The Art and Science of Rotating Field Machines Design: A Practical Approach

The Art and Science of Rotating Field Machines Design: A Practical Approach

by

Vlado Ostović

Professor of Electrical Engineering and Consultant

Formerly

Chief Engineer for Solving Special Electromagnetic and Thermal Problems in Large Electric Machines at ABB Kraftwerke AG, Mannheim, Germany

Springer

Vlado Ostović
Weinheim
Germany

ISBN 978-3-319-39079-6 ISBN 978-3-319-39081-9 (eBook)
DOI 10.1007/978-3-319-39081-9

Library of Congress Control Number: 2016942030

Printed on acid-free paper

This Springer imprint is published by Springer Nature
The registered company is Springer International Publishing AG Switzerland

To Thea, Thomas, and Marko
without whom this book would not have been possible

Preface

Rotating field electric machine is one of the most complex devices in the whole electrical engineering, because physical quantities in it are both a function of space and time, the medium in which they spread is nonlinear and all three dimensions of electromagnetic, mechanical, and thermal fields determine its performance. When fed from AC source(s), rotating field machine windings generate flux density waves which spread at different velocities through the machine, produce time dependent forces between stator and rotor, and dissipate losses which increase the temperature throughout the machine.

Only in an electric machine Faraday's law of electromagnetic induction can show its real nature: the tendency of space to oppose any change of magnetic field. When in a vicinity of a stationary coil the magnetic field is changed, the voltage is induced in the coil, which drives current through it, the magnetic field of which compensates for the primary change of the field. If the coil can move, or rotate, and if it is placed in a space in which the magnetic field rotates, not only the voltage will be induced in it, as is the case with a stationary coil. The coil will start to rotate in the direction of the field, trying to oppose the change of the concatenated flux created by the rotating field. After infinitely long time, neglecting mechanical losses, the coil will reach the synchronous speed and concatenate the magnetic field which in that case does not change relative to it.

Besides generating electromagnetic torques, forces in an electric machine create mechanical stress on its components. Electromagnetic and mechanical losses increase the machine temperature and must be taken out from it by means of a gaseous or fluid coolant. The amount of losses and related temperature increase is crucial when determining the machine rated power, which is a pure thermal quantity, defined as the power transferred to a load at which the dissipated losses increase the machine temperature to the value allowed by the winding class of insulation. Therefore, to design an electric machine is primarily a thermal task which is successfully completed when the losses accompanying the electromechanical energy conversion do not increase the hot spot temperature above the given value.

An exact description of electromagnetic, fluid flow, and thermal effects which simultaneously take place in an electric machine would require enormous computational power, which would barely justify the accuracy of results and time consumed. Therefore, it is designer's job to simplify the bulky mathematical apparatus when solving a particular problem, building a simple physical model of a machine, and solving it with regular mathematical tools. In order to make adequate physical assumptions for a particular problem, the designer has to understand well the tool he is using, which helps him avoid application errors.

Like in almost no other field of engineering, when designing an electric machine there is a high need for proper physical interpretation of mathematical solutions, and especially of conditions under which these are derived. Mathematics is a good servant, but a poor master. This can be illustrated by the example of the probably most misinterpreted equation in electrical engineering, which states that electrical current i through a capacitor is equal to the time derivative of charge Q

$$i = \frac{dQ}{dt}$$

At the latest since the Millikan–Fletcher experiment in the year 1910 the engineering community is aware of the *discrete* character of electrical charge: the amount of electricity can only change in steps of $e^{-} = 1.6 \cdot 10^{-19}$ As, which is the charge of electron. In terms of mathematical analysis, a derivative of a discrete function is not defined; therefore, the expression dQ/dt is mathematically undetermined. At time instants at which the amount of charge in a closed volume V changes, its time derivative is not defined; at all other time instants the derivative of the amount of charge is equal to zero, Fig. 1. Accordingly, an electric current defined as a time rate of change of the amount of charge is either undetermined, or equal to zero.

Such definition of electrical current is a result of misinterpretation of Maxwell's concept of displacement current in dielectrics. Knowing that the divergence of the curl of any vector is equal to zero, and applying operator div to the Ampère's circuital law, one obtains

$$div\left(curl\vec{H}\right) = div\left(\vec{\Gamma} + \frac{\partial \vec{D}}{\partial t}\right) = 0$$

with H denoting the magnetic field strength, Γ the current density, and D the electric displacement. Setting for

$$div\vec{D} = \rho = \frac{Q}{V} = \frac{\sum_{i=1}^{n} q_i}{V}$$

with ρ standing for charge density, and V for volume, one can further write

$$V \cdot div\vec{\Gamma} + \frac{dQ}{dt} = 0$$

which leads to a mathematically correct, yet physically false construction

$$i = \frac{dQ}{dt}$$

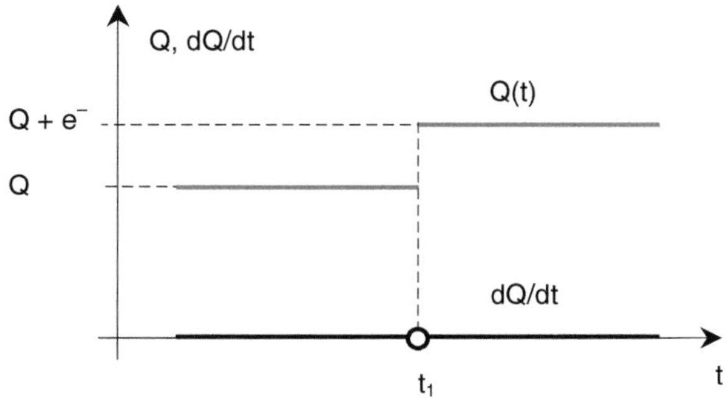

Fig. 1 Illustrating expression dQ/dt: time rate of change of the amount of charge e^- in a closed volume V is not defined at time instant at which the change takes place; at all other time instants the time rate of change is equal to zero

When correctly interpreted, the displacement current density Γ_d, defined, e.g. for a parallel plate capacitor as

$$\vec{\Gamma}_d = \frac{\partial\vec{D}}{dt} = \varepsilon\frac{\partial\vec{E}}{dt} = \varepsilon\frac{\partial}{\partial t}(-\nabla\varphi) = \frac{\varepsilon}{d}\frac{du}{dt}$$

helps one express the current i_c through a capacitor by means of voltage drop u across its plates as

$$i_c = S\frac{\varepsilon}{d}\frac{du}{dt} = C\frac{du}{dt}$$

where S denotes the area of plates, d the distance between plates, ε the permittivity of the medium and C the capacitance. The current through a capacitor is proportional to the time derivative of voltage drop across its plates. *Electric current is definitely not equal to a time derivative of charge,* because the amount of charge Q is a discrete quantity and therefore it is not a derivable function.

Any discussion about physically proper description of electric current inevitably leads to the question about its real nature. The answer to this question is astonishingly simple: electric current is a state of continuum characterized by thermal, light, chemical, and magnetic effects. Whereas thermal, light, and chemical effects do not always accompany electric current, magnetic field is *always* generated when a current flows. Therefore, electric current can be defined as the source of magnetic field (Fig. 2).

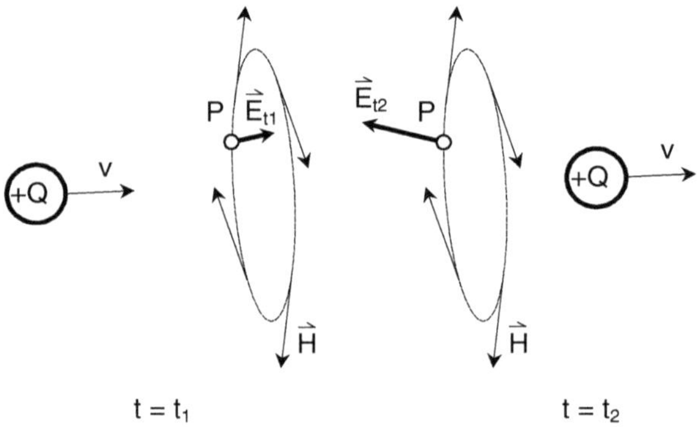

Fig. 2 Illustrating effects of motion of charge +Q, which travels at finite velocity v, as a source of magnetic field. The charge +Q creates at point P the electric field E_{t1} at time instant t_1, and the electric field E_{t2} at time instant t_2. The change of electric field distribution in time interval $[t_1, t_2]$, which spreads through space with the speed of light, gives rise to displacement current density $\varepsilon \frac{\partial E}{\partial t}$, which is a source of magnetic field H

The emphasis in the definition of electric current is on the state of continuum, which, according to the currently valid physical model of the universe, spreads with the maximum possible speed through space. For example, when a body with mass m changes its position, a new distribution of gravitational field is generated which spreads with the speed of light through the universe. In the same manner when a charge Q changes its position in space, the effect of the change—a new distribution of electromagnetic field—spreads with the speed of light throughout the whole universe.

Since every change of electric field distribution is accompanied by a creation of magnetic field, a charge which changes its position creates magnetic field in space. Independent of how slow a charge moves, the electric field distribution created at its every new position spreads with the speed of light through space. Therefore, one can state that no matter how fast or slow a charge moves, the state of continuum described by electric current spreads at the speed of light.

Previously discussed topics illustrate the tenor of the book: thorough understanding of physical (electromagnetic, mechanical, and thermal) processes in rotating field electric machines by using sophisticated computational tools, along with correct physical interpretation of mathematical results.

A reader who expects to find a compilation of recipes on how to design a rotating field electric machine in this book will be disappointed, because this is a book on "why", rather than on "how". Alexander Gray, one of the pioneers in the field of electric machines design, stated more than one hundred years ago "Since the design of electrical machinery is as much an art as a science, no list of formulae or collection of data is sufficient to enable one to become a successful designer." The art in this statement relates to balanced trade-offs of often contradictory requests. Technical specifications for a particular machine can be fulfilled in many different ways, none of them being absolutely the best. There exists no recipe on how to design a perfect electric machine—it is more a question of experience and manufacturing capabilities, with a strong influence of engineering talent. Nevertheless, one should always keep in mind the simplicity of the solution, as a dominating design goal which guarantees longer mean times between failures and decreases the machine price.

One of the basic ideas followed in this book is a wide use of lumped circuit approach when solving field problems. Electrical engineers think in terms of circuits. Ohm's law, one of the most pronounced laws in the nature, states that the output is proportional to the input. Just like for electric circuits, similar laws of proportionality can be written for pressure and mass flow, temperature and heat flow, as well as for magnetic flux and MMF drop.

Despite various nonlinearities which dominate the physics of electric machines, an emphasis in the book is given to an analytical solution of a particular problem. An analytical solution helps designer to get insight into the nature of processes in a machine and estimate the influence of each parameter on its overall performance. A numerical solution, on the other hand, is valid only for a particular set of parameters. A single numerical solution might deliver correct numbers, but it does not help the designer to understand the relationships between the quantities involved.

The book is organized in eight chapters. In the first seven chapters, a thorough analysis of electromagnetic and thermal effects which dominate the operation of electric machines is given. In the last chapter important steps in the design of induction and synchronous machines are sketched. In each chapter several case studies are presented, dealing with a topic relevant to the chapter contents.

Properties of permanent magnet machines are discussed thoroughly in the first chapter and in the Appendix. Considering inferior performance of a permanent magnet AC machine as compared to a wound rotor synchronous machine, the design procedure for the former machine type is not carried out separately in the book. Nevertheless, permanent magnet machines have been referred to in several case studies. The interested reader can obtain the performance curves of a permanent magnet machine simply by setting the field current of an equivalent wound rotor synchronous machine down to its no-load value.

Paraphrasing the statement "Never trust a statistics you didn't make yourself," one might find it a good idea not to use commercial machine design software packages, the vendors of which consider the applied methods and procedures their intellectual property. When applying such software packages it could happen that the user, without being aware of assumptions and limitations made when creating such a package, obtains physically meaningless results, like current flowing from a conductor into air, overall power factor larger than one, or even sum of all power components different from zero, etc. Having this in mind, the intention of this book is to support those engineers who want to design an electric machine standing on their own two feet.

Last but not least, the text of the book was checked, rechecked, and checked again in order to eliminate possible errors. If despite the careful preparation certain expression turns out to be incomplete or misleading, one should keep in mind that this was done on purpose, because those readers, who desperately look for an error in a work like this, should also be satisfied after having found it.

Weinheim, July 2016 Vlado Ostović

Contents

List of Symbols

A	Current sheet (A/m)
B	Flux density (T)
C	Cross-sectional (–); Constant of integration (–); System matrix (–)
D	Electric displacement field (As/m^2); Diameter (m)
DE	Driving end (–)
E	Electric field strength (V/m); Matrix of coefficients for eccentric air gap (–)
F	Force (N)
G	Conductance (–)
H	Magnetic field strength (A/m)
I	Current (A)
M	Torque (Nm)
N	Number of slots (–)
NDE	Non-driving end (–)
P	Power (W); Wetted perimeter (m); Product (–)
P′	Heat per surface area (Ws/m^2)
PM	Permanent magnet (–)
Q	Charge (As); Heat energy (Ws); Volume flow rate (m^3/s); Quotient (–)
R	Radius (m); Resistance (–)
Re	Reynolds number (–)
S	Sum (–); Apparent power (VA); Surface area (m^2)
T	Period length (s)
U	Voltage (V)
V	Volume (m^3)
W	Energy (Ws)
a	Cosine terms of Fourier series (–)
air	Related to air (–)
b	Sine terms of Fourier series (–); Conductor width (m); Vector of unknowns in equations of eccentric air gap (–)
c	Specific heat (Ws/kg K); Parameter (–)
cyl	Cylindrical (–)

d	Conductor width (m); Slot width (m)
eff	Effective (–)
f	Factor (–); Frequency (Hz)
i	Current (A)
k	Factor (–)
l	Length (m)
leak	Related to leakage
m	Mass (kg); Number of phases (–)
main	Related to main (–)
p	Power (W); Pressure (N/m^2); Number of pole pairs (–)
q	Specific losses (W/m^3); Number of slots per pole and phase (–)
r	Number of principal poles per fundamental pole of a fractional slot winding (–); Saliency ratio (–)
s	Slot opening (m)
t	Time (s)
u	Voltage (V); Auxiliary function (–)
v	Velocity (m/s)
w	Number of turns (–)
x	Circumferential coordinate (m)
y	Coil pitch (–)
Δ	Difference (–)
Γ	Current density (A/m^2)
ϑ	Temperature (°C)
$\Delta\vartheta$	Temperature difference (K)
Ω	Angular velocity (mechanical) (s^{-1})
α	Angle of flux lines to the normal of a surface; heat transfer coefficient (W/m^2 K)
β	Temperature coefficient of electric resistance (K^{-1}); Auxiliary function (–)
δ	Air gap width (m)
ε	Permittivity (As/Vm); Eccentricity (m)
γ	Rotor angle (–); Auxiliary function (–)
φ	Potential (–)
κ	Electric conductivity ($1/\Omega$ m)
λ	Thermal conductivity (W/m K); Slot specific permeance (–); Friction factor (–)
μ	Permeability (Vs/Am)
ν	Kinematic viscosity (m^2/s)
ρ	Charge density (As/m^3); Mass density (kg/m^3); Specific electric resistance (Ωm)
τ	Angle (–); Step(–)
ξ	Shift (m)
ζ	Correction factor for hydraulic resistance (–); Pressure loss coefficient (–)

Chapter 1
Introduction

Contents

Fundamental electromagnetic, fluid flow, and thermal laws governing the design philosophy of rotating field electric machines are discussed, and the importance of simultaneous consideration of various engineering disciplines thereby is emphasized. The application of basic laws of electromagnetism in electric machines is illustrated with examples of iron–air boundary and current imaging. Solutions of the fundamental heat transfer equation for various combinations of electric and thermal parameters are discussed, and electrothermal conductance is introduced. Performance of permanent magnet excitation is analyzed. It is shown that constant magnetization is the Achilles heel of a permanent magnet excited synchronous machine and the main reason for its inferiority to wound rotor synchronous machine. Lumped element parameters in electric, magnetic, thermal, and fluid flow circuits are introduced.

V. Ostović, *The Art and Science of Rotating Field Machines Design*,
DOI 10.1007/978-3-319-39081-9_1

1.1 General Considerations

To design an electric machine means to determine the combination of geometric parameters, winding configuration, and implemented materials, which ensures a flawless operation of the designed machine throughout the whole period of its service life. Considering numerous technical and economical criteria which have to be satisfied in the process of machine design, one comes to a conclusion that there are many ways to design a good machine. On the other hand, there are even more ways to design a machine, the performance of which does not fulfill given specifications, the most pronounced of them being thermal overloading of insulation. The process of electromechanical energy conversion is accompanied with a generation of electromagnetic and mechanical losses in the machine's active and passive components. On their way through machine parts the losses create temperature gradients and hot spots, the latter denoting locally overheated areas.

Considering thermal limits of winding insulation, the main criterion that a successfully designed electric machine has to fulfill is that the hot spot temperature in it does not exceed the amount allowed for the given class of insulation. Proper thermal design is crucial for the operation of an electric machine; a thermally poor designed machine must either operate derated, delivering less power than foreseen, or it fails completely because of overheating of its windings.

Losses generated in the whole *volume* of active part can be transferred to a coolant only on heat exchange *surfaces*. Considering constant losses per volume and denoting by x the machine's linear dimension, the total losses increase with the machine volume, i.e., proportional to x^3, whereas the heat exchange surface can only increase proportionally to x^2. The ratio of total losses to the heat exchange surface, being proportional to the temperature drop on the heat exchange surface, increases proportionally to the machine size, i.e., to the quotient $x^3/x^2 = x$. For this simple reason it is easier to cool small than large electric machines. Without changing the machine cooling type, current density in a small induction motor may reach 10 A/mm^2; it has, however, to be reduced to some 2–4 A/mm^2 in a large machine with the same class of insulation. Consequently, the importance of cooling increases proportionally to the machine size and dominates design of large electric machines.

The simple scaling law for heating not only is typical for machines but also can be observed in the whole nature. An Antarctic penguin is bigger than an Australian: Considering identical metabolism, body temperature, and produced energy per body volume, an Antarctic penguin can cover a higher temperature difference to its environment than an Australian only by being bigger, i.e., by having a larger volume to body surface area.

1.2 Stationary Coils and Rotating Magnetic Field

Constant current creates constant magnetic field at each point in space around a current-carrying conductor. Time-dependent current creates time-dependent magnetic field. Considering linear and homogenous magnetic properties of medium surrounding the conductor, the direction of magnetic field at any point in space is determined only by geometry and independent of the amount of current. Time-dependent current creates at every point in a homogenous medium a magnetic field with an amplitude dependent on current and with constant direction in space. In other words, magnetic field around a current-carrying conductor *pulsates* when the current *alternates*. One should note that electric current, as a *time-dependent* quantity, can only alternate, i.e., change its amount in time. Magnetic field, as a *spatial* quantity, can both change its magnitude and direction in space. As shown in Fig. 1.1, the direction of a pulsating magnetic field at point P at time instant t_1, when the current is positive, is collinear with axis *a–a′*. After one half of period of current *i*, at time instant $t_1 + T/2$ when the current has an opposite sign, the direction of magnetic field changes by 180°.

Sinusoidal current $i(t) = I_{\max} \cos \omega t$ flowing through the coil in Fig. 1.1 produces pulsating magnetic field, the direction of which varies from one point in the space to another. A pulsating physical quantity is a synonym for *standing wave*. A standing wave emerges from two traveling waves with equal amplitudes, B_+ and B_-, which travel (rotate) at the same speed in opposite directions as shown in Fig. 1.2.

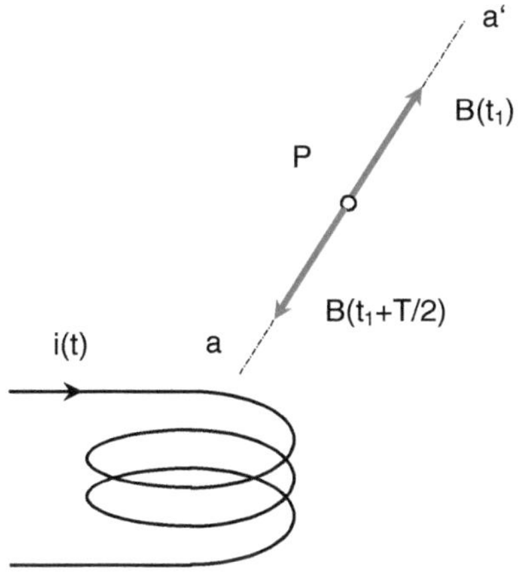

Fig. 1.1 Illustrating the direction of magnetic field at an arbitrary point *P* at two time instants which are one half of period of current shifted to each other

The axis of pulsating field B_{puls} acting at point P in Fig. 1.2 is a–a', as shown in Fig. 1.1. For a given sinusoidal coil current the amplitude of field B_{puls} changes periodically from 0 over B_{max} down to $-B_{max}$ and back to 0. The axis a–a' of field B_{puls} acting at point P does not change its slope and position in space. The actual field B_{puls} which pulsates along the axis a–a' can be replaced by a sum of two fields having constant amplitudes B_+ and B_- which rotate in opposite directions with angular velocities $+\Omega$ and $-\Omega$. The relationship between the amplitudes of rotating and pulsating field is $B_+ = B_- = B_{max}/2$. The rotating field component denoted by B_+ is called *positive sequence*, and the rotating field component denoted by B_- is the *negative sequence.*

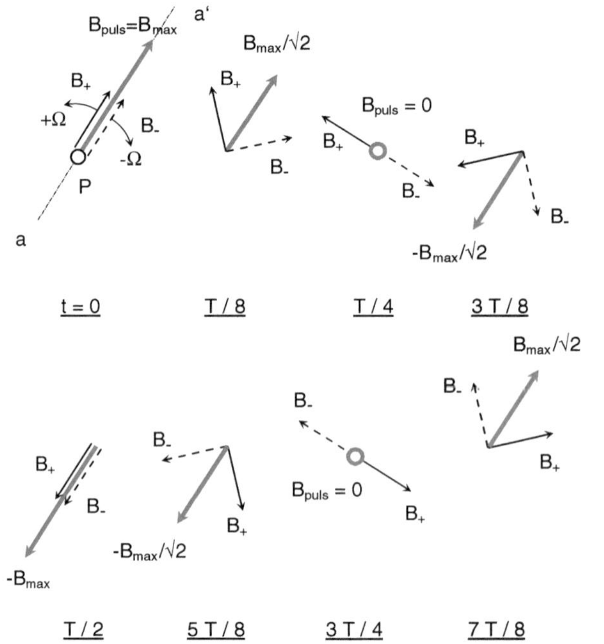

Fig. 1.2 Representation of a pulsating field created by sinusoidal current with angular frequency ω as a sum of two waves of rotating fields B_+ and B_-

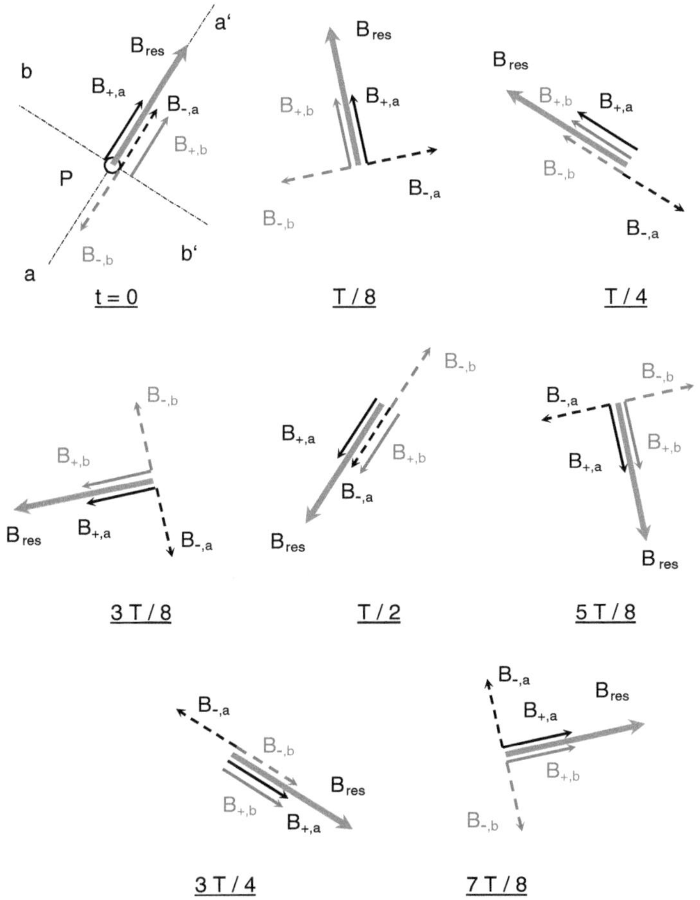

Fig. 1.3 Generation of rotating field with two perpendicular coils carrying 90° out of phase alternating currents

As illustrated in Fig. 1.2, a single coil carrying sinusoidal current produces at each point in space two fields with equal amplitudes, which rotate in opposite directions. *Pure rotating field* is obtained when one of the components—either negative or positive sequence—is fully eliminated. This can only be done by introducing another coil(s) which carry current(s) phase shifted to the current in the first coil. The effect of additional coil(s) is illustrated in Fig. 1.3, in which the magnetic field created by currents in two coils is represented, the lines of flux of which are perpendicular to each other at point P.

The line *a–a′* shown in Fig. 1.3 represents the axis of pulsating field created by the first coil and the line *b–b′* analogously for the second coil. For the purpose of simplicity, it will be further assumed that the two amplitudes of magnetic field at point P created by the two currents are equal to each other and that the phase shift between the two coil currents is equal to 90°. All components of field created by the current in the first coil are drawn black, and those created by the second coil are gray. The resulting field B_{res} is represented with thick gray arrow.

One notes in Fig. 1.3 that the resulting field has a constant amplitude and completes one revolution in space after one period T of the coil currents. The resulting field rotates at an angular velocity $\vec{\Omega}$, which is a *vector* perpendicular to the plane of rotation. Spatial quantity $\vec{\Omega}$ should not be confused with the *scalar* of angular frequency ω equal to $2\pi f$, with f denoting the reciprocal of the period T of coil currents, $f = 1/T$. In other words, ω is *not* the magnitude of $\vec{\Omega}$.

Analyzing the generation of rotating field illustrated in Fig. 1.3, one can state that one of the components of pulsating field created by a single coil—in this case its negative sequence—was fully eliminated by introducing additional source of pulsating field, the negative sequence component of which is 180° phase shifted to the negative sequence component of the first coil. At the same time, the amplitude of the negative sequence component of the second coil must be equal to the amplitude of the negative sequence component of the first coil. One can state that *pure rotating field is obtained by superimposing two pulsating fields* created by stationary current-carrying coils if the following conditions are satisfied:

- The axes of coils are perpendicular to each other;
- Coil currents are 90° out of phase.

In a polyphase system, the rules above are adapted in order to consider the number of phases.

In the process of creating the rotating field shown in Fig. 1.3 each phase contributes with its full positive sequence component to the resulting positive sequence field, whereas the amplitude of positive sequence component per phase is equal to 50 % of the amplitude of total field created by single phase.

It will be shown in Chap. 2 how the previously discussed basic connection of coils has to be modified in order to extend the space in which rotating field is generated from the single point P in Fig. 1.1 to a larger volume, e.g., the air gap of an electric machine. If suitable objects, such as coils, ferromagnetic components, and permanent magnets, are brought into this space, the rotating field can interact with them, and force F [N] and torque M [Nm] can be generated which tend to move these objects in the direction of rotating field. In order to move objects, the rotating field has to perform in each revolution the mechanical work W_{mech} equal to

$$W_{mech} = \int_0^{2\pi} M(\gamma)\mathrm{d}\gamma = \int_0^T p(t)\mathrm{d}t = P \cdot T \tag{1.1}$$

Graphical interpretation of Eq. 1.1 is shown in Fig. 1.4. Only the constant torque component M_{ave} shown in Fig. 1.4 performs mechanical work, because the average value of all other torque components in one revolution, i.e., in angle interval $[0, 2\pi]$, is equal to zero.

Equation (1.1) has far-reaching effects on design of rotating field electric machines, one of which being articulated in the statement that permanent electromechanical energy conversion can only take place if torque-generating components of magnetic field rotate at the same speed. Replacing differential $d\gamma$ in Eq. 1.1 with $\Omega\,dt$, and replacing the expression $M(\gamma) \cdot \Omega$ by instantaneous power $p(t)$, one obtains as a result that mechanical energy is proportional to the average torque. Constant T in Eq. 1.1 denotes the time interval within which the rotor completes one revolution from 0 to 2π.

The contribution of a pulsating component of torque to the total energy converted from electrical to mechanical form and vice versa is equal to 0. At a given speed of rotation, only active power creates mechanical work; higher harmonics of active power create torques with an average value (mechanical work) equal to zero.

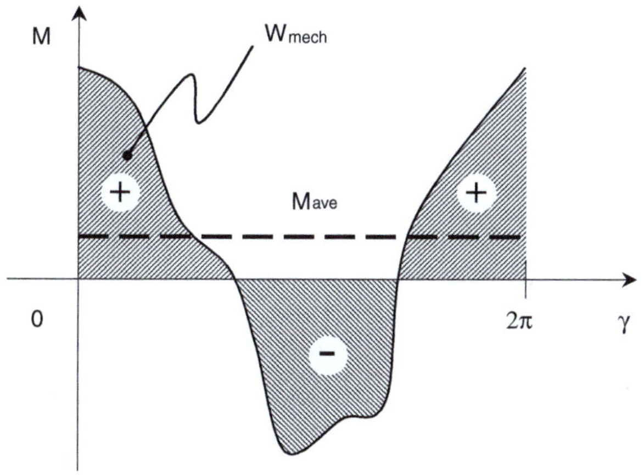

Fig. 1.4 Torque as a function of angle. The area below the curve $M(\gamma)$ is equal to the mechanical work performed

1.3 Electromagnetic Field Equations and Boundary Conditions; Field Distribution in Heteropolar Machines

The source of magnetic field is electric current. Magnetic field lines form closed curves and concatenate electric field lines. The sourcelessness of magnetic field can be formulated as [1]

$$\operatorname{div} \vec{B} = 0 \tag{1.2}$$

which states that the magnetic flux through a closed surface is equal to zero. This simple condition and Ampère's circuital law for a closed contour C

$$\oint_C \vec{H} \cdot d\vec{l} = \sum i \tag{1.3}$$

are the only two physical laws that help one determine the flux distribution in the complex geometry of electric machines.

Vast majority of rotating field electric machines are built as heteropolar, i.e., all flux lines in their active part close in a plane perpendicular to the machine shaft. Thus, the machine's active part can be analyzed two-dimensionally without losing accuracy. In the end winding zone, on the other hand, all three spatial components of magnetic field are equally important when determining analytical solutions of field distribution, which help find leakage reactances, forces, losses, etc.

As illustrated in [2], Eqs. 1.2 and 1.3 applied to the machine's iron–air boundary without current sheet (surface current density) help one determine the ratio of angles of flux lines to the boundary surface normal line as (Fig. 1.5)

$$\frac{\tan \alpha_{\mathrm{air}}}{\tan \alpha_{\mathrm{Fe}}} = \frac{1}{\mu_{r,\mathrm{Fe}}} \tag{1.4}$$

where the iron relative permeability, $\mu_{r,\mathrm{Fe}}$, equals typically to 10^3–10^4. Therefore, one can state that lines of flux enter the iron surface on the air side of air–iron boundary almost perpendicularly. The change of direction of the line of flux shown in Fig. 1.5 on the air–iron boundary is a consequence of different permeabilities of air and iron.

Equation 1.4 helps one find magnetic field distribution of a current-carrying conductor in front of an iron half-space, as shown in Fig. 1.6. The problem is solved by applying the method of current imaging in which the region with high iron permeance and without current-carrying conductor(s) is replaced by fictitious current-carrying conductors in air [3]. As a result, Ampère's circuital law can be applied on both sides of the air–iron boundary on which the lines of flux do not break as a consequence of different relative permeabilities.

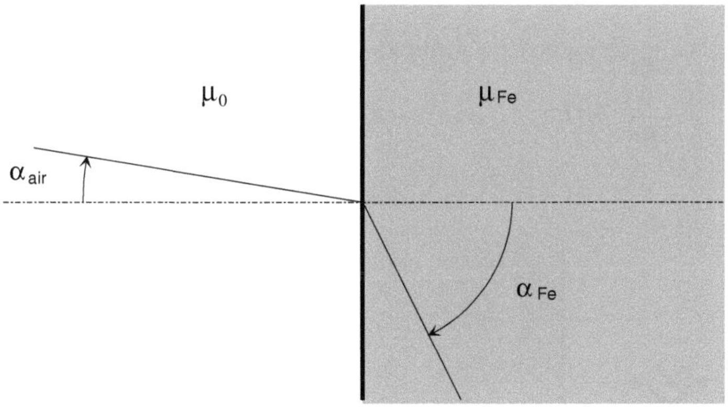

Fig. 1.5 Refraction of magnetic flux on iron–air border

Boundary conditions for a particular problem are a means to replace the influence of the rest of the world on electromagnetic state within the problem region by applying simple physical relationships on the boundary of the problem region. Typical boundary condition can be defined as a line of flux parallel to the border; in this case, no flux goes out of the problem region into the rest of the world.

In Fig. 1.6b a flux line created by current I flowing through a conductor at a distance d from iron half-space is shown. According to Eq. 1.4, the lines of flux in air on the air–iron boundary are practically perpendicular to iron. On the iron side the lines of flux choose the shortest way and end up almost parallel to the boundary.

The actual electromagnetic DC excitation and problem geometry shown in Fig. 1.6a are replaced by boundary conditions as illustrated in Fig. 1.6b, c, in which the whole problem region has only one permeability, μ_0. This way the Ampère's law can be directly applied in the whole problem region. Field distribution on the air side shown in Fig. 1.6a is represented in Fig. 1.6b as a sum of the field distribution created by the original current I and the field distribution created by a fictitious current $I_{r,\mathrm{Fe}}$ flowing through a fictitious conductor placed in iron at the same distance d from the air–iron boundary as the actual conductor on the air side. The resulting magnetic field strength H [A/m] on the air side of the air–iron boundary can be expressed as follows:

$$\vec{H}_{\text{air}} = \frac{I}{2R\pi}\left(-\vec{a}_y \cos\alpha + \vec{a}_x \sin\alpha\right) + \frac{I_{r,\text{Fe}}}{2R\pi}\left(\vec{a}_y \cos\alpha + \vec{a}_x \sin\alpha\right) = \\ = \frac{1}{2R\pi}\left[\vec{a}_x\left(I_{r,\text{Fe}} + I\right)\sin\alpha + \vec{a}_y\left(I_{r,\text{Fe}} - I\right)\cos\alpha\right] \tag{1.5}$$

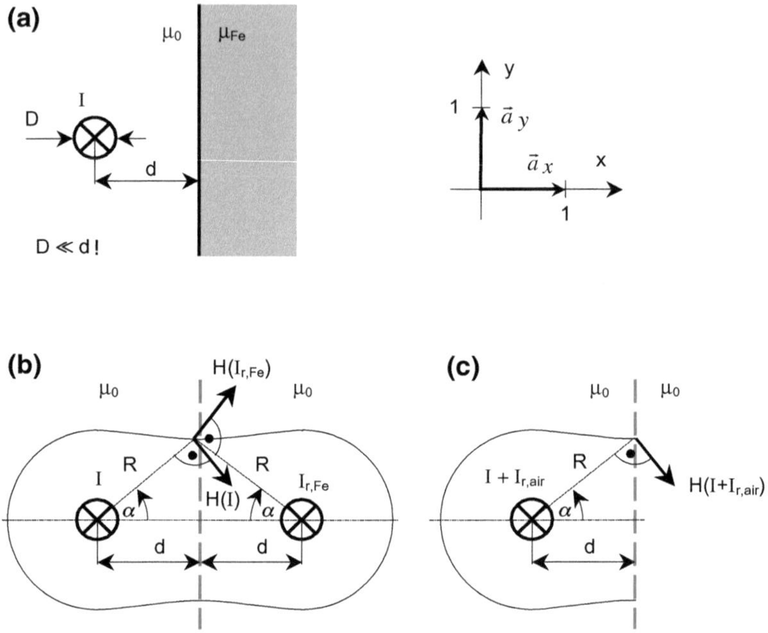

Fig. 1.6 Current-carrying conductor in front of iron half-space (**a**) and its equivalents (**b** and **c**) in case of DC excitation

the normal (x in Fig. 1.6) and tangential (y in Fig. 1.6) components of which can be written as

$$H_{x,\text{air}} = \frac{1}{2R\pi}\left(I_{r,\text{Fe}} + I\right)\sin\alpha \tag{1.6}$$

$$H_{y,\text{air}} = \frac{1}{2R\pi}\left(I_{r,\text{Fe}} - I\right)\cos\alpha \tag{1.7}$$

One can express in the same manner the field distribution on the iron side of the border by the field of fictitious current $I_{r,\mathrm{air}}$ flowing through the conductor additionally to the actual current I, as shown in Fig. 1.6c:

$$\vec{H}_{\mathrm{Fe}} = \frac{I + I_{r,\mathrm{air}}}{2R\pi}\left(\vec{a}_x \sin\alpha - \vec{a}_y \cos\alpha\right) \tag{1.8}$$

Applying boundary conditions [2] for normal and tangential components of the magnetic field strength on the air–iron surface without current sheet:

$$\mu_{\mathrm{Fe}} H_{x,\mathrm{Fe}} = \mu_0 H_{x,\mathrm{air}} \tag{1.9}$$

one obtains

$$\mu_{\mathrm{Fe}}\left(I + I_{r,\mathrm{air}}\right) = \mu_0\left(I + I_{r,\mathrm{Fe}}\right) \tag{1.10}$$

as well as

$$H_{y,\mathrm{Fe}} = H_{y,\mathrm{air}} \tag{1.11}$$

and

$$I_{r,\mathrm{air}} = -I_{r,\mathrm{Fe}} \tag{1.12}$$

Fictitious currents $I_{r,\mathrm{Fe}}$ and $I_{r,\mathrm{air}}$ are now

$$I_{r,\mathrm{Fe}} = -I_{r,\mathrm{air}} = \frac{\mu_{r,\mathrm{Fe}} - 1}{\mu_{r,\mathrm{Fe}} + 1} I \tag{1.13}$$

whereas the normal (x in Fig. 1.6) and tangential (y in Fig. 1.6) components of magnetic field strength in air, Eqs. 1.6 and 1.7, can be written as follows:

$$H_{x,\mathrm{air}} = \frac{I}{2R\pi} \cdot \frac{2\mu_{r,\mathrm{Fe}}}{\mu_{r,\mathrm{Fe}} + 1} \sin\alpha \tag{1.14}$$

$$H_{y,\mathrm{air}} = -\frac{I}{2R\pi} \cdot \frac{2}{\mu_{r,\mathrm{Fe}} + 1} \cos\alpha \tag{1.15}$$

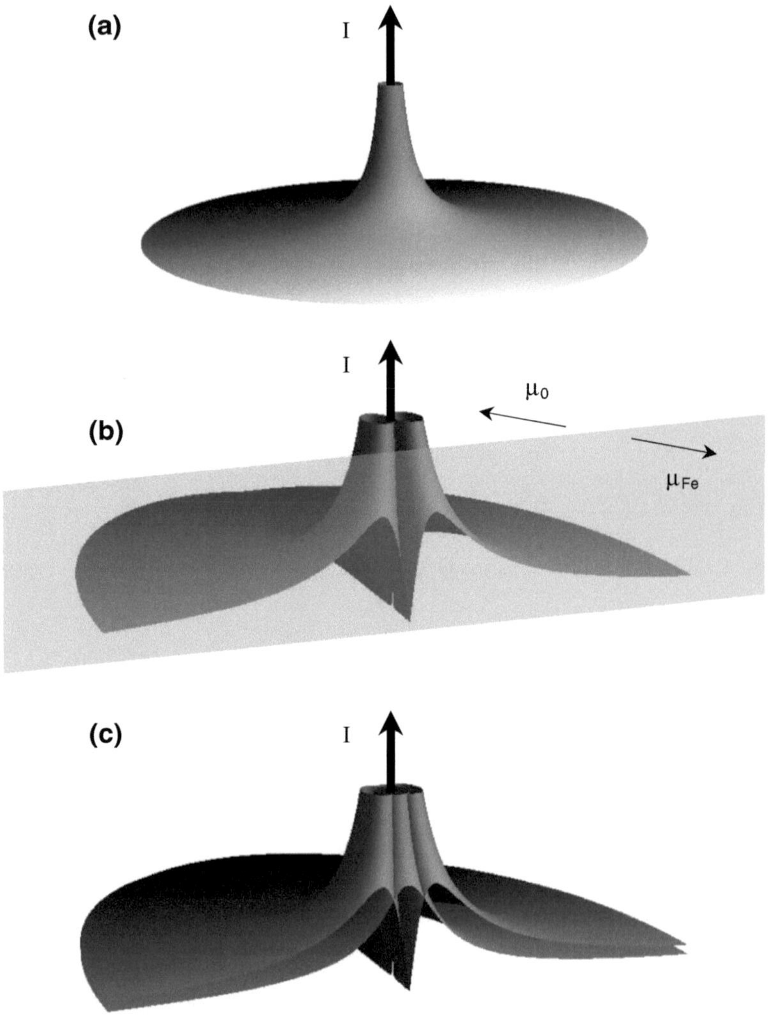

Fig. 1.7 Magnetic field strength H created by a current-carrying conductor (**a**) in air alone; (**b**) in air in front of iron half-space; **c** comparison of the two distributions in (**a** and **b**)

By substituting the expression for fictitious current $I_{r,\text{air}}$ (Eq. 1.13) into Eq. 1.8, one can express the normal and tangential components of magnetic field strength in iron as

$$H_{x,\text{Fe}} = \frac{I}{2R\pi} \cdot \frac{2}{\mu_{r,\text{Fe}} + 1} \sin \alpha \tag{1.16}$$

$$H_{y,\mathrm{Fe}} = -\frac{I}{2R\pi} \cdot \frac{2}{\mu_{r,\mathrm{Fe}} + 1} \cos \alpha \tag{1.17}$$

Except for small values of angle α (both positive and negative), the normal component of the magnetic field strength in air, defined in Eq. 1.14, gets almost doubled in the vicinity of iron with $\mu_{r,\mathrm{Fe}} \gg$, as a consequence of practically halving the length of lines of flux in air. Due to its high relative permeability, the MMF drop in iron is negligible as compared to the MMF drop in air. The tangential component of magnetic field strength in air, defined in Eq. 1.15, is radically reduced, since the lines of flux enter the iron surface almost perpendicularly, see Eq. 1.4. Whereas the normal component of magnetic field strength in air is $\mu_{r,\mathrm{Fe}}$ times larger than in iron, the tangential components are equal in both media.

Graphical interpretation of results obtained by Eqs. 1.14 and 1.15 for a relative permeability $\mu_{r,\mathrm{Fe}}$ equal to 1000 is shown in Figs. 1.7a–c.

In Fig. 1.7a the amplitude of magnetic field strength of a current-carrying conductor in a medium with a relative permeability equal to 1 as a function of radial distance and angle is shown. The amplitude of magnetic field strength at a given point is inversely proportional to the distance from the center of the conductor to that point.

In Fig. 1.7b the magnetic field strength distribution in air created by a current-carrying conductor parallel to a ferromagnetic half-space with $\mu_{r,\mathrm{Fe}} = 1000$ is shown. According to Eq. 1.8, the magnetic field strength in iron is almost equal to zero. Therefore, the magnetic field strength distribution on the air–iron boundary is discontinuous, which is shown as a vertical cut in Fig. 1.7b.

The amplitudes of the magnetic field strength in air without and with ferromagnetic half-space are compared in Fig. 1.7c. Magnetic field created by a conductor in front of the iron half-space is equal to zero along the plane through the conductor perpendicular to the magnetic half plane. Outside of the very narrow neighborhood of this perpendicular plane, the magnetic field strength with ferromagnetic half-space is almost twice as big as the field strength without it.

As a conclusion, one can state that the vicinity of ferromagnetic media almost doubles the magnetic field in air created by current-carrying conductors. This is a very important fact which has to be considered when analyzing forces acting on the end winding.

1.4 Fluid Flow and Heat Transfer in Electric Machines; Types of Cooling

Thermal design of an electric machine is governed by fundamental equation relating the temperature rise $\Delta\vartheta$ [K] of a coolant with volume V [m^3], mass density ρ [kg/m^3], and specific heat c [Ws/(kg K)] at a given pressure p [N/m^2] to the amount of heat ΔQ [Ws] absorbed

$$\Delta Q = P \cdot \Delta t = V \cdot \rho \cdot c \cdot \Delta \vartheta \tag{1.18}$$

with P denoting the power of losses (heat) [W] and Δt (s) the length of time interval in which the heat gets absorbed by the coolant. Introducing volume flow rate $\dot{V}\,[\mathrm{m}^3/\mathrm{s}]$

$$\dot{V} = \frac{V}{\Delta t} \tag{1.19}$$

one can express the temperature rise $\Delta\vartheta$ of the coolant as

$$\Delta \vartheta = \frac{P}{c \cdot \rho \cdot \dot{V}} \tag{1.20}$$

In Eq. 1.20 the global physical relationship between coolant thermal properties, its volume flow rate, and the absorbed heat is described. Typical values for coolant temperature rise in electrical machines vary between 15 and 40 K. Having these values in mind, one can determine the necessary coolant volume flow rate $\dot{V}$ delivered by a fan or some other form of coolant mover. A coolant mover/fan transports fresh coolant from heat exchanger to the machine, and warm coolant back to the heat exchanger. In order to perform this function, the coolant mover has to create certain pressure difference. In terms of electric circuits, pressure difference corresponds to electrical voltage and volume flow rate to current. Coolant mover is the source, and hydraulic resistances are the loads.

Fluid flow in loss-free case is described by Bernoulli's equation which relates the pressure drop on a hydraulic resistance Δp [N/m^2] to the coolant velocity

$$\Delta p = \frac{\rho}{2} \cdot v^2 \tag{1.21}$$

assuming that the term related to geodesic component of energy, $\rho \cdot g \cdot h$, may be neglected. ρ in equation above denotes mass density [kg/m^3] and v the coolant velocity [m/s]. In reality, the coolant flow in an electric machine is lossy due to two reasons: turbulence and friction. Every change of cooling channel cross section and/or direction of coolant flow changes pressure drop across it. On the other hand, cooling channels in a machine are not perfectly smooth, and the coolant pressure has to be increased in order to reach the required volume flow rate. How important the smoothness of the heat exchange surface for the heat flow can be, shows a simple experiment with an almost perfectly smooth rubber disposable glove: Even without wind, the temperature of the hand wearing it is sensitively lower than the temperature of the hand without a rubber glove. Roughness of the heat exchange surface of the hand is in this case the means to increase the temperature drop on the hand and reduce the loss of body heat on it.

The increase of pressure drop along a cooling channel is taken into account by correction factor ζ, which can be a function of the coolant velocity v:

$$\Delta p = \zeta(v) \cdot \frac{\rho}{2} \cdot v^2 \tag{1.22}$$

Hydraulic resistance, defined as a ratio between pressure drop and volume flow rate, is nonlinear. The dependence of pressure drop across a coolant mover on the generated volume flow rate is also nonlinear. Therefore, in order to solve hydraulic circuit of an electric machine, one of the methods for solution of nonlinear circuits has to be applied, the result of which is the coolant velocity distribution in the machine's hydraulic resistances.

Heat transfer takes place both in electrically active and passive components of electric machines. Dominating mechanisms of heat transfer in electric machines are convection and conduction; heat transfer by means of radiation plays a minor role.

Coolant velocity determines the heat transfer coefficient α [W/(m^2 K)], the crucial parameter for heat transfer on a boundary surface by means of *heat convection*. For a total area S_{B} [m^2] of boundary surface through which heat power P [W] is transferred from a solid body to a coolant, the difference between the body and coolant temperature $\Delta\vartheta$ [K] according to the Newton's law of cooling is equal to

$$\Delta\vartheta = \frac{P}{\alpha(v) \cdot S_B} \tag{1.23}$$

The minimum value of the heat transfer coefficient is equal to 6–7 W/(m^2 K) and corresponds to air buoyancy in the vicinity of a hot body due to difference in air density.

Heat is transferred through solid portions of electric machines by means of *heat conduction*, which is analogous to electric conduction. The amount of heat P [W] passing through a body with length l [m], cross-sectional area S_{C} [m^2], and thermal conductivity λ_{th} [W/(m K)] creates across the body a temperature drop $\Delta\vartheta$ [K] equal to

$$\Delta\vartheta = \frac{P \cdot l}{\lambda_{\mathrm{th}} \cdot S_C} \tag{1.24}$$

According to the Wiedemann–Franz law, thermal conductivity λ_{th} of a metal is proportional to its electrical conductivity κ [1/(Ω m)]

$$\lambda_{\mathrm{th}} = \kappa \cdot c \cdot T \tag{1.25}$$

with T denoting the absolute temperature of the body [K] and c a parameter.

When the thermal conductivity is not equal in all three spatial directions, as is the case in lamination where thermal conductivity in axial direction is substantially smaller than in radial and tangential, the temperature distribution can be found by solving Poisson's partial differential equation of elliptic type

$$\lambda_{\text{th}}^{x}\frac{\partial^2\vartheta}{\partial x^2}+\lambda_{\text{th}}^{y}\frac{\partial^2\vartheta}{\partial y^2}+\lambda_{\text{th}}^{z}\frac{\partial^2\vartheta}{\partial z^2}+q(x,y,z)=0 \tag{1.26}$$

with $q(x, y, z)$ denoting the spatial distribution of specific losses [W/m^3].

Heat radiation follows the Stefan–Boltzmann law according to which a body with absolute surface temperature T_s radiates at absolute temperature T_e into its surroundings the amount of heat P' per square meter of its surface [W/m^2] equal to

$$P' = c\cdot\left[\left(\frac{T_s}{100}\right)^4-\left(\frac{T_e}{100}\right)^4\right] \tag{1.27}$$

with c equal to 5.8 W/(K^4 m^2) for an absolutely black body and about to 5 W/(K^4 m^2) for the surface of a conventional electric machine. For the purpose of practical computations, the equivalent heat transfer coefficient for radiation, α_r, is introduced, the value of which varies between 5 and 7 W/(m^2 K). By using Eq. 1.6, one obtains the total heat transfer coefficient for an electric machine without cooling in the range between 11 and 14 W/(m^2 K).

Cooling type of an electric machine is a compromise between the need to take the heat out of the machine and available means for implementation of cooling.

Natural cooling requires no fan, and the turning rotor sets the air in the machine in motion. In addition, radiation takes the heat from the machine casing. This type of cooling is typical for micro- and small machines, as well as for special purpose machines.

Self-cooling indicates that there is a fan on the machine shaft, which creates pressure difference and fluid flow within the machine. Due to its simplicity, this is the most widely spread cooling method for constant speed small, medium, and large electric machines.

External cooling means that either an auxiliary motor runs the fan, or that the machine is cooled by means of another cooling medium instead of air, transported e.g., by a pump. Again, electric machine of any size can be cooled this way, and not only air, but also hydrogen, oil, water, etc., can be used as a coolant. The cooling circuit is more complex than in the case of self-cooling and often requires a heat exchanger.

1.5 Electric, Magnetic, and Thermal Properties of Materials for Electric Machines; Classes of Insulation

Performance of an electric machine is not only dependent on how well it is designed, but also on physical properties of materials used for its construction. Physical properties are, as a rule, dependent on temperature; sometimes, they also can show effects of aging.

Electrical properties of materials are described by their electrical conductance (metals) and dielectric strength (insulators). Here the emphasis will be given to the electrical conductance κ $[1/(\Omega\ \mathrm{m})]$ and its reciprocal, the specific electric resistance ρ_{el} $[\Omega\ \mathrm{m}]$. The specific electric resistance at a temperature ϑ, $\rho_{el,\vartheta}$, is a linear function of temperature

$$\rho_{el,\vartheta} = \rho_{el,0}[1 + \beta(\vartheta - \vartheta_0)] \tag{1.28}$$

with $\rho_{el,0}$ denoting the specific electric resistance at the reference temperature ϑ_0 and β the temperature coefficient of electric resistance [1/K].

When current I flows through a conductor with length l [m] and cross-sectional area S_C [m^2], the losses P [W] in the amount of

$$P = I^2 \cdot \rho \cdot \frac{l}{S_C} \tag{1.29}$$

are dissipated. Introducing current density Γ [A/m^2] defined as

$$\Gamma = \frac{I}{S_C} \tag{1.30}$$

one can express the losses in Eq. 1.29 as

$$P = \Gamma^2 \cdot \rho \cdot l \cdot A_C \tag{1.31}$$

and the specific losses, or losses per volume P/V [W/m^3]

$$\frac{P}{V} = \Gamma^2 \cdot \rho \tag{1.32}$$

Current density of 2.5 MA/m^2 creates in a copper conductor with $\rho_{Cu} = 1/56{,}000{,}000\ \Omega$ m at 20 °C the losses per volume in the amount of 111.6 kW/m^3. The same constant current density creates in iron with $\rho_{Fe} = 1/7{,}000{,}000\ \Omega$ m the losses per volume in the amount of 893 kW/m^3.

In Eq. 1.18 energy balance in a body with volume V is described in case that the whole generated power P increases its temperature. In a more general case, illustrated in Fig. 1.8, only the portion $\mathrm{m} \cdot \mathrm{c} \cdot \mathrm{d}\vartheta$ of generated energy $P \cdot \mathrm{d}t$ remains in the body and increases its temperature. The rest of generated energy equal to $S_B \cdot \alpha \cdot (\vartheta - \vartheta_0) \cdot \mathrm{d}t$ is transferred through the surface of the body to the surroundings by means of heat convection:

$$P\mathrm{d}t = m \cdot c d\vartheta + S_B \cdot \alpha \cdot (\vartheta - \vartheta_0)\mathrm{d}t \tag{1.33}$$

If the losses P in Eq. 1.33 are generated by current I at temperature ϑ, Eq. 1.29 can be further written as

$$\begin{aligned} P = I^2 \cdot \frac{l}{S_C} \cdot \rho_{el,\vartheta} = I^2 \cdot \frac{l}{S_C} \cdot \rho_{el,0}[1 + \beta(\vartheta - \vartheta_0)] = \\ = I^2 \cdot R_0(1 - \beta \cdot \vartheta_0) + I^2 \cdot R_0 \cdot \beta \cdot \vartheta \end{aligned} \tag{1.34}$$

with R_0 denoting the conductor resistance at the reference temperature ϑ_0.

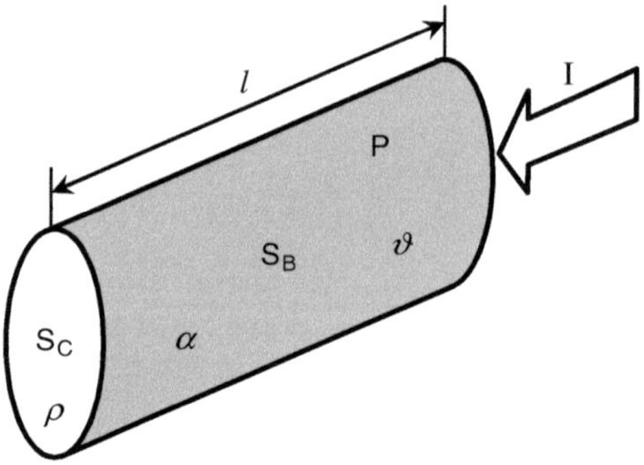

Fig. 1.8 Current-carrying conductor in which losses P are dissipated, a part of which is transferred through its boundary surface S_B to surroundings by means of heat convection. The remaining losses increase the conductor temperature

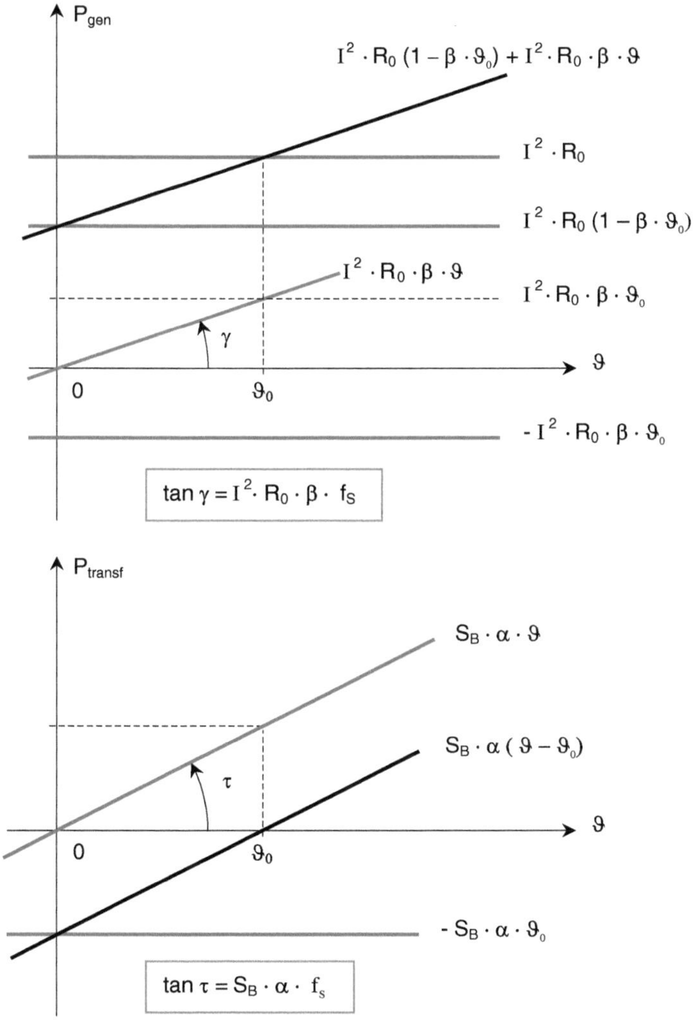

Fig. 1.9 Components of the generated P_{gen} (*above*) and transferred P_{transf} (*below*) power as functions of conductor temperature ϑ; f_S [K/W] is the scale factor which makes the argument of tangent dimensionless

After substituting Eq. 1.34 in Eq. 1.33 one obtains

$$\left[I^2 \cdot R_0(1-\beta\cdot\vartheta_0)+S_B\cdot\alpha\cdot\vartheta_0+\vartheta\cdot\left(I^2\cdot R_0\cdot\beta - S_B\cdot\alpha\right)\right]\mathrm{d}t = m\cdot c\mathrm{d}\vartheta \quad (1.35)$$

Relationships between quantities in Eq. 1.35 are visualized in Fig. 1.9. By substituting 0 for the conductor temperature ϑ in Eq. 1.34, one obtains that the first term in square brackets of Eq. 1.35 is equal to the losses *generated* at temperature $\vartheta = 0$, $P_{\vartheta=0}$:

$$P_{\vartheta=0} = I^2 \cdot R_0(1 - \beta \cdot \vartheta_0) \tag{1.36}$$

and the second term to the losses *transferred* from the conductor at temperature $\vartheta = \vartheta_0$ to the surroundings at temperature $\vartheta = 0$:

$$P_{0\to\vartheta_0} = S_B \cdot \alpha \cdot \vartheta_0 \tag{1.37}$$

The third term in square brackets of Eq. 1.35 is equal to the product of temperature ϑ and electrothermal conductance G_{eth} [W/K], which is defined as:

$$G_{\text{eth}} = I^2 \cdot R_0 \cdot \beta - S_B \cdot \alpha = \frac{\tan\gamma - \tan\tau}{fs} \tag{1.38}$$

with $\tan\gamma$, $\tan\tau$, and f_S defined in Fig. 1.9.

The electrothermal conductance G_{eth} is a crucial quantity which determines behavior of an electrothermal system in Fig. 1.8 as a function of electrical and thermal parameters, as shown in Fig. 1.10. If G_{eth} is positive, more losses are generated than can be dissipated and the temperature rate of change increases; if G_{eth} is negative, less losses are generated than can be dissipated and the temperature rate of change decreases. In limit case, when $G_{\text{eth}} = 0$, only the temperature-dependent portion of losses, $I^2 \cdot R_0 \cdot \beta \cdot \vartheta$, is transferred to the surroundings; the constant component of losses $I^2 \cdot R_0\ (1 - \beta \cdot \vartheta_0)$ increases the temperature of the body at a constant rate of change (adiabatic).

The energy balance relationship described by Eq. 1.33 is a first-order linear differential equation with constant coefficients, which can be rewritten as

$$(P_{\vartheta=0} + P_{0\to\vartheta_0} + \vartheta \cdot G_{\text{eth}})\mathrm{d}t = m \cdot c\mathrm{d}\vartheta \tag{1.39}$$

and the general solution of which can be expressed as

$$t = \frac{m \cdot c}{G_{\text{eth}}} \ln(\vartheta \cdot G_{\text{eth}} + P_{\vartheta=0} + P_{0\to\vartheta_0}) + C \tag{1.40}$$

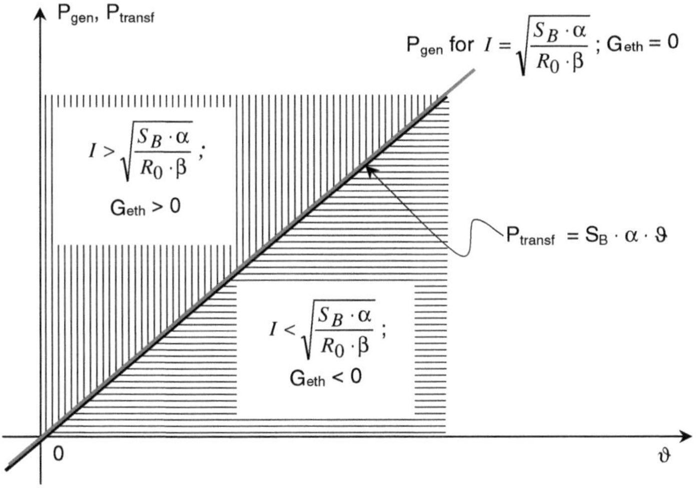

Fig. 1.10 The influence of conductor current I on the amount of electrothermal conductance G_{eth} and on the relationship between the generated and transferred power

The constant of integration C in differential Eq. (1.40) can be determined from initial condition $\vartheta = \vartheta_0$ at $t = 0$:

$$C = -\frac{m \cdot c}{G_{\text{eth}}} \ln(\vartheta_0 \cdot G_{\text{eth}} + P_{\vartheta=0} + P_{0 \to \vartheta_0}) \tag{1.41}$$

By substituting Eq. 1.41 in 1.40, one obtains

$$t = \frac{m \cdot c}{G_{\text{eth}}} \ln \frac{\vartheta \cdot G_{\text{eth}} + P_{\vartheta=0} + P_{0 \to \vartheta_0}}{\vartheta_0 \cdot G_{\text{eth}} + P_{\vartheta=0} + P_{0 \to \vartheta_0}} \tag{1.42}$$

and after rearranging:

$$\vartheta = \left(\vartheta_0 + \frac{P_{\vartheta=0} + P_{0 \to \vartheta_0}}{G_{\text{eth}}} \right) \cdot e^{\frac{G_{\text{eth}}}{m \cdot c} t} - \frac{P_{\vartheta=0} + P_{0 \to \vartheta_0}}{G_{\text{eth}}} \tag{1.43}$$

The character of solution of Eq. 1.43 depends on the sign of electrothermal conductance G_{eth}. Three typical responses can be distinguished:

(a) $\underline{G_{\text{eth}} > 0}$: If the current I dissipates more losses than that can be transferred to the surroundings through the conductor surface S_B at a given heat transfer coefficient α, the temperature will increase faster than linearly, because the conductor specific electric resistance increases too. The system behavior is typical for positive feedback: The output (temperature) increases the input (losses), which again increases the output. At certain time instant the temperature exceeds the melting point of the conductor material, which then melts. This process takes place in thermal fuses, the triggering time of which shortens rapidly as the current exceeds its rated value.

Time instant $t_{\max}$ at which the body temperature reaches the value of $\vartheta_{\max}$ is obtained by setting $\vartheta = \vartheta_{\max}$ in Eq. 1.42:

$$t_{\max} = \frac{m \cdot c}{I^2 \cdot R_0 \cdot \beta - S_B \cdot \alpha} \ln \frac{I^2 \cdot R_0 + (\vartheta_{\max} - \vartheta_0) \cdot (I^2 \cdot R_0 \cdot \beta - S_B \cdot \alpha)}{I^2 \cdot R_0} \tag{1.44}$$

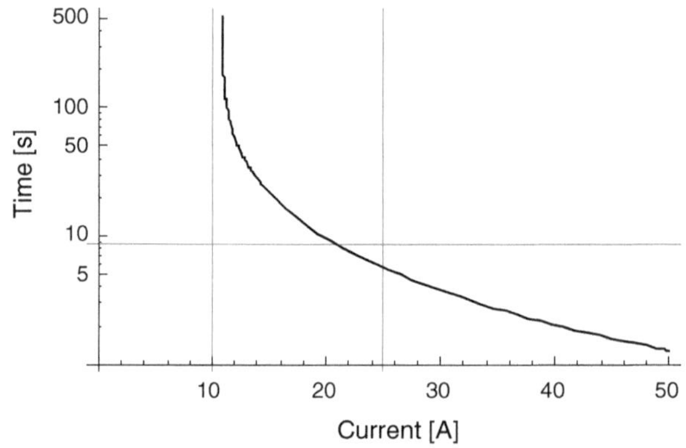

Fig. 1.11 Characteristic of a 10 A fuse

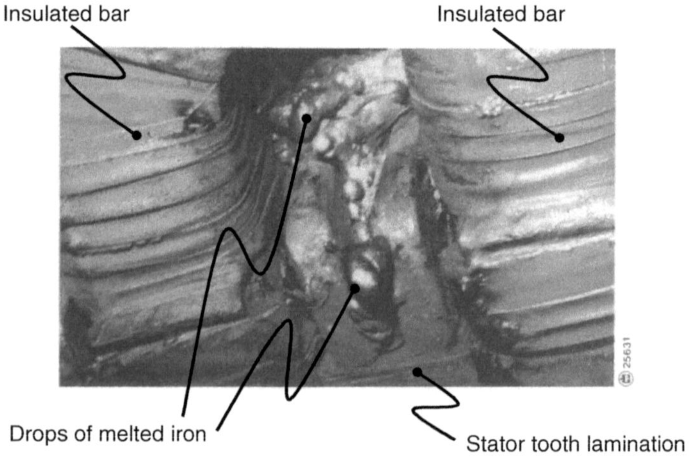

Fig. 1.12 Melted iron lamination on stator tooth tip of a large electric machine as a consequence of too high local temperatures (hot spot). From: H. Kugler "Schaeden an Turbogeneratoren," Der Maschinenschaden 49 (1976), Issue 6, pp. 221–235, photo courtesy of Allianz Deutschland AG

Equation 1.44 is illustrated by the fuse curve in Fig. 1.11, in which the characteristic of a 10 A fuse is shown. According to [4], this fuse must blow not later than 8.5 s (horizontal line) when overloaded with 2.5 times rated current.

Electrothermal conductance G_{eth} can be defined for any physical entity in which heat is produced, not only for a current-carrying conductor. For example, losses in iron lamination of large electric machines are in normal operation taken out from the volume where they are dissipated, and the iron core temperature remains below its critical value, because G_{eth} is negative. If locally dissipated losses increase due to lamination short circuit in such an extent that for a given volume G_{eth} becomes positive, a *hot spot* is created. The hot spot temperature increases until it reaches the melting point of iron. As a result of the burning of stator core lamination, iron becomes liquid, as illustrated in Fig. 1.12. In this figure one recognizes insulated stator bars on both sides of the tooth, along with tooth lamination in the lower portion of the figure. In the upper portion of this figure numerous drops of melted iron can be seen. Extremely high temperature of melted iron destroyed stator bar insulation in its vicinity, which ended in stator winding earth fault. Typical time necessary to develop the damage shown in Fig. 1.12 can reach couple of months.

(b) $\underline{G_{\text{eth}} = 0}$: The limit of expression (1.43) in this case yields

$$\lim_{G_{\text{eth}}\to 0}(\vartheta) = \frac{P_{\vartheta=0} + P_{0\to\vartheta_0}}{m \cdot c} t + \vartheta_0 \tag{1.45}$$

which is nothing but the equation of adiabatic heating

$$P_{\vartheta_0} \cdot t = m \cdot c \cdot (\vartheta - \vartheta_0) \tag{1.46}$$

where

$$P_{\vartheta_0} = P_{\vartheta=0} + P_{0\to\vartheta_0} \tag{1.47}$$

The temperature difference $\vartheta - \vartheta_0$ in Eq. 1.46 increases linearly over time.

(c) $\underline{G_{\text{eth}} < 0}$: Electric machines are designed in such a manner that the electrothermal conductance G_{eth} *is always negative*. Following Eq. 1.43, the temperature reaches its steady-state value of

$$\vartheta_{t\to\infty} = -\frac{P_{\vartheta=0} + P_{0\to\vartheta_0}}{G_{\text{eth}}} = -\frac{I^2 \cdot R_0(1 - \beta \cdot \vartheta_0) + S_B \cdot \alpha \cdot \vartheta_0}{I^2 \cdot R_0 \cdot \beta - S_B \cdot \alpha} \tag{1.48}$$

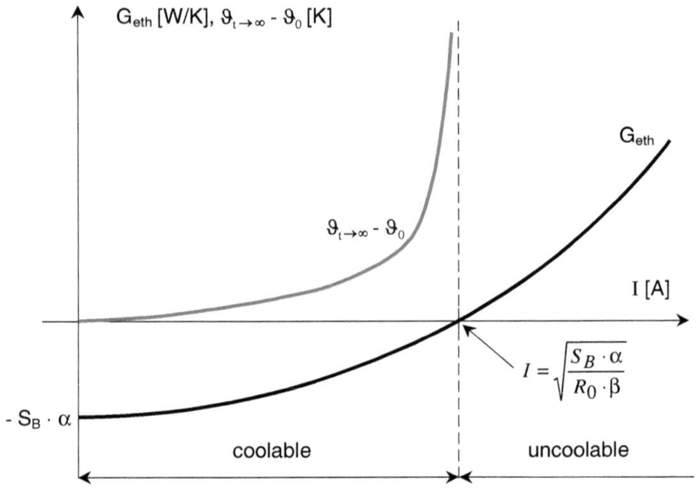

Fig. 1.13 Dependence of electrothermal conductance G_{eth} (Eq. 1.38) and temperature rise $\vartheta_{t\to\infty}-\vartheta_0$ (Eq. 1.49) on conductor current I. The conductor is coolable only for negative values of G_{eth}

or, after rearrangement:

$$\vartheta_{t\to\infty} = \frac{I^2 \cdot R_0}{S_B \cdot \alpha - I^2 \cdot R_0 \cdot \beta} + \vartheta_0 \tag{1.49}$$

Temperature rise is characterized by thermal time constant τ_{th}, which is defined as

$$\tau_{th} = -\frac{m \cdot c}{G_{eth}} = \frac{m \cdot c}{S_B \cdot \alpha - I^2 \cdot R_0 \cdot \beta} \tag{1.50}$$

The larger the current I, the higher the steady-state temperature $\vartheta_{t\to\infty}$ and the steeper the temperature increase, i.e., the shorter the thermal time constant τ_{th}.
The dependence of electrothermal conductance G_{eth} and temperature rise $\vartheta_{t\to\infty}-\vartheta_0$ on conductor current I is shown in Fig. 1.13. The conductor is coolable as long as G_{eth} is negative.
In Fig. 1.14 typical curve forms $\vartheta(t)$ for negative, zero, and positive values of electrothermal conductance G_{eth} are shown. A finite value of temperature rise $\vartheta_{t\to\infty}-\vartheta_0$ can be reached only with negative electrothermal conductance G_{eth}, i.e., only if more heat can be taken from the surface of a body than produced inside of it.

Magnetic materials used in active parts of electric machines are characterized by their magnetizing curve, AC losses, and electrical conductivity. Whereas in electric circuits a clear boundary line can be drawn between good (metals) and poor

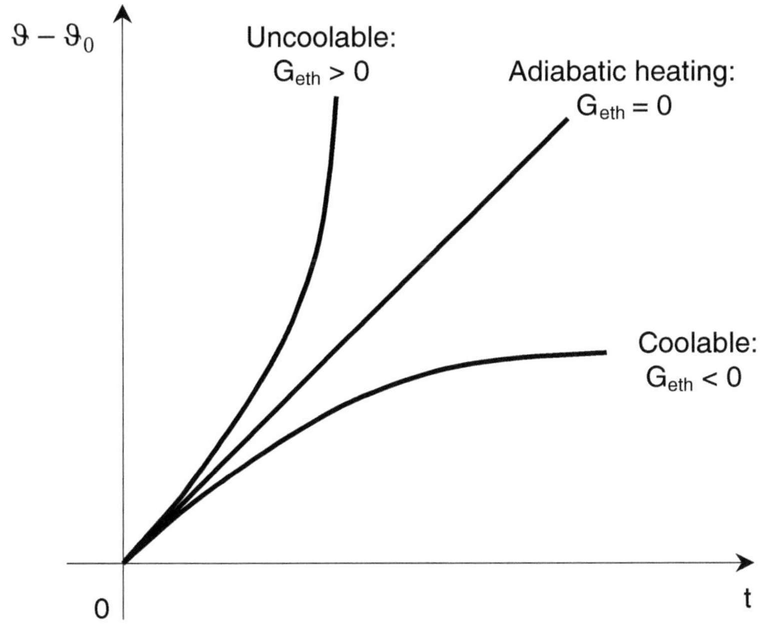

Fig. 1.14 Temperature increase of a body for various values of G_{eth}

(insulators) conductors, the relationships in magnetic field are much more vague. Good magnetic material has a relative permeability of several thousands, which is much less than the ratio between the electrical conductivity of an electrical conductor and air. Therefore, magnetic flux cannot be restricted to a magnetic material, as is the electric current channeled in a conductor. Accordingly, magnetic flux in an electric machine is not limited to its yoke and teeth only: It spreads parallel to stator and rotor yoke as well.

Depending on the character of electric source, magnetic circuit of an electric machine can be either current/MMF or flux driven, as in Fig. 1.15.

When the coil is supplied from a DC source, as shown in Fig. 1.15a, the MMF in the amount of $U_= \cdot w/R_{el}$ is imposed in the magnetic circuit, and the resulting flux Φ is determined by the *B–H* curve of the iron core. Similar situation prevails in magnetic circuits with permanent magnets which create an MMF equal to the product of coercive force and magnet thickness. If more coils share the same magnetic circuit, the resulting flux is obtained from the resulting MMF and *B–H* curve of the iron core. This situation is typical for an uncompensated DC machine, as well as for a synchronous machine, in which the load ampere-turns modify the air gap flux density created by the field winding through the mechanism of armature reaction.

If the coil is connected to an AC source, as shown in Fig. 1.15b, the amplitude of flux Φ is according to the Faraday's law determined by the amplitude and frequency of applied voltage. The amount of coil current follows from the *B–H* curve of iron core for a given flux. If another coil is placed in the magnetic circuit and connected to an impedance, a voltage will be induced between its terminals and a current in the coil will flow. The ampere-turns of the second coil will create their own flux, which is superimposed to the previously generated flux; i.e., they will try to modify the flux created by the applied AC voltage.

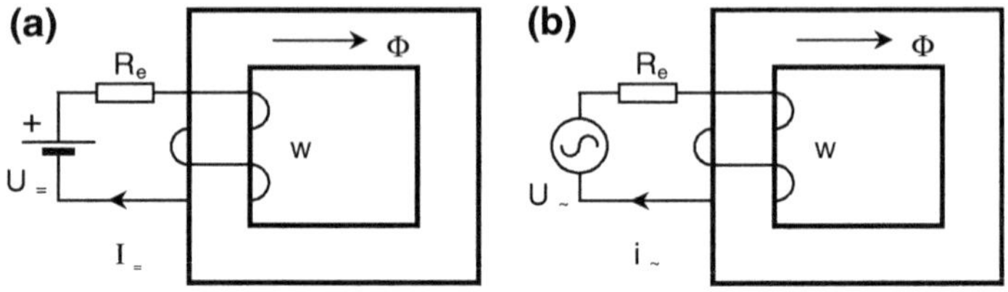

Fig. 1.15 Current/MMF (**a**) and flux (**b**) driven magnetic circuit

In order to preserve the validity of Faraday's law, the applied voltage $U_\sim$ in Fig. 1.15b supplies additional component of current, the ampere-turns of which compensate for the ampere-turns of the second coil, and the flux Φ remains unchanged. This behavior is typical for transformers and induction machines, in which the fundamental component of flux at load is equal as at no load; i.e., there is no armature reaction.

AC flux generates hysteresis and eddy current losses in a laminated core in the amount of a couple of W/kg at 50 Hz and 1.5 T [5]. Typical AC lamination losses of 4 W/kg correspond to a loss density of about 30 kW/m^3. Hysteresis losses dominate at industrial frequencies if the AC flux spreads in direction of lamination. This is the most common case in heteropolar machines, the active part of which carries flux density with only radial and tangential components. When the flux from end winding region axially penetrates into the stator lamination, it does not experience the laminated structure. Eddy currents generated by the axial component of flux density flow freely in lamination, uninterrupted by insulation between single iron sheets. In order to minimize additional eddy current losses caused by the axial component of flux, front and end edges of stator lamination in large electric machines are often punched with steadily increasing inner diameter, thus reaching a stair-like form, as shown in Fig. 1.16. This way the axial component of end winding flux density faces a kind of laminated iron when penetrating the stator lamination, which helps reduce eddy current losses and minimize risk of lamination burning [6].

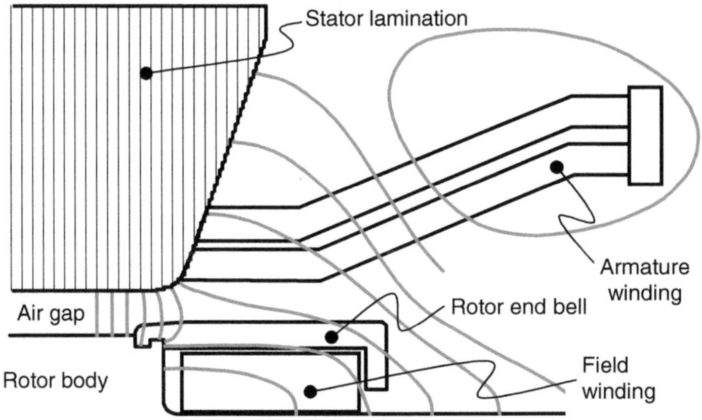

Fig. 1.16 Flux lines (*gray*) distribution in the end winding zone of a large electric machine

Due to skin effect, losses in solid iron components are not uniformly distributed throughout their whole volume. Therefore, the loss density is quantified rather per surface than per volume. Using analytical tools developed in Chap. 5, one can find that a tangential component of flux density in the air gap with an amplitude of 15 mT, which corresponds to magnetic field strength of 12 kA/m, creates a current density on the rotor solid iron surface with an rms value of 13 A/mm^2, taking 200 π [s^{-1}] for angular frequency, iron relative permeability of 1000, and electrical conductivity of iron equal to 3,000,000 [1/(Ωm)]. The loss surface density in this case equals to 26 kW/m^2. Taking a typical value of 150 W/(m^2 K) for heat transfer coefficient α, one obtains a temperature rise of 173 K by applying Eq. 1.23.

Not only the components of active part of a machine are endangered by excessive local losses, but the supporting structure can also be exposed to high alternating magnetic fields. The electromagnetic loading of passive components in electric machines can be illustrated by an example of loss generation in stator yoke wedges, as in Fig. 1.17.

Neglecting radial components of flux density in the stator yoke in Fig. 1.17, one can claim that peak values of tangential components of field strength are equal, i.e., $H_y = H_w = H_a$ (no current sheet on boundary surfaces). Due to nonlinear B–H curve of stator lamination, every increase of stator yoke flux causes a faster than linear increase of field strength. AC field strength H_w generates eddy current losses in solid steel wedges, which are proportional to the square of H_w. For this reason the stator yoke flux density in machines with stator yoke wedges is limited to values typically below 1.8 T, whereas the upper limit for tooth flux densities is determined only by the quality of magnetic material.

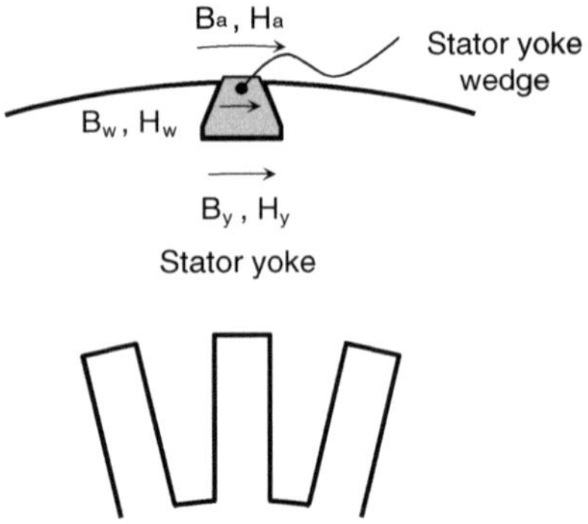

Fig. 1.17 Flux density inside and outside the stator yoke

Permanent magnets have been always attractive for electric machine manufacturers due to their ability to generate magnetic flux without dissipating electric losses, as is the case with a coil carrying DC. The advantage of loss-free generation of flux in permanent magnets is, however, accompanied with numerous disadvantages, out of which:

- Limited amount of accumulated energy per volume;
- Performance deterioration with increasing temperatures;
- Unmodifiable coercive force and residual flux density and, therefore, no direct field-weakening possibility in DC machines for the purpose of speed increase above the no-load speed, as well as no means to act against armature reaction in synchronous machines;
- Hazard of field source loss due to demagnetization;
- Eddy current losses;
- Magnet price;
- Negative environmental impact during manufacturing process of rare earth magnets—production of toxic and radioactive waste [7].

are among the most pronounced.

One of the dominating parameters which determine the performance of an electric machine is the amount of magnetic energy accumulated in its air gap, the partial derivative with respect to angle of which is equal to the electromagnetic torque. In order to illustrate the capability of permanent magnet excitation, the values of energy density created by various types of surface-mounted magnets and by current excitation are compared in Table 1.1 and in Fig. 1.18.

Table 1.1 Comparison of magnetic properties and energy densities of various permanent magnet types and current excitation [8]

	Magnet type			Current excitation: B_δ =		
	Ferrite	AlNiCo	NdFeB	0.5 T	0.7 T	0.9 T
B_r @ 20 °C [T]	0.405	1.05	1.24	N/A	N/A	N/A
$H_{c,B}$ @ 20 °C [kA/m]	260	112	940	N/A	N/A	N/A
k_B [%/K]	−0.2	−0.025	−0.1	N/A	N/A	N/A
k_H [%/K]	0.3	−0.025	−0.6	N/A	N/A	N/A
B_r @ 75 °C [T]	0.36	1.04	1.172	N/A	N/A	N/A
$H_{c,B}$ @ 75 °C [kA/m]	231	110.5	888	N/A	N/A	N/A
$(BH)_{max}$ @ 20 °C [kJ/m^3]	26.3	72	291.4	100	195	322
$(BH)_{max}$ @75 °C [kJ/m^3]	20.9	70	260	100	195	322
$(BH)_{3/4}$ @75 °C [kJ/m^3]	15.6	53	195	100	195	322

Typical data for representatives of surface-mounted ferrite, AlNiCo, and NdFeB magnets are given in columns 2–4 of Table 1.1 for magnet temperatures of 20 and 75 °C [8]. The value of $(BH)_{max}$ corresponds to the point of maximum accumulated energy in magnets, at which a conventional machine never operates, because at this operating point the magnet flux density is equal to one half of the residual flux density B_r only.

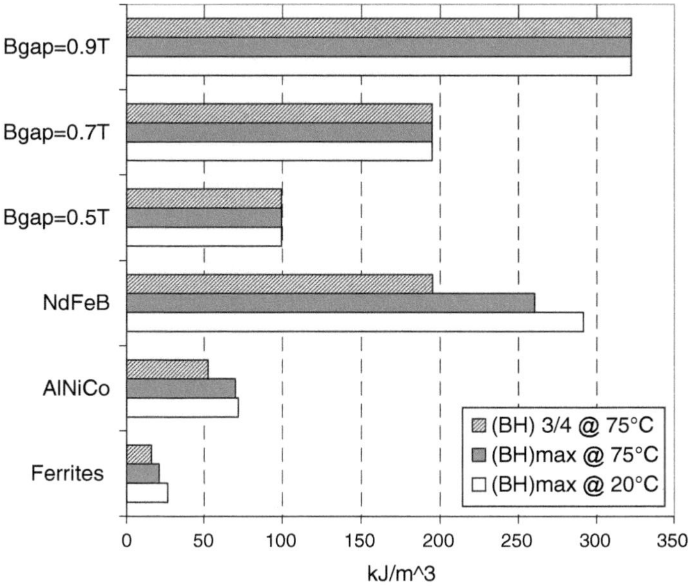

Fig. 1.18 Maximum energy density created by various magnet types and current excitation. Shaded bars reflect the most realistic operating conditions

A more realistic operating point is located above ¾ B_r, denoted by $(BH)_{3/4}$, at which, however, the accumulated energy is reduced down to about $(0.75 \cdot 0.25)/0.5^2 = 0.75$ of its maximum value $(BH)_{max}$.

By comparing the values of magnetic energy density in the air gap of current excited electric machines (columns 5–7 of Table 1.1) and in permanent magnets, one concludes that at usual operating temperatures even so-called high energy density NdFeB magnets are characterized by modest values of accumulated energy. Magnetic energy density stored in NdFeB magnets at 75 °C is below average, and it is not higher than energy density in the air gap of low- to medium-utilized induction or wound field synchronous machines with $B_\delta \leq 0.7$ T (see Table 1.1). Being plagued by such a low energy density in magnets, permanent magnet machines cannot be considered serious competitors to wound field synchronous machines when comparing torque densities in the two machine types.

The physical reason for inferior performance of permanent magnets is obvious: as any other autarkic source of energy, e.g., a DC battery, a permanent magnet has a *limited* capability of energy storage. Just as a battery, a permanent magnet cannot make more than 50 % of accumulated energy available to external magnetic circuit-the rest of it remains stored in magnet internal permeance.

The *B–H* curve of a conventional permanent magnet in the second quadrant is linear with slope $\mu_0\mu_r$, connecting points $(-H_c; 0)$ and $(0, B_r)$.

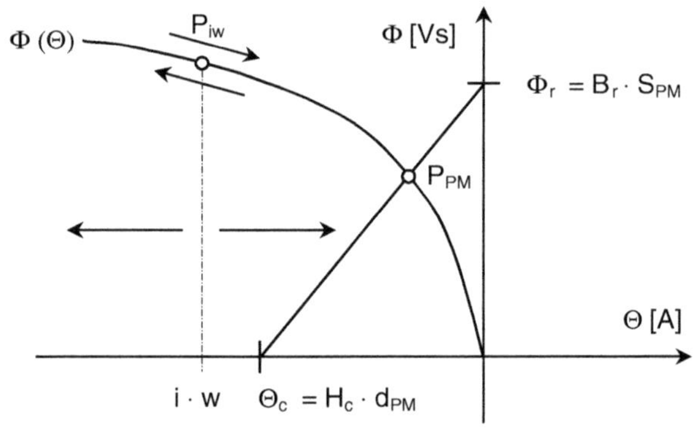

Fig. 1.19 Permanent magnet (*solid line*) and current excitation (*dashed-dotted line*) in a magnetic circuit. Whereas the operating point P_{PM} in a magnetic circuit with permanent magnet is fixed, the operating point P_{iw} in a current excited circuit slides along the magnetic circuit characteristic $\Phi(\Theta)$

By multiplying the values on *H*-axis of the magnetization curve with magnet thickness d_{PM}, and values on *B*-axis with magnet cross-sectional area S_{PM}, a line in the (Θ, Φ) coordinate system is obtained, which can be interpreted as a source characteristic in a magnetic circuit, as in Fig. 1.19.

Consequently, a permanent magnet can be formally represented as an MMF source with no-load ampere-turns $\Theta_c = H_c \cdot d_{PM}$, and an in-series-connected reluctance $R_{PM} = H_c \cdot d_{PM} / (B_r \cdot S_{PM})$, or as a flux source with short-circuit flux $\Phi_r = B_r \cdot S_{PM}$ and a parallel-connected permeance $G_{PM} = B_r \cdot S_{PM}/(H_c \cdot d_{PM})$. In both cases the linear characteristic of permanent magnet in Fig. 1.19 reflects in the best manner its limited capability as a source of magnetic flux: The internal magnet reluctance R_{PM} is a bottleneck which restricts the maximum amount of flux the magnet can deliver.

On the other hand, a current-carrying coil generates ampere-turns $i \cdot w$ *without any internal permeance*, so that the maximum amount of flux is limited only by the rest of the magnetic circuit. Whereas the operating point P_{iw} in a magnetic circuit excited by current-carrying conductor moves at will along the magnetic circuit characteristic $\Phi(\Theta)$ in Fig. 1.19, in case of permanent magnet excitation the operating point P_{PM} is fixed because of restrictions imposed by the magnet.

In terms of source and load description, a current-carrying coil is an *ideal* source in a magnetic circuit, whereas a permanent magnet with its internal permeance is a *real* source, as in Fig. 1.20. Magnet dimensions and *B–H* curve determine the maximum magnetic energy which it can deliver to a magnetic circuit; a current-carrying coil does not suffer under such restrictions and can deliver arbitrary amount of energy to a magnetic circuit. This property is utilized in superconducting magnets, in which current in a coil creates flux densities unthinkable for permanent magnets.

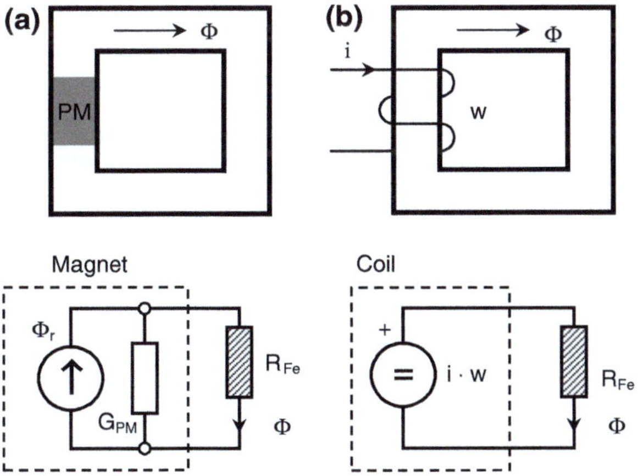

Fig. 1.20 Permanent magnet acts as a real source in a magnetic circuit (**a**), whereas a current-carrying coil is an ideal source, without internal reluctance (**b**)

As stated previously, a permanent magnet can formally be represented either as a MMF source with $\Theta_c = H_c \cdot d_{\mathrm{PM}}$ and in-series-connected reluctance $R_{\mathrm{PM}} = H_c \cdot d_{\mathrm{PM}}/(B_r \cdot S_{\mathrm{PM}})$, or as a flux source $\Phi_r = B_r \cdot S_{\mathrm{PM}}$ connected parallel to a permeance $G_{\mathrm{PM}} = B_r \cdot S_{\mathrm{PM}}/(H_c \cdot d_{\mathrm{PM}})$. Physically correct is, however, the representation in terms of flux source and parallel-connected permeance, in which magnetic energy $W_{\mathrm{PM}} = ½\, \Theta_{\mathrm{PM}} \cdot \Phi_{\mathrm{PM}}$ is stored, as in Fig. 1.21a. According to the representation with an MMF source, as in Fig. 1.21b, magnetic energy can only be stored in a loaded permanent magnet, which is apart from the physical reality.

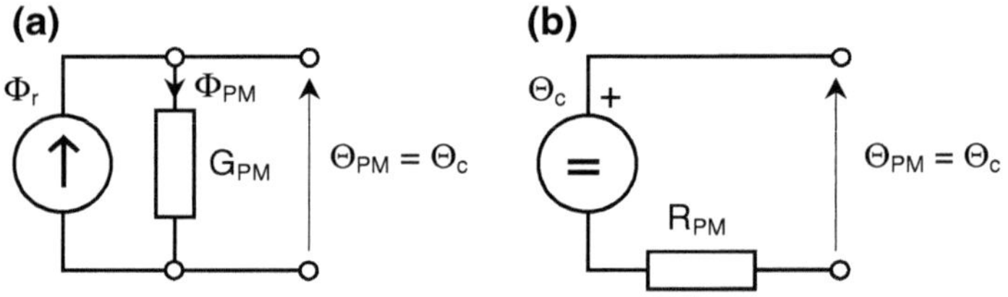

Fig. 1.21 Permanent magnet representation allowing for stored energy (**a**) and not allowing for stored energy (**b**) at no load

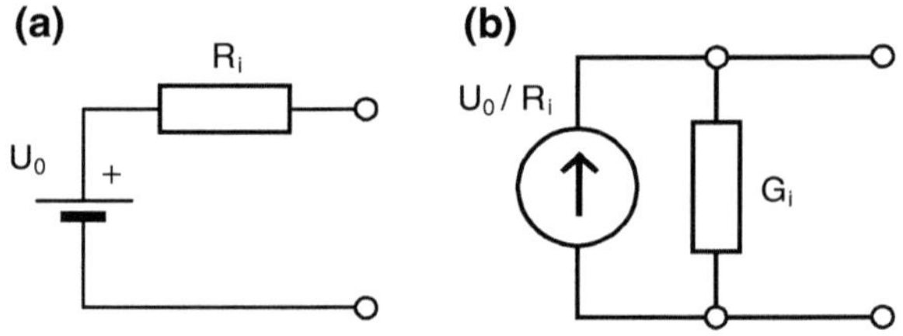

Fig. 1.22 Real voltage source dissipating no losses at no load (**a**) and its Norton current equivalent in which short-circuit losses are dissipated at no load (**b**)

For the sake of completeness, similar considerations in terms of energy consumption can be made for equivalent voltage and current sources in electric circuits, as in Fig. 1.22. Following Helmholtz's representation of a complex network with only two elements between two arbitrary terminals, one can create the physically founded equivalent circuit as shown in Fig. 1.22a. U_0 is here the open-circuit voltage, and R_i is the inner resistance of the source. Interestingly, the procedure of determination of parameters U_0 and R_i, introduced by Helmholtz in [9], is called Thevenin's theorem, although Thevenin was not born yet at the time Helmholtz published the results of his work!

Following Norton's theorem, a voltage source with no-load voltage U_0 and internal resistance R_i shown in Fig. 1.22a can be formally replaced by a current source U_0/R_i and internal conductance $G_i = 1/R_i$, as illustrated in Fig. 1.22b. Although the conversion from one source form to another in Fig. 1.22 is mathematically correct, the two sources are not identical from the point of view of energy balance. Whereas a real voltage source at no load does not dissipate energy, since the current in it is equal to zero, a current source dissipates at no load the losses in the amount of short-circuit losses of the voltage source U_0^2/R_i. This is another example of how carefully the results of mathematical operations have to be interpreted in order not to lose the physical insight into the problem.

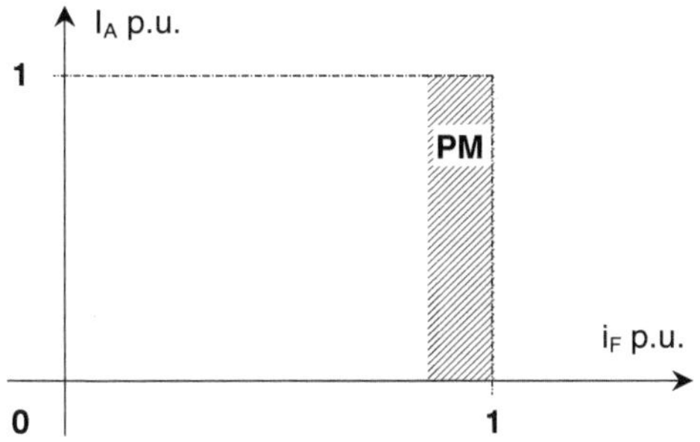

Fig. 1.23 Operating regions of wound field and PM DC machine

Operating regions of permanent magnet and wound field DC machines are compared in Fig. 1.23. Whereas there exists no electromagnetic limitation for a wound field machine to operate at any point within area delimited by ($0 < i_F < i_{F,\mathrm{rated}}$) and ($0 < I_A < I_{A,\mathrm{rated}}$) in Fig. 1.23, the operating point of a permanent magnet machine can move only within the shaded area in this figure, the width of which is determined by the magnet temperature. Therefore, the application of permanent magnet excitation is limited to DC motors without flux weakening.

Not only the efficiency of a permanent magnet excited DC motor is better than that of a wound field machine but also negative effects of armature reaction in a permanent magnet machine are negligible in the former, as illustrated in Fig. 1.24.

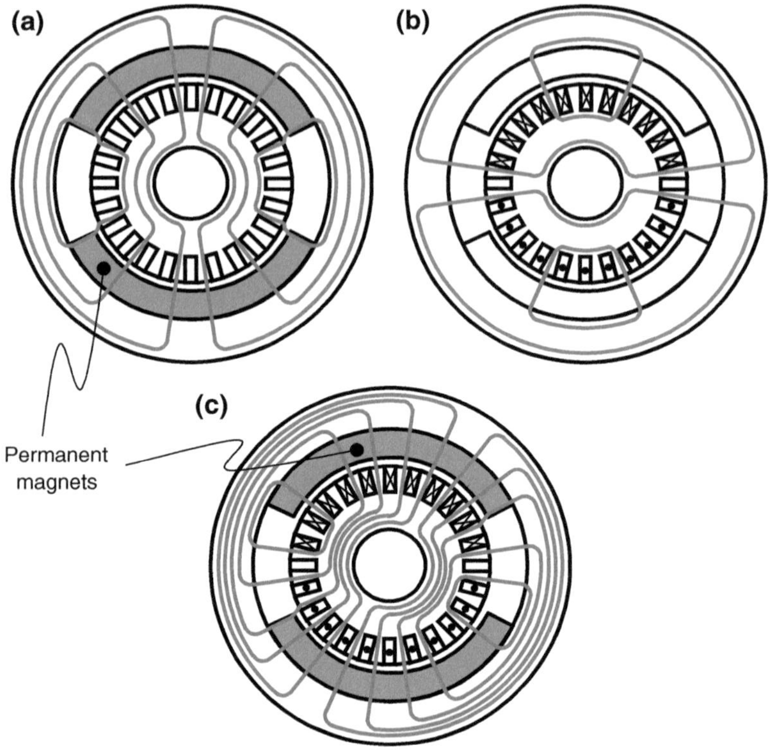

Fig. 1.24 Flux lines distribution in a permanent magnet DC motor: **a** at no load; **b** due to load ampere-turns, without magnets; **c** at load

Considering very low relative permeability of permanent magnets (typically slightly above 1), the flux created by armature reaction ampere-turns is too weak to leave any considerable trace on machine performance. Aside from uncontrollability in the speed range above the no-load speed, a DC permanent magnet motor is more advantageous than its wound rotor counterpart.

As opposed to a permanent magnet excited DC machine, where armature current has negligible influence on main flux, the armature reaction ampere-turns play a decisive role in the performance of a synchronous machine. In order to compensate for the influence of armature reaction, the field winding in a wound rotor synchronous machine has to be dimensioned for about a triple no-load field current. Since the magnetization of permanent magnets only can change (decrease) due to temperature, there is no means for it to act against armature reaction in a permanent magnet excited synchronous machine. As a consequence, the performance of a permanent magnet synchronous machine is inferior to that of a wound field synchronous machine, as illustrated in Fig. 1.25.

In Fig. 1.25 typical V-curves of a synchronous machine are shown, i.e., the dependence of armature current on the field current for constant active power as a parameter. In addition, constant power factor curves cos φ = 1 and cos φ = 0 (leading and lagging) are drawn in this figure.

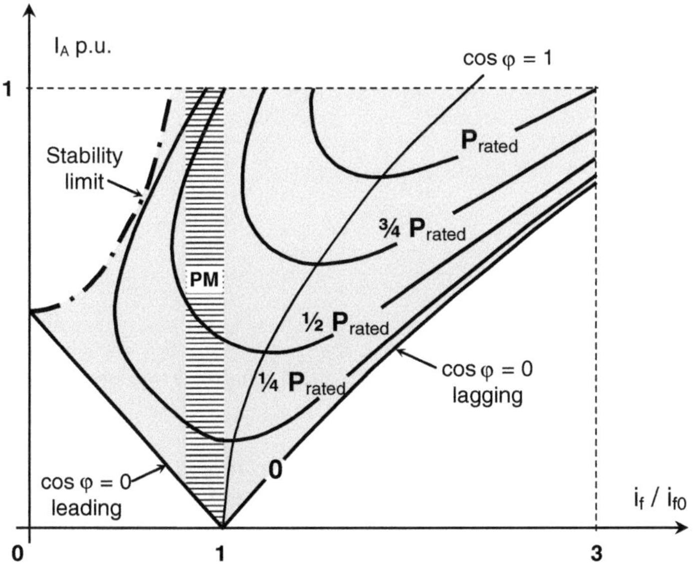

Fig. 1.25 V-curves of a wound field synchronous machine and the operating region of permanent magnet excited synchronous machine, denoted ▤

The operating point of a wound field synchronous machine can be located anywhere within the gray-colored region in Fig. 1.25. In order to operate at the rated power defined by the class of insulation, the wound field synchronous machine has to be overexcited.

A permanent magnet synchronous machine can operate only at no-load field level, i.e., it cannot be overexcited. Therefore, it reaches the thermal limit posed by its class of insulation at a power level which is (significantly) lower than the rated power. Permanent magnet synchronous machine operates underexcited at a poor power factor and has a poor torque to volume ratio.

Rated power factor cos φ_r of a permanent magnet synchronous machine is a function of its synchronous reactance X_s and magnet temperature ϑ_{PM} only. The magnet temperature ϑ_{PM} determines the rate of change $c\vartheta$ of the induced voltage. The value of cos φ_r can be found by means of machine voltage diagram, Fig. 1.26.

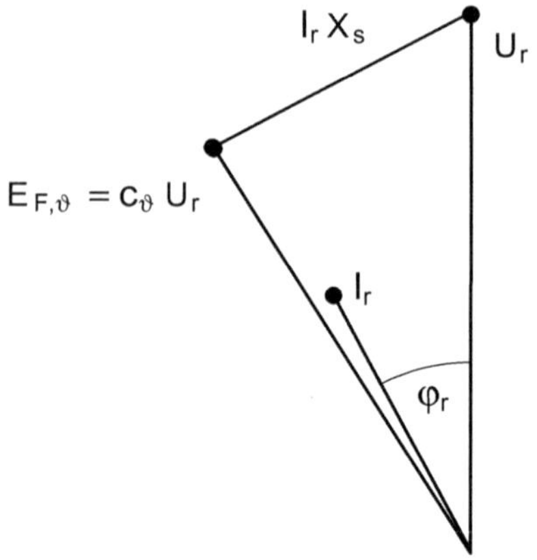

Fig. 1.26 Voltage diagram of a permanent magnet synchronous machine at a temperature ϑ_{PM} of the magnets. U_r is the terminal (rated) voltage, I_r the rated current, $E_{F,\vartheta}$ the induced voltage at a temperature ϑ, and c_ϑ the coefficient of change of induced voltage due to magnet temperature change

Replacing machine synchronous reactance X_s by its p.u. value x_s, and applying the law of cosines to components of the voltage diagram in Fig. 1.26, one obtains the rated power factor cos φ_r of a permanent magnet excited synchronous machine as

$$\cos\varphi_r = \sqrt{1 - \left(\frac{1 + x_s^2 - c_\vartheta^2}{2x_s}\right)^2} \tag{1.51}$$

The rated power factor of a permanent magnet excited synchronous machine decreases as the synchronous reactance and/or magnet temperature increases, Fig. 1.27.

Stator copper losses I^2R are equal to

$$I^2R = \left(\frac{P}{U}\right)^2 \frac{R}{m\cos^2\varphi_r} \tag{1.52}$$

i.e., they are proportional to the reciprocal of the power factor squared. The dependence of p.u. stator copper losses $p_{Cu,s}$, defined as

$$p_{Cu,s} = \frac{1}{\cos^2\varphi_r} \tag{1.53}$$

on p.u. synchronous reactance x_s at rated current I_r is shown in Fig. 1.27. One recognizes rapid increase of stator copper losses as a function of increasing magnet temperature and machine synchronous reactance. Having in mind that most permanent magnet machines are fed from an inverter, the deteriorating influence of

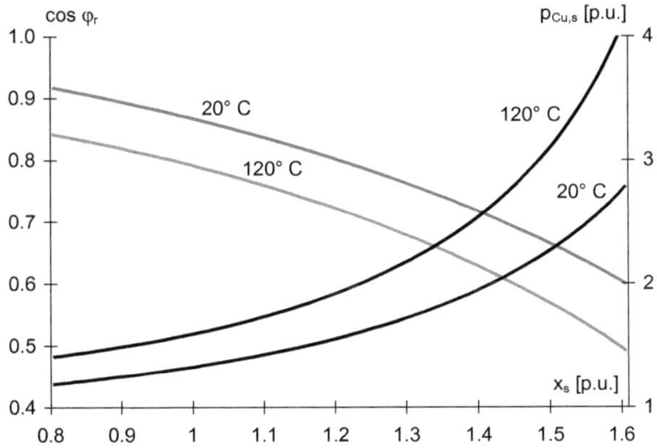

Fig. 1.27 Rated power factor of a permanent magnet synchronous machine cos φ_r (*gray*) and relative stator I^2R losses $p_{Cu,s}$ (*black*) as functions of the machine p.u. synchronous reactance x_s at two magnet temperatures

poor power factor of a permanent magnet machine spreads to the inverter and increases its rating and losses.

Case Study 1.1: A 2.85 MW, 14.5 rpm, 1.874 MNm PM generator for wind applications has a no-load voltage of 400 V, rated power factor 0.875, stator copper losses 147 kW, stator iron losses 13.5 kW, magnet losses 5 kW, rotor iron losses 19 kW, and efficiency of 93.5 %. The torque per air gap volume at rated point equals to $1874/(4.8^2 \cdot \pi \cdot 0.865/4) = 119.7$ kNm/m^3.

A wound rotor synchronous generator with identical active part dimensions has an open-circuit characteristic as shown in Fig. 1.28. For the purpose of comparison, the firm point of PM generator excitation which corresponds to 1 p.u. of the field current is also shown in this figure.

Considering equal losses in the field and armature winding of the wound rotor machine, the ratio between the rated torque M_{WR} of the wound rotor generator and M_{PM} of the PM generator is equal to

$$\frac{M_{\mathrm{WR}}}{M_{\mathrm{PM}}} = \frac{c_M \Phi_{\mathrm{WR}} I_{\mathrm{WR}} \cos \varphi_{\mathrm{WR}}}{c_M \Phi_{\mathrm{PM}} I_{\mathrm{PM}} \cos \varphi_{\mathrm{PM}}} = \frac{1.4}{0.875} = 1.6$$

for identical machine dimensions (c_{M}) and the same stator current ($I_{\mathrm{WR}} = I_{\mathrm{PM}}$) in both machines. The torque density of the wound rotor generator is, accordingly,

$$\left(\frac{M}{V}\right)_{\mathrm{WR}} = 1.6 \cdot \left(\frac{M}{V}\right)_{\mathrm{PM}} = 191.5 \frac{\mathrm{kNm}}{\mathrm{m}^3}$$

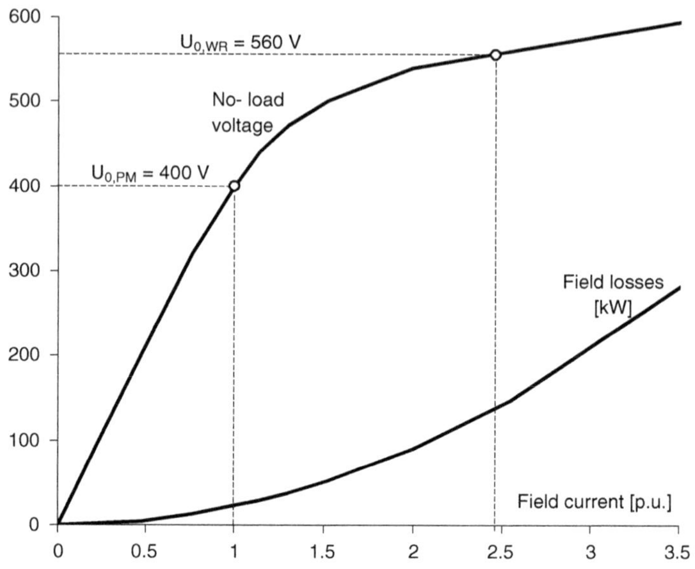

Fig. 1.28 No-load voltage and field losses of the wound rotor generator. Field current, which generates the same losses as the rated armature current, induces the no-load voltage of 560 V, i.e., 40 % higher than the induced voltage at 1 p.u. of the field current. Accordingly, the main flux in the wound rotor machine can be up to 40 % higher than in the PM machine

The PM and wound rotor machine data are summarized in the following table

	PM rotor	Wound rotor
Rated power [MW]	2.85	4.56
Rated speed [rpm]	14.5	14.5
Rated torque [kNm]	1874	2998
Torque per volume at rated point [kNm/m^3]	119.7	191.5
No-load air gap flux [p.u.]	1	1.4*
Stator copper losses [kW]	147	147
Stator iron losses [kW]	13.5	13.5
Rotor copper losses [kW]	–	147
PM losses [kW]	5	–
Rotor iron losses [kW]	19	–
Rated power factor cos φ	0.875 leading	1
Efficiency [%]	93.5	93.3
Total copper weight [kg]	5880	11,980
Magnet weight [kg]	1480	–

*For equal field and armature losses

The wound rotor synchronous machine in this case study, designed to have equal stator and rotor copper losses at rated point, is capable of generating 2.85 · 1.6 = 4.56 MW rated power from the volume from which a PM machine delivers not more than 2.85 MW at a given speed. Considering the same amount of stator copper losses in both machines and laminated rotor iron of the wound rotor generator,

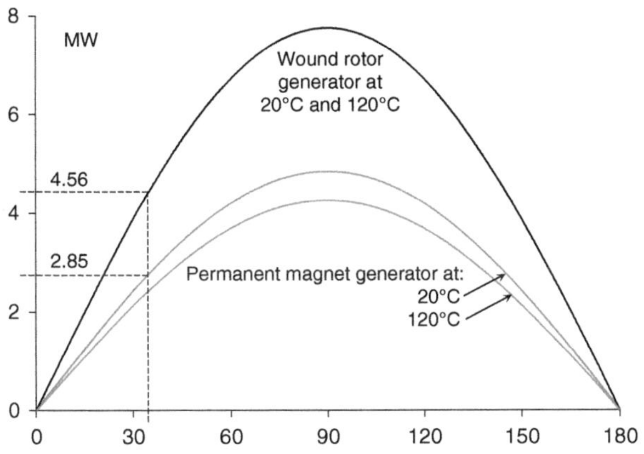

Fig. 1.29 Output power of a wound rotor and permanent magnet generator with equal active volume as a function of load angle [°] at rated speed of rotation

the total losses in the wound rotor generator are equal to $2 \cdot 147 + 13.5 = 307.5$ kW, which gives the efficiency of the wound rotor generator of 93.3 %.

The output power-load angle curves of both generators are shown in Fig. 1.29.

One should keep in mind that in the given application (wind generator), the generator active part weight has an impact on the tower construction. Therefore, the wound rotor generator not only delivers more torque from a given volume, but also helps reduce the weight and mechanical stresses on the tower significantly when built for the same power as its PM counterpart. Last but not least, the wound rotor generator does not require magnets, the price of which lies typically between 1/3 and ½ of the PM generator price.

Thermal properties of materials in electric machines depend on various physical parameters. Whereas solid parts in a machine are described by their thermal conductivity λ_{th} [W/(m K)], more information is needed in order to determine thermal properties of fluids, such as:

Mass density ρ [kg/m^3];
Specific heat c [Ws/(kg K)];
Kinematic viscosity ν [m^2/s].

Thermal properties of a coolant flowing through a cooling channel, in particular its heat transfer coefficient α, are a function of dimensionless Reynolds number Re, which is defined as

$$\mathrm{Re} = \frac{\bar{v} \cdot d_{\mathrm{h}}}{\nu} \tag{1.54}$$

with $\bar{v}$ denoting the average flow speed [m/s], ν the kinematic viscosity, and d_{h} the hydraulic diameter of the cooling channel defined as

Table 1.2 Comparison of thermal properties of various coolants

Property	Air	H_2, 1 ata	H_2, 2 ata	Water	Oil
Mass density	1	1/14.4	1/7.2	860	750
Thermal conductivity	1	7.7	7.1	23	5.3
Specific heat per unit mass	1	14.1	14.1	4.1	1.9
Specific heat per unit volume	1	1	2	3500	1400
Kinematic viscosity	1	7.3	3.6	1/16	2.2
Heat transfer coefficient for the same speed	1	1.7	2.75	570	22.2
Heat transfer coefficient for the same mass flow rate	1	11.8	11.8	2.35	1/32
Speed for the same heat transfer coefficient	1	½	¼	1/3100	1/60
Hydraulic resistance for the same heat transfer coefficient	1	½	1/8	1/6200	1/520

$$d_{\mathrm{h}} = \frac{4 \cdot S}{P} \tag{1.55}$$

with S standing for the cross sectional area and P for the wetted perimeter of the channel. Reynolds number determines the character of fluid flow. At a Reynolds number below a critical value, fluid motion will ultimately be laminar; for large Reynolds numbers, the flow will be turbulent. Turbulent fluid flow is characterized by higher heat transfer coefficients than laminar. The critical Reynolds number for fluid flow through a straight pipe with a smooth wall and a circular cross-sectional spreads in the range between 1900 and 2700.

In Table 1.2 the benchmarking thermal data for various types of coolant are given as p.u. values of air at normal pressure [10].

Class of insulation is the crucial technological parameter which determines the rated power of an electric machine. Comparing two identically manufactured machines operating under identical electrical, mechanical, and thermal conditions and having different classes of insulation, the machine built with an insulation material for higher operating temperature has a higher rated power. Electrical insulation usually contains organic materials, the aging of which eventually renders it unfit to perform its electrical and mechanical function. Aging is a function of time and temperature; its rate increases rapidly as temperature increases [11]. Although it is not possible to exactly predict the life expectancy of a particular insulation material, a rule of thumb can be used which tells that a permanent increase of machine temperature for 8 K halves the life expectancy of its insulation. However, as any other statistical quantity, life expectancy is not a precise number, and therefore, it has to be handled with reserve. Besides, temperature distribution in electric machines is not uniform, with *hot spots* denoting areas with maximum temperatures.

Class of insulation restricts the hot spot temperature at steady state. It is described with a Latin capital letter which gives the maximum permanently allowed hot spot temperature according to the following scheme:

Class of insulation	Maximum hot spot temperature [°C]
A	105
E	120
B	130
F	155
H	180

Inorganic materials, such as mica, ceramics, glass, and quartz, withstand temperatures above 180 °C and belong to C class of insulation. Large electric machines are usually built in class F, but operated in B in order to decrease the thermal stress of insulation.

Given the class of insulation, the rated power of an electric machine is a function of environmental temperature and the same machine delivers less power in tropical than in polar conditions. The more efficient cooling of a machine, the more torque it can deliver at the rated point.

1.6 Lumped Element Presentation of Electric, Magnetic, Thermal, and Fluid Flow Circuits

In Table 1.3 the lumped element presentation of electromagnetic, thermal and fluid flow quantities for an element with length ℓ and cross-sectional S is shown.

Besides electric conductivity κ and magnetic permeability μ, the following physical parameters and quantities are used in Table 1.3:

- α [W/(m^2 K)]: heat transfer coefficient;
- λ_{th} [W/(m K)]: thermal conductivity;
- ρ [kg/m^3]: mass density;
- d_h [m]: hydraulic diameter;
- λ_{ff} [−]: coefficient of increase of hydraulic resistance due to friction;
- ζ [−]: coefficient of increase of hydraulic resistance due to change of cross section.

Some of these parameters are nonlinear functions of relevant physical quantities, such as:

- Relative permeability, μ_r, dependent on flux Φ through element;
- Coefficient of increase of hydraulic resistance due to friction, λ_{ff}, dependent on volume flow rate Q [12].

In addition, the I^2R losses are a function of temperature. For all these reasons, the use of equivalent magnetic, thermal, and fluid flow circuits leads generally to a solution procedure for a system of nonlinear algebraic equations.

Considering field quantities in Table 1.3 as input and output to and out of a circuit element, one can state that except for fluid flow computation, the output is

Table 1.3 Analog quantities and parameters of lumped element presentation

Electric	Magnetic	Thermal		Fluid flow
		Convection	Conduction	
U [V]	Θ [At]	$\Delta\vartheta$ [K]	$\Delta\vartheta$ [K]	Δp [N/m^2]
I [A]	Φ [Vs]	P [W]	P [W]	Q [m^3/s]
$U = I \cdot l/(\kappa S)$	$\Theta = \Phi \cdot l/(\mu S)$	$\Delta\vartheta = P/(\alpha S)$	$\Delta\vartheta = P \cdot l/(\lambda_{th} S)$	$\Delta p = Q^2 \cdot \rho(\lambda_{ff}\, l/d_h + \zeta)/(2\, S^2)$
R_{el} [Ω] $= l/(\kappa S)$	R_{mg} [H^{-1}] $= l/(\mu S)$	R_{cv} [K/W] $= 1/(\alpha S)$	R_{cd} [K/W] $= l/(\lambda_{th} S)$	R_h [kg/m^{4}s] $= Q \cdot \rho\, (\lambda_{ff}\, l/d_h + \zeta)/(2\, S^2)$

Table 1.4 Kirchhoff's current and voltage law in various fields

	Electric	Magnetic	Thermal	Fluid flow
Current law	$\Sigma\, I = 0$	$\Sigma\, \Phi = 0$	$\Sigma\, P = 0$	$\Sigma\, \Delta p = 0$
Voltage law	$\Sigma\, U = 0$	$\Sigma\, \Theta = 0$	$\Sigma\, \vartheta = 0$	$\Sigma\, Q = 0$

proportional to the input, whereas the coefficient of proportionality is either constant, or a function of input quantity. In this sense, the simple logic of Ohm's law for DC helps solving not only electric, but also magnetic and thermal circuits. In case of fluid flow, the output (pressure drop Δp) is proportional to the square of input (volume flow Q), which in no sense influences the nature of lumped element presentation of fluid flow circuits. In Table 1.4 the implementation of Kirchhoff's laws on quantities in Table 1.3 is illustrated.

References

1. Maxwell JC (1891) A treatise on electricity and magnetism. Clarendon Press, London
2. Ostovic V (1994) Computer-Aided Analysis of Electric Machines: a Mathematica Approach. Prentice Hall, London
3. Berberovic S (1998) Teorijska elektrotehnika-Odabrani primjeri. Graphis, Zagreb
4. VDE 0635/1.59
5. Beckley P (2002) Electrical steels for rotating machines. The Institution of Electrical Engineers, London
6. Kugler H (1976) Schaeden an Turbogeneratoren. Der Maschinenschaden 49(6):221–235
7. Margonelli L (2009) Clean energy's dirty little secret. Retrieved from www.theantlantic.com
8. Yantai Zhenghai Magnetic Material Co., Ltd. (2006) Catalogue of permanent magnets. Shandong, P.R. China
9. Helmholtz H (1853) Ueber einige Gesetze zur Verteilung elektrischer Stroeme in koerperlichen Leitern mit Anwendung auf die tierisch-elektrischen Versuche. Annalen der Physik und Chemie (89/6): 211–233
10. Goetze F (1942) Neuzeitliche Loesungen der Kuehlungs- und Belueftungsfrage bei elektrischen Maschinen. Elektrotechnik und Maschiennbau 60:197–206
11. Montsinger VM (1930) Loading transformers by temperature. A.I.E.E Transac 4:776–790
12. Richter H (1971) Rohrhydraulik. Springer, Berlin

Chapter 2
Windings

Contents

V. Ostović, *The Art and Science of Rotating Field Machines Design*,
DOI 10.1007/978-3-319-39081-9_2

Diversity of winding topologies in rotating field machines helps them adapt to arbitrary load characteristic and source properties. Windings of a rotating field machine determine to the largest extent its electromechanical performance. Spatial distribution of conductors along the air gap circumference and connection of coils to particular phases are a clue to the desired machine features. Not less important is the end winding geometry, as a function of the number of winding layers, its design (wave, lap, concentric, tooth wound, etc.), number of poles, etc. Spatial distribution of winding determines its current sheet, MMF, and flux density curves, all of them being periodical functions of the air gap circumferential coordinate. Spatial Fourier analysis is therefore employed as a mathematical tool for predicting the winding performance. Both integer and fractional slot AC, as well as DC field windings are analyzed. Harmonics generated by toothed air gap and their role in electric machines and magnetic gearboxes is illustrated. The mechanism of voltage generation is illustrated on example of rotating field and rotating coil excitation. Spatial spectra of air gap MMF created by regular and discontinuous squirrel cage windings are calculated. Winding failures are analyzed in various types of windings.

2.1 Active Part and End Winding Zone, Air Gap Winding Versus Coils in Slots, Slot Fill Factor

Electric machines are built in such a manner as to have stator and rotor field components in the air gap firmly concatenated with each other. The distance between current-carrying stator and rotor conductors in the end winding region is substantially larger than in its active part. Therefore, the contribution of portions of conductors in the end winding region to the total torque is negligible as compared to the effects in the active part.

On the other hand, it is not unusual that the leakage reactance of the end winding portions of coils, and, even more pronounced, their resistance, is of the same order of magnitude as those of the active part. For this reason the end winding zone is considered a ballast, and numerous tricks of the trade have been tried out in order to diminish its influence on machine performance. However, when modifying the end winding topology in the manner it is done e.g. in tooth-wound machines, one should always try not to deteriorate too much the parameters of windings in the machine's active part and not to worsen its overall characteristics, e.g., by increasing its Carter factor and air gap leakage inductance.

Following the fundamental relation for force acting on a current-carrying conductor in magnetic field $F = B\,\ell\,I$, one would tend to place the machine windings into air gap with flux density B. This well-known equation for force in magnetic field is, however, only a special case of a more general relation, which tells that force is equal to the rate of change of energy stored in magnetic field, $\partial W_{mg}/\partial x$. Therefore, a current-carrying conductor does not have to be directly exposed to magnetic field in order to generate force in it; it is only necessary that the current-carrying conductor is a part of a turn which concatenates external field with flux density B on the place of the conductor. Accordingly, a current-carrying conductor in a slot in which the radial flux density is equal to zero generates the same force as if it were placed in the air gap above the slot, where the radial flux density is equal to B [1].

Placing conductors into slots, instead into air gap, brings electromagnetic and mechanical advantages: Coils in slots require substantially less current in order to generate certain flux level (shorter air gap!), and force does not act directly on conductors, but on teeth. On the other hand, inductances in a machine with a shorter air gap (conductors in slots) are larger than in a machine with a wider gap (conductors in air gap), keeping all other machine parameters unchanged. Besides, transfer of I^2R losses to the cooling air generates less temperature gradient if conductors are directly exposed to the air, than if they are placed in slots.

Conductors in slots are usually insulated against lamination and against each other. Their potential to the ground can reach dozens of kV. Uninsulated conductors (bars) are employed in squirrel cage induction machines and in damper cages of synchronous machines. Insulated conductors are manufactured either as round, or rectangular. Coils with rectangular conductors are more labor-intensive to manufacture than coils with round conductors.

Slot fill factor is an important parameter which determines the machine's rated power. The slot fill factor is equal to the ratio between total conductor area in a slot and slot area. Obviously, the highest slot fill factor (close to one) is achieved by using uninsolated conductors, followed by solid rectangular conductors in rectangular slots with form-wound coils (about 0.7, depending on the rated voltage). Slot fill factor of a coil with random-wound conductors is worse due to two reasons: There is always a space between round conductors in a slot, even when they are ideally aligned parallel to each other. Besides, round conductors are very hard to bring parallel to each other in a slot, as shown in Fig. 2.1a—their position in a slot is more or less chaotic, as shown in Fig. 2.1b. These two reasons limit typical slot fill factor for coils with random-wound conductors to the amounts between 0.4 and 0.5.

Windings of electric machines are usually impregnated with resin, which, among others, improves heat transfer from conductors in slot to air gap and lamination.

The influence of slot fill factor f_{Cu} on the amount of machine's rated power will be illustrated by the following example: a slot with area S_{slot} should generate the MMF of $w \cdot I$ ampere-turns when fed by current I. The I^2R losses in a conductor are equal to

$$P_{cond} = I^2 \rho \frac{l_{ax}}{S_{cond}} \tag{2.1}$$

where S_{cond} denotes the cross-sectional area of a single conductor, and ℓ_{ax} its axial length in the slot. The total I^2R losses generated by w conductors, P_{coil}, are w times larger

$$P_{coil} = w \cdot I^2 \rho \frac{l_{ax}}{S_{cond}} = (w \cdot I)^2 \rho \frac{l_{ax}}{S_{coil}} = (w \cdot I)^2 \rho \frac{l_{ax}}{f_{Cu} \cdot S_{slot}} \tag{2.2}$$

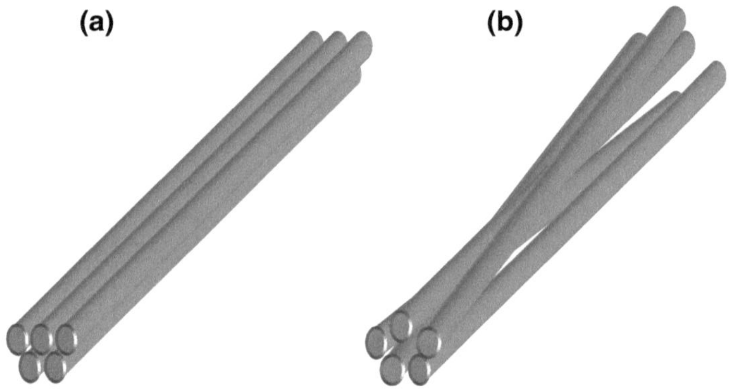

Fig. 2.1 Position of round conductors in a slot: idealized (**a**), and actual (**b**)

where $S_{coil} = w \cdot S_{cond}$. Temperature drop $\Delta\vartheta$ across the insulation layer with thickness d is equal to

$$\Delta\vartheta = \frac{P}{\lambda}\frac{d}{A} \tag{2.3}$$

with P denoting the heat power, λ the thermal conductivity of insulation, and S the area of surface through which the heat spreads. The average temperature drop across insulation can be evaluated by using equivalent slot geometry, as shown in Fig. 2.2.

The equivalent slot as shown in Fig. 2.2c carries a single fictitious conductor, the area of which is equal to the area of actual coil, $S_{coil} = w \cdot h$. The equivalent insulation with thickness d is uniformly distributed around the coil and has the total area S_{ins} of

$$S_{ins} = 2d(h + w) = S_{slot} - S_{coil} = (1 - f_{Cu})S_{slot} \tag{2.4}$$

from which one can write

$$d = S_{slot} \frac{1 - f_{Cu}}{2(h + w)} \tag{2.5}$$

Temperature drop $\Delta\vartheta$ across the insulation layer with thickness d can now be written as

$$\Delta\vartheta = \frac{P_{coil}}{\lambda \cdot l_{ax}} S_{slot} \frac{1 - f_{Cu}}{4(h + w)^2} \tag{2.6}$$

After inserting expression (2.2) for coil losses into Eq. (2.6), one obtains

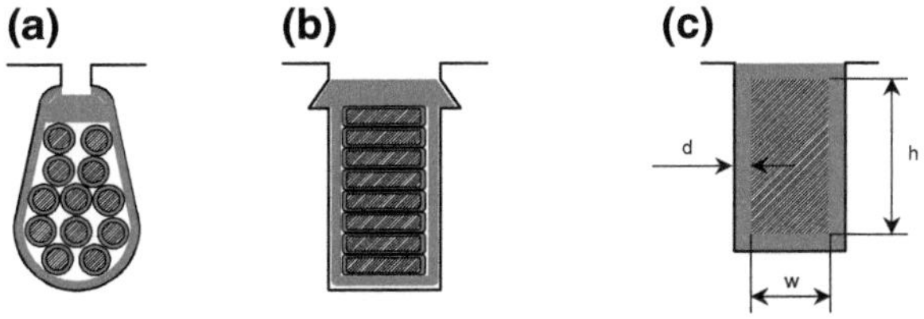

Fig. 2.2 Slot with round (**a**) and rectangular (**b**) conductors and equivalent slot geometry for thermal computations (**c**)

$$\Delta\vartheta = \frac{(w \cdot I)^2}{f_{Cu}} \frac{\rho}{\lambda} \frac{1 - f_{Cu}}{4(h + w)^2} \tag{2.7}$$

For a given temperature drop $\Delta\vartheta$ across the insulation the slot ampere-turns are proportional to

$$w \cdot I \propto \sqrt{\frac{f_{Cu}}{1 - f_{Cu}}} \tag{2.8}$$

which means that slot ampere-turns created by rectangular conductors with $f_{Cu} = 0.7$ are by a factor of

$$\sqrt{\frac{0.7}{1 - 0.7}} \Big/ \sqrt{\frac{0.5}{1 - 0.5}} = 1.53 \tag{2.9}$$

larger than slot ampere-turns created by round conductors with $f_{\mathrm{Cu}} = 0.5$, considering identical equivalent slot geometry, as shown in Fig. 2.2.

2.2 Single- and Double-Layer Windings, Coil Pitch, Skewing, Feasibility

Polyphase windings for rotating field machines are usually built to satisfy *phase symmetry condition*, which requires identical winding distribution for each phase, and *pole symmetry condition*, which specifies identical winding distribution under each (fundamental) pole. From the point of view of placement in slots, a winding can be carried out as a single layer, double layer, or mixed [2].

Integer slot windings create identical air gap flux density distribution under each pole pair. *Fractional slot windings* produce air gap flux density distributions the fundamental interval of which is larger than one pole pair. An integer slot winding is a special case of fractional slot winding—the coil distribution in an integer slot winding repeats under every pole, whereas the coil pattern in a fractional slot winding spreads over a *fundamental pole*, which comprises an odd number of machine poles.

Fig. 2.3 Coil of a single-layer winding

Integer slot windings can be designed to generate the air gap flux density distribution spectra in which the strongest harmonics are the fundamental and the slot harmonics of the order $N/p \pm 1$. Therefore, integer slot windings can be utilized in both induction and synchronous machines. Fractional slot windings generate spectra in which arbitrary harmonics can dominate. As such, they are suitable for synchronous machines only.

Probably the simplest winding configuration imaginable is the one in which all conductors in a slot belong to a single coil, or a **single-layer winding**, as shown in Fig. 2.3. The total number of coils of a single-layer winding is $N/2$, N denoting the number of slots, since each coil occupies two slots.

Depending on the form of coil ends, single-layer windings are manufactured either as *concentric*, or *distributed*. Both topologies can generate identical MMF distribution as long as the conductor placement in slots is identical.

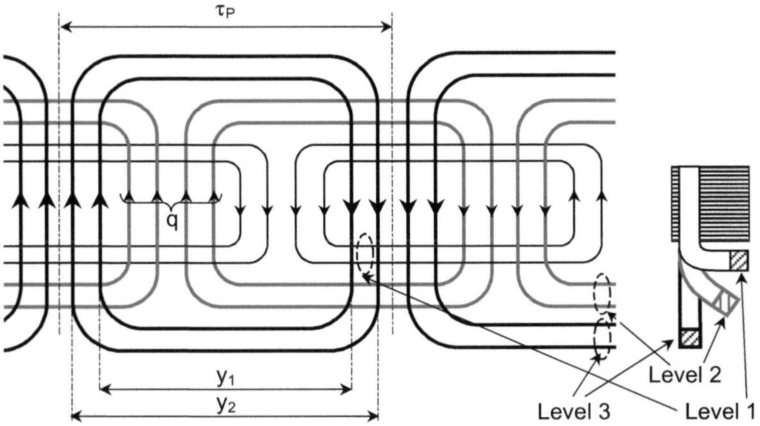

Fig. 2.4 Single-layer concentrated three-phase winding with coil ends manufactured in 3 levels and $q = 4$ slots per pole and phase. Assuming that all coils are wound in the same way, the arrows show direction of slot ampere-turns at time instant when the current in one phase is maximal

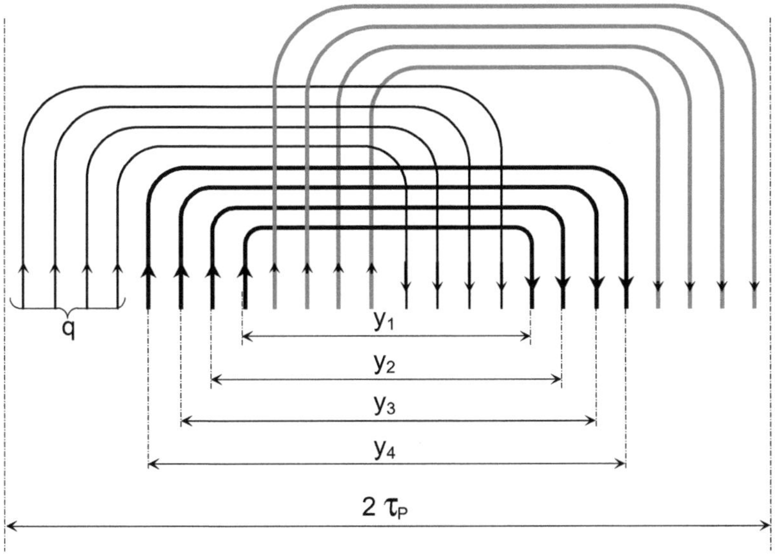

Fig. 2.5 Single-layer concentrated three-phase winding with coil ends manufactured in 3 levels and $q = 4$ slots per pole and phase. Stator with such winding is separable every 2 poles

The most important parameter of a coil is its *pitch*, or the circumferential distance in air gap given in number of teeth between the left-hand side and the right-hand side of a coil. *Winding pitch*, on the other hand, depends on several parameters and can be equal to the coil pitch for some winding topologies. Coils in concentric windings have different pitches and can spread either through three (Figs. 2.4 and 2.5), or two levels (planes) in the end winding zone (Fig. 2.6). The innermost coil of a concentric winding has the shortest pitch (e.g., y_1 as shown in Fig. 2.5), and the outermost coil the largest geometric pitch (y_4 as shown in Fig. 2.5). The winding pitch y of a single-layer concentric winding, defined as the arithmetic mean of all coil pitches, is less or equal to the pole pitch τ_p.

The *zone width q* of a single-layer concentrated winding (Fig. 2.5) is equal to the number of slots under one pole which belong to the same phase, i.e., which generate the same ampere-turns. The zone width is expressed either as a number of slots per pole and phase, $q = N/(2 \cdot p \cdot m)$, or as an electrical angle π/m, m denoting the number of phases.

If the stator outer diameter is too large, its lamination can be segmented in circumferential direction and wound with single-layer windings with 3 coil end levels as shown in Fig. 2.5.

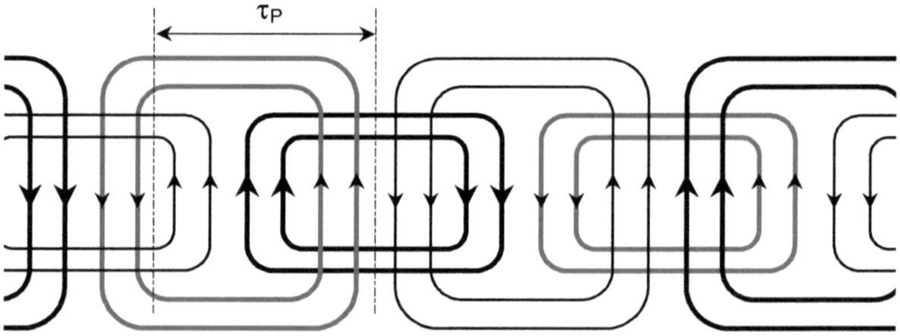

Fig. 2.6 Single-layer concentrated three-phase winding with coil ends manufactured in 2 levels and $q = 2$ slots per pole and phase

Single-layer windings with coil ends manufactured in 2 levels have shorter end windings than those with 3-level coils. Besides, they make use of only two different coil shapes. However, their end coils are concatenated in such a way that they do not allow for segmentation without cutting a coil group belonging to one phase, as shown in Fig. 2.6.

Field windings of synchronous machines are manufactured with concentric coils, as shown in Fig. 2.7. Coil ends are placed next to each other in axial direction and supported against centrifugal forces. The outermost slots are sometimes manufactured with smaller height in order to reduce the level of saturation in poles.

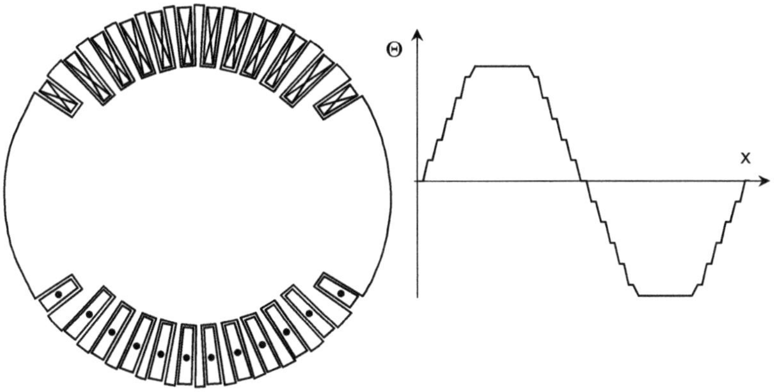

Fig. 2.7 Single-layer winding with concentric coils

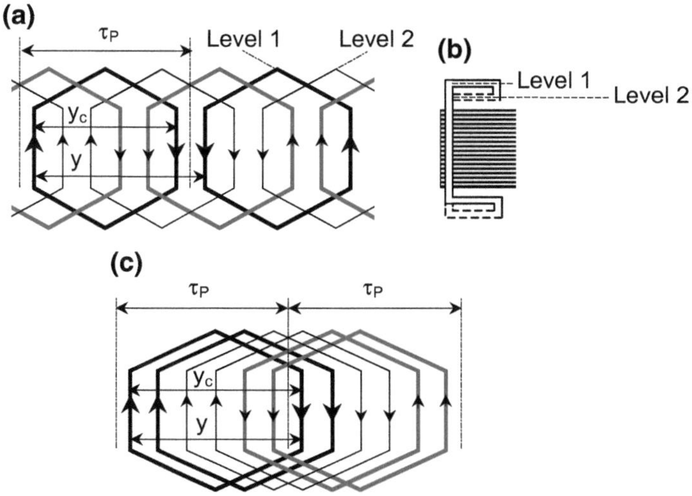

Fig. 2.8 Single-layer distributed three-phase lap windings (**a**), (**c**) and their end winding form (**b**). Winding shown in (**a**) has coil pitch y_c different from, and in (**c**) equal to the winding pitch τ_p. Winding (**c**) allows for segmentation in circumferential direction

Distributed single-layer windings can be manufactured as *wave* or *lap* (Fig. 2.8). Wave windings require less welding, because one phase group (parallel circuit) can be manufactured in a single production stage. All coils of a distributed single-layer lap winding have pitch y_c, which is not necessarily equal to the pole pitch τ_p.

A single-layer distributed winding can be short-pitched (chorded) for the amount of 0, 1, 3, 5, 7, …, etc., slots if the number of slots per pole and phase q is even, as shown in Fig. 2.9. If q is odd, as shown in Fig. 2.10, a single-layer distributed winding can be short-pitched for 0, 2, 4, 6, 8, …, etc., slots. Coil pitch chording for 1 slot has no influence on the winding pitch and, therefore, no influence on the air gap MMF distribution. Chording for more than 1 slot results in phase interleaving (Figs. 2.9 and 2.10). Chorded coils in windings with odd values of q cannot be symmetrically distributed in a two-pole interval. As a result, even harmonics in the air gap MMF distribution are generated, as shown in Fig. 2.10.

Coils of a single-layer distributed winding get themselves in the way because a slot occupied by conductors of one coil cannot be used by conductors of other coils. Such a rigid structure of single-layer windings limits substantially the field of possible applications.

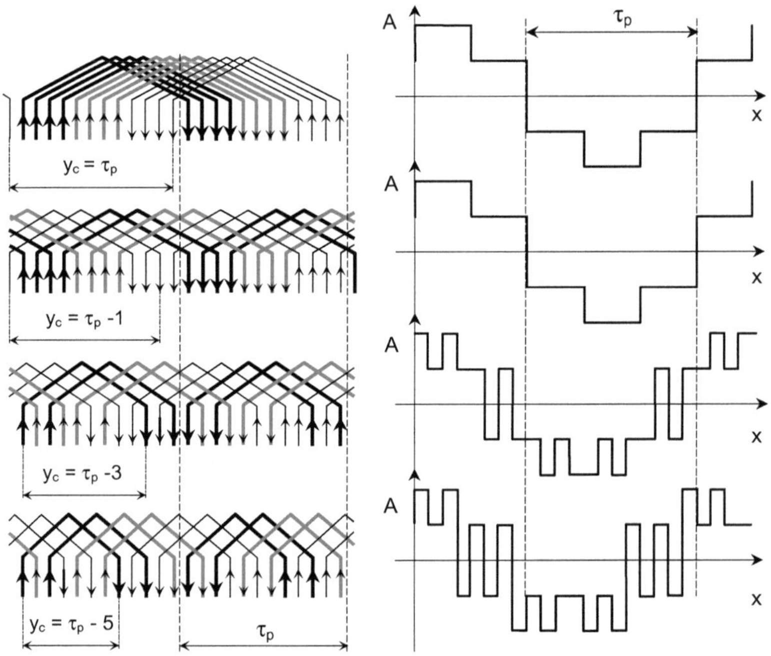

Fig. 2.9 Single-layer distributed three-phase lap windings with $q = 4$ slots per pole and phase and different coil pitch y_c (*left*) and current sheet A at a time instant of maximal current in one phase (*right*). The lower two windings are chorded for more than one slot and, therefore, interleaved

Double-layer windings offer much more freedom in creating air gap MMF distribution than single-layer windings. Each slot of a double-layer winding carries conductors of two coils, which belong either to separate phases (so-called *mixed slots*), or to the same phase (*monoslots*). One side of a coil of symmetrical double-layer winding lies in the bottom layer of one slot, and the other coil side in the top layer of another slot (Fig. 2.11). Thus the number of coils of a double-layer winding equals to the number of slots in which it is placed.

Since the two layers in a slot belong to separate coils, the coil pitch can be selected arbitrarily. Higher harmonics of air gap MMF created by a double-layer winding can be controlled more precisely than for a single-layer winding, which makes the double-layer winding the mostly spread winding type at all.

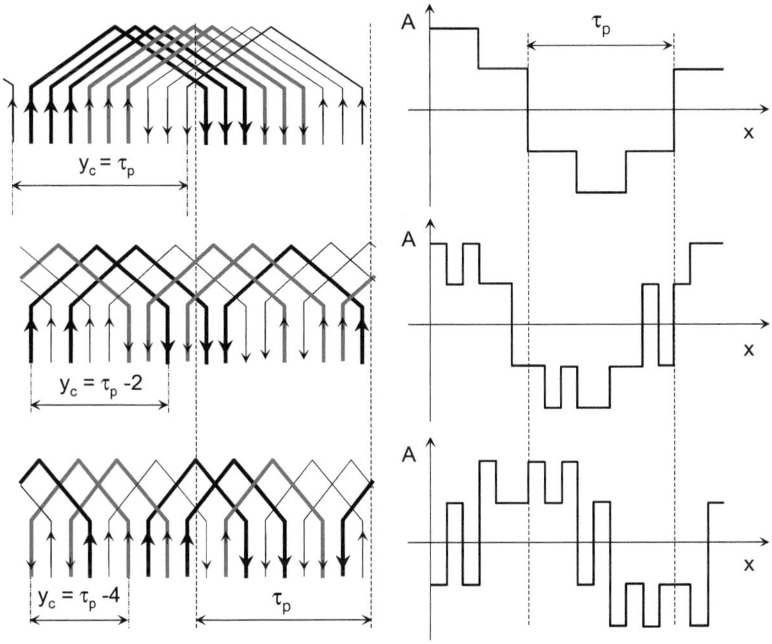

Fig. 2.10 Single-layer distributed three-phase lap windings with $q = 3$ slots per pole and phase and different coil pitch y_c (*left*) and current sheet A at a time instant of maximal current in one phase (*right*). The lower two windings are interleaved and asymmetrically distributed over two poles. As a result, even harmonics of current sheet and air gap MMF are generated

On the other hand, the slot fill factor of a double-layer winding is not as good as in case of a single-layer one, because the two layers have to be insulated against each other. This aspect is especially important in medium voltage machines, which anyway have thick coil insulation.

A comparison of properties of single and double-layer winding is given in Table 2.1.

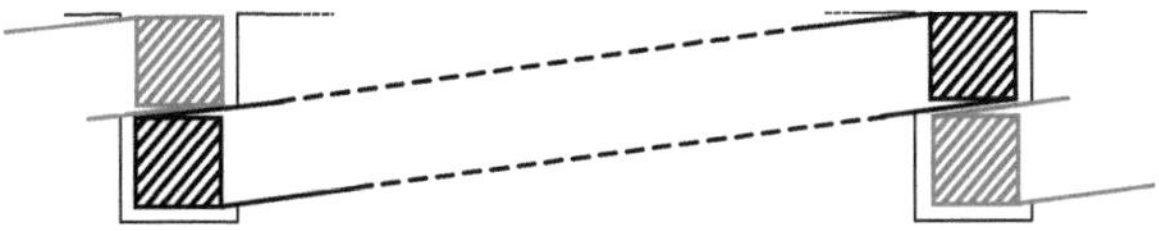

Fig. 2.11 Coil placement in a double-layer winding

Table 2.1 Comparison of winding-type properties

	Single layer	Double layer
Number of coils	*N*/2 for *N* slots	*N* for *N* slots
Coil pitch	Dependent on type	Arbitrary
Coil form	Dependent on type	Equal for all coils
Slot utilization	High	Moderate
End winding	Dependent on type	Compact
Mechanical strength	Moderate	High
Application	Predominantly small and medium machines	Unlimited

Mixed-layer windings combine single- and double-layer windings, as shown in Fig. 2.12. Coils of a mixed-layer winding are inserted into slots in accordance with their affiliation to a particular phase. In Fig. 2.12 the outermost coils can have twice the number of turns than the other ones if all slots have the same area. Alternatively, the slots for outermost coils can be made smaller than others, and all slots can have the same area. Windings as shown in Fig. 2.12 are suitable for automated production because first all coils of the phase A are inserted, after that all coils of the phase B and at the end all coils of the phase C.

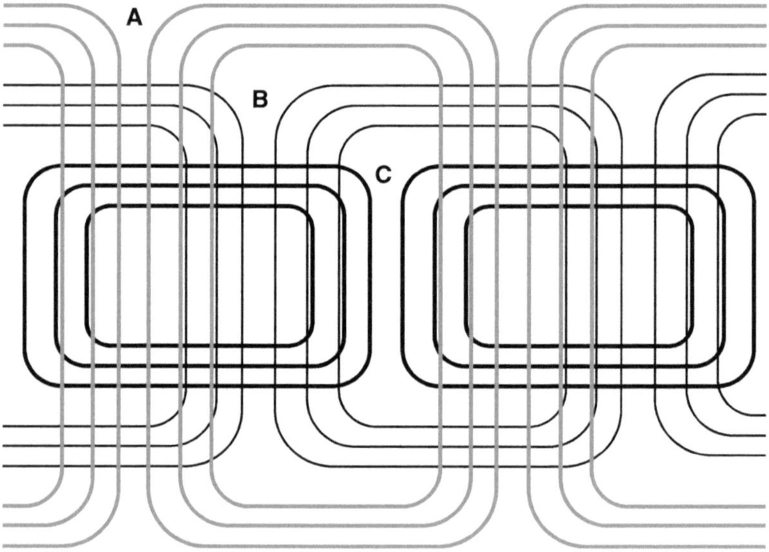

Fig. 2.12 Coil placement in a mixed-layer winding

2.3 Current Sheet and Air Gap MMF

The fundamental idea governing physical and engineering sciences is to build mathematical models of physical phenomena, which reflect all effects of interest for a particular phenomenon, and which are interpreted as physical laws. The simpler the model, the wider the area of its applications. Accordingly, an analytical model gives a significantly deeper insight into a particular phenomenon than its numerical counterpart. Therefore, a numerical model can at best complete machine analytical model, but in no case replace it.

Electric machines are complex 3D structures, the operation of which is based on magnetic circuits with nonlinear magnetic materials, in which electromagnetic quantities permanently change their values and direction in space. A detailed model of an electric machine, which would allow for all physical peculiarities at any time instant and at any point in the machine volume, could only be built as a numerical approximation. As such, it could deliver a quantitative solution to a particular problem at a given accuracy, but it would not allow a qualitative insight into physics of machine operation.

Obviously, the level of complexity has to be reduced if an analytical model of an electric machine should be built which represents the crucial machine properties. Effects neglected in such a model have marginal influence on the accuracy of results.

Analytical model of an electric machine is based on space-time representation of electromagnetic quantities in it, whereas the spatial coordinate is placed in the middle of the air gap and spreads in circumferential direction. Consequently, only the radial component of air gap flux density is considered. It is also assumed that the machine is infinitely long, which legalizes disregarding of axial components of field in its active part.

Winding distribution and affiliation of coils to a particular phase are expressed by means of *current sheet* A, an auxiliary quantity with a meaning of linear current density, defined as

$$A = \frac{i \cdot w}{b} \tag{2.10}$$

with b denoting the width of zone in the air gap in which w conductors, each of which carrying current i, are placed. The sign of current sheet is determined by the orientation of coil and by the current direction.

2.4 Spatial Harmonics in Air Gap MMF, Slot-Opening Factor, Winding Factors

The overwhelming majority of rotating field machines is built as heteropolar, with active part excited by coils the conductors of which are placed between two ferromagnetic structures schematically represented in Fig. 2.13. In general, the current sheet distribution $A(x)$ along the circumference does not necessarily have to be symmetrical with respect to the axis of winding.

In Fig. 2.13, the **x** and • marks within conductors denote products of coil orientation and current sign and, as such, can be interpreted as positive or negative. Orientation of a coil along with sign of current through it determines the sign of current sheet, MMF, and flux density. Therefore, not the current I, but the ampere-turns $I \cdot w$ are the primordial machine spatial quantity, since they carry both the information on the amount and direction of air gap fields.

If the relative permeability μ_r of stator and rotor iron is very large, μ_r >>, the whole MMF drop $\Theta(x)$ created by coil ampere-turns is spent on air gap and can be defined as

$$\Theta(x) = \int A(x)\mathrm{d}x \tag{2.11}$$

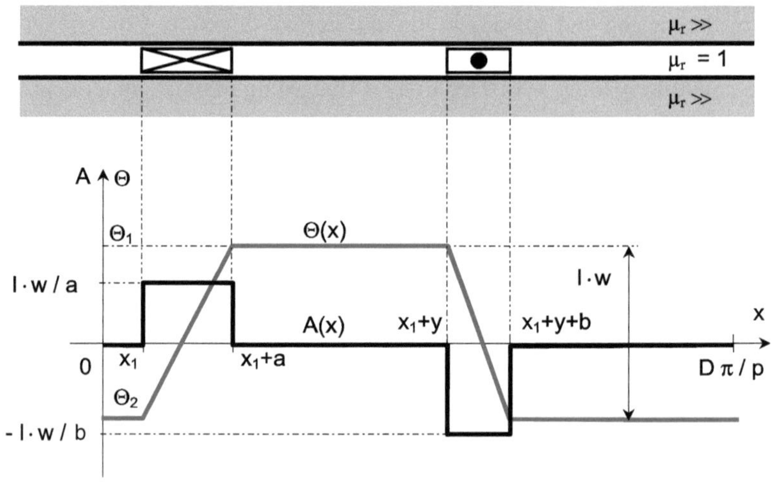

Fig. 2.13 Current sheet distribution $A(x)$ and its integral, the MMF distribution $\Theta(x)$ of an arbitrary coil

with x denoting the air gap circumferential coordinate and $A(x)$ the current sheet created by current I.

Constant MMFs Θ_1 and Θ_2 in Fig. 2.13 are evaluated as

$$\Theta_1 = \frac{2D\pi + p \cdot (a - 2y - b)}{2D\pi} \cdot I \cdot w; \quad \Theta_2 = p \cdot \frac{a - 2y - b}{2D\pi} \cdot I \cdot w \tag{2.12}$$

Air gap quantities are periodical functions of circumferential coordinate x with period length $2\tau_p$, τ_p being the pole pitch:

$$\tau_p = \frac{D\pi}{2p} \tag{2.13}$$

Spatial distribution $\Theta(x)$ can be expressed in terms of Fourier series as

$$\Theta(x) = \sum_{n=1}^{\infty} \Theta_n \sin n \frac{\pi}{\tau_p} x \tag{2.14}$$

with Θ_n denoting the amplitude of the nth harmonic of the air gap MMF. Fourier coefficients of MMF distribution in Fig. 2.13 are

$$a_0 = 0 \tag{2.15}$$

because there is no homopolar flux (div $\mathbf{B}$ = 0), and

$$a_n = \tau_p \cdot I \cdot w \frac{a \cdot \left[\cos n\frac{\pi}{\tau_p}(x_1 + y) - \cos n\frac{\pi}{\tau_p}(x_1 + y + b)\right] + b \cdot \left[\cos n\frac{\pi}{\tau_p}(x_1 + a) - \cos n\frac{\pi}{\tau_p}x_1\right]}{n^2 ab\pi} \tag{2.16}$$

$$b_n = \tau_p \cdot I \cdot w \frac{a \cdot \left[\sin n\frac{\pi}{\tau_p}(x_1 + y) - \sin n\frac{\pi}{\tau_p}(x_1 + y + b)\right] + b \cdot \left[\sin n\frac{\pi}{\tau_p}(x_1 + a) - \sin n\frac{\pi}{\tau_p}x_1\right]}{n^2 ab\pi} \tag{2.17}$$

Conventional windings are placed in slots, the width of which in the air gap is called *slot opening d*. Single coil per pole pair placed in slots on one side of air gap as illustrated in Fig. 2.14 generates the MMF distribution which can be expressed in terms of Fourier series with coefficients a_n and b_n:

$$a_n = \frac{2}{\pi}\frac{Iw}{n}\frac{\sin n\frac{d}{\tau_p}\frac{\pi}{2}}{n\frac{d}{\tau_p}\frac{\pi}{2}} \cdot \cos n\frac{\pi}{2} \cdot \sin n\frac{y}{\tau_p}\frac{\pi}{2} \tag{2.18}$$

$$b_n = \frac{2}{\pi}\frac{Iw}{n}\frac{\sin n\frac{d}{\tau_p}\frac{\pi}{2}}{n\frac{d}{\tau_p}\frac{\pi}{2}} \cdot \sin n\frac{\pi}{2} \cdot \sin n\frac{y}{\tau_p}\frac{\pi}{2} \tag{2.19}$$

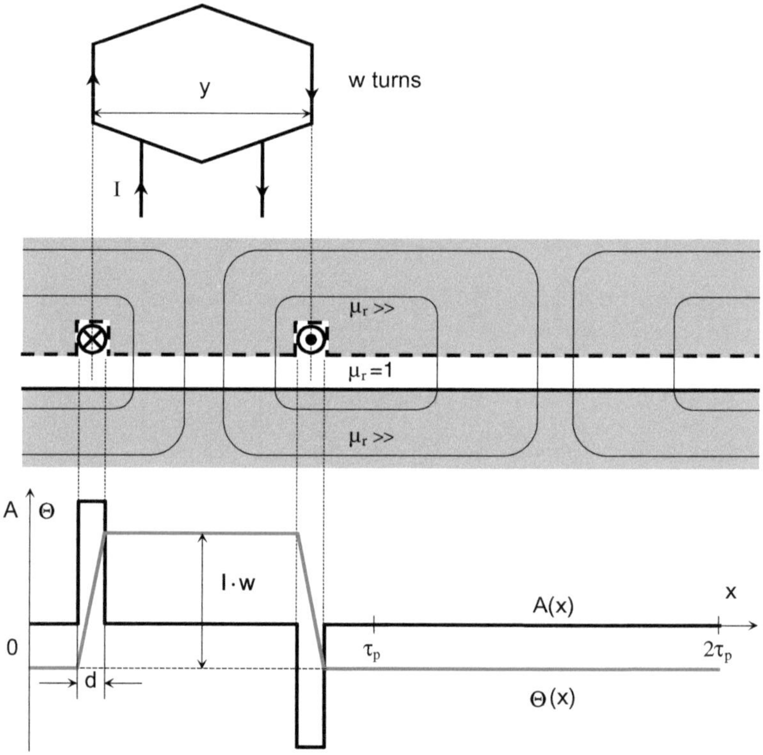

Fig. 2.14 Current sheet distribution $A(x)$ and its integral, the MMF distribution $\Theta(x)$ of single coil per pole pair

Fourier coefficients a_n and b_n are equal to the product of three factors, each of which is smaller or equal to one:

(a) *Slot-opening factor* $f_{o,n}$, defined for the nth harmonic as

$$f_{o,n} = \frac{\sin n\frac{d}{\tau_p}\frac{\pi}{2}}{n\frac{d}{\tau_p}\frac{\pi}{2}} \tag{2.20}$$

The slot-opening factor is a function of the ratio between the slot opening *d* and pole pitch τ_p. As long as the slot opening d is much smaller than the pole pitch τ_p, the slot-opening factor for the fundamental, $f_{o,1}$, is almost equal to one. Large slot openings, typical for tooth-wound machines, cause low slot-opening factors, thus diminishing resulting MMF and deteriorating machine performance.

(b) A *trigonometric indicator*, defined for cosine terms a_n as $\cos(n \cdot \pi/2)$ and for sine terms b_n as $\sin(n \cdot \pi/2)$. The indicator for cosine terms, $\cos(n \cdot \pi/2)$, is equal to zero for odd harmonics, and indicator for sine terms, $\sin(n \cdot \pi/2)$, is equal to zero for even harmonics. In other words, the phase shift of odd harmonics relative to the origin is equal to zero, and the phase shift of even harmonics relative to the origin is $\pm\pi/2$.

Every odd harmonic from Fourier approximation of MMF distribution has at least one zero crossing point identical with the zero crossing point of the fundamental. Accordingly, every odd harmonic has an extreme at the same point on the circumference where the fundamental is extreme. The character of extreme of adjacent odd harmonics alters permanently between maximum and minimum, as given by the indicator $\sin(n \cdot \pi/2)$. For this reason, the resultant effect of (odd) slot harmonics of MMF on the fundamental of air gap flux density is negligible (see discussion to Fig. 2.37).

(c) *Coil pitch factor* $f_{p,n}$, defined for the *n*th harmonic as

$$f_{p,n} = \sin n \frac{y}{\tau_{ps}} \frac{\pi}{2} \tag{2.21}$$

where τ_{ps} denotes the pole pitch expressed in number of slots

$$\tau_{ps} = \frac{N}{2p} \tag{2.22}$$

Coil pitch factor quantifies the MMF derogation due to pitching of coils. In extreme case when the product $n \cdot y/\tau_{ps}$ is an even number, the amplitude of the *n*th harmonic is equal to zero. The influence of coil pitch on the amplitude of the *n*th harmonic of air gap MMF can be illustrated by the example of the fifth harmonic of MMF created by a coil with pitch of $y/\tau_p = 4/5$, as shown in Fig. 2.15.

Fig. 2.15 MMF distributions created by a full-pitch coil per pole, $y = \tau_p$ (**a**), and a short-pitch coil per pole, $y = 4/5\ \tau_p$ (**b**). The fifth harmonic in the MMF distribution of the coil below vanishes

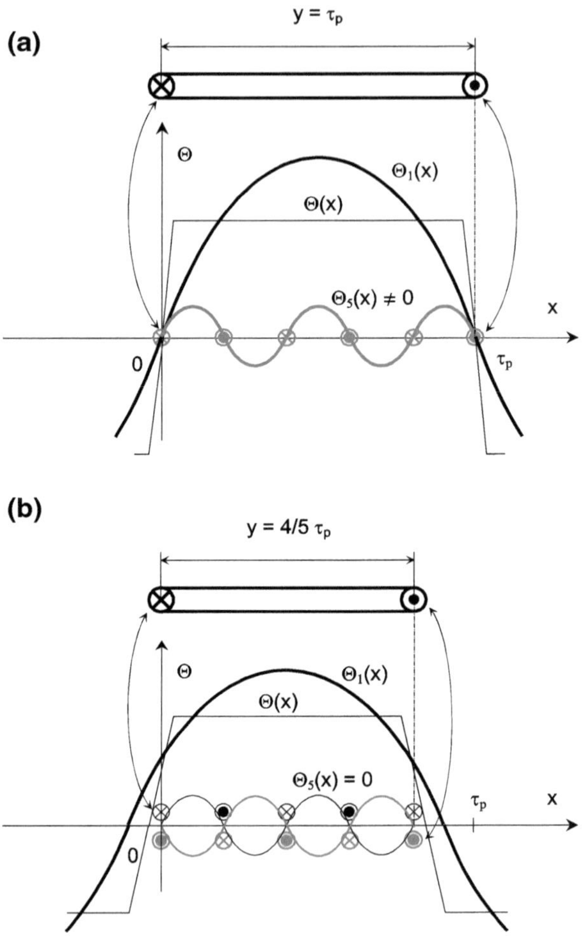

The full-pitch coil in Fig. 2.15a creates trapezoidal MMF distribution $\Theta(x)$. One can imagine that the fifth spatial harmonic of the MMF distribution $\Theta(x)$ is created by a fictitious winding, shown gray in Fig. 2.15a, with a coil pitch equal to $\tau_p/5$. At $x = 0$ and $x = \tau_p$, the direction of current and coil orientation of the real and the fictitious winding coincide, which means that the left and the right half of the real coil support each other when creating the fifth harmonic component of the MMF.

The ampere-turns of the left-hand side of the short-pitch coil in Fig. 2.15b with $y = 4/5\ \tau_p$ would create the fifth harmonic of the MMF identical to that drawn with solid black line, originated from fictitious winding with $y = \tau_p/5$. The same fictitious winding would, however, require the ampere-turns on the right-hand side of the coil (at $x = 4/5\ \tau_p$) acting in the *opposite* direction than the real ones created by the coil right portion.

If the ampere-turns of the fictitious winding with $y = \tau_p/5$ should coincide at $x = 4/5\ \tau_p$ with those created by the right-hand side of the real coil, the winding distribution drawn gray in Fig. 2.15b would be valid. This means that the fictitious coil should generate ampere-turns at $x = 0$ *opposite* to the real coil! Obviously, the sum of the black and gray distributions representing fictitious 5th harmonic is equal to zero when $y = 4/5\ \tau_p$.

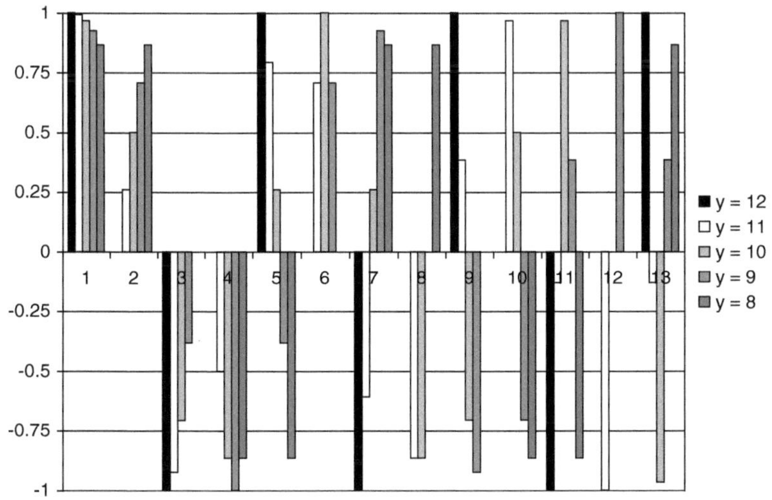

Fig. 2.16 p.u. values of coil pitch factors for higher spatial harmonics in a machine with 12 slots per pole and various values of coil pitch

Pitch factors in p.u. of the fundamental for various spatial harmonics in a machine with 12 slots per pole and various coil pitch values are shown in Fig. 2.16. Coil pitch shortening is a very powerful means for the minimization, or even elimination, of particular harmonics from the MMF distribution. The price paid is a (slight) decrease of the useful fundamental harmonic.

Absolute value of pitch factor for any spatial harmonic n for full-pitch coils ($y = \tau_p$) is equal to 1. By applying the coil shortening, the amplitudes of pitch factors become periodically dependent on harmonic order. One should note that, independently of the coil pitch to pole pitch ratio y/τ_p, spatial harmonics with order $N/p \pm 1$ have always the same pitch factor as the fundamental.

The amplitude c_n of nth harmonic of air gap MMF created by one coil per pole pair can be expressed by means of Eqs. (2.18) and (2.19) as

$$c_n = \sqrt{a_n^2 + b_n^2} = \frac{2}{\pi}\frac{Iw}{n}\frac{\sin n\frac{d}{\tau_p}\frac{\pi}{2}}{n\frac{d}{\tau_p}\frac{\pi}{2}} \cdot \sin n\frac{y}{\tau_p}\frac{\pi}{2} \tag{2.23}$$

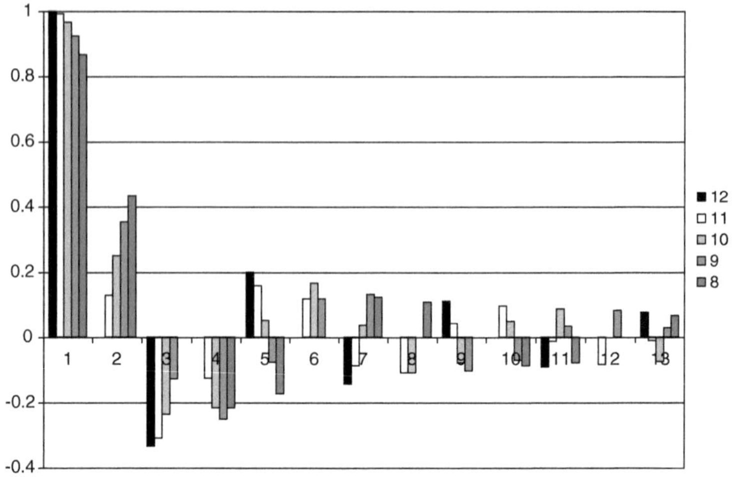

Fig. 2.17 p.u. amplitudes of MMF harmonics created by a single coil per pole pair in a machine with 12 slots per pole. The coil pitch varies between 8 and 12

One coil per pole pair with $y < \tau_p$ *generates both even and odd harmonics of air gap MMF, the amplitudes of which are approximately proportional to the reciprocal of their order*. This property is illustrated in Fig. 2.17, in which p.u. amplitudes of the first 13 harmonics created by a single coil per pole pair for variable values of coil pitch and for $\tau_p = 12$ are shown.

In Fig. 2.17 one can see that only a full-pitch coil per pole pair generates air gap MMF spectrum without even harmonics; as long as the coil is chorded, it acts as a source of both even and odd harmonics. Since a single full-pitch coil per pole pair creates identical MMF spectrum as a full-pitch coil per pole, one can state that even harmonics created by two full-pitch coils per pole pair act against each other.

Case Study 2.1: A 20-pole, 3-phase machine has 30 stator slots. Air gap diameter is 1200 mm and stator slot opening $d = 56$ mm. Stator winding has $y = 1$ (tooth-wound). Slot-opening and coil pitch factors of the stator winding along with % amplitudes of harmonics are listed in Table 2.2.

Table 2.2 Parameters of winding in Case Study 2.1

n	1	2	3	4	5	6	7	8
$f_{o,n}$	.99	.94	.87	.78	.67	.55	.42	.29
$f_{p,n}$	.87	.87	0	−.87	−.87	0	.87	.87
%	100	47.8	0	19.8	13.6	0	6.1	3.7

One recognizes in Table 2.2 that the stator winding generates extremely strong 2. and 4. harmonic of the air gap MMF. These harmonics increase substantially the air gap leakage inductance of the selected winding and create additional components of eddy current losses in solid rotor parts, along with pulsating torques on the shaft.

It is easy to demonstrate that not only two fully pitched, but also two identically chorded coils per pole pair generate even harmonics the sum of which is always equal to zero, because

$$\Theta_n \sin\left[n\frac{\pi}{\tau_p}(x-\tau_p)\right] = \Theta_n \sin\left(n\frac{\pi}{\tau_p}x - n\pi\right) = \Theta_n \sin n\frac{\pi}{\tau_p}x \tag{2.24}$$

for n even, and

$$\Theta_n \sin\left[n\frac{\pi}{\tau_p}(x-\tau_p)\right] = -\Theta_n \sin n\frac{\pi}{\tau_p}x \tag{2.25}$$

for n odd. Since the ampere-turns of coils under one (N) pole, $(I \cdot w)_N$, have an opposite sign than the ampere-turns under adjacent (S) pole, $(I \cdot w)_S$, odd harmonics created by coils of one pole support odd harmonics created by coils of adjacent pole. Even harmonics created by coils of one pole, on the other hand, act against even harmonics created by coils of adjacent pole.

Vast majority of windings in heteropolar machines is built in such a manner that the winding distribution pattern repeats on pole basis, see Fig. 2.18. As a result, only odd harmonics of air gap MMF can be generated, the amplitudes of which are

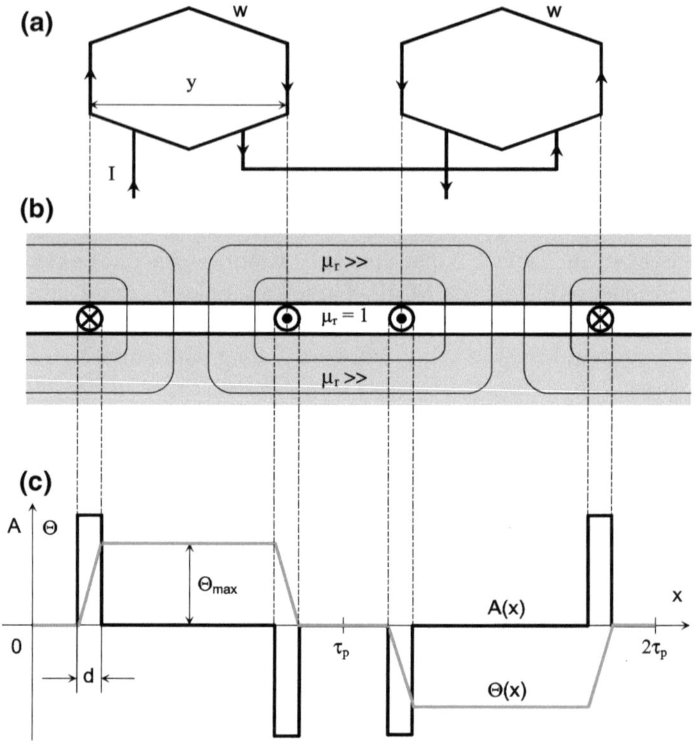

Fig. 2.18 Schematic representation (**a**) of coils in the air gap (**b**), along with the current sheet $A(x)$ and MMF $\Theta(x)$ distributions created by positive current I flowing through the coils (**c**). For distributions $A(x)$ and $\Theta(x)$ it is irrelevant whether the coils are placed in slots, or directly in the air gap

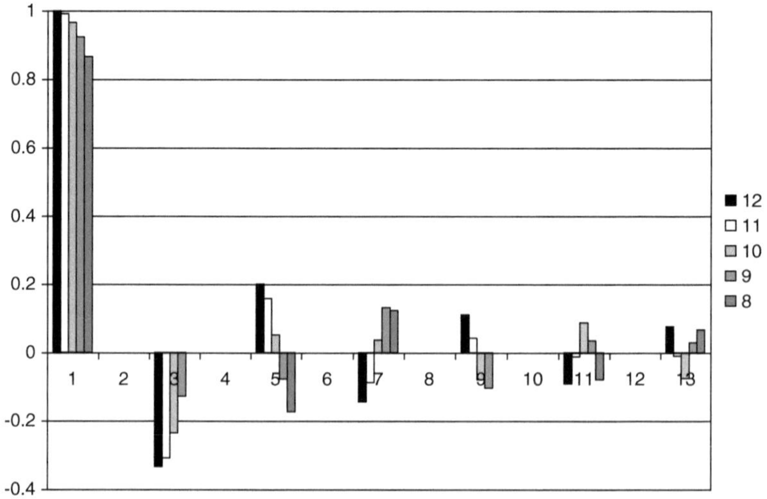

Fig. 2.19 p.u. amplitudes of MMF harmonics created by one coil per pole in a machine with 12 slots per pole. The coil pitch varies between 8 and 12

$$a_n = \frac{4}{\pi}\frac{Iw}{n}\frac{\sin n\frac{d}{\tau_p}\frac{\pi}{2}}{n\frac{d}{\tau_p}\frac{\pi}{2}} \cdot \sin n\frac{\pi}{2} \cdot \sin n\pi \cdot \sin n\frac{y}{\tau_p}\frac{\pi}{2} \equiv 0 \tag{2.26}$$

because n is odd, and

$$b_n = \frac{4}{\pi}\frac{Iw}{n}\frac{\sin n\frac{d}{\tau_p}\frac{\pi}{2}}{n\frac{d}{\tau_p}\frac{\pi}{2}} \cdot \sin n\frac{\pi}{2} \cdot \sin n\frac{y}{\tau_p}\frac{\pi}{2} \tag{2.27}$$

MMF spectra of one coil per pole with identical data as in Fig. 2.15 are shown in Fig. 2.19.

Pole symmetry (identical winding distribution under each pole) eliminates even harmonics in air gap MMF spectrum. Even harmonics in air gap MMF spectrum are a sign of pole asymmetry, caused typically by winding short turns. If p [p.u.] turns under one pole are short-circuited, Fourier coefficients of air gap MMF spectrum can be expressed as

$$a_n = \frac{2}{\pi}\frac{Iw}{n}\frac{\sin n\frac{d}{\tau_p}\frac{\pi}{2}}{n\frac{d}{\tau_p}\frac{\pi}{2}} \cdot \sin n\frac{y}{\tau_p}\frac{\pi}{2} \cdot \left[\cos n\frac{\pi}{2} - (1-p)\cos 3n\frac{\pi}{2}\right] \tag{2.28}$$

$$b_n = \frac{2}{\pi}\frac{Iw}{n}\frac{\sin n\frac{d}{\tau_p}\frac{\pi}{2}}{n\frac{d}{\tau_p}\frac{\pi}{2}} \cdot \sin n\frac{y}{\tau_p}\frac{\pi}{2} \cdot \left[\sin n\frac{\pi}{2} - (1-p)\sin 3n\frac{\pi}{2}\right] \tag{2.29}$$

The amplitude c_n of nth harmonic is equal to

$$c_n = \frac{2}{\pi}\frac{Iw}{n}\frac{\sin n\frac{d}{\tau_p}\frac{\pi}{2}}{n\frac{d}{\tau_p}\frac{\pi}{2}} \cdot \sin n\frac{y}{\tau_p}\frac{\pi}{2} \cdot \sqrt{1 - 2\cdot(1-p)\cdot\cos n\pi + (1-p)^2} \tag{2.30}$$

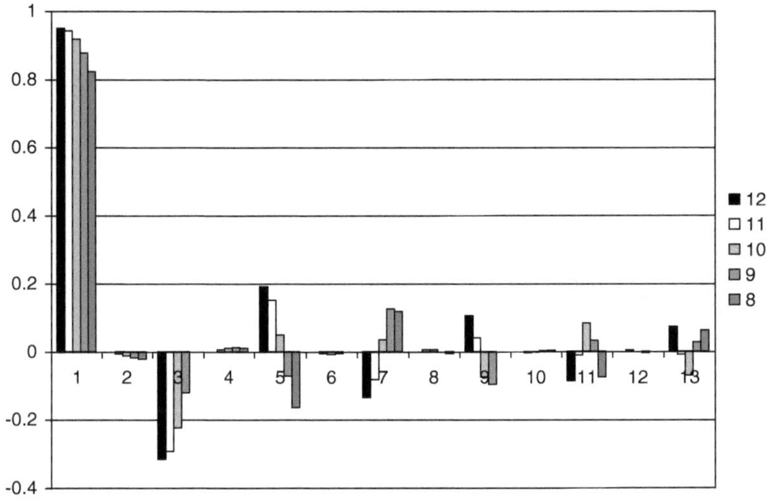

Fig. 2.20 p.u. amplitudes of MMF harmonics created by one coil per pole in a machine with 12 slots per pole for pole asymmetry of p = 0.1 p.u. The coil pitch is varied between 8 and 12

The influence of pole asymmetry on amplitudes of air gap MMF harmonics in a machine with 12 slots per pole is shown in Fig. 2.20. The machine is assumed to have 10 % less turns under one pole ($p = 0.1$ p.u.) than under another. Besides general decrease of amplitudes of odd harmonics (less ampere-turns!), one recognizes the appearance of even harmonics in Fig. 2.20, the amplitudes of which are dependent on coil pitch y.

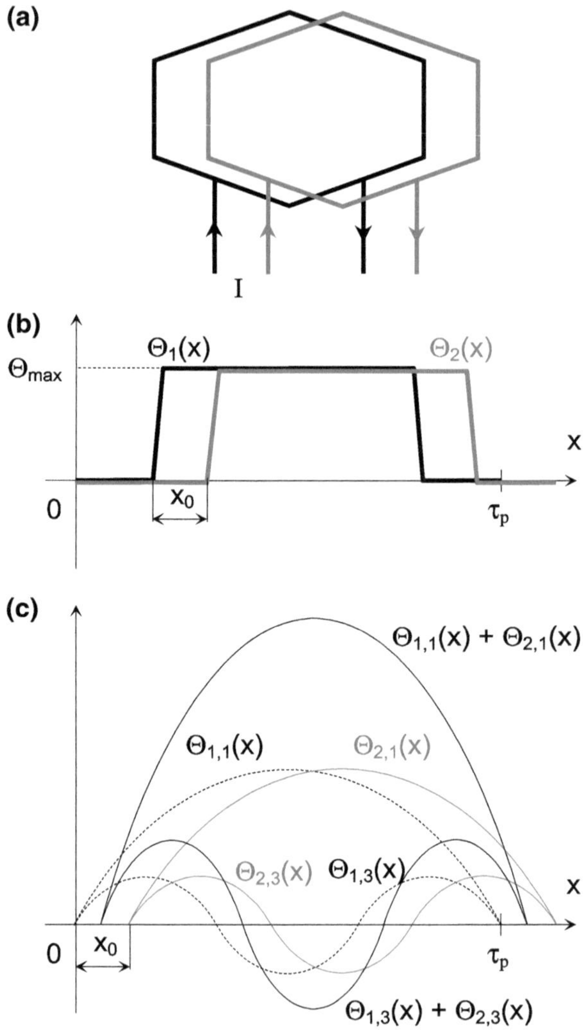

Fig. 2.21 Two coils shifted for x_0 to each other (**a**), their MMF distributions (**b**), and the fundamental and third components of MMF created by each coil (**c**)

Phase windings of electric machines are seldom wound with only one coil per pole. Putting phase coils into adjacent slots under a pole is a further means to suppress higher spatial harmonics. In Fig. 2.21 two coils shifted for the amount of x_0 are shown along with their air gap MMF distributions.

The shift x_0 of the two coils along circumferential coordinate in Fig. 2.21 does not have the same impact on all spatial harmonics of their MMFs, because the higher the order of harmonics n, the bigger the spatial shift $x_{0,n}$

$$x_{0,n} = n \cdot x_0 \tag{2.31}$$

The amplitude $\Theta_{res,1,max}$ of the fundamental component of the resulting MMF in Fig. 2.21c, $\Theta_{res,1}(x) = \Theta_{1,1}(x) + \Theta_{2,1}(x)$ can be evaluated by using the trigonometric identity

$$\sum_{j=1}^{k} \cos[\alpha + (j-1) \cdot \delta] = \frac{\cos\left(\alpha + \frac{k-1}{2}\delta\right) \cdot \sin\frac{k \cdot \delta}{2}}{\sin\frac{\delta}{2}} \tag{2.32}$$

and setting for $\alpha = 0$, $k = 2$:

$$1 + \cos\delta = \frac{\cos\frac{\delta}{2} \cdot \sin\delta}{\sin\frac{\delta}{2}} = 2\cos^2\frac{\delta}{2} \tag{2.33}$$

which helps one define $\Theta_{res,1,max}$

$$\Theta_{res,1,max} = \Theta_{1,max}\sqrt{(1+\cos\alpha_0)^2 + \sin^2\alpha_0} = \Theta_{1,max}\sqrt{2 \cdot (1+\cos\alpha_0)} \tag{2.34}$$

as

$$\Theta_{res,1,max} = 2 \cdot \Theta_{1,max}\cos\frac{\alpha_0}{2} \tag{2.35}$$

with $x_0 = R \cdot \alpha_0$, R denoting the air gap radius.

The ratio between the trigonometric and algebraic sum of amplitudes of MMFs created by adjacent coils under one pole is called the *zone factor*, which in case of two coils per zone can be written for the fundamental harmonic as

$$\frac{\Theta_{res,1,max}}{2 \cdot \Theta_{1,max}} = \cos\frac{\alpha_0}{2}$$

and for a spatial harmonic of the order n

$$\frac{\Theta_{res,n,max}}{2 \cdot \Theta_{n,max}} = \cos n\frac{\alpha_0}{2}$$

One recalls that the expression for the zone winding factor of the nth harmonic, $f_{z,n}$:

$$f_{z,n} = \frac{\sin \frac{nq\alpha_0}{2}}{q \sin \frac{n\alpha_0}{2}} \tag{2.36}$$

turns into $\cos(n \cdot \alpha_0/2)$ for $q = 2$.

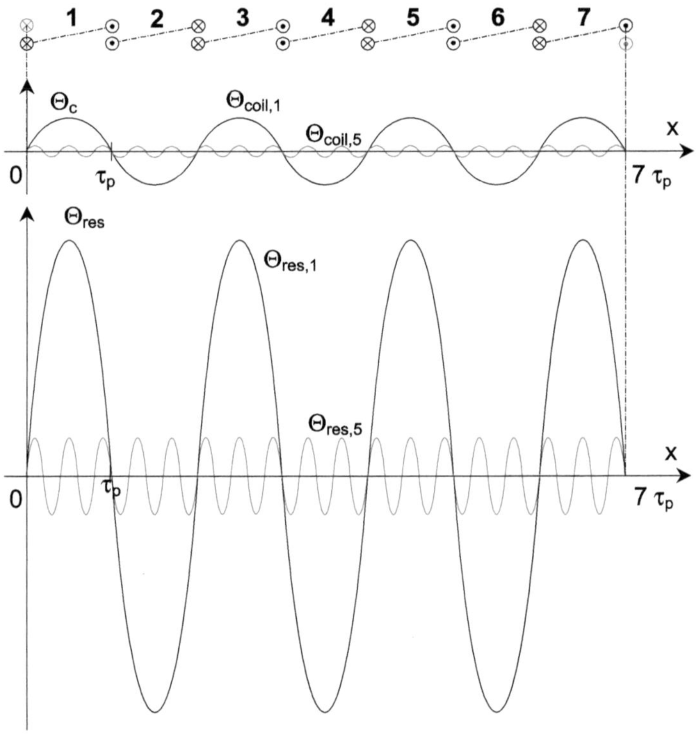

Fig. 2.22 Symmetrical integer slot winding with one coil per pole and phase: coil MMF distribution Θ_c repeats from pole to pole, making the resulting MMF equal to the algebraic sum of coil MMFs, both for the fundamental $\Theta_{coil,1}$ and for the fifth harmonic $\Theta_{coil,5}$

In high-polarity electric machines there is often not enough space for more than one coil per phase under one pole. Both previously discussed countermeasures against high spatial harmonics—chording the coil pitch, or connecting more coils in series under each pole—obviously cannot be applied in this case. Therefore, other means for control of spatial higher harmonics has to be employed: the undesirable harmonics are not compensated within one pole, but within several poles. The winding pattern repeats every *fundamental pole*, the width of which is an odd multiple of the pole pitch τ_p. One refers to *fractional slot winding* in which the number of slots per pole and phase q is not an integer.

In order to illustrate the principle of higher harmonics minimization in a fractional slot winding the air gap MMF distribution in a 84-pole, 3-phase machine with 252 slots and $q = 252/(84 \cdot 3) = 1$ (Fig. 2.22) is compared with air gap MMF distribution in an 84-pole, 3-phase machine with 288 slots, $q = 288/(84 \cdot 3) = 8/7$, as shown in Fig. 2.23. In particular, the influence of winding connection on the resulting fundamental and fifth harmonic is analyzed.

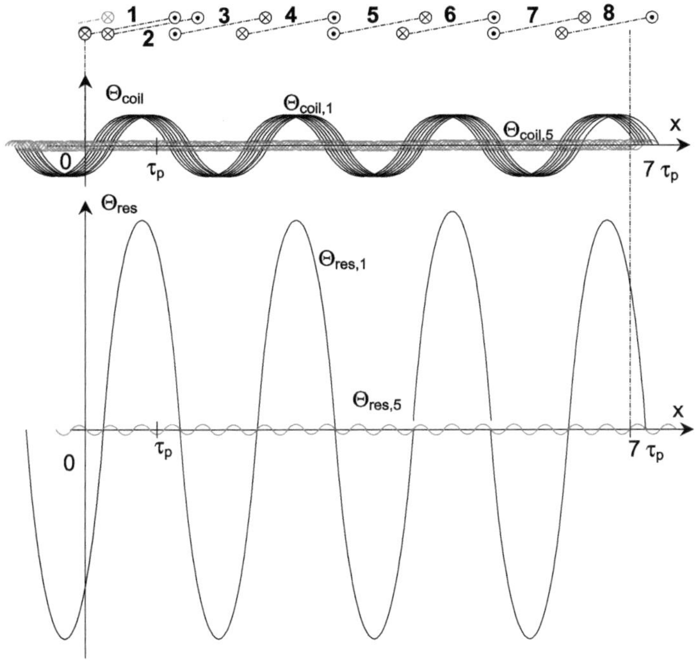

Fig. 2.23 Symmetrical fractional slot winding with $q = 8/7$ coil per pole and phase. The fundamental and 5 harmonic components of all coils are shown

For $q = 1$, as shown in Fig. 2.22, the resulting MMF is equal to the algebraic sum of MMFs created by all coils. In the resulting MMF the ratio between the amplitude of the fifth harmonic and the fundamental term is equal to the ratio between the amplitude of the fifth harmonic and the fundamental term of a coil.

The MMF distribution in a machine with $q = 8/7$, shown in Fig. 2.23, is for the fifth spatial harmonic completely different from that in a machine with $q = 1$. In a machine with fractional slot winding the period length of the winding distribution corresponds to the fundamental pole. A fundamental pole of the winding in Fig. 2.23 with $q = 8/7$ includes seven machine poles. Each phase of the analyzed winding has eight coils per fundamental pole, denoted by numbers 1–8. The coils are connected to each other in such a manner as to maximize the resulting fundamental, and minimize higher harmonics. The sequence of connection of coils in Fig. 2.23 is $1 \to 4 \to 6 \to 8 \to -3 \to -5 \to -7 \to 2$, with negative signs standing for a reversely connected coils. This sequence of coil connections shifts fundamentals slightly to each other. On the other hand, the fifth harmonic components are significantly shifted to each other. Ultimately, the resulting fundamental component is slightly lower than the algebraic sum of coil MMFs, whereas the resulting fifth harmonic is drastically reduced.

2.5 Air Gap Permeance, Carter Factor, Air Gap Flux Density Distribution

Air gap flux density is the crucial quantity in an electric machine, since it determines both the induced voltage and the torque, the two most important machine attributes. Spatial distribution of air gap flux density, including all higher harmonics, determines time dependencies of induced voltages, torque, and radial forces. Therefore, special attention has to be paid to the proper shaping of the air gap flux density distribution.

Neglecting the MMF drop in iron, one can write for the air gap flux density distribution $B_\delta(x)$:

$$B_\delta(x) = \mu_0 \cdot H_\delta(x) = \mu_0 \frac{\Theta(x)}{\delta(x)} \tag{2.37}$$

i.e., the amount of flux density at a particular point in air gap is proportional to the MMF and inversely proportional to the air gap width at that point.

Windings of electric machines are usually placed in slots separated by teeth, which provide mechanical support. The price for mechanical fixation of windings by putting them into slots is a loss of flux expressed by the Carter factor.

In a machine with **single-slotted air gap** and constant excitation over slots and teeth, the air gap flux density $B(x)$ is minimal along slot centerline (Fig. 2.24). The flux density distribution $B(x)$ can be represented in terms of its average value and fundamental harmonic due to slotting as

$$B_\delta(x) = B_0 + B_s \cos\frac{\pi}{\tau_s}x \tag{2.38}$$

where

$$B_0 = \frac{B_{max} + B_{min}}{2}; \quad B_s = \frac{B_{max} - B_{min}}{2} \tag{2.39}$$

B_s denotes here the amplitude of the fundamental component of air gap flux density due to slotting, and B_0 its average value over one slot pitch τ_s.

The Carter factor k_C is defined as

$$k_C = \frac{\tau_s}{\tau_s - \gamma \cdot \delta} \tag{2.40}$$

and the effective air gap width δ_{eff}:

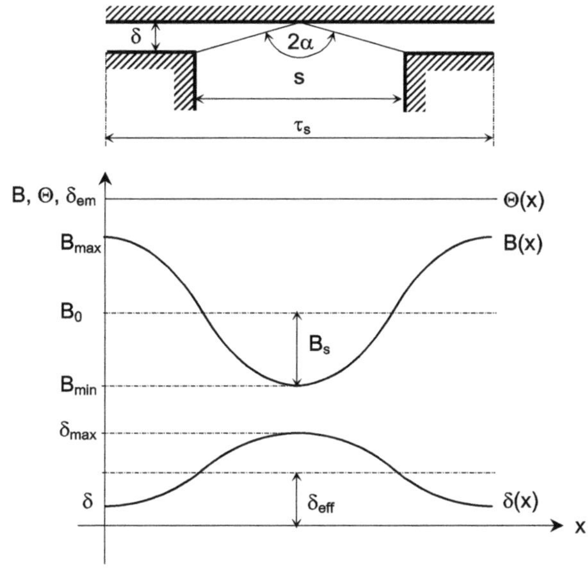

Fig. 2.24 Dominating components of air gap flux density $B(x)$ and electromagnetic air gap width $\delta_{em}(x)$ at constant MMF

$$\delta_{eff} = k_C \cdot \delta \tag{2.41}$$

where

$$\gamma = \frac{4}{\pi}\left[\frac{s}{2\delta}\arctan\frac{s}{2\delta} - \ln\sqrt{1+\left(\frac{s}{2\delta}\right)^2}\right] \tag{2.42}$$

Not the geometrical distance δ between smooth and slotted surface, but the effective air gap width $\delta_{eff} = k_C \cdot \delta$ determines the ampere-turns demand for a given flux density. Carter factor is a function of the ratio between the slot opening s and slot pitch τ_s, and the ratio between the air gap width δ and slot pitch τ_s. By keeping the slot-opening constant for a given slot pitch, the Carter factor decreases as the air gap width increases. On the other hand, by keeping the air gap width for a given slot pitch constant, the Carter factor increases as the slot opening increases.

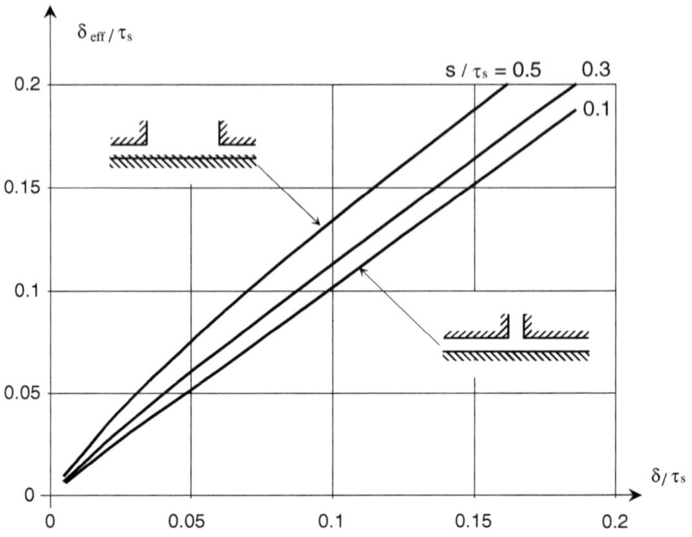

Fig. 2.25 Dependence of normalized effective air gap width δ_{eff}/τ_s on normalized air gap width δ/τ_s with normalized slot opening s/τ_s as a parameter

The analysis of Carter factor can be simplified by normalizing the air gap width and the slot opening to the slot pitch of 1 p.u. Introduce first the normalized air gap width δ_N:

$$\delta_N = \frac{\delta}{\tau_s} \tag{2.43}$$

and the normalized slot width s_N:

$$s_N = \frac{s}{\tau_s} \tag{2.44}$$

The Carter factor can be now written as

$$k_C = \frac{1}{1 - \gamma \cdot \delta_N} \tag{2.45}$$

and the auxiliary function γ

$$\gamma = \frac{4}{\pi}\left[\frac{s_N}{2\delta_N}\arctan\frac{s_N}{2\delta_N} - \ln\sqrt{1 + \left(\frac{s_N}{2\delta_N}\right)^2}\right] \tag{2.46}$$

The dependence of the normalized effective air gap width on normalized air gap width with normalized slot opening as a parameter is shown in Fig. 2.25. One recognizes in this figure that the difference between effective and geometrical air gap width becomes constant for increasing δ/τ_s, which means that the Carter factor in that case slowly decreases.

Introducing the auxiliary quantity u

$$u = \frac{s}{2\delta} + \sqrt{1 + \left(\frac{s}{2\delta}\right)^2} = \frac{s_N}{2\delta_N} + \sqrt{1 + \left(\frac{s_N}{2\delta_N}\right)^2} \tag{2.47}$$

one can define the ratio β (see Fig. 2.24) as

$$\beta = \frac{(1-u)^2}{2\cdot(1+u^2)} = \sin^2\frac{\alpha}{2} \tag{2.48}$$

which also can be written as

$$\beta = \frac{B_s}{B_{max}} = \frac{B_{max} - B_{min}}{2B_{max}} \tag{2.49}$$

Now one can write for flux densities B_0 and B_s

$$B_0 = \frac{B_{max} + B_{min}}{2} = (1-\beta)B_{max}; \quad B_s = \frac{B_{max} - B_{min}}{2} = \beta B_{max} \tag{2.50}$$

as well as for their ratio

$$\frac{B_s}{B_0} = \frac{\beta}{1-\beta} \tag{2.51}$$

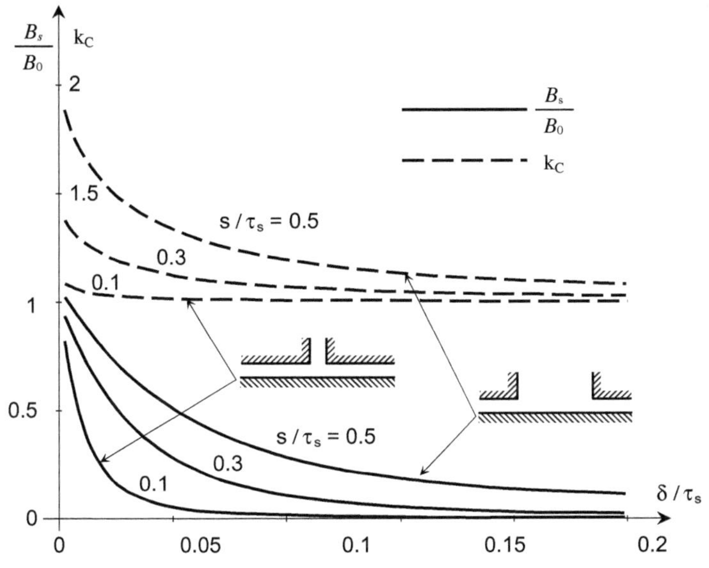

Fig. 2.26 Ratio B_s/B_0 (*solid*) and Carter factor (*dashed*) as functions of normalized air gap width δ/τ_s with normalized slot width s/τ_s as a parameter

The dependence of Carter factor and the ratio between the amplitudes B_0 and B_s on the normalized air gap width for various values of normalized slot-opening width is shown in Fig. 2.26. For small values of normalized air gap width and large values of normalized slot-opening width, the magnitude of the fundamental slot harmonic B_s can be almost as large as the magnitude of the constant term B_0!

Pulsating component of air gap flux density has thermal and mechanical consequences: Eddy current losses in conducting media are proportional to the square of flux density amplitude, as is the radial (attractive) force between stator and rotor. Therefore, special attention has to be paid to the design of air gap geometry and to minimization of slot harmonics in the flux density distribution.

In Fig. 2.24 the electromagnetic air gap width $\delta_{em}(x)$ was introduced, which stands for the length of flux line at a given circumferential coordinate x. The maximum value δ_{max} of the electromagnetic air gap width corresponds to the minimum value B_{min} of air gap flux density $B(x)$:

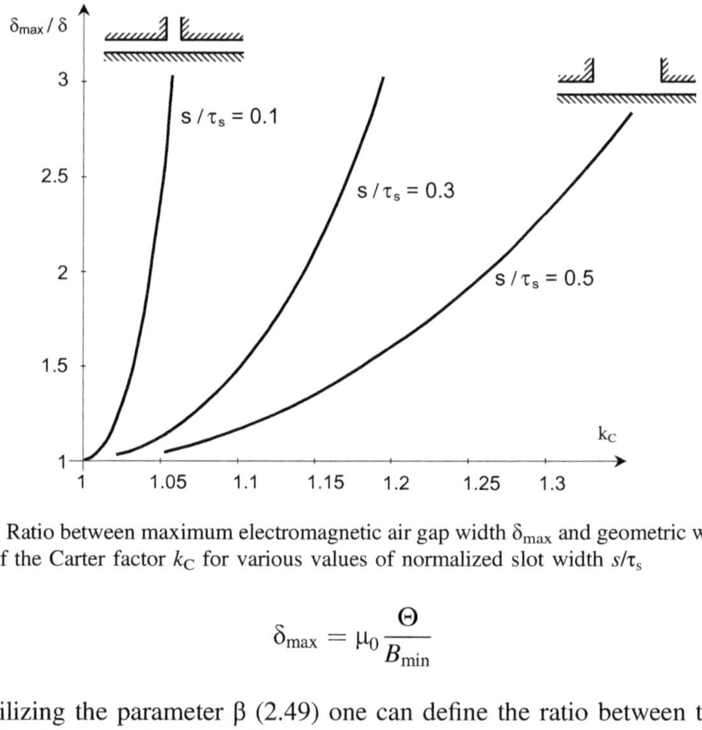

Fig. 2.27 Ratio between maximum electromagnetic air gap width δ_{max} and geometric width δ as a function of the Carter factor k_C for various values of normalized slot width s/τ_s

$$\delta_{max} = \mu_0 \frac{\Theta}{B_{min}} \tag{2.52}$$

By utilizing the parameter β (2.49) one can define the ratio between the maximum and minimum air gap width as

$$\frac{\delta_{max}}{\delta} = \frac{1}{1 - 2\beta} \tag{2.53}$$

The ratio between maximum electromagnetic air gap width δ_{max} and geometric width δ as a function of the Carter factor k_C for various values of normalized slot width s/τ_s is shown in Fig. 2.27.

If the air gap flux density is generated by a $2p$-pole winding placed in N slots, the air gap width $\delta(x)$ can be represented with a constant term and a fundamental harmonic with period length τ_s:

$$\delta(x) = \frac{\delta_{max} + \delta}{2} + \frac{\delta_{max} - \delta}{2} \cos\frac{2\pi}{\tau_s} x = \delta_0 + \delta_1 \cos\frac{N}{p}\frac{\pi}{\tau_p} x \tag{2.54}$$

with τ_s denoting the *slot pitch*

$$\tau_s = \frac{D\pi}{N} = \frac{2p}{N}\tau_p \tag{2.55}$$

and D the average air gap diameter. One should note that the terms δ_0 and δ_1 are both functions of Carter factor and can be of the same order of magnitude.

If the **air gap of an electric machine is doubly slotted** (Fig. 2.28), the resulting Carter factor k_C is equal to the product of the Carter factor for the stator $k_{C,s}$ and for the rotor $k_{C,r}$

$$k_C = k_{C,s} \cdot k_{C,r} \tag{2.56}$$

where both $k_{C,s}$ and $k_{C,r}$ are calculated assuming that the opposite side of air gap has no slots. Here the principle of reciprocity can be applied, since for the computation of Carter factor in electrically excited machines the source of the field is placed outside the air gap. Both stator and rotor windings face the same air gap geometry in Fig. 2.28 and, therefore, the same effective air gap width δ_{eff}.

The air gap width $\delta(x)$ of a doubly slotted machine in Fig. 2.28 can be represented with a constant term, the fundamental harmonic representing the stator slotting, with period length $\tau_{s,s}$, and the fundamental harmonic representing the rotor slotting, with period length $\tau_{s,r}$

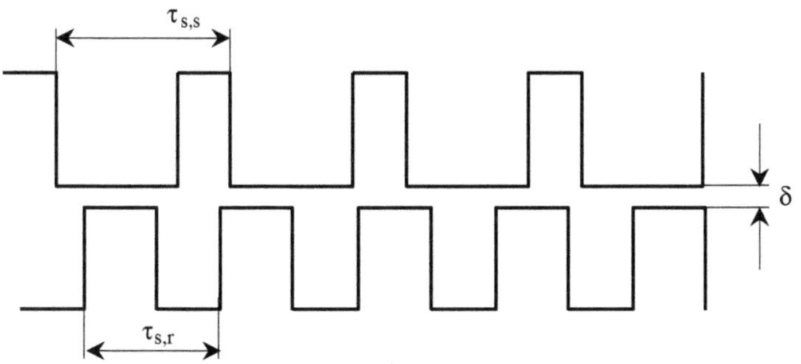

Fig. 2.28 Doubly slotted air gap of an electric machine

$$\delta(x) = \delta_0 + \delta_{1,s} \cos \frac{N_s}{p} \frac{\pi}{\tau_p} x + \delta_{1,r} \cos \frac{N_r}{p} \frac{\pi}{\tau_p} x \tag{2.57}$$

with N_s standing for the number of stator, and N_r for the number of rotor slots.

In order to increase the heat exchange surface and improve the cooling performance, electric machines are often built with radial cooling channels through which the cooling air is blown from inner to outer, or from outer to inner portions of machines. If the lamination contains N_{cc} radial cooling channels, it is built of $N_{cc} + 1$ lamination stacks separated by the cooling channels. Radial cooling channels extend the machine axial length for the amount of $w_{cc} \cdot N_{cc}$ and deteriorate the lamination fill factor. The rate of deterioration can be expressed by using the axial Carter factor $k_{C,ax}$ defined as

$$k_{C,ax} = \frac{l_{ax} - N_{cc} w_{cc}}{l_{ax} - N_{cc} w'_{cc} + 2\delta} \tag{2.58}$$

with

$$w'_{cc} \approx \frac{w_{cc}}{1 + 5 \frac{\delta}{w_{cc}}} \tag{2.59}$$

for the same number of stator and rotor radial cooling channels, and

$$w'_{cc} \approx \frac{w_{cc}}{1 + 2.5 \frac{\delta}{w_{cc}}} \tag{2.60}$$

for different numbers of stator and rotor cooling channels. l_{ax} in Eq. 2.58 denotes the axial length of active part, i.e., the distance between the beginning of the first and the end of the last lamination stack. The resulting Carter factor for a machine with radial cooling channels can be expressed as

$$k_C = k_{C,s} \cdot k_{C,r} \cdot k_{C,ax} \tag{2.61}$$

The principle of reciprocity used for the determination of Carter factor for doubly slotted air gap is disturbed when the excitation on one side of the air gap is relocated from iron structure into the air gap, as is the case in surface-mounted PM machines, as shown in Fig. 2.29.

Permanent magnet in the air gap of the machine in Fig. 2.29a faces teeth and slots on the other side of mechanical air gap δ_m relatively close to its surface. As a consequence, the air gap flux density and the flux density in the magnet below a slot differ significantly from their values below a tooth. Large pulsations of air gap flux density are an indicator of a large Carter factor. When excited by a coil in slots as shown in Fig. 2.29b, the same air gap geometry generates lower amplitudes of flux density pulsations underneath the slots, which is an indicator for a lower Carter factor.

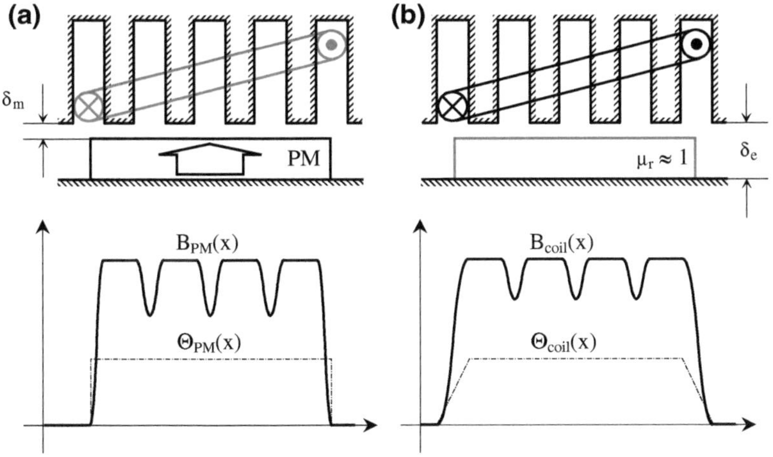

Fig. 2.29 Qualitative air gap flux density distribution in a surface-mounted PM machine created by magnets (**a**) and armature winding (**b**)

2.5.1 Uneven Air Gap and Homopolar Flux

Consider a two-pole machine with a winding only on one side of the air gap, as shown in Fig. 2.30. Winding current i creating spatially alternating flux density in the active part flows through end winding conductors in the given sectional (z, r) plane in the same direction, as shown in Fig. 2.30b, c. Flux lines created by end winding MMF go through bearing shields and shaft (if these are made of magnetic material) into the active part. The total flux through the shaft is equal to zero as long as the magnetic circuit is perfectly symmetrical, as shown in Fig. 2.30b. If there is an asymmetry in the machine's magnetic circuit, such as uneven gap caused e.g. by rotor eccentricity, the total shaft flux Φ_s is not equal to zero any more, as shown in Fig. 2.30c, and the homopolar flux is generated.

Homopolar flux Φ_s can be evaluated by means of the simplified magnetic equivalent circuit of the machine in Fig. 2.31b. One recognizes in this circuit the bridge structure created by gap permeances $G_{sh,\delta,N}$, $G_{sh,\delta,S}$, $G_{\delta,N}$, and $G_{\delta,S}$. Flux Φ_s is equal to

$$\Phi_s = 2iw\frac{G_{shaft}\left(G_{\delta,N}G_{sh,\delta,S} - G_{\delta,S}G_{sh,\delta,N}\right)}{G_{shaft}\left(G_{\delta,N} + G_{\delta,S}\right) + \left(G_{sh,\delta,N} + G_{sh,\delta,S}\right)\left(G_{\delta,N} + G_{\delta,S} + G_{Shaft}\right)} \tag{2.62}$$

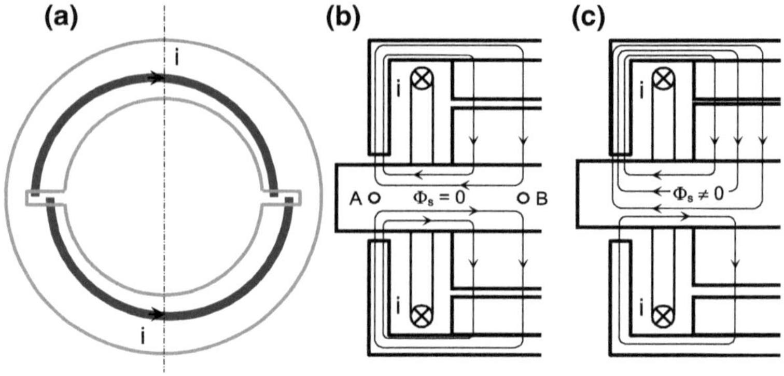

Fig. 2.30 Two-pole machine with stator winding only: **a** schematical representation of the end winding MMF with sectional (z, r) plane denoted by a *dash-dot line*; **b** distribution of fluxes in the (z, r) plane of a machine with perfectly symmetrical magnetic circuit and bearing shields made of magnetic material; **c** distribution of fluxes in the (z, r) plane of a machine with uneven air gap(s) and bearing shields made of magnetic material

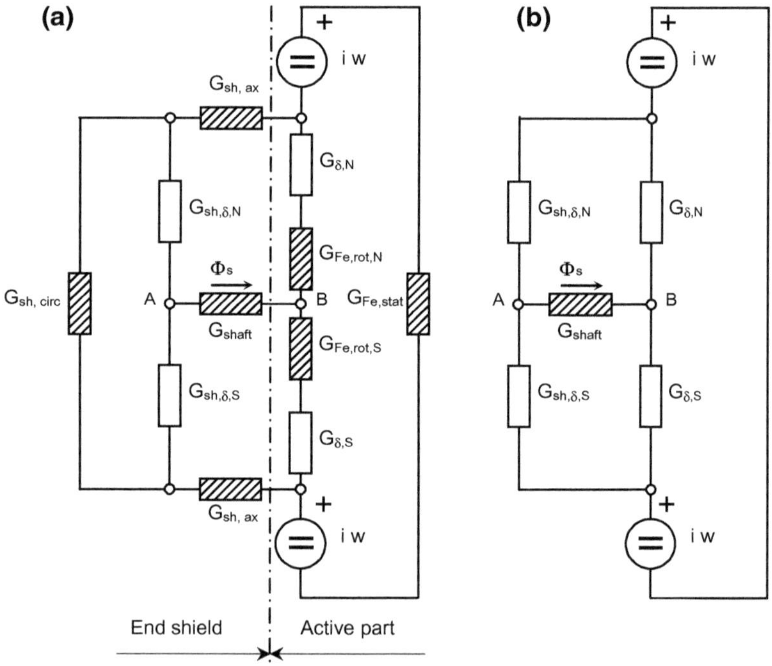

Fig. 2.31 Magnetic equivalent circuit of the portion of the machine in Fig. 2.30b, c: detailed (**a**) and reduced to its most significant components (**b**). Index *sh* relates to the bearing shield, *ax* to axial and *circ* to circumferential direction. Index δ stands for air gap, *rot* for rotor, *stat* for stator, *Fe* for iron, *N* for the N-pole, and *S* for the S-pole. Ampere-turns Iw denote the MMF per pole created by the particular winding

As long as the magnetic circuit in Fig. 2.31 is balanced, the shaft flux is equal to zero. Any unbalance in the bridge of air gap permeances in Fig. 2.31b results in shaft flux. Shaft flux can be constant, or time-dependent. Time dependence of shaft flux is caused either by time-dependent MMFs, or by variation of gap permeances due to rotor motion. Time dependent shaft flux induces shaft voltage, which can cause bearing currents.

2.5.2 *Flux Density Distribution in Eccentric Air Gap of a Slotless Machine*

The most common reason for nonuniformity of air gap is the rotor eccentricity shown in Fig. 2.32, which can be single or both sided (DE and/or NDE), as well as static or dynamic. The eccentric air gap width varies with periodicity of $2R\pi$, R being the average radius of the gap:

$$\delta = \delta_0 + \varepsilon \cos \frac{x - x_r}{R} \tag{2.63}$$

The air gap flux density distribution in a $2p$-pole machine with rotor eccentricity can be expressed as

$$\left(\delta_0 + \varepsilon \cos \frac{x - x_r}{R}\right) \sum_{i=0,1,2,\dots}^{\infty} B_i \cos i \frac{p}{R}(x - x_i) = \mu_0 \sum_{j=1,3,\dots}^{\infty} \Theta_j \cos j \frac{p}{R} x \tag{2.64}$$

or

$$\begin{aligned} &\delta_0 \sum_{i=0,1,2,\dots}^{\infty} B_i \cos i \frac{p}{R}(x - x_i) + \frac{\varepsilon}{2} \sum_{i=0,1,2,\dots}^{\infty} B_i \cos \frac{(ip+1)x - (ipx_i + x_r)}{R} \\ &+ \frac{\varepsilon}{2} \sum_{i=0,1,2,\dots}^{\infty} B_i \cos \frac{(ip-1)x - (ipx_i - x_r)}{R} = \mu_0 \sum_{j=1,3,\dots}^{\infty} \Theta_j \cos j \frac{p}{R} \end{aligned} \tag{2.65}$$

Fig. 2.32 Rotor eccentricity

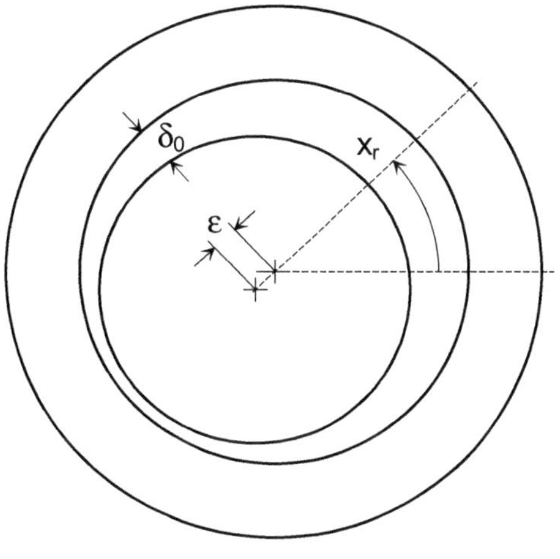

The solution of Eq. 2.61 depends strongly on the number of pole pairs p, since for $p = 1$ some terms exist which disappear at other pole pairs. Physical reason for different behavior of 2-pole machine is obvious, because the rotor eccentricity repeats with the lowest possible periodicity.

(a) $p = 1$

For a 2-pole machine Eq. 2.61 can be further written as

$$\begin{aligned}
&\delta_0\left(B_0 + B_1\cos\frac{x-x_1}{R} + B_2\cos 2\frac{x-x_2}{R} + B_3\cos 3\frac{x-x_3}{R} + B_4\cos 4\frac{x-x_4}{R} + \cdots\right)\\
&\quad + \varepsilon_+\left(B_0\cos\frac{x-x_\mathrm{r}}{R} + B_1\cos\frac{2x-x_1-x_\mathrm{r}}{R} + B_2\cos\frac{3x-2x_2-x_\mathrm{r}}{R}\right.\\
&\qquad\left. + B_3\cos\frac{4x-3x_3-x_\mathrm{r}}{R} + \cdots\right)\\
&\quad + \varepsilon_-\left(B_0\cos\frac{x+x_\mathrm{r}}{R} + B_1\cos\frac{-x_1+x_\mathrm{r}}{R} + B_2\cos\frac{x-2x_2+x_\mathrm{r}}{R}\right.\\
&\qquad\left. + B_3\cos\frac{2x-3x_3+x_\mathrm{r}}{R} + \cdots\right)\\
&= \mu_0\left(\Theta_1\cos\frac{p}{R}x + \Theta_3\cos 3\frac{p}{R}x + \Theta_5\cos 5\frac{p}{R}x + \cdots\right)
\end{aligned} \tag{2.66}$$

with $\varepsilon_+ = \varepsilon/2$ and $\varepsilon_- = \varepsilon/2$.

The unknowns in Eq. 2.66 are amplitudes B_i of air gap flux density harmonics and their spatial shifts x_i. Since spatial harmonics B_i are orthogonal to each other, one can derive an infinite number of trigonometric identities from Eq. 2.66—one identity for each harmonic:

$$\delta_0 B_0 + \frac{\varepsilon}{2}B_1\cos\frac{x_1-x_\mathrm{r}}{R} = 0 \tag{2.67}$$

$$\varepsilon B_0\cos\frac{x-x_\mathrm{r}}{R} + \delta_0 B_1\cos\frac{x-x_1}{R} + \frac{\varepsilon}{2}B_2\cos\frac{x-2x_2+x_\mathrm{r}}{R} = \mu_0\Theta_1\cos\frac{x}{R} \tag{2.68}$$

$$\frac{\varepsilon}{2}B_1\cos\frac{2x-x_1-x_\mathrm{r}}{R} + \delta_0 B_2\cos 2\frac{x-x_2}{R} + \frac{\varepsilon}{2}B_3\cos\frac{2x-3x_1+x_\mathrm{r}}{R} = 0 \tag{2.69}$$

$$\begin{aligned}\frac{\varepsilon}{2}B_2\cos\frac{3x-2x_2-x_r}{R}+\delta_0B_3\cos 3\frac{x-x_3}{R}\\+\frac{\varepsilon}{2}B_4\cos\frac{3x-4x_4+x_r}{R}=\mu_0\Theta_3\cos 3\frac{x}{R}\end{aligned}\tag{2.70}$$

$$\frac{\varepsilon}{2}B_3\cos\frac{4x-3x_3-x_r}{R}+\delta_0B_4\cos 4\frac{x-x_4}{R}+\frac{\varepsilon}{2}B_5\cos\frac{4x-5x_5+x_r}{R}=0\tag{2.71}$$

$$\begin{aligned}\frac{\varepsilon}{2}B_4\cos\frac{5x-4x_4-x_r}{R}+\delta_0B_5\cos 5\frac{x-x_5}{R}\\+\frac{\varepsilon}{2}B_6\cos\frac{5x-6x_6+x_r}{R}=\mu_0\Theta_5\cos 5\frac{x}{R}\end{aligned}\tag{2.72}$$

etc. Each air gap flux harmonic can be resolved along axes $x = 0$ and $x = R\pi/2$, which results in cosine

$$\begin{aligned}
&\varepsilon B_0\cos\frac{x_r}{R}+\delta_0B_1\cos\frac{x_1}{R}+\frac{\varepsilon}{2}B_2\cos\frac{2x_2-x_r}{R}=\mu_0\Theta_1\\
&\frac{\varepsilon}{2}B_1\cos\frac{x_1+x_r}{R}+\delta_0B_2\cos\frac{2x_2}{R}+\frac{\varepsilon}{2}B_3\cos\frac{3x_1-x_r}{R}=0\\
&\frac{\varepsilon}{2}B_2\cos\frac{2x_2+x_r}{R}+\delta_0B_3\cos\frac{3x_3}{R}+\frac{\varepsilon}{2}B_4\cos\frac{4x_4-x_r}{R}=\mu_0\Theta_3\\
&\frac{\varepsilon}{2}B_3\cos\frac{3x_3+x_r}{R}+\delta_0B_4\cos\frac{4x_4}{R}+\frac{\varepsilon}{2}B_5\cos\frac{5x_5-x_r}{R}=0\\
&\frac{\varepsilon}{2}B_4\cos\frac{4x_4+x_r}{R}+\delta_0B_5\cos\frac{5x_5}{R}+\frac{\varepsilon}{2}B_6\cos\frac{6x_6-x_r}{R}=\mu_0\Theta_5
\end{aligned}$$

etc., as well as in sine terms

$$\begin{aligned}
&\varepsilon B_0\sin\frac{x_r}{R}+\delta_0B_1\sin\frac{x_1}{R}+\frac{\varepsilon}{2}B_2\sin\frac{2x_2-x_r}{R}=0\\
&\frac{\varepsilon}{2}B_1\sin\frac{x_1+x_r}{R}+\delta_0B_2\sin\frac{2x_2}{R}+\frac{\varepsilon}{2}B_3\sin\frac{3x_1-x_r}{R}=0\\
&\frac{\varepsilon}{2}B_2\sin\frac{2x_2+x_r}{R}+\delta_0B_3\sin\frac{3x_3}{R}+\frac{\varepsilon}{2}B_4\sin\frac{4x_4-x_r}{R}=0\\
&\frac{\varepsilon}{2}B_3\sin\frac{3x_3+x_r}{R}+\delta_0B_4\sin\frac{4x_4}{R}+\frac{\varepsilon}{2}B_5\sin\frac{5x_5-x_r}{R}=0\\
&\frac{\varepsilon}{2}B_4\sin\frac{4x_4+x_r}{R}+\delta_0B_5\sin\frac{5x_5}{R}+\frac{\varepsilon}{2}B_6\sin\frac{6x_6-x_r}{R}=0
\end{aligned}$$

etc. Denoting by $s_r = \sin(x_r/R)$ and $c_r = \cos(x_r/R)$ one can define the matrix $\underline{E}$

$$
\underline{E} = \begin{bmatrix}
\delta_0 & \frac{\varepsilon}{2}c_r & -\frac{\varepsilon}{2}s_r & 0 & 0 & 0 & 0 & 0 & 0 & \cdots \\
\varepsilon c_r & \delta_0 & 0 & \frac{\varepsilon}{2}c_r & -\frac{\varepsilon}{2}s_r & 0 & 0 & 0 & 0 & \cdots \\
\varepsilon s_r & 0 & \delta_0 & -\frac{\varepsilon}{2}s_r & \frac{\varepsilon}{2}c_r & 0 & 0 & 0 & 0 & \cdots \\
0 & \frac{\varepsilon}{2}c_r & -\frac{\varepsilon}{2}s_r & \delta_0 & 0 & \frac{\varepsilon}{2}c_r & \frac{\varepsilon}{2}s_r & 0 & 0 & \cdots \\
0 & \frac{\varepsilon}{2}s_r & \frac{\varepsilon}{2}c_r & 0 & \delta_0 & -\frac{\varepsilon}{2}s_r & \frac{\varepsilon}{2}c_r & 0 & 0 & \cdots \\
0 & 0 & 0 & \frac{\varepsilon}{2}c_r & -\frac{\varepsilon}{2}s_r & \delta_0 & 0 & \frac{\varepsilon}{2}c_r & \frac{\varepsilon}{2}s_r & \cdots \\
0 & 0 & 0 & \frac{\varepsilon}{2}s_r & \frac{\varepsilon}{2}c_r & 0 & \delta_0 & -\frac{\varepsilon}{2}s_r & \frac{\varepsilon}{2}c_r & \cdots \\
0 & 0 & 0 & 0 & 0 & \frac{\varepsilon}{2}c_r & -\frac{\varepsilon}{2}s_r & \delta_0 & 0 & \cdots \\
0 & 0 & 0 & 0 & 0 & \frac{\varepsilon}{2}s_r & \frac{\varepsilon}{2}c_r & 0 & \delta_0 & \cdots \\
\cdots & \cdots & \cdots & \cdots & \cdots & \cdots & \cdots & \cdots & \cdots & \cdots
\end{bmatrix} \tag{2.73}
$$

and the vector of unknowns $\underline{b}$ as

$$
\underline{b} = \begin{bmatrix} B_0 & B_1 \cos\frac{x_1}{R} & B_1 \sin\frac{x_1}{R} & B_2 \cos\frac{2x_2}{R} & B_2 \sin\frac{2x_2}{R} & B_3 \cos\frac{3x_3}{R} & \cdots \end{bmatrix}^{\mathrm{T}} \tag{2.74}
$$

which along with the vector of applied MMF harmonics $\underline{\theta}$

$$
\underline{\theta} = \mu_0 [0 \quad \Theta_1 \quad 0 \quad 0 \quad 0 \quad \Theta_3 \quad 0 \quad 0 \quad 0 \quad \Theta_5 \quad 0 \quad 0 \quad 0 \quad \Theta_7 \quad \ldots]^{\mathrm{T}} \tag{2.75}
$$

build the system of algebraic equations

$$
\underline{E} \cdot \underline{b} = \underline{\theta} \tag{2.76}
$$

the solution of which are the magnitudes and phase shifts of air gap flux density harmonics in a machine with eccentric rotor.

In order to illustrate the influence of eccentricity on air gap flux density, a two-pole machine with full-pitch coil and with eccentric rotor was analyzed.

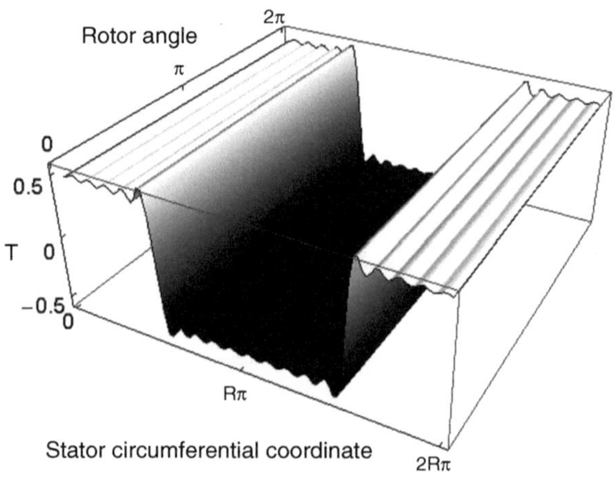

Fig. 2.33 Air gap flux density in a two-pole machine with centric rotor created by the first 19 harmonics of MMF of a full-pitch coil

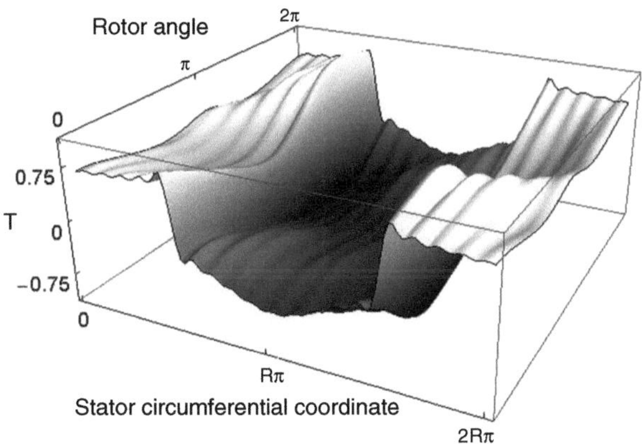

Fig. 2.34 Air gap flux density in a two-pole machine with eccentric rotor created by the first 19 harmonics of MMF of a full-pitch coil

The air gap flux density distribution evaluated for the first 19 harmonics of excitation by a full-pitch coil with an amplitude of the fundamental term of 500 At and for a homogenous air gap width of 1 mm is shown in Fig. 2.33. Apart from small deviations caused by representation of MMF with Fourier series having a finite number of terms, the air gap flux density is constant under one pole, independent of the rotor position. The amplitude of the fundamental term of air gap flux density is equal to μ_0 500/.001 = 0.628 *T*. The air gap flux density spectrum contains only the terms present in the MMF spectrum, i.e., the 1., 3., 5., etc., harmonics.

If the rotor is eccentric with a maximum eccentricity ε of 0.5 mm, i.e., 50 % of the air gap width, the shape of the air gap flux density distribution becomes dependent on the rotor to stator angle. In addition to terms existing in the MMF distribution, the air gap flux density spectrum contains harmonics generated by variable air gap width due to eccentricity. In particular, the terms with order 0, 2, 4, 6, etc., in the air gap flux density spectrum are generated in addition to odd terms. The air gap flux density distribution under the same conditions as in Fig. 2.33, but with an eccentricity ε of 0.5 mm (or 50 %) is shown in Fig. 2.34.

Amplitudes of harmonics of air gap flux density in Fig. 2.34 as functions of rotor angle are shown in Fig. 2.35. Constant term B_0, as a measure of shaft flux in Fig. 2.32, becomes negative after a rotor shift of $\pi/2$ because the opposite polarity of stator MMF is a source of more flux. Accordingly, the shaft flux in Fig. 2.32 becomes also negative.

One notes in Fig. 2.35 that the amplitudes of all harmonics of air gap flux density in a machine with eccentric rotor pulsate as a function of the rotor angle.

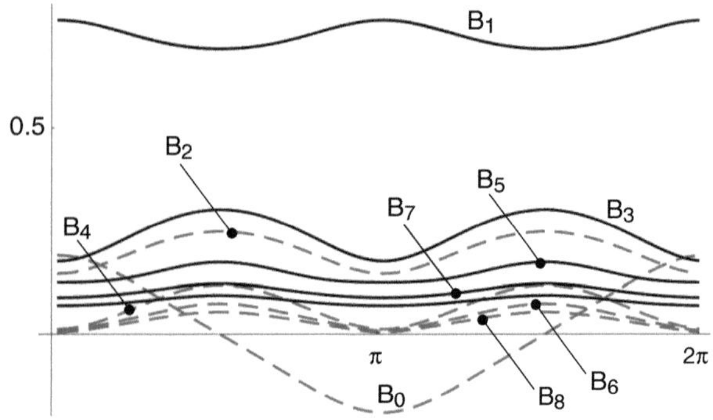

Fig. 2.35 Amplitudes of air gap flux density harmonics as functions of rotor angle in a two-pole machine with eccentric rotor

Consequently, one can write for the amplitude B_0:

$$B_0 = B_{0,\max} \cos \frac{\pi}{\tau_p} x_r \tag{2.77}$$

as well as for the amplitude of the ith harmonic B_i in Eq. 2.60:

$$B_i = B_{i,\text{const}} + B_{i,\max} \cos\left(2\frac{\pi}{\tau_p} x_r \pm \pi\right) \tag{2.78}$$

(b) $p > 1$

For a machine with more than one pole pair, Eq. 2.61 can be written as

$$\begin{aligned}
&\delta_0 \Big[B_0 + B_1\left(\cos p\frac{x}{R}\cos p\frac{x_1}{R} + \sin p\frac{x}{R}\sin p\frac{x_1}{R}\right) + B_2\left(\cos 2p\frac{x}{R}\cos 2p\frac{x_1}{R} + \sin 2p\frac{x}{R}\sin 2p\frac{x_1}{R}\right)\\
&\quad + B_3\left(\cos 3p\frac{x}{R}\cos 3p\frac{x_1}{R} + \sin 3p\frac{x}{R}\sin 3p\frac{x_1}{R}\right) + B_4\left(\cos 4p\frac{x}{R}\cos 4p\frac{x_1}{R} + \sin 4p\frac{x}{R}\sin 4p\frac{x_1}{R}\right) + \cdots\Big]\\
&\quad + \frac{\varepsilon}{2}\Big[B_0\left(\cos\frac{x}{R}\cos\frac{x_r}{R} + \sin\frac{x}{R}\sin\frac{x_r}{R}\right) + B_1\left(\cos\frac{p+1}{R}x\cos\frac{px_1+x_r}{R} + \sin\frac{p+1}{R}x\sin\frac{px_1+x_r}{R}\right)\\
&\quad + B_2\left(\cos\frac{2p+1}{R}x\cos\frac{2px_2+x_r}{R} + \sin\frac{2p+1}{R}x\sin\frac{2px_2+x_r}{R}\right) + B_3\left(\cos\frac{3p+1}{R}x\cos\frac{3px_3+x_r}{R} + \sin\frac{3p+1}{R}x\sin\frac{3px_3+x_r}{R}\right) + \cdots\Big]\\
&\quad + \frac{\varepsilon}{2}\Big[B_0\left(\cos\frac{x}{R}\cos\frac{x_r}{R} + \sin\frac{x}{R}\sin\frac{x_r}{R}\right) + B_1\left(\cos\frac{p-1}{R}x\cos\frac{px_1-x_r}{R} + \sin\frac{p-1}{R}x\sin\frac{px_1-x_r}{R}\right)\\
&\quad + B_2\left(\cos\frac{2p-1}{R}x\cos\frac{2px_2-x_r}{R} + \sin\frac{2p-1}{R}x\sin\frac{2px_2-x_r}{R}\right) + B_3\left(\cos\frac{3p-1}{R}x\cos\frac{3px_3-x_r}{R} + \sin\frac{3p-1}{R}x\sin\frac{3px_3-x_r}{R}\right) + \cdots\Big]\\
&= \mu_0\left(\Theta_1\cos p\frac{x}{R} + \Theta_3\cos 3p\frac{x}{R} + \Theta_5\cos 5p\frac{x}{R} + \cdots\right)
\end{aligned} \tag{2.79}$$

The orders of harmonics which multiply δ_0 in Eq. 2.79 are 0, p, $2p$, $3p$, …, etc., and of harmonics which multiply $\varepsilon/2$ are 1, $p \pm 1$, $2p \pm 1$, $3p \pm 1$, …, etc. The orders of harmonics of the air gap flux density in a machine with eccentric rotor as functions of the number of machine pole pairs are given in Table 2.3.

Table 2.3 The orders of harmonics multiplying air gap width components δ_0, ε_+, and ε_- after Eq. 2.66 in a machine with eccentric rotor as functions of the number of pole pairs p

δ_0	p	1	2	3	4
	$2p$	2	4	6	8
	$3p$	3	6	9	12
	$4p$	4	8	12	16
ε_+	$p + 1$	2	3	4	5
	$2p + 1$	3	5	7	9
	$3p + 1$	4	7	10	13
	$4p + 1$	5	9	13	17
ε_-	$p - 1$	0	1	2	3
	$2p - 1$	1	3	5	7
	$3p - 1$	2	5	8	11
	$4p - 1$	3	7	11	15

The strongest interaction between harmonics occurs in a two-pole machine with eccentric rotor. The higher the number of poles, the weaker the influence of the rotor eccentricity, e.g., for an 8-pole machine, there exists no single harmonic common for the sets of terms multiplying δ_0, ε_+, and ε_-. In a two-pole machine, on the other hand, every harmonic of the order of 2 and above appears in all three sets of terms multiplying δ_0, ε_+, and ε_-. Only in a two-pole machine the constant term B_0 can appear, see Table 2.3. Therefore, the rotor eccentricity can be a source of significant shaft flux only in a two-pole machine.

2.5.3 *Flux Density Distribution in the Air Gap of a Single-Slotted Machine*

Consider a single-slotted machine the air gap width of which is described by a constant term and an infinite series of harmonics with amplitudes δ_i:

$$\delta(x) = \delta_0 + \sum_{i=1,3,5,\ldots}^{\infty} \delta_i \cos i \frac{N}{p} \frac{\pi}{\tau_p} x \tag{2.80}$$

If the slot geometry is identical for all N slots, the order i of harmonics is an odd number. In most practical cases the influence of slot harmonics with order above 1 is negligible; therefore, the air gap width will be represented with a constant term and the fundamental slot harmonic of the order of N/p:

$$\delta(x) = \delta_0 + \delta_1 \cos \frac{N}{p} \frac{\pi}{\tau_p} x \tag{2.81}$$

For a salient pole machine the number of poles per pole pair $N/p = 2$.

The air gap flux density distribution $B(x)$ created by conventional pole-symmetric winding (only odd harmonics of MMF!), the axis of which is shifted for the amount of x_c relative to the center of the first slot (Fig. 2.36), satisfies equation

$$B(x)\cdot\left(\delta_0+\delta_1\cos\frac{N}{p}\frac{\pi}{\tau_p}x\right)=\mu_0\cdot\sum_{n=1,3,5,\ldots}^{\infty}\Theta_n\cos n\frac{\pi}{\tau_p}(x-x_c) \tag{2.82}$$

Machines with integer numbers of slots per pole have **even number of slots per pole pair N/p**. The air gap flux density distribution $B(x)$ in that case contains only odd terms

$$B(x)=\sum_{n=1,3,5,\ldots}^{\infty}B_n\cos n\frac{\pi}{\tau_p}(x-x_c) \tag{2.83}$$

and satisfies equation

$$\begin{aligned}\delta_0\cdot\sum_{i=1,3,5,\ldots}^{\infty}B_i\cos i\frac{\pi}{\tau_p}(x-x_c)+\delta_1\cdot\sum_{j=1,3,5,\ldots}^{\infty}B_j\cos j\frac{\pi}{\tau_p}(x-x_c)\cdot\cos\frac{N}{p}\frac{\pi}{\tau_p}x\\=\mu_0\cdot\sum_{n=1,3,5,\ldots}^{\infty}\Theta_n\cos n\frac{\pi}{\tau_p}(x-x_c)\end{aligned} \tag{2.84}$$

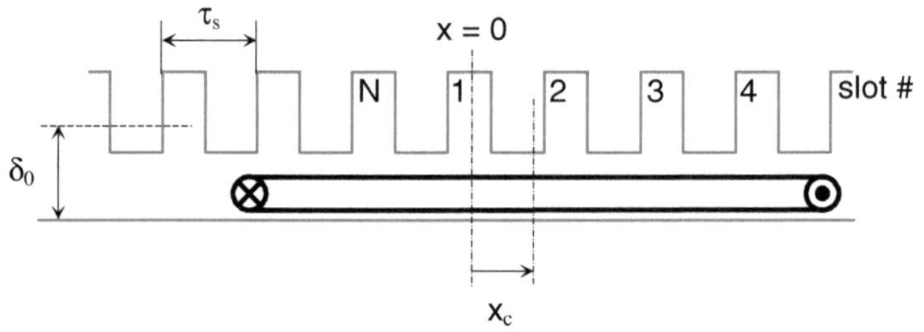

Fig. 2.36 Single-slotted air gap with a coil

The order i of flux density distribution due to constant air gap width δ_0 in Eq. 2.84 must always be equal to the order n of MMF distribution, $i = n$. The order j of flux density distribution due to slotting in Eq. 2.84 must satisfy the condition

$$j=n\pm\frac{N}{p} \tag{2.85}$$

in order to generate flux density components with the same order n as the MMF in Eq. 2.84. Note that for $n < N/p$ the order j can be negative, which means nothing but a 180° phase shift of the particular term. By applying condition in Eq. 2.85, one can write Eq. 2.84 further as

$$\begin{aligned}
&\delta_0 \cdot \sum_{i=1,3,5,\ldots}^{\infty} B_i \cos i\frac{\pi}{\tau_\mathrm{p}}(x - x_\mathrm{c}) + \frac{\delta_1}{2}\sum_{j_+=1,3,5,\ldots}^{\infty} B_{j_+} \cos\left[\left(j_+ + \frac{N}{p}\right)\cdot x - j_+ \cdot x_\mathrm{c}\right]\frac{\pi}{\tau_\mathrm{p}} \\
&\quad + \frac{\delta_1}{2}\sum_{j_-=1,3,5,\ldots}^{\infty} B_{j_-} \cos\left[\left(j_- - \frac{N}{p}\right)\cdot x - j_- \cdot x_\mathrm{c}\right]\frac{\pi}{\tau_\mathrm{p}} \\
&= \mu_0 \cdot \sum_{n=1,3,5,\ldots}^{\infty} \Theta_n \cos n\frac{\pi}{\tau_\mathrm{p}}(x - x_\mathrm{c})
\end{aligned} \tag{2.86}$$

The nth (odd) harmonic of applied MMF creates in a single-slotted air gap of an electric machine *three components of flux density with the same order n*:

- B_n (odd), in the amount of $\mu_0\ \Theta_n/\delta_0$,
- $B_{n-N/p}$ (odd), in the amount of 2 $\mu_0\ \Theta_n/\delta_1$, and
- $B_{n+N/p}$ (odd), in the amount of 2 $\mu_0\ \Theta_n/\delta_1$

which satisfy equation

$$\begin{aligned}
\mu_0\Theta_n \cos n\frac{\pi}{\tau_\mathrm{p}}(x - x_\mathrm{c}) = {} & \delta_0 B_n \cos n\frac{\pi}{\tau_\mathrm{p}}(x - x_\mathrm{c}) \\
& + \frac{\delta_1}{2} B_{n-\frac{N}{p}} \cos\frac{\pi}{\tau_\mathrm{p}}\left[nx - \left(n + \frac{N}{p}\right)x_\mathrm{c}\right] \\
& + \frac{\delta_1}{2} B_{n+\frac{N}{p}} \cos\frac{\pi}{\tau_\mathrm{p}}\left[nx - \left(n - \frac{N}{p}\right)x_\mathrm{c}\right]
\end{aligned} \tag{2.87}$$

Relationship between the three components of MMF drop and coil MMF harmonic of the order n can be visualized by means of Fig. 2.37. In this figure the nth harmonic of applied air gap MMF and nth, $(n - N/p)$th, and $(n + N/p)$th harmonics of MMF drop are resolved into components along the axis α, coincident with centerline of slot 1, and along the axis β, shifted for $90°_\mathrm{el}$ to α.

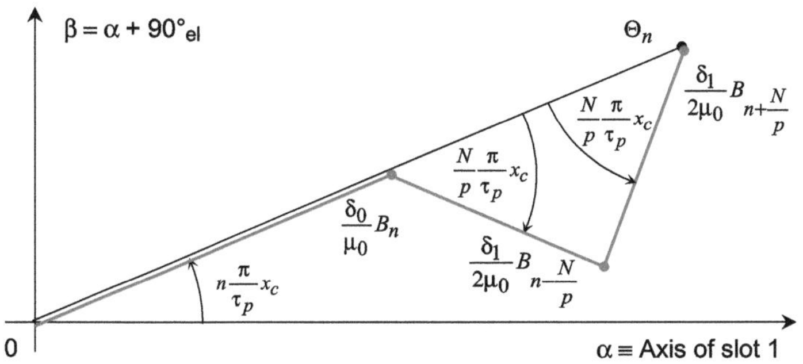

Fig. 2.37 Illustrating relationship between applied MMF and flux density harmonic components in a machine with single-slotted air gap for a given rotor shift x_c

Due to slotting, the nth harmonic of applied MMF created by coil current contributes to the amplitudes of $(n - N/p)$th and $(n + N/p)$th harmonics of air gap flux density along with corresponding fluxes $\Phi_{n-N/p}$ and $\Phi_{n+N/p}$. The fluxes $\Phi_{n-N/p}$ and $\Phi_{N/p+n}$ create MMF drops $\vartheta_{n-N/p}$ and $\vartheta_{n+N/p}$ across the air gap. One distinguishes here between the *applied* MMF Θ_n and its effects, the flux density harmonics B_n, $B_{n-N/p}$, and $B_{n+N/p}$, and the *evoked* MMF harmonics $\vartheta_{n-N/p}$ and $\vartheta_{n+N/p}$, along with all their effects. In order to emphasize this difference, capital letters are used for applied quantities, and small letters for evoked quantities.

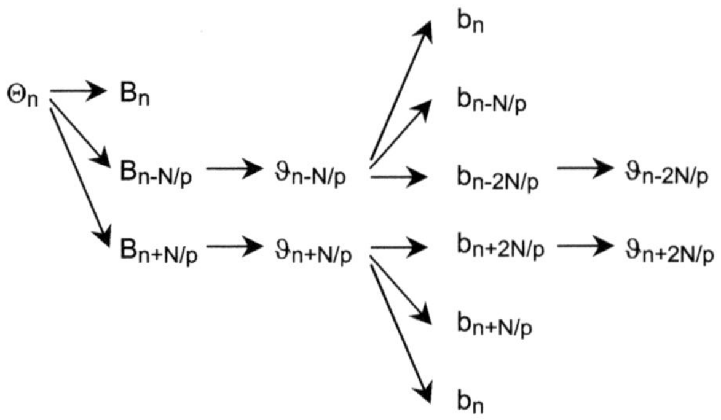

Fig. 2.38 The beginning of the chain of generation of flux density harmonics from a single harmonic of applied MMF in a single-slotted machine

The evoked MMF harmonic $\vartheta_{n-N/p}$ creates flux density harmonics $b_{n-N/p}$, $b_{n-2N/p}$, and b_n, whereas $\vartheta_{n+N/p}$ is a source of $b_{n+N/p}$, $b_{n+2N/p}$, and b_n harmonics, etc. The initial terms of the chain of generation of air gap flux harmonics from a single MMF harmonic in a slotted air gap are shown in Fig. 2.38. Since the number of slots per pole pair N/p is even, an odd harmonic of MMF can create only odd harmonics of air gap flux density.

Table 2.4 Harmonics of flux density B created by particular harmonics of MMF Θ in a machine with even number of slots per pole pair $N/p = \nu$

		B														
		1	3	5	…	$\nu - 5$	$\nu - 3$	$\nu - 1$	$\nu + 1$	$\nu + 3$	$\nu + 5$	…	$2\nu - 3$	$2\nu - 1$	$2\nu + 1$	$2\nu + 3$
	1	Xx						Xx	Xx					Xx	Xx	
	3		Xx				Xx			Xx			Xx			Xx
	5			Xx		Xx					Xx					
	…															
	$\nu - 5$			Xx		Xx										
	$\nu - 3$		Xx				Xx						Xx			
	$\nu - 1$	Xx						Xx						Xx		
Θ	$\nu + 1$	Xx							Xx						Xx	
	$\nu + 3$		Xx							Xx						Xx
	$\nu + 5$			Xx							Xx					
	…															
	$2\nu - 3$		Xx				Xx						Xx			
	$2\nu - 1$	Xx						Xx						Xx		
	$2\nu + 1$	Xx							Xx						Xx	
	$2\nu + 3$		Xx							Xx						Xx

All harmonics of flux density contain components both due to applied (denoted by X) and evoked (denoted by x) harmonics of MMF

The nonuniformity of air gap width caused by slotting is a source of higher spatial harmonics in the spectrum of flux density for sinusoidal applied MMF. The mechanism of generation of higher spatial harmonics of air gap flux density due to nonuniform air gap width (cosine terms!) is analogous to the mechanism of creation of higher time harmonics of current in a magnetic circuit with nonlinear B–H curve. Independent of whether a time or spatial problem is analyzed, the nonlinearity is always a source of higher harmonics in the system.

In Table 2.4 the orders of components of air gap flux density distribution in a slotted machine with even number of slots per pole pair are given as functions of orders of applied MMF. The fundamental component of applied MMF creates the fundamental component of air gap flux density, along with harmonics of the order $N/p - 1$ and $N/p + 1$. These two harmonics correspond to evoked harmonics of air gap MMF drops $\vartheta_{n-N/p}$ and $\vartheta_{n+N/p}$ and both of them generate, among other, the fundamental components of flux density due to higher harmonics of MMF.

Using Eq. 2.86 and Table 2.4, one can relate the amplitude of nth harmonic of air gap flux density to corresponding applied and evoked MMFs as

$$\begin{aligned} B_n \cos n\frac{\pi}{\tau_\mathrm{p}}(x - x_\mathrm{c}) = {} & \frac{\mu_0}{\delta_0}\Theta_n \cos n\frac{\pi}{\tau_\mathrm{p}}(x - x_\mathrm{c}) \\ & + 2\frac{\mu_0}{\delta_1}\sum_{i=1}^{\infty}\Theta_{i\frac{N}{p}-n}\cos\frac{\pi}{\tau_\mathrm{p}}\left[nx - \left(n + i\frac{N}{p}\right)x_\mathrm{c}\right] \\ & + 2\frac{\mu_0}{\delta_1}\sum_{i=1}^{\infty}\Theta_{i\frac{N}{p}+n}\cos\frac{\pi}{\tau_\mathrm{p}}\left[nx - \left(n - i\frac{N}{p}\right)x_\mathrm{c}\right] \end{aligned} \tag{2.88}$$

Salient pole machine has two poles (teeth) per pole pair, $N/p = 2$ and $\tau_\mathrm{p} = \tau_\mathrm{s}$ (Fig. 2.39). The fundamental harmonic of applied MMF creates with air gap width three components of flux density of the same order 1 (Eq. 2.87):

- B_1 with constant air gap width component in the amount of $\mu_0\, \Theta_1/\delta_0$,
- B_1 with fundamental component of air gap width in the amount of $2\mu_0\Theta_1/\delta_1$, and
- B_3 with fundamental component of air gap width in the amount of $2\mu_0\Theta_1/\delta_1$

which satisfy equation

$$\Theta_1 = \frac{\delta_0}{\mu_0}B_1 - \frac{\delta_1}{2\mu_0}B_1 - \frac{\delta_1}{2\mu_0}B_3 \tag{2.89}$$

The fundamental component of air gap flux density B_1 can be found by applying Eq. 2.88, here rewritten for $N/p = 2$:

$$B_1 = \frac{\mu_0}{\delta_0}\Theta_1 + 2\frac{\mu_0}{\delta_1}\sum_{i=1}^{\infty}(\Theta_{2i-1} - \Theta_{2i+1}) \tag{2.90}$$

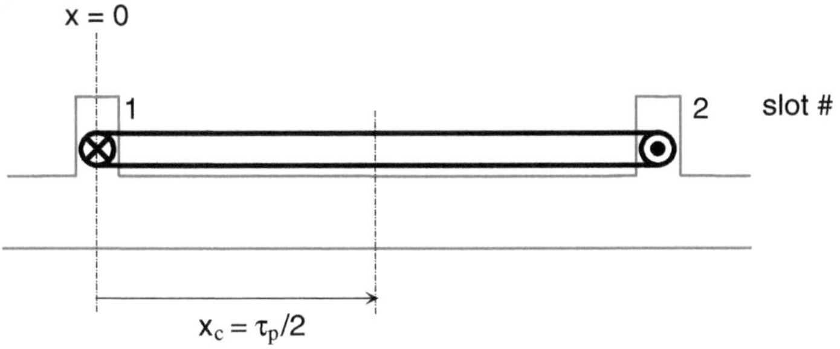

Fig. 2.39 Salient pole machine with a coil

Components of air gap flux density harmonics along perpendicular axes α and β in Fig. 2.37 satisfy simultaneously two algebraic equations (see also Eq. 2.86):

$$\begin{aligned} B_{n,\alpha} \cos n\frac{\pi}{\tau_p} x_c + \frac{\delta_1}{2\delta_0} B_{n-,\alpha} \cos\left(n + i\frac{N}{p}\right) x_c \\ + \frac{\delta_1}{2\delta_0} B_{n+,\alpha} \cos\left(n - i\frac{N}{p}\right) x_c = \frac{\mu_0 \cdot \Theta_n}{\delta_0} \cos n\frac{\pi}{\tau_p} x_c \end{aligned} \quad (2.91)$$

and

$$\begin{aligned} B_{n,\beta} \sin n\frac{\pi}{\tau_p} x_c + \frac{\delta_1}{2\delta_0} B_{n-,\beta} \sin\left(n + i\frac{N}{p}\right) x_c \\ + \frac{\delta_1}{2\delta_0} B_{n+,\beta} \sin\left(n - i\frac{N}{p}\right) x_c = \frac{\mu_0 \cdot \Theta_n}{\delta_0} \sin n\frac{\pi}{\tau_p} x_c \end{aligned} \quad (2.92)$$

Considering the first r odd harmonics of air gap flux density and MMF, one can write after rearranging of Eqs. 2.91 and 2.92 two matrix equations for the components along the axes α and β as

$$\underline{C}_\delta \cdot \underline{B}_\alpha = \underline{\vartheta}_\alpha \quad (2.93)$$

and

$$\underline{C}_\delta \cdot \underline{B}_\beta = \underline{\vartheta}_\beta \quad (2.94)$$

The vectors ϑ_α and ϑ_β representing the applied MMFs can be written as

$$\underline{\vartheta}_\alpha = \frac{\mu_0}{\delta_0} \cdot \left[\Theta_1 \cos\tfrac{\pi}{\tau_p} x_c \;\; \Theta_3 \cos 3\tfrac{\pi}{\tau_p} x_c \;\; \ldots \;\; \Theta_r \cos r\tfrac{\pi}{\tau_p} x_c\right]^T \quad (2.95)$$

$$\underline{\vartheta}_\beta = \frac{\mu_0}{\delta_0} \cdot \left[\Theta_1 \sin\tfrac{\pi}{\tau_p} x_c \;\; \Theta_3 \sin 3\tfrac{\pi}{\tau_p} x_c \;\; \ldots \;\; \Theta_r \sin r\tfrac{\pi}{\tau_p} x_c\right]^T \quad (2.96)$$

and the system matrix C_δ

Column $\mathrm{N/p}-1$
$\downarrow$

$$\underline{C}_\delta = \begin{bmatrix}
1 & 0 & \cdots & 0 & 0 & \cdots & 0 & r & r & 0 & \cdots & \cdots & 0 & 0 & 0 & 0 & \cdots & 0 & 0 & \cdots & 0 & 0 \\
0 & 1 & \cdots & 0 & 0 & \cdots & r & 0 & 0 & r & \cdots & \cdots & 0 & 0 & 0 & 0 & \cdots & 0 & 0 & \cdots & 0 & 0 \\
\cdots & \cdots & \cdots & \cdots & \cdots & \cdots & \cdots & \cdots & \cdots & \cdots & \cdots & \cdots & \cdots & \cdots & \cdots & \cdots & \cdots & \cdots & \cdots & \cdots & \cdots & \cdots \\
0 & 0 & \cdots & 1 & r & \cdots & 0 & 0 & 0 & 0 & \cdots & \cdots & r & 0 & 0 & 0 & \cdots & 0 & 0 & \cdots & 0 & 0 \\
0 & 0 & \cdots & r & 1 & \cdots & 0 & 0 & 0 & 0 & \cdots & \cdots & 0 & r & 0 & 0 & \cdots & 0 & 0 & \cdots & 0 & 0 \\
\cdots & \cdots & \cdots & \cdots & \cdots & \cdots & \cdots & \cdots & \cdots & \cdots & \cdots & \cdots & \cdots & \cdots & \cdots & \cdots & \cdots & \cdots & \cdots & \cdots & \cdots & \cdots \\
0 & r & \cdots & 0 & 0 & \cdots & 1 & 0 & 0 & 0 & \cdots & \cdots & 0 & 0 & r & 0 & \cdots & 0 & 0 & \cdots & 0 & 0 \\
r & 0 & \cdots & 0 & 0 & \cdots & 0 & 1 & 0 & 0 & \cdots & \cdots & 0 & 0 & 0 & r & \cdots & 0 & 0 & \cdots & 0 & 0 \\
r & 0 & \cdots & 0 & 0 & \cdots & 0 & 0 & 1 & 0 & \cdots & \cdots & 0 & 0 & 0 & 0 & \cdots & r & 0 & \cdots & 0 & 0 \\
0 & r & \cdots & 0 & 0 & \cdots & 0 & 0 & 0 & 1 & \cdots & \cdots & 0 & 0 & 0 & 0 & \cdots & 0 & r & \cdots & 0 & 0 \\
\cdots & \cdots & \cdots & \cdots & \cdots & \cdots & \cdots & \cdots & \cdots & \cdots & \cdots & \cdots & \cdots & \cdots & \cdots & \cdots & \cdots & \cdots & \cdots & \cdots & \cdots & \cdots \\
\cdots & \cdots & \cdots & \cdots & \cdots & \cdots & \cdots & \cdots & \cdots & \cdots & \cdots & \cdots & \cdots & \cdots & \cdots & \cdots & \cdots & \cdots & \cdots & \cdots & \cdots & \cdots \\
0 & 0 & \cdots & r & 0 & \cdots & 0 & 0 & 0 & 0 & \cdots & \cdots & 1 & 0 & 0 & 0 & \cdots & 0 & 0 & \cdots & r & 0 \\
0 & 0 & \cdots & 0 & r & \cdots & 0 & 0 & 0 & 0 & \cdots & \cdots & 0 & 1 & 0 & 0 & \cdots & 0 & 0 & \cdots & 0 & r \\
0 & 0 & \cdots & 0 & 0 & \cdots & r & 0 & 0 & 0 & \cdots & \cdots & 0 & 0 & 1 & 0 & \cdots & 0 & 0 & \cdots & 0 & 0 \\
0 & 0 & \cdots & 0 & 0 & \cdots & 0 & r & 0 & 0 & \cdots & \cdots & 0 & 0 & 0 & 1 & \cdots & 0 & 0 & \cdots & 0 & 0 \\
\cdots & \cdots & \cdots & \cdots & \cdots & \cdots & \cdots & \cdots & \cdots & \cdots & \cdots & \cdots & \cdots & \cdots & \cdots & \cdots & \cdots & \cdots & \cdots & \cdots & \cdots & \cdots \\
0 & 0 & \cdots & 0 & 0 & \cdots & 0 & 0 & r & 0 & \cdots & \cdots & 0 & 0 & 0 & 0 & \cdots & 1 & 0 & \cdots & 0 & 0 \\
0 & 0 & \cdots & 0 & 0 & \cdots & 0 & 0 & 0 & r & \cdots & \cdots & 0 & 0 & 0 & 0 & \cdots & 0 & 1 & \cdots & 0 & 0 \\
\cdots & \cdots & \cdots & \cdots & \cdots & \cdots & \cdots & \cdots & \cdots & \cdots & \cdots & \cdots & \cdots & \cdots & \cdots & \cdots & \cdots & \cdots & \cdots & \cdots & \cdots & \cdots \\
0 & 0 & \cdots & 0 & 0 & \cdots & 0 & 0 & 0 & 0 & \cdots & \cdots & r & 0 & 0 & 0 & \cdots & 0 & 0 & \cdots & 1 & 0 \\
0 & 0 & \cdots & 0 & 0 & \cdots & 0 & 0 & 0 & 0 & \cdots & \cdots & 0 & r & 0 & 0 & \cdots & 0 & 0 & \cdots & 0 & 1
\end{bmatrix} \tag{2.97}$$

if the number of slots per pole $N/(2p)$ is even, and

Column $\mathrm{N/p}-1$
$\downarrow$

$$\underline{C}_\delta = \begin{bmatrix}
1 & 0 & \cdots & 0 & 0 & 0 & \cdots & 0 & r & r & 0 & \cdots & 0 & 0 & 0 & 0 & \cdots & 0 & 0 & 0 & \cdots & 0 & 0 \\
0 & 1 & \cdots & 0 & 0 & 0 & \cdots & r & 0 & 0 & r & \cdots & 0 & 0 & 0 & 0 & \cdots & 0 & 0 & 0 & \cdots & 0 & 0 \\
\cdots & \cdots & \cdots & \cdots & \cdots & \cdots & \cdots & \cdots & \cdots & \cdots & \cdots & \cdots & \cdots & \cdots & \cdots & \cdots & \cdots & \cdots & \cdots & \cdots & \cdots & \cdots & \cdots \\
0 & 0 & \cdots & 1 & 0 & r & \cdots & 0 & 0 & 0 & 0 & \cdots & r & 0 & 0 & 0 & \cdots & 0 & 0 & 0 & \cdots & 0 & 0 \\
0 & 0 & \cdots & 0 & 1+r & 0 & \cdots & 0 & 0 & 0 & 0 & \cdots & 0 & r & 0 & 0 & \cdots & 0 & 0 & 0 & \cdots & 0 & 0 \\
0 & 0 & \cdots & r & 0 & 1 & \cdots & 0 & 0 & 0 & 0 & \cdots & 0 & 0 & r & 0 & \cdots & 0 & 0 & 0 & \cdots & 0 & 0 \\
\cdots & \cdots & \cdots & \cdots & \cdots & \cdots & \cdots & \cdots & \cdots & \cdots & \cdots & \cdots & \cdots & \cdots & \cdots & \cdots & \cdots & \cdots & \cdots & \cdots & \cdots & \cdots & \cdots \\
0 & r & \cdots & 0 & 0 & 0 & \cdots & 1 & 0 & 0 & 0 & \cdots & 0 & 0 & 0 & 0 & \cdots & 0 & 0 & 0 & \cdots & 0 & 0 \\
r & 0 & \cdots & 0 & 0 & 0 & \cdots & 0 & 1 & 0 & 0 & \cdots & 0 & 0 & 0 & 0 & \cdots & r & 0 & 0 & \cdots & 0 & 0 \\
r & 0 & \cdots & 0 & 0 & 0 & \cdots & 0 & 0 & 1 & 0 & \cdots & 0 & 0 & 0 & 0 & \cdots & 0 & r & 0 & \cdots & 0 & 0 \\
0 & r & \cdots & 0 & 0 & 0 & \cdots & 0 & 0 & 0 & 1 & \cdots & 0 & 0 & 0 & 0 & \cdots & 0 & 0 & r & \cdots & 0 & 0 \\
\cdots & \cdots & \cdots & \cdots & \cdots & \cdots & \cdots & \cdots & \cdots & \cdots & \cdots & \cdots & \cdots & \cdots & \cdots & \cdots & \cdots & \cdots & \cdots & \cdots & \cdots & \cdots & \cdots \\
0 & 0 & \cdots & r & 0 & 0 & \cdots & 0 & 0 & 0 & 0 & \cdots & 1 & 0 & 0 & 0 & \cdots & 0 & 0 & 0 & \cdots & r & 0 \\
0 & 0 & \cdots & 0 & r & 0 & \cdots & 0 & 0 & 0 & 0 & \cdots & 0 & 1 & 0 & 0 & \cdots & 0 & 0 & 0 & \cdots & 0 & r \\
0 & 0 & \cdots & 0 & 0 & r & \cdots & 0 & 0 & 0 & 0 & \cdots & 0 & 0 & 1 & 0 & \cdots & 0 & 0 & 0 & \cdots & 0 & 0 \\
0 & 0 & \cdots & 0 & 0 & 0 & \cdots & 0 & 0 & 0 & 0 & \cdots & 0 & 0 & 0 & 1 & \cdots & 0 & 0 & 0 & \cdots & 0 & 0 \\
\cdots & \cdots & \cdots & \cdots & \cdots & \cdots & \cdots & \cdots & \cdots & \cdots & \cdots & \cdots & \cdots & \cdots & \cdots & \cdots & \cdots & \cdots & \cdots & \cdots & \cdots & \cdots & \cdots \\
0 & 0 & \cdots & 0 & 0 & 0 & \cdots & 0 & r & 0 & 0 & \cdots & 0 & 0 & 0 & 0 & \cdots & 1 & 0 & 0 & \cdots & 0 & 0 \\
0 & 0 & \cdots & 0 & 0 & 0 & \cdots & 0 & 0 & r & 0 & \cdots & 0 & 0 & 0 & 0 & \cdots & 0 & 1 & 0 & \cdots & 0 & 0 \\
0 & 0 & \cdots & 0 & 0 & 0 & \cdots & 0 & 0 & 0 & r & \cdots & 0 & 0 & 0 & 0 & \cdots & 0 & 0 & 1 & \cdots & 0 & 0 \\
\cdots & \cdots & \cdots & \cdots & \cdots & \cdots & \cdots & \cdots & \cdots & \cdots & \cdots & \cdots & \cdots & \cdots & \cdots & \cdots & \cdots & \cdots & \cdots & \cdots & \cdots & \cdots & \cdots \\
0 & 0 & \cdots & 0 & 0 & 0 & \cdots & 0 & 0 & 0 & 0 & \cdots & r & 0 & 0 & 0 & \cdots & 0 & 0 & 0 & \cdots & 1 & 0 \\
0 & 0 & \cdots & 0 & 0 & 0 & \cdots & 0 & 0 & 0 & 0 & \cdots & 0 & r & 0 & 0 & \cdots & 0 & 0 & 0 & \cdots & 0 & 1
\end{bmatrix} \tag{2.98}$$

if the number of slots per pole $N/(2p)$ is odd. The coefficient r is defined as

$$r = \frac{\delta_1}{2 \cdot \delta_0} \tag{2.99}$$

and the vectors of α- and β-axes components of resulting flux density as

$$\underline{B}_\alpha = \begin{bmatrix} B_{1,\alpha} & B_{3,\alpha} & \cdots & B_{\frac{N}{p}-1,\alpha} & B_{\frac{N}{p}+1,\alpha} & \cdots & B_{r-2,\alpha} & B_{r,\alpha} \end{bmatrix}^{\mathrm{T}} \tag{2.100}$$

$$\underline{B}_\beta = \begin{bmatrix} B_{1,\beta} & B_{3,\beta} & \cdots & B_{\frac{N}{p}-1,\beta} & B_{\frac{N}{p}+1,\beta} & \cdots & B_{r-2,\beta} & B_{r,\beta} \end{bmatrix}^{\mathrm{T}} \tag{2.101}$$

In order to illustrate the mechanism of generation of air gap flux harmonics in single-slotted machines, a machine with 12 slots per pole is excited with sinusoidal air gap MMF containing only the fundamental harmonic in the amount of 1 p.u. and after that with only 5. harmonic in the amount of 0.2 p.u., as shown in Figure 2.40. The excitation with 1 p.u. of fundamental harmonic of MMF creates in this machine the 1., 11., 13., 23., 25, etc., harmonics of air gap flux density. Without contribution of evoked harmonics of MMF, the fundamental harmonic of flux density would be equal to 1 p.u.; in reality, however, it is slightly increased due to action of slot harmonics. The excitation with only fifth harmonic in the amount of 0.2 p.u. results in 5., 7., 17., 19., etc., harmonics of air gap flux density corresponding to the scheme in Table 2.4.

In Fig. 2.41 the air gap flux density distributions in three machines with full-pitch coils and various air gap geometries are shown. The machine denoted by "0" in Fig. 2.41 has no teeth and/or poles (smooth air gap), the machine denoted by "2" has two salient poles on one side of air gap and $r = \delta_1/(2\delta_0) = 0.288$, whereas the machine denoted by "12" has 12 teeth per pole pair on one side of air gap and $r = 0.331$.

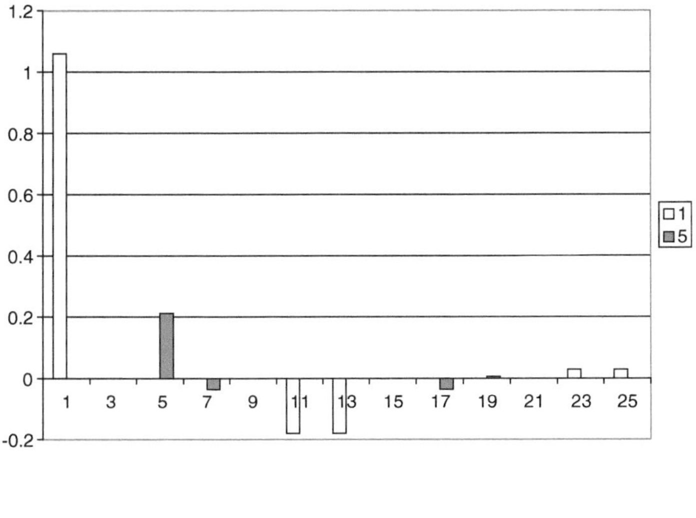

Fig. 2.40 Air gap flux density in p.u. in a machine with 12 slots per pole pair excited by the fundamental (*white bars*) and 5 harmonic (*gray bars*) of MMF

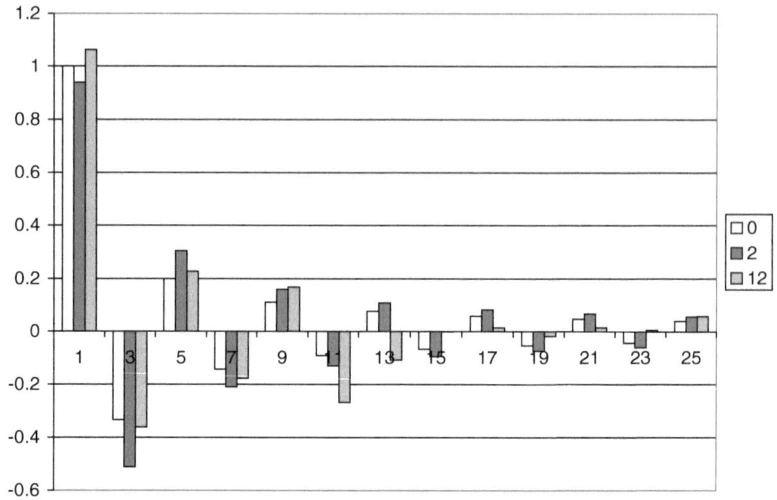

Fig. 2.41 Air gap flux density in p.u. created by current through full-pitch coil in machines with different air gap geometries and even numbers of slots per pole pair

The spectrum of air gap flux density in a machine with smooth air gap in Fig. 2.41 is identical to the spectrum of applied MMF. The amplitude of a particular harmonic of air gap flux density depends solely on the amplitude of air gap MMF with the same order. In a salient pole machine the amplitudes of flux density harmonics are lower due to relatively large interpolar space (Carter factor!). The machine with 12 teeth per pole pair illustrates in the best manner the mechanism of generation of a particular flux density harmonic through evoked harmonics. The fundamental harmonic of air gap flux density is higher than in the machine with smooth air gap, as a consequence of contribution of 11. and 13. air gap MMF harmonics. Besides, the amplitudes of 11. and 13. harmonics of air gap flux density in this machine are significantly larger than the amplitudes of their next neighbors, since very strong fundamental of the MMF contributes to their creation.

One should note that slot harmonics of the order $N/p \pm 1$ are 180° shifted to the fundamental at the axis of slot 1. If the fundamental is oriented in the positive direction of the axis of the first slot, the $(N/p - 1)$th and $(N/p + 1)$th slot harmonics are oriented in its negative direction.

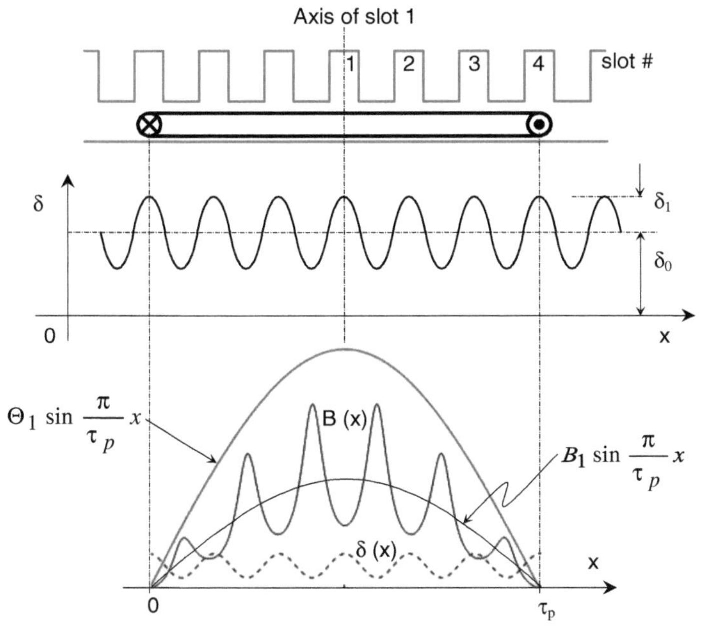

Fig. 2.42 Slotted air gap represented with a constant term and fundamental harmonic of air gap width along with the air gap flux density distribution created by the fundamental harmonic of MMF

Different phase shifts of adjacent slot harmonics are illustrated in Fig. 2.43, in which air gap flux density distribution in a single-slotted machine with 12 teeth per pole pair created by the fundamental component of MMF after Fig. 2.42 is shown.

The air gap flux density distribution in Fig. 2.43 with 12 slots per pole pair has the fundamental harmonic with amplitude B_1 and two slot harmonics with orders $N/p + 1 = 13$ and $N/p - 1 = 11$, as shown in Fig. 2.42

$$B_{\frac{N}{p}\pm1}(x) = \Theta_1 \frac{\mu_0}{2\delta_1}\left[\cos\left(\frac{N}{p}-1\right)\frac{\pi}{\tau_p}x - \cos\left(\frac{N}{p}+1\right)\frac{\pi}{\tau_p}x\right] = B_{\frac{N}{p}-1}(x) + B_{\frac{N}{p}+1}(x) \tag{2.102}$$

Slot harmonics with orders $N/p \pm 1$ are at minimum along the axis of slot 1, where the fundamental is maximum, since the air gap width at this position is maximum (slot!).

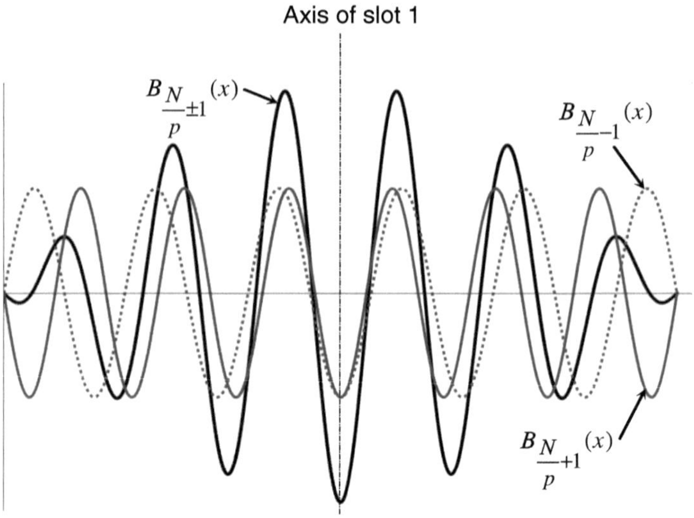

Fig. 2.43 Air gap flux density under one pole in a machine with $N/p = 12$ created by the fundamental component of MMF and fundamental harmonic of air gap width of the order N/p. In addition, two components of air gap flux density with orders $N/p - 1 = 11$ and $N/p + 1 = 13$ are shown

In special case of $N/p = 2$ (salient pole machine), the fundamental component of MMF ($n = 1$) creates

- A portion of the fundamental component ($j = n = 1$) of air gap flux density when acting on constant air gap with width δ_0,
- A portion of the fundamental component ($j = N/p - n = 1$) of air gap flux density when acting on a variable air gap with width δ_1, and
- A portion of the third harmonic component ($j = N/p + n = 3$) of air gap flux density when acting on a variable air gap with width δ_1.

which can be described by equation

$$\begin{aligned}\mu_0\Theta_1 \sin\frac{\pi}{\tau_p}x = \delta_0 B_1 \sin\frac{\pi}{\tau_p}x + \frac{\delta_1}{2}B_1 \sin(1-2)\frac{\pi}{\tau_p}\left(x - x_{0,1}\right)\\ + \frac{\delta_1}{2}B_3 \sin(3-2)\frac{\pi}{\tau_p}\left(x - x_{0,3}\right)\end{aligned} \tag{2.103}$$

as illustrated in Fig. 2.39.

In machines with **odd number of slots per pole pair N/p** the air gap flux density distribution $B(x)$ contains both odd and even harmonics

$$B(x) = \sum_{n=0,1,2,\ldots}^{\infty} B_n \cos n \frac{\pi}{\tau_p}(x - x_c) \tag{2.104}$$

in order to satisfy equation

$$B(x) \cdot \left(\delta_0 + \delta_1 \cos \frac{N}{p}\frac{\pi}{\tau_p} x\right) = \mu_0 \cdot \sum_{n=1,2,3,\ldots}^{\infty} \Theta_n \cos n \frac{\pi}{\tau_p}(x - x_c) \tag{2.105}$$

in which the MMF as well can have both odd and even terms. Note that the air gap flux density spectrum contains a constant term B_0, as a result of interaction between the *N/p*th harmonic of MMF and *N/p*th harmonic of air gap width. Flux lines of air gap flux density B_0 spread in axial direction and modify iron core flux densities in the whole machine in the manner illustrated in Fig. 2.44.

Since iron core losses are proportional to the square of maximum flux density, the shift of operating point of the machine's magnetic circuit due to homopolar flux results in increased iron core losses in the amount of $(B_0/B_m)^2$ [p.u.].

The relationship between MMF and flux density can be further written as

$$\delta_0 \cdot \sum_{i=0,1,2,\ldots}^{\infty} B_i \cos i \frac{\pi}{\tau_p}(x - x_c) + \delta_1 \cdot \sum_{j=0,1,2,\ldots}^{\infty} B_j \cos j \frac{\pi}{\tau_p}(x - x_c) \cdot \cos \frac{N}{p}\frac{\pi}{\tau_p} x$$
$$= \mu_0 \cdot \sum_{n=1,2,3,\ldots}^{\infty} \Theta_n \cos n \frac{\pi}{\tau_p}(x - x_c) \tag{2.106}$$

Since *N/p* is an odd number, odd harmonics of air gap MMF can create only even harmonics of air gap flux density, whereas even harmonics of MMF create odd harmonics of flux density when interacting with the fundamental harmonic of air gap width δ_1. Following the same logic, odd harmonics of air gap MMF can create only odd harmonics of air gap flux density, and even harmonics of air gap MMF can create only even harmonics of air gap flux density when interacting with the constant term of the air gap width δ_0. Therefore,

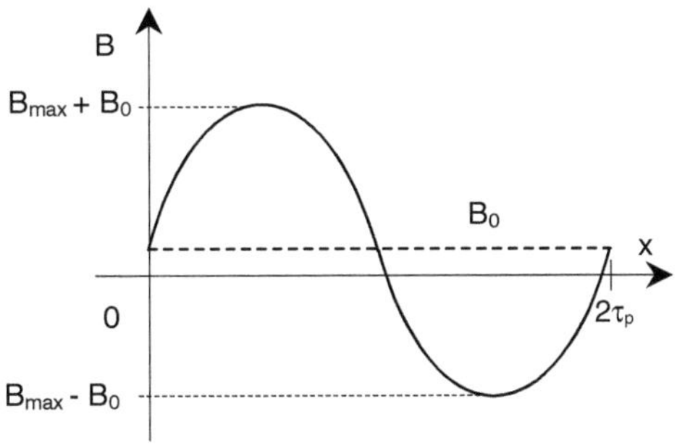

Fig. 2.44 Shift of flux density in iron core due to homopolar flux

$$\delta_0 \cdot \sum_{i=0,1,2,\ldots}^{\infty} B_i \cos i \frac{\pi}{\tau_p}(x - x_c)$$
$$+ \frac{\delta_1}{2} \sum_{j_+=0,1,2,\ldots}^{\infty} B_{j_+} \cos\left[\left(j_+ + \frac{N}{p}\right) \cdot x - j_+ \cdot x_c\right] \frac{\pi}{\tau_p}$$
$$+ \frac{\delta_1}{2} \sum_{j_-=0,1,2,\ldots}^{\infty} B_{j_-} \cos\left[\left(j_- + \frac{N}{p}\right) \cdot x - j_- \cdot x_c\right] \frac{\pi}{\tau_p} \tag{2.107}$$
$$= \mu_0 \cdot \sum_{n=1,2,3,\ldots}^{\infty} \Theta_n \cos n \frac{\pi}{\tau_p}(x - x_c)$$

If the applied air gap MMF distribution contains only odd terms, its nth harmonic with an amplitude of Θ_n creates in a machine with an odd number of slots per pole pair N/p again *three spatial harmonics of air gap flux density*:

- B_n (odd), in the amount of $\mu_0\ \Theta_n/\delta_0$,
- $B_{N/p-n}$ (even), in the amount of $2\ \mu_0\ \Theta_n/\delta_1$, and
- $B_{N/p+n}$ (even), in the amount of $2\ \mu_0\ \Theta_n/\delta_1$.

with corresponding fluxes Φ_ν, $\Phi_{N/p-n}$, and $\Phi_{N/p+n}$. The chain of generation of higher harmonics of air gap flux density is identical to the one shown in Fig. 2.38.

By using Eq. 2.107 and Table 2.5, one can relate the amplitude of nth harmonic of air gap flux density to corresponding applied and evoked MMFs as

$$B_n \cos n \frac{\pi}{\tau_p}(x - x_c) = \frac{\mu_0}{\delta_0} \Theta_n \cos n \frac{\pi}{\tau_p}(x - x_c)$$
$$+ 2\frac{\mu_0}{\delta_1} \sum_{i=1}^{\infty} \Theta_{i\frac{N}{p}-n} \cos \frac{\pi}{\tau_p}\left[nx - \left(n + i\frac{N}{p}\right)x_c\right] \tag{2.108}$$
$$+ 2\frac{\mu_0}{\delta_1} \sum_{i=1}^{\infty} \Theta_{i\frac{N}{p}+n} \cos \frac{\pi}{\tau_p}\left[nx - \left(n - i\frac{N}{p}\right)x_c\right]$$

Table 2.5 Harmonics of flux density B created by particular harmonics of MMF Θ in a machine with odd number of slots per pole pair $N/p = \nu$

		B																
		0	1	2	3	4	5	…	ν − 5	ν − 4	ν − 3	ν − 2	ν − 1	ν	ν + 1	ν + 2	ν + 3	ν + 4	ν + 5
Θ	0	x												x					
	1		Xx										Xx		Xx				
	2			x								x				x			
	3				Xx						Xx						Xx		
	4					x				x								x	
	5						Xx		Xx										Xx
	…																		
	ν − 4					Xx				Xx								Xx	
	ν − 3				x						x						x		
	ν − 2			Xx								Xx				Xx			
	ν − 1		x										x		x				
	ν	Xx												Xx					
	ν + 1		x										x		x				
	ν + 2			Xx								Xx				Xx			
	ν + 3				x						x						x		
	ν + 4					Xx				Xx								Xx	

X denotes harmonics of flux density created by applied, and x by evoked harmonics of MMF for n odd, and

$$B_n \cos n\frac{\pi}{\tau_p}(x-x_c) = \frac{\delta_1}{2\mu_0}\sum_{i=1}^{\infty}\Theta_{i\frac{N}{p}-n}\cos\frac{\pi}{\tau_p}\left[nx-\left(n+i\frac{N}{p}\right)x_c\right] + \frac{\delta_1}{2\mu_0}\sum_{i=1}^{\infty}\Theta_{i\frac{N}{p}+n}\cos\frac{\pi}{\tau_p}\left[nx-\left(n-i\frac{N}{p}\right)x_c\right] \tag{2.109}$$

for n even, except for $n = 0$. The constant term of air gap flux density ($n = 0$) representing homopolar flux is equal to

$$B_0 = \frac{\delta_1}{2\mu_0}\sum_{i=1}^{\infty}\Theta_{i\frac{N}{p}}\cos\frac{\pi}{\tau_p}\left(i\frac{N}{p}x_c\right) \tag{2.110}$$

Components of air gap flux harmonics along perpendicular axes α and β satisfy simultaneously two algebraic equations:

$$B_{n,\alpha}\cos n\frac{\pi}{\tau_p}x_c + \frac{\delta_1}{2\delta_0}B_{n-,\alpha}\cos\left(n+i\frac{N}{p}\right)x_c + \frac{\delta_1}{2\delta_0}B_{n+,\alpha}\cos\left(n-i\frac{N}{p}\right)x_c = \frac{\mu_0\cdot\Theta_n}{\delta_0}\cos n\frac{\pi}{\tau_p}x_c \tag{2.111}$$

and

$$B_{n,\beta}\sin n\frac{\pi}{\tau_p}x_c + \frac{\delta_1}{2\delta_0}B_{n-,\beta}\sin\left(n+i\frac{N}{p}\right)x_c + \frac{\delta_1}{2\delta_0}B_{n+,\beta}\sin\left(n-i\frac{N}{p}\right)x_c = \frac{\mu_0\cdot\Theta_n}{\delta_0}\sin n\frac{\pi}{\tau_p}x_c \tag{2.112}$$

for n odd, and

$$B_{n,\alpha}\cos n\frac{\pi}{\tau_p}x_c + \frac{\delta_1}{2\delta_0}B_{n-,\alpha}\cos\left(n+i\frac{N}{p}\right)x_c + \frac{\delta_1}{2\delta_0}B_{n+,\alpha}\cos\left(n-i\frac{N}{p}\right)x_c = 0 \tag{2.113}$$

$$B_{n,\beta}\sin n\frac{\pi}{\tau_p}x_c + \frac{\delta_1}{2\delta_0}B_{n-,\beta}\sin\left(n+i\frac{N}{p}\right)x_c + \frac{\delta_1}{2\delta_0}B_{n+,\beta}\sin\left(n-i\frac{N}{p}\right)x_c = 0 \tag{2.114}$$

for n even (no harmonics of applied MMF).

Denoting by index α the air gap flux components along the axis of the 1. slot, and by β the air gap flux components 90° ahead, one can write a system of linear equations which relate the air gap flux harmonics to the air gap MMF harmonics as

$$\underline{C}_\delta \cdot \underline{B}_{\alpha,\beta} = \underline{\theta}_{\alpha,\beta} \tag{2.115}$$

where

$$\underline{C}_\delta = \begin{bmatrix}
1 & 0 & 0 & 0 & 0 & \cdots & 0 & 0 & 0 & 0 & r & r & 0 & 0 & 0 & 0 & \cdots & 0 & 0 & \cdots & 0 & 0 \\
0 & 1 & 0 & 0 & 0 & \cdots & 0 & 0 & r & 0 & 0 & 0 & r & 0 & 0 & 0 & \cdots & 0 & 0 & \cdots & 0 & 0 \\
0 & 0 & 1 & 0 & 0 & \cdots & 0 & 0 & 0 & r & 0 & 0 & 0 & r & 0 & 0 & \cdots & 0 & 0 & \cdots & 0 & 0 \\
0 & 0 & 0 & 1 & 0 & \cdots & r & 0 & 0 & 0 & 0 & 0 & 0 & 0 & r & 0 & \cdots & 0 & 0 & \cdots & 0 & 0 \\
0 & 0 & 0 & 0 & 1 & \cdots & 0 & r & 0 & 0 & 0 & 0 & 0 & 0 & 0 & r & \cdots & 0 & 0 & \cdots & 0 & 0 \\
\cdots & \cdots & \cdots & \cdots & \cdots & \cdots & \cdots & \cdots & \cdots & \cdots & \cdots & \cdots & \cdots & \cdots & \cdots & \cdots & \cdots & \cdots & \cdots & \cdots & \cdots & \cdots \\
0 & 0 & 0 & r & 0 & \cdots & 1 & 0 & 0 & 0 & 0 & 0 & 0 & 0 & 0 & 0 & \cdots & 0 & 0 & \cdots & 0 & 0 \\
0 & 0 & 0 & 0 & r & \cdots & 0 & 1 & 0 & 0 & 0 & 0 & 0 & 0 & 0 & 0 & \cdots & 0 & 0 & \cdots & 0 & 0 \\
0 & r & 0 & 0 & 0 & \cdots & 0 & 0 & 1 & 0 & 0 & 0 & 0 & 0 & 0 & 0 & \cdots & 0 & 0 & \cdots & 0 & 0 \\
0 & 0 & r & 0 & 0 & \cdots & 0 & 0 & 0 & 1 & 0 & 0 & 0 & 0 & 0 & 0 & \cdots & 0 & 0 & \cdots & 0 & 0 \\
r & 0 & 0 & 0 & 0 & \cdots & 0 & 0 & 0 & 0 & 0 & 0 & 0 & 0 & 0 & 0 & \cdots & 0 & 0 & \cdots & 0 & 0 \\
r & 0 & 0 & 0 & 0 & \cdots & 0 & 0 & 0 & 0 & 0 & 1 & 0 & 0 & 0 & 0 & \cdots & 0 & 0 & \cdots & 0 & 0 \\
0 & r & 0 & 0 & 0 & \cdots & 0 & 0 & 0 & 0 & 0 & 0 & 1 & 0 & 0 & 0 & \cdots & 0 & 0 & \cdots & 0 & 0 \\
0 & 0 & r & 0 & 0 & \cdots & 0 & 0 & 0 & 0 & 0 & 0 & 0 & 1 & 0 & 0 & \cdots & 0 & 0 & \cdots & 0 & 0 \\
0 & 0 & 0 & r & 0 & \cdots & 0 & 0 & 0 & 0 & 0 & 0 & 0 & 0 & 1 & 0 & \cdots & 0 & 0 & \cdots & 0 & 0 \\
0 & 0 & 0 & 0 & r & \cdots & 0 & 0 & 0 & 0 & 0 & 0 & 0 & 0 & 0 & 1 & \cdots & 0 & 0 & \cdots & 0 & 0 \\
\cdots & \cdots & \cdots & \cdots & \cdots & \cdots & \cdots & \cdots & \cdots & \cdots & \cdots & \cdots & \cdots & \cdots & \cdots & \cdots & \cdots & \cdots & \cdots & \cdots & \cdots & \cdots \\
0 & 0 & 0 & 0 & 0 & \cdots & 0 & 0 & 0 & 0 & 0 & 0 & 0 & 0 & 0 & 0 & \cdots & 1 & 0 & \cdots & r & 0 \\
0 & 0 & 0 & 0 & 0 & \cdots & 0 & 0 & 0 & 0 & 0 & 0 & 0 & 0 & 0 & 0 & \cdots & 0 & 1 & \cdots & 0 & r \\
\cdots & \cdots & \cdots & \cdots & \cdots & \cdots & \cdots & \cdots & \cdots & \cdots & \cdots & \cdots & \cdots & \cdots & \cdots & \cdots & \cdots & \cdots & \cdots & \cdots & \cdots & \cdots \\
0 & 0 & 0 & 0 & 0 & \cdots & 0 & 0 & 0 & 0 & 0 & 0 & 0 & 0 & 0 & 0 & \cdots & r & 0 & \cdots & 1 & 0 \\
0 & 0 & 0 & 0 & 0 & \cdots & 0 & 0 & 0 & 0 & 0 & 0 & 0 & 0 & 0 & 0 & \cdots & 0 & r & \cdots & 0 & 1
\end{bmatrix} \tag{2.116}$$

and

$$\underline{B}_{\alpha,\beta} = \begin{bmatrix} B_0 & B_{1,\alpha} & B_{1,\beta} & B_{2,\alpha} & B_{2,\beta} & \ldots & B_{n-1,\alpha} & B_{n-1,\beta} & B_{n,\alpha} & B_{n,\beta} \end{bmatrix}^T \tag{2.117}$$

Here again the coefficient r was used

$$r = \frac{\delta_1}{2 \cdot \delta_0} \tag{2.118}$$

The vector of applied MMFs has only odd terms

$$
\underline{\Theta}_{\alpha,\beta} = \frac{\mu_0}{\delta_0}\begin{bmatrix} 0 \\ \Theta_1 \cos\frac{\pi}{\tau_p} x_c \\ \Theta_1 \sin\frac{\pi}{\tau_p} x_c \\ 0 \\ 0 \\ \Theta_3 \cos 3\frac{\pi}{\tau_p} x_c \\ \Theta_3 \sin 3\frac{\pi}{\tau_p} x_c \\ \cdots \\ \cdots \\ \Theta_{n-2} \cos(n-2)\frac{\pi}{\tau_p} x_c \\ \Theta_{n-2} \sin(n-2)\frac{\pi}{\tau_p} x_c \\ 0 \\ 0 \\ \Theta_n \cos n\frac{\pi}{\tau_p} x_c \\ \Theta_n \sin n\frac{\pi}{\tau_p} x_c \end{bmatrix} \tag{2.119}
$$

The previous considerations can be illustrated by means of Fig. 2.45, in which the air gap flux density spectrum in a tooth-wound machine with $N/p = 3$ and in a machine with $N/p = 11$ are shown for a coil shift of $x_{\text{coil}} = 0$. The spectrum of applied MMF in the machine with $N/p = 3$ contains both odd and even harmonics, and in the machine with $N/p = 11$ only odd harmonics were present. In both cases, however, even harmonics of air gap flux density, including the homopolar flux, are generated. The amplitudes of air gap flux density harmonics in Fig. 2.45 are shown along with their phase shifts.

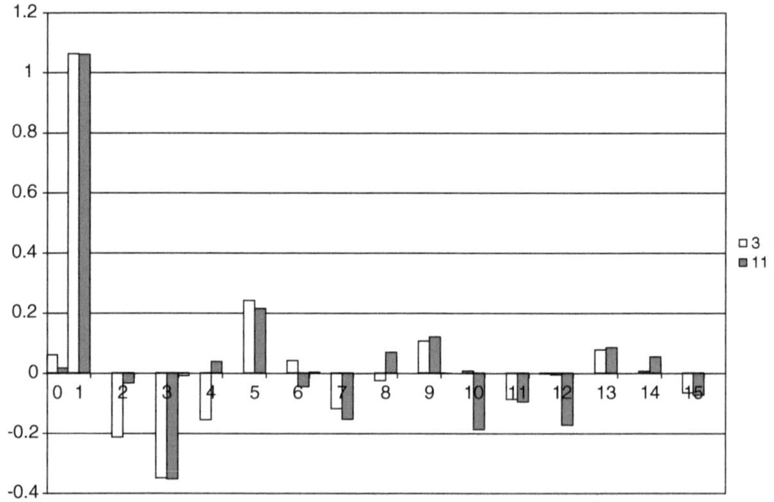

Fig. 2.45 Air gap flux density harmonics in p.u. created by current-carrying coil in machines with 3 and 11 slots per pole pair

Slot harmonics with order $N/p \pm 1$ are in both machines even and have relatively large amplitudes. Besides, a homopolar flux is generated, which is proportional to B_0.

The two components of air gap flux density with period lengths of $\tau_p/(N/p \pm 1)$ are sources of higher harmonics in the induced voltage, pulsating torque on the shaft, and vibrations due to pulsating radial forces. For these reasons, the amplitudes of the two components of flux density must be reduced.

Pitch factor for slot harmonics of order $N/p \pm 1$ is equal to (see Eq. 2.21):

$$\begin{aligned} f_{\mathrm{p},\frac{N}{p}\pm 1} &= \sin\left(\frac{N}{p} \pm 1\right)\frac{y}{\tau_{\mathrm{ps}}}\frac{\pi}{2} = \sin\left(\frac{N}{p}\frac{y}{\frac{N}{2p}}\frac{\pi}{2} \pm \frac{y}{\tau_{\mathrm{ps}}}\frac{\pi}{2}\right) = \sin\left(y\pi \pm \frac{y}{\tau_{\mathrm{ps}}}\frac{\pi}{2}\right) \\ &= \pm\sin\frac{y}{\tau_{\mathrm{ps}}}\frac{\pi}{2} \end{aligned} \quad (2.120)$$

which is the pitch factor for the fundamental. Since the coil pitch is selected in such a manner as not to decrease too much the fundamental, the pitch factor for air gap slot harmonics with orders $N/p \pm 1$ is high. Pitching the coils cannot fight slot harmonics.

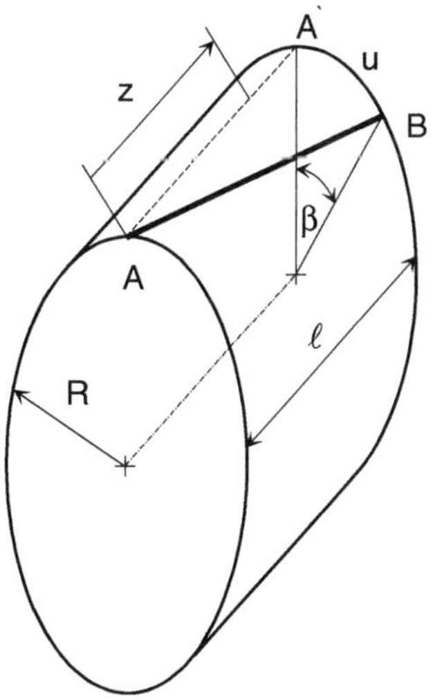

Fig. 2.46 Axial skewing of a conductor for an angle β

One recalls that higher harmonics of the MMF can be reduced on the coil, pole, and/or fundamental pole basis. In all three cases, the coil parameters in *circumferential* direction are modified. In heteropolar machines, the coil geometry offers a possibility of impacting the air gap quantities by *skewing* the coils in *axial* direction, as shown in Fig. 2.46.

Effects of skewing in Fig. 2.46 can be illustrated in the circumferential–axial (x, z) plane in the manner shown in Fig. 2.47. The black drawn coil has a pitch y and is skewed for the amount of u

$$u = \beta \cdot R \tag{2.121}$$

The fundamental component of air gap flux density can be written as

$$B_1(x) = B_{\max} \sin \frac{\pi}{\tau_p} x \tag{2.122}$$

The amount of flux concatenated by the skewed coil in Fig. 2.47 is equal to

$$\Phi_1 = \int_0^l \int_{x_0 + \frac{u}{l}z}^{x_0 + \frac{u}{l}z + y} B_{\max} \sin \frac{\pi}{\tau_p} x \mathrm{d}x \mathrm{d}z = \frac{l\tau_p^2}{u\pi^2} \cdot \left[-\sin \frac{\pi}{\tau_p} x_0 + \sin \frac{\pi}{\tau_p}(u + x_0) + \sin \frac{\pi}{\tau_p}(y + x_0) - \sin \frac{\pi}{\tau_p}(u + y + x_0) \right] \tag{2.123}$$

or, after some trigonometric simplifications

$$\Phi_1 = \frac{2}{\pi} l \tau_p B_{\max} \cdot \frac{\sin \frac{\beta_{el}}{2}}{\frac{\beta_{el}}{2}} \cdot \sin\left(\frac{y}{\tau_p}\frac{\pi}{2}\right) \cdot \sin\left(\frac{2x_0 + u + y}{2}\frac{\pi}{\tau_p}\right) \tag{2.124}$$

As expected, the flux concatenated between the fundamental component of air gap flux density and a coil is a function of coil shift x_0. The concatenated flux is reduced by two factors: the pitch factor, and the skewing factor $f_{sk,1}$

$$f_{sk,1} = \frac{\sin \frac{\beta_{el}}{2}}{\frac{\beta_{el}}{2}} \tag{2.125}$$

where

$$\beta_{el} = p \cdot \beta = p \cdot \frac{u}{R} \tag{2.126}$$

The skewing factor is a measure for loss of flux due to skewing and is always less than one. For an arbitrary nth harmonic, the concatenated flux is equal to

$$\Phi_n = \frac{2}{\pi} l \tau_p B_{\max} \cdot \frac{\sin n \frac{\beta_{el}}{2}}{n \frac{\beta_{el}}{2}} \cdot \sin\left(n \frac{y}{\tau_p}\frac{\pi}{2}\right) \cdot \sin\left(\frac{2nx_0 + u + y}{2}\frac{\pi}{\tau_p}\right) \tag{2.127}$$

and the skewing factor $f_{sk,n}$:

$$f_{\mathrm{sk},n} = \frac{\sin n \frac{\beta_{\mathrm{el}}}{2}}{n \frac{\beta_{\mathrm{el}}}{2}} \tag{2.128}$$

If the machine is skewed for one slot pitch

$$u = \frac{2R\pi}{N}; \quad \beta_{\mathrm{el}} = p \cdot \frac{u}{R} = p\frac{2\pi}{N} \tag{2.129}$$

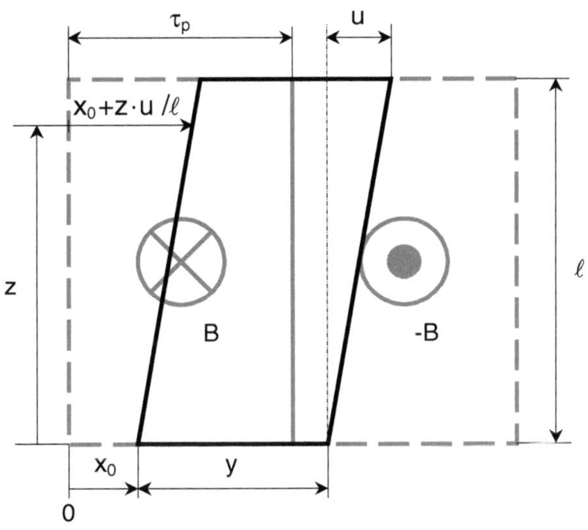

Fig. 2.47 Representation of a skewed coil in the (x, z) plane

the skewing factor becomes

$$f_{\mathrm{sk},n} = \frac{\sin np\frac{\pi}{N}}{np\frac{\pi}{N}} \tag{2.130}$$

The skewing factor for slot harmonics of the order $n = N/p \pm 1$ can now be expressed as

$$f_{\mathrm{sk},\frac{N}{p}\pm 1} = \frac{\sin\left(\frac{N}{p} \pm 1\right)p\frac{\pi}{N}}{\left(\frac{N}{p} \pm 1\right)p\frac{\pi}{N}} = \pm\frac{\sin p\frac{\pi}{N}}{p\frac{\pi}{N}}\frac{1}{\frac{N}{p} \pm 1} = \pm f_{\mathrm{sk},1} \cdot \frac{1}{\frac{N}{p} \pm 1} \tag{2.131}$$

By skewing the slots for one slot pitch, the amplitude of the nth harmonic of concatenated flux decreases to $1/(N/p \pm 1)$ of the amplitude of the fundamental. Slot skewing is the only efficient means to decrease the amplitudes of slot harmonics in the air gap flux density distribution.

2.5.4 Magnetic Gears

Relationships between the orders of harmonic in a single-slotted air gap, summarized in Tables 2.4 and 2.5, can be illustrated by the example of magnetic gears, as shown in Fig. 2.48. The magnetic gear in this figure has three concentric shells, each of which can freely rotate. The outer and inner shells carry excitation, usually permanent magnets, and the intermediate shell is used to modulate the permeance between the two.

Denoting by n the number of pole pairs of the inner shell, by k the order of lowest harmonic of the air gap permeance created by the intermediate shell, and by ℓ the number of pole pairs of the outer shell, one can express the condition for interaction between harmonics given in Tables 2.4 and 2.5 as

$$k = n \pm l \tag{2.132}$$

If the sign of ℓ in Eq. 2.132 is positive, the inner and outer field must rotate in opposite directions in order to build a torque with the given number of poles of the intermediate shell. If the sign is negative, both inner and outer shells rotate in the same direction.

Gear ratio i, i.e., the ratio between speeds of outer and inner shells is obviously equal to

$$i = \frac{n}{l} \tag{2.133}$$

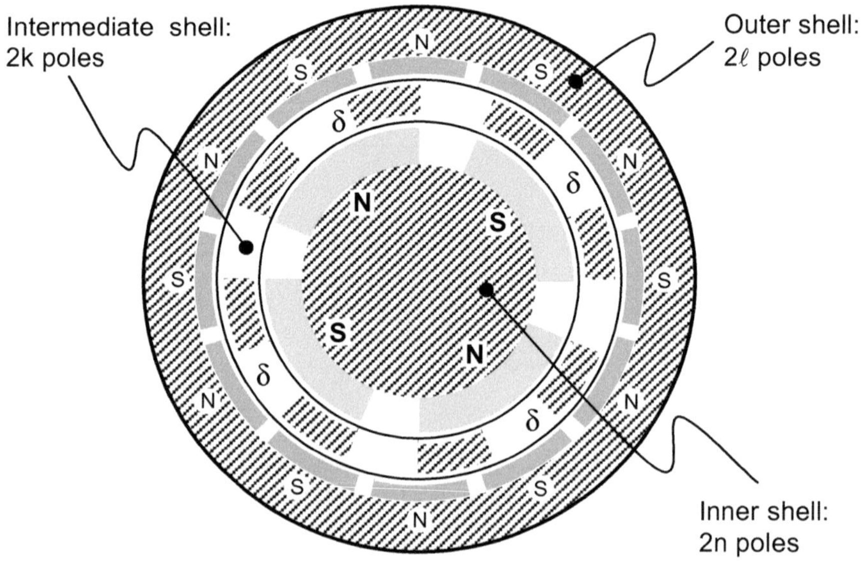

Fig. 2.48 Magnetic gear. *Shaded area* iron; *gray area* magnets; *white area* non-magnetic

2.5.5 Flux Density Distribution in the Air Gap of a Double-Slotted Machine

Consider a double-slotted machine, the air gap width of which is described as

$$\delta(x) = \delta_0 + \sum_{i=1,3,5,\ldots}^{\infty} \delta_{s,i} \cos i \frac{N_s}{p} \frac{\pi}{\tau_p} x + \sum_{j=1,3,5,\ldots}^{\infty} \delta_{r,j} \cos j \frac{N_r}{p} \frac{\pi}{\tau_p} (x - x_0) \quad (2.134)$$

where index "s" is related to stator, and "r" to rotor quantities, and x_0 is the rotor shift. A double-slotted machine with excitation on both sides of the air gap is shown in Fig. 2.49.

Due to their minor influence on machine performance, higher harmonics of stator and rotor air gap widths can be neglected. With this assumption, one can write for air gap quantities

$$\begin{aligned} B(x, x_0) \cdot \left[\delta_0 + \delta_{s,1} \cos \frac{N_s}{p} \frac{\pi}{\tau_p} x + \delta_{r,1} \cos \frac{N_r}{p} \frac{\pi}{\tau_p} (x - x_0)\right] \\ = \mu_0 \cdot \left[\sum_{n=1,3,5,\ldots}^{\infty} \Theta_{s,n} \cos n \frac{\pi}{\tau_p} x + \sum_{k=1,3,5,\ldots}^{\infty} \Theta_{r,k} \cos k \frac{\pi}{\tau_p} (x - x_0)\right] \end{aligned} \quad (2.135)$$

where

$$B(x, x_0) = \sum_{i=0,1,2,\ldots}^{\infty} B_i \cos i \frac{\pi}{\tau_p} (x - x_i) \quad (2.136)$$

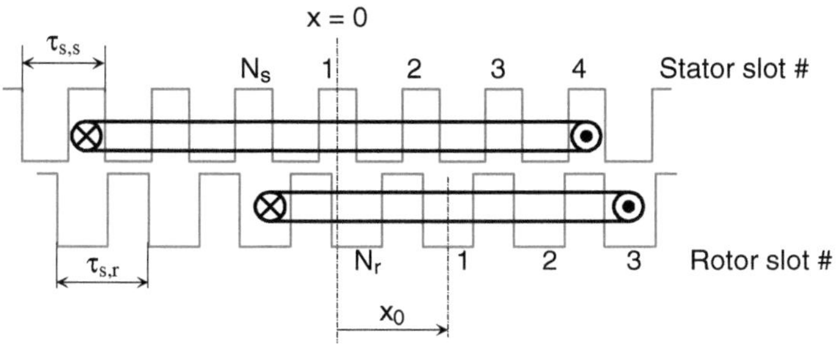

Fig. 2.49 Double-slotted air gap with stator and rotor excitation

The nth harmonic of total (stator plus rotor) applied MMF can be written as

$$\Theta_n = \Theta_{n,\max} \cos n \frac{\pi}{\tau_p}(x - \xi_n) \tag{2.137}$$

where

$$\Theta_{n,\max} = \sqrt{\Theta_{s,n}^2 + \Theta_{r,n}^2 + 2\Theta_{s,n}\Theta_{r,n} \cos n \frac{\pi}{\tau_p} x_0} \tag{2.138}$$

and

$$\xi_n = \frac{\tau_p}{\pi} \arctan \varepsilon_n; \quad \varepsilon_n = \frac{\Theta_{r,n} \sin n \frac{\pi}{\tau_p} x_0}{\Theta_{s,n} + \Theta_{r,n} \cos n \frac{\pi}{\tau_p} x_0} \tag{2.139}$$

The shift ξ_n denotes the position of maximum of total MMF. Constant term in the flux density spectrum B_0 is generated when either N_s/p or N_r/p is odd. Accordingly

$$\begin{aligned} \delta_0 & \sum_{i=0,1,2,\ldots}^{\infty} B_i \cos i \frac{\pi}{\tau_p}(x - \xi_i) \\ & + \delta_{s,1} \sum_{j=0,1,2,\ldots}^{\infty} B_j \cos j \frac{\pi}{\tau_p}(x - \xi_j) \cdot \cos \frac{N_s}{p} \frac{\pi}{\tau_p} x \\ & + \delta_{r,1} \sum_{k=0,1,2,\ldots}^{\infty} B_k \cos k \frac{\pi}{\tau_p}(x - \xi_k) \cdot \cos \frac{N_r}{p} \frac{\pi}{\tau_p}(x - x_0) \\ & = \mu_0 \sum_{n=1,3,5,\ldots}^{\infty} \Theta_{n,\max} \cos n \frac{\pi}{\tau_p}(x - \xi_n) \end{aligned} \tag{2.140}$$

or

$$\begin{aligned} & \delta_0 \sum_{i=0,1,2,\ldots}^{\infty} B_i \cos i \frac{\pi}{\tau_p}(x - \xi_i) + \frac{\delta_{s,1}}{2} \sum_{j=0,1,2,\ldots}^{\infty} B_j \cos \frac{\pi}{\tau_p}\left[\left(\frac{N_s}{p} + j\right) \cdot x - j \cdot \xi_j\right] \\ & + \frac{\delta_{s,1}}{2} \sum_{j=0,1,2,\ldots}^{\infty} B_j \cos \frac{\pi}{\tau_p}\left[\left(\frac{N_s}{p} - j\right) \cdot x + j \cdot \xi_j\right] + \frac{\delta_{r,1}}{2} \sum_{k=0,1,2,\ldots}^{\infty} B_k \cos \frac{\pi}{\tau_p}\left[\left(\frac{N_r}{p} + k\right) \cdot x - k \cdot \left(\xi_k + \frac{N_r}{p} x_0\right)\right] \\ & + \frac{\delta_{r,1}}{2} \sum_{k=0,1,2,\ldots}^{\infty} B_k \cos \frac{\pi}{\tau_p}\left[\left(\frac{N_r}{p} - k\right) \cdot x + k \cdot \left(\xi_k - \frac{N_r}{p} x_0\right)\right] = \mu_0 \sum_{n=1,3,5,\ldots}^{\infty} \Theta_{n,\max} \cos n \frac{\pi}{\tau_p}(x - \xi_n) \end{aligned} \tag{2.141}$$

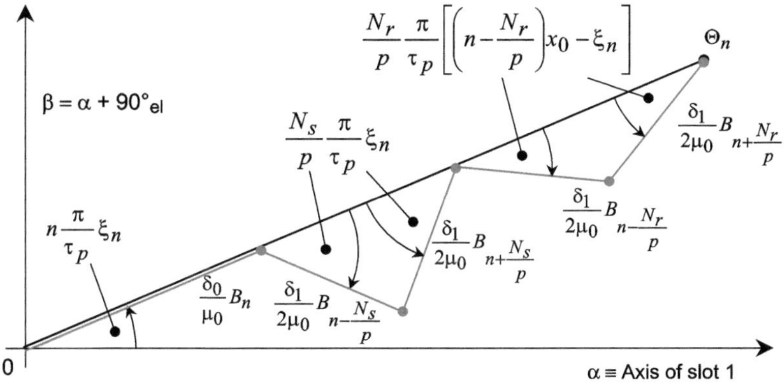

Fig. 2.50 Illustrating the relationship between MMF and flux density harmonic components in a machine with double-slotted air gap

Assume for the purpose of simplicity that both N_s/p and N_r/p are even. The nth (odd) harmonic of applied MMF creates *five components of flux density with the same order n* (Fig. 2.50):

- B_n (odd), in the amount of $\mu_0\ \Theta_n/\delta_0$,
- $B_{n-N_s/p}$ (odd), in the amount of $2\ \mu_0\ \Theta_n/\delta_1$,
- $B_{n+N_s/p}$ (odd), in the amount of $2\ \mu_0\ \Theta_n/\delta_1$,
- $B_{n-N_r/p}$ (odd), in the amount of $2\ \mu_0\ \Theta_n/\delta_1$, and
- $B_{n+N_r/p}$ (odd), in the amount of $2\ \mu_0\ \Theta_n/\delta_1$.

which satisfy equation

$$\begin{aligned}\mu_0\Theta_{n,\max}\cos n\frac{\pi}{\tau_p}(x-\xi_n) &= \delta_0 B_n\cos n\frac{\pi}{\tau_p}(x-\xi_n)+\frac{\delta_{s,1}}{2}B_{n-\frac{N_s}{p}}\cos\frac{\pi}{\tau_p}\left[nx-\left(n+\frac{N_s}{p}\right)\xi_n\right]\\ &+\frac{\delta_{s,1}}{2}B_{n+\frac{N_s}{p}}\cos\frac{\pi}{\tau_p}\left[nx-\left(n-\frac{N_s}{p}\right)\xi_n\right]+\frac{\delta_{r,1}}{2}B_{n-\frac{N_r}{p}}\cos\frac{\pi}{\tau_p}\left[nx-\left(n-\frac{N_r}{p}\right)\left(\xi_n+\frac{N_r}{p}x_0\right)\right]\\ &+\frac{\delta_{r,1}}{2}B_{n+\frac{N_r}{p}}\cos\frac{\pi}{\tau_p}\left[nx-\left(n-\frac{N_r}{p}\right)\left(\xi_n-\frac{N_r}{p}x_0\right)\right]\end{aligned} \tag{2.142}$$

The nth harmonic of total (stator plus rotor) applied MMF contributes to the nth, $N_s/p - n$, $N_s/p + n$, $N_r/p - n$, and $N_r/p + n$ harmonics of the air gap flux density, since these harmonics, when modulated with corresponding slot harmonics, have the same order equal to n.

Table 2.6 Initial harmonics of flux density *B* created by particular harmonics of MMF Θ in a double-slotted machine with even number of slots per pole pair ν on one side of the air gap and μ. on the other

		B																
		1	3	5	…	ν − 5	ν − 3	ν − 1	ν + 1	ν + 3	ν + 5	…	μ − 5	μ − 3	μ − 1	μ + 1	μ + 3	μ + 5
Θ	1	**Xx**						Xx	Xx						*Xx*	*Xx*		
	3		**Xx**				Xx			Xx				*Xx*			*Xx*	
	5			**Xx**		Xx					Xx		*Xx*					*Xx*
	…																	
	ν − 5			Xx		**Xx**					*Xx*							
	ν − 3		Xx				**Xx**			*Xx*								
	ν − 1	Xx						**Xx**	*Xx*									
	ν + 1	Xx						*Xx*	**Xx**									
	ν + 3		Xx				*Xx*			**Xx**								
	ν + 5			Xx		*Xx*					**Xx**							
	…																	
	μ − 5			*Xx*									**Xx**					
	μ − 3		*Xx*											**Xx**				
	μ − 1	*Xx*													**Xx**			
	μ + 1	*Xx*														**Xx**		
	μ + 3		*Xx*														**Xx**	
	μ + 5			*Xx*														**Xx**

X denotes harmonics of flux density created by applied, and x by evoked harmonics of MMF

Initial terms of the chain of generation of flux density from a single harmonic of total applied MMF is shown in Fig. 2.51.

By using the scheme in Fig. 2.51, one can generate Table 2.6 in which the relationship between applied and evoked MMF harmonics on one side and air gap flux density harmonics on the other is given. Here symbol ν is used to denote the number of slots per pole pair on one side of the air gap, and μ on the other. Normal font letters denote harmonics due to ν, and italic styled those due to μ. Harmonics denoted by bold letters contain components due to both ν and μ.

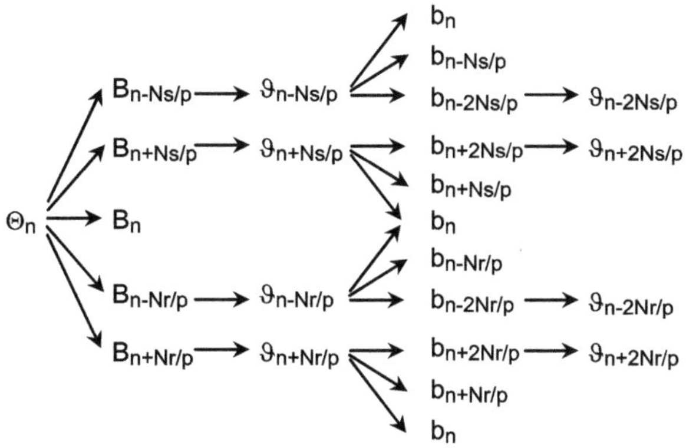

Fig. 2.51 The beginning of the chain of generation of flux density harmonics from a single harmonic of total applied MMF in a double-slotted machine

Using Eq. 2.142 and Table 2.6, one can relate the amplitude of *n*th harmonic of air gap flux density to corresponding applied and evoked MMFs as

$$\begin{aligned} B_n \cos n\frac{\pi}{\tau_p}(x-\xi_n) &= \frac{\mu_0}{\delta_0}\Theta_{n,\max}\cos n\frac{\pi}{\tau_p}(x-\xi_n) \\ &+ 2\frac{\mu_0}{\delta_{s,1}}\sum_{i=1}^{\infty}\Theta_{i\frac{N_s}{p}-n}\cos\frac{\pi}{\tau_p}\left[nx-\left(n+i\frac{N_s}{p}\right)\xi_n\right] \\ &+ 2\frac{\mu_0}{\delta_{s,1}}\sum_{i=1}^{\infty}\Theta_{i\frac{N_s}{p}+n}\cos\frac{\pi}{\tau_p}\left[nx-\left(n-i\frac{N_s}{p}\right)\xi_n\right] \\ &+ 2\frac{\mu_0}{\delta_{r,1}}\sum_{i=1}^{\infty}\Theta_{i\frac{N_r}{p}-n}\cos\frac{\pi}{\tau_p}\left[nx-\left(n+i\frac{N_r}{p}\right)\left(\xi_n+\frac{N_r}{p}x_0\right)\right] \\ &+ 2\frac{\mu_0}{\delta_{r,1}}\sum_{i=1}^{\infty}\Theta_{i\frac{N_r}{p}+n}\cos\frac{\pi}{\tau_p}\left[nx-\left(n-i\frac{N_r}{p}\right)\left(\xi_n-\frac{N_r}{p}x_0\right)\right] \end{aligned} \tag{2.143}$$

Components of air gap flux created by applied MMF harmonics along perpendicular axes α and β in Fig. 2.50 satisfy for $i = 1$ simultaneously two algebraic equations

$$B_{n,\alpha} \cos n \frac{\pi}{\tau_p} \xi_n + \frac{\delta_1}{2\delta_0} B_{n-,\alpha} \cos\left(n + \frac{N_s}{p}\right) \xi_n + \frac{\delta_1}{2\delta_0} B_{n+,\alpha} \cos\left(n - \frac{N_s}{p}\right) \xi_n$$
$$+ \frac{\delta_1}{2\delta_0} B_{n-,\alpha} \cos\left(n + \frac{N}{p}\right)\left(\xi_n + \frac{N_r}{p} x_0\right)$$
$$+ \frac{\delta_1}{2\delta_0} B_{n+,\alpha} \cos\left(n - \frac{N}{p}\right)\left(\xi_n - \frac{N_r}{p} x_0\right) = \frac{\mu_0 \cdot \Theta_n}{\delta_0} \cos n \frac{\pi}{\tau_p} \xi_n \tag{2.144}$$

and

$$B_{n,\alpha} \sin sn \frac{\pi}{\tau_p} \xi_n + \frac{\delta_1}{2\delta_0} B_{n-,\alpha} \sin\left(n + \frac{N_s}{p}\right) \xi_n + \frac{\delta_1}{2\delta_0} B_{n+,\alpha} \sin\left(n - \frac{N_s}{p}\right) \xi_n$$
$$+ \frac{\delta_1}{2\delta_0} B_{n-,\alpha} \sin\left(n + \frac{N}{p}\right)\left(\xi_n + \frac{N_r}{p} x_0\right)$$
$$+ \frac{\delta_1}{2\delta_0} B_{n+,\alpha} \sin\left(n - \frac{N}{p}\right)\left(\xi_n - \frac{N_r}{p} x_0\right) = \frac{\mu_0 \cdot \Theta_n}{\delta_0} \sin n \frac{\pi}{\tau_p} \xi_n \tag{2.145}$$

By applying the principle of orthogonality of trigonometric functions through the separation of sine and cosine terms, one can derive from Eq. 2.143 a system of $2n + 1$ algebraic equations (one for the constant term plus two for each harmonic) for air gap quantities in a doubly slotted machine:

$$\left(\underline{C}_{s,\delta} + \underline{C}_{r,\delta}\right) \cdot \underline{B}_{\alpha,\beta} = \underline{\vartheta}_{\alpha,\beta} \tag{2.146}$$

where the solution vector of air gap flux density harmonics $B_{\alpha,\beta}$ is defined as

$$\underline{B}_{\alpha,\beta} = \begin{bmatrix} B_0 & B_{1,\alpha} & B_{1,\beta} & B_{2,\alpha} & B_{2,\beta} & \ldots & B_{n-1,\alpha} & B_{n-1,\beta} & B_{n,\alpha} & B_{n,\beta} \end{bmatrix} \tag{2.147}$$

By making use of substitutions

$$r_s = \frac{\delta_{s,1}}{2 \cdot \delta_0}; \quad r_r = \frac{\delta_{r,1}}{2 \cdot \delta_0} \tag{2.148}$$

one can define the system matrices $C_{s,\delta}$ (Eq. 2.149) and $C_{r,\delta}$ (Eq. 2.150) as

$$
\underline{C}_{s,\delta} = \begin{bmatrix}
1 & 0 & 0 & 0 & 0 & \cdots & 0 & 0 & 0 & 0 & r_s & r_s & 0 & 0 & 0 & 0 & \cdots & 0 & 0 & \cdots & 0 & 0 \\
0 & 1 & 0 & 0 & 0 & \cdots & 0 & 0 & r_s & 0 & 0 & 0 & r_s & 0 & 0 & 0 & \cdots & 0 & 0 & \cdots & 0 & 0 \\
0 & 0 & 1 & 0 & 0 & \cdots & 0 & 0 & 0 & r_s & 0 & 0 & 0 & r_s & 0 & 0 & \cdots & 0 & 0 & \cdots & 0 & 0 \\
0 & 0 & 0 & 1 & 0 & \cdots & r_s & 0 & 0 & 0 & 0 & 0 & 0 & 0 & r_s & 0 & \cdots & 0 & 0 & \cdots & 0 & 0 \\
0 & 0 & 0 & 0 & 1 & \cdots & 0 & r_s & 0 & 0 & 0 & 0 & 0 & 0 & 0 & r_s & \cdots & 0 & 0 & \cdots & 0 & 0 \\
\cdots & \cdots & \cdots & \cdots & \cdots & \cdots & \cdots & \cdots & \cdots & \cdots & \cdots & \cdots & \cdots & \cdots & \cdots & \cdots & \cdots & \cdots & \cdots & \cdots & \cdots & \cdots \\
0 & 0 & 0 & r_s & 0 & \cdots & 1 & 0 & 0 & 0 & 0 & 0 & 0 & 0 & 0 & 0 & \cdots & 0 & 0 & \cdots & 0 & 0 \\
0 & 0 & 0 & 0 & r_s & \cdots & 0 & 1 & 0 & 0 & 0 & 0 & 0 & 0 & 0 & 0 & \cdots & 0 & 0 & \cdots & 0 & 0 \\
0 & r_s & 0 & 0 & 0 & \cdots & 0 & 0 & 1 & 0 & 0 & 0 & 0 & 0 & 0 & 0 & \cdots & 0 & 0 & \cdots & 0 & 0 \\
0 & 0 & r_s & 0 & 0 & \cdots & 0 & 0 & 0 & 1 & 0 & 0 & 0 & 0 & 0 & 0 & \cdots & 0 & 0 & \cdots & 0 & 0 \\
r_s & 0 & 0 & 0 & 0 & \cdots & 0 & 0 & 0 & 0 & 0 & 0 & 0 & 0 & 0 & 0 & \cdots & 0 & 0 & \cdots & 0 & 0 \\
r_s & 0 & 0 & 0 & 0 & \cdots & 0 & 0 & 0 & 0 & 0 & 1 & 0 & 0 & 0 & 0 & \cdots & 0 & 0 & \cdots & 0 & 0 \\
0 & r_s & 0 & 0 & 0 & \cdots & 0 & 0 & 0 & 0 & 0 & 0 & 1 & 0 & 0 & 0 & \cdots & 0 & 0 & \cdots & 0 & 0 \\
0 & 0 & r_s & 0 & 0 & \cdots & 0 & 0 & 0 & 0 & 0 & 0 & 0 & 1 & 0 & 0 & \cdots & 0 & 0 & \cdots & 0 & 0 \\
0 & 0 & 0 & r_s & 0 & \cdots & 0 & 0 & 0 & 0 & 0 & 0 & 0 & 0 & 1 & 0 & \cdots & 0 & 0 & \cdots & 0 & 0 \\
0 & 0 & 0 & 0 & r_s & \cdots & 0 & 0 & 0 & 0 & 0 & 0 & 0 & 0 & 0 & 1 & \cdots & 0 & 0 & \cdots & 0 & 0 \\
\cdots & \cdots & \cdots & \cdots & \cdots & \cdots & \cdots & \cdots & \cdots & \cdots & \cdots & \cdots & \cdots & \cdots & \cdots & \cdots & \cdots & \cdots & \cdots & \cdots & \cdots & \cdots \\
0 & 0 & 0 & 0 & 0 & \cdots & 0 & 0 & 0 & 0 & 0 & 0 & 0 & 0 & 0 & 0 & \cdots & 1 & 0 & \cdots & r_s & 0 \\
0 & 0 & 0 & 0 & 0 & \cdots & 0 & 0 & 0 & 0 & 0 & 0 & 0 & 0 & 0 & 0 & \cdots & 0 & 1 & \cdots & 0 & r_s \\
\cdots & \cdots & \cdots & \cdots & \cdots & \cdots & \cdots & \cdots & \cdots & \cdots & \cdots & \cdots & \cdots & \cdots & \cdots & \cdots & \cdots & \cdots & \cdots & \cdots & \cdots & \cdots \\
0 & 0 & 0 & 0 & 0 & \cdots & 0 & 0 & 0 & 0 & 0 & 0 & 0 & 0 & 0 & 0 & \cdots & r_s & 0 & \cdots & 1 & 0 \\
0 & 0 & 0 & 0 & 0 & \cdots & 0 & 0 & 0 & 0 & 0 & 0 & 0 & 0 & 0 & 0 & \cdots & 0 & r_s & \cdots & 0 & 1
\end{bmatrix} \tag{2.149}
$$

$$
\underline{C}_{r,\delta} = \begin{bmatrix}
1 & 0 & 0 & 0 & 0 & \cdots & 0 & 0 & 0 & 0 & r_r & r_r & 0 & 0 & 0 & 0 & \cdots & 0 & 0 & \cdots & 0 & 0 \\
0 & 1 & 0 & 0 & 0 & \cdots & 0 & 0 & r_r & 0 & 0 & 0 & r_r & 0 & 0 & 0 & \cdots & 0 & 0 & \cdots & 0 & 0 \\
0 & 0 & 1 & 0 & 0 & \cdots & 0 & 0 & 0 & r_r & 0 & 0 & 0 & r_r & 0 & 0 & \cdots & 0 & 0 & \cdots & 0 & 0 \\
0 & 0 & 0 & 1 & 0 & \cdots & r_r & 0 & 0 & 0 & 0 & 0 & 0 & 0 & r_r & 0 & \cdots & 0 & 0 & \cdots & 0 & 0 \\
0 & 0 & 0 & 0 & 1 & \cdots & 0 & r_r & 0 & 0 & 0 & 0 & 0 & 0 & 0 & r_r & \cdots & 0 & 0 & \cdots & 0 & 0 \\
\cdots & \cdots & \cdots & \cdots & \cdots & \cdots & \cdots & \cdots & \cdots & \cdots & \cdots & \cdots & \cdots & \cdots & \cdots & \cdots & \cdots & \cdots & \cdots & \cdots & \cdots & \cdots \\
0 & 0 & 0 & r_r & 0 & \cdots & 1 & 0 & 0 & 0 & 0 & 0 & 0 & 0 & 0 & 0 & \cdots & 0 & 0 & \cdots & 0 & 0 \\
0 & 0 & 0 & 0 & r_r & \cdots & 0 & 1 & 0 & 0 & 0 & 0 & 0 & 0 & 0 & 0 & \cdots & 0 & 0 & \cdots & 0 & 0 \\
0 & r_r & 0 & 0 & 0 & \cdots & 0 & 0 & 1 & 0 & 0 & 0 & 0 & 0 & 0 & 0 & \cdots & 0 & 0 & \cdots & 0 & 0 \\
0 & 0 & r_r & 0 & 0 & \cdots & 0 & 0 & 0 & 1 & 0 & 0 & 0 & 0 & 0 & 0 & \cdots & 0 & 0 & \cdots & 0 & 0 \\
r_r & 0 & 0 & 0 & 0 & \cdots & 0 & 0 & 0 & 0 & 0 & 0 & 0 & 0 & 0 & 0 & \cdots & 0 & 0 & \cdots & 0 & 0 \\
r_r & 0 & 0 & 0 & 0 & \cdots & 0 & 0 & 0 & 0 & 0 & 1 & 0 & 0 & 0 & 0 & \cdots & 0 & 0 & \cdots & 0 & 0 \\
0 & r_r & 0 & 0 & 0 & \cdots & 0 & 0 & 0 & 0 & 0 & 0 & 1 & 0 & 0 & 0 & \cdots & 0 & 0 & \cdots & 0 & 0 \\
0 & 0 & r_r & 0 & 0 & \cdots & 0 & 0 & 0 & 0 & 0 & 0 & 0 & 1 & 0 & 0 & \cdots & 0 & 0 & \cdots & 0 & 0 \\
0 & 0 & 0 & r_r & 0 & \cdots & 0 & 0 & 0 & 0 & 0 & 0 & 0 & 0 & 1 & 0 & \cdots & 0 & 0 & \cdots & 0 & 0 \\
0 & 0 & 0 & 0 & r_r & \cdots & 0 & 0 & 0 & 0 & 0 & 0 & 0 & 0 & 0 & 1 & \cdots & 0 & 0 & \cdots & 0 & 0 \\
\cdots & \cdots & \cdots & \cdots & \cdots & \cdots & \cdots & \cdots & \cdots & \cdots & \cdots & \cdots & \cdots & \cdots & \cdots & \cdots & \cdots & \cdots & \cdots & \cdots & \cdots & \cdots \\
0 & 0 & 0 & 0 & 0 & \cdots & 0 & 0 & 0 & 0 & 0 & 0 & 0 & 0 & 0 & 0 & \cdots & 1 & 0 & \cdots & r_r & 0 \\
0 & 0 & 0 & 0 & 0 & \cdots & 0 & 0 & 0 & 0 & 0 & 0 & 0 & 0 & 0 & 0 & \cdots & 0 & 1 & \cdots & 0 & r_r \\
\cdots & \cdots & \cdots & \cdots & \cdots & \cdots & \cdots & \cdots & \cdots & \cdots & \cdots & \cdots & \cdots & \cdots & \cdots & \cdots & \cdots & \cdots & \cdots & \cdots & \cdots & \cdots \\
0 & 0 & 0 & 0 & 0 & \cdots & 0 & 0 & 0 & 0 & 0 & 0 & 0 & 0 & 0 & 0 & \cdots & r_r & 0 & \cdots & 1 & 0 \\
0 & 0 & 0 & 0 & 0 & \cdots & 0 & 0 & 0 & 0 & 0 & 0 & 0 & 0 & 0 & 0 & \cdots & 0 & r_r & \cdots & 0 & 1
\end{bmatrix} \tag{2.150}
$$

The vector $\vartheta_{\alpha,\beta}$ representing applied MMFs is equal to

$$
\underline{\vartheta}_{\alpha,\beta} = \frac{\mu_0}{\delta_0}
\begin{bmatrix}
0 \\
\Theta_{s,1} + \Theta_{r,1}\cos\frac{\pi}{\tau_p}x_0 \\
\Theta_{r,1}\sin\frac{\pi}{\tau_p}x_0 \\
\Theta_{s,2} + \Theta_{r,2}\cos 2\frac{\pi}{\tau_p}x_0 \\
\Theta_{s,2} + \Theta_{r,2}\sin 2\frac{\pi}{\tau_p}x_0 \\
\Theta_{s,3} + \Theta_{r,3}\cos 3\frac{\pi}{\tau_p}x_0 \\
\Theta_{r,3}\sin 3\frac{\pi}{\tau_p}x_0 \\
\cdots \\
\cdots \\
\Theta_{s,n-2} + \Theta_{r,n-2}\cos(n-2)\frac{\pi}{\tau_p}x_0 \\
\Theta_{r,n-2}\sin(n-2)\frac{\pi}{\tau_p}x_0 \\
\Theta_{s,n-1} + \Theta_{r,n-1}\cos(n-1)\frac{\pi}{\tau_p}x_0 \\
\Theta_{r,n-1}\sin(n-1)\frac{\pi}{\tau_p}x_0 \\
\Theta_{s,n} + \Theta_{r,n}\cos n\frac{\pi}{\tau_p}x_0 \\
\Theta_{r,n}\sin n\frac{\pi}{\tau_p}x_0
\end{bmatrix}
\tag{2.151}
$$

In Fig. 2.52 the results of analysis of air gap flux density in a machine with 18 stator slots per pole pair, $r_s = \delta_{s,1}/(2\delta_0) = r_r = \delta_{r,1}/(2\delta_0) = 0.33$, and

(a) 2 rotor poles;
(b) 14 rotor slots per pole pair; and
(c) 20 rotor slots per pole pair

are shown.

The machine with 2 rotor (salient) poles has a spectrum similar to that of a machine with smooth stator and salient poles, as shown in Fig. 2.33—both fundamental and 3. harmonic of the air gap flux density are lower than in case of completely smooth air gap, as a consequence of 180° phase shift of rotor slot harmonics. In addition, stator slot harmonics (17. and 19.) are stronger than their next neighbors.

In the machine with 14 rotor slots per pole pair, both stator (17. and 19.) and rotor (13. and 15.) slot harmonics are strong as compared to their next neighbors.

The air gap flux density in the machine with 20 rotor slots has a very strong peak at the common slot harmonic of the order 19 (=20 − 1 = 18 + 1). This high peak is typical for all machines with slot combinations $N_s - N_r = \pm 2p$.

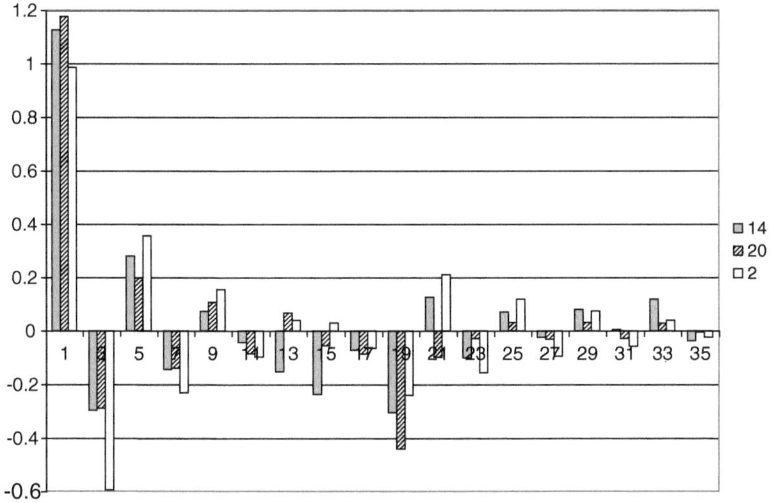

Fig. 2.52 Air gap flux density harmonics in p.u. created by currents through a stator and a rotor full-pitch coil in machines with doubly slotted air gap, 18 stator slots per pole pair and 2, 14, and 20 rotor slots per pole pair

The spectrum of air gap flux density harmonics in a machine with equal number of stator and rotor slots is shown in Fig. 2.53, along with the spectrum of air gap flux density harmonics in a machine with the same number of stator teeth and smooth rotor.

Various factors affecting the amplitude of the nth harmonic of MMF can be expressed in terms of the winding factor for the nth harmonic $f_{w,n}$, defined as

$$f_{w,n} = f_{o,n} \cdot f_{p,n} \cdot f_{z,n} \cdot f_{sk,n} \tag{2.152}$$

By dividing the winding factor for the nth harmonic through the Carter factor, one obtains the *excitation efficacy* factor $f_{ee,n}$ for the nth harmonic

$$f_{ee,n} = \frac{f_{w,n}}{k_C} \tag{2.153}$$

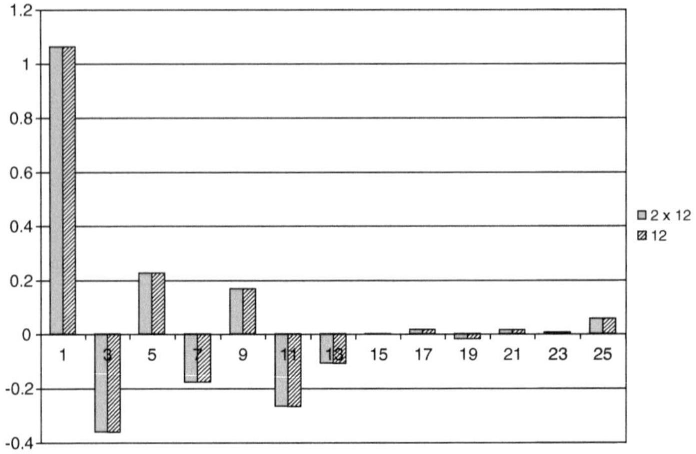

Fig. 2.53 Air gap flux density harmonics in p.u. created by currents through a stator and a rotor full-pitch coil in a machine with doubly slotted air gap, 12 stator slots per pole pair and 12 rotor slots per pole pair (denoted by "2 × 12"). The spectrum of air gap flux density in a machine with 12 stator teeth and smooth rotor (denoted by "12") is shown as reference. The two spectra differ only slightly from each other

Physical meaning of the excitation efficacy factor becomes obvious when one analyzes the relationship between the nth harmonic of flux density and MMF which created it:

$$B_n = \mu_0 \frac{\Theta_n}{k_C \cdot \delta} = I \cdot w \cdot \frac{\sin\frac{n\pi}{2}}{n} \cdot \frac{\mu_0}{\delta} \cdot \frac{4}{\pi} \cdot \frac{f_{w,n}}{k_C} = I \cdot w \cdot \frac{\sin\frac{n\pi}{2}}{n} \cdot \frac{\mu_0}{\delta} \cdot \frac{4}{\pi} \cdot f_{ee,n} \qquad (2.154)$$

The excitation efficacy factor determines how much flux density the ampere-turns $I \cdot w$ can generate in an air gap with a width δ.

2.5.6 *Flux Density Distribution in Eccentric Air Gap of a Single-Slotted Machine*

Air gap width in a single-slotted machine with eccentric rotor can be described by equation

$$\delta = \delta_0 + \delta_1 \cos \frac{N}{p} \frac{\pi}{\tau_p} (x - x_r) + \varepsilon \cos \frac{\pi}{\tau_p} (x - x_r) \qquad (2.155)$$

Air gap flux density created by odd harmonics of MMF contains harmonics of the order 0, 1, 2, 3,…, etc., as discussed in previous sections. In special case of salient pole machine with p = 1, the air gap flux density can be expressed as

$$\delta = \delta_0 + \delta_1 \cos 2\frac{\pi}{\tau_p}(x - x_r) + \varepsilon \cos\frac{\pi}{\tau_p}(x - x_r) \tag{2.156}$$

The fundamental spatial component of MMF Θ_1 creates homopolar flux density B_0, the amplitude of which varies with period length of 2 τ_p, along with the fundamental B_1 second B_2, and third harmonic B_3 of air gap flux density. The amplitudes of all harmonics with order equal to or larger than 1 pulsate with period length of τ_p.

2.5.7 The Influence of Saturation

In the previous sections it was assumed that the MMF drop across iron core is negligible and only those additional harmonics of air gap flux density were calculated, which originated in non-even air gap width. In reality, however, the MMF drop across iron core cannot be neglected (Fig. 2.54).

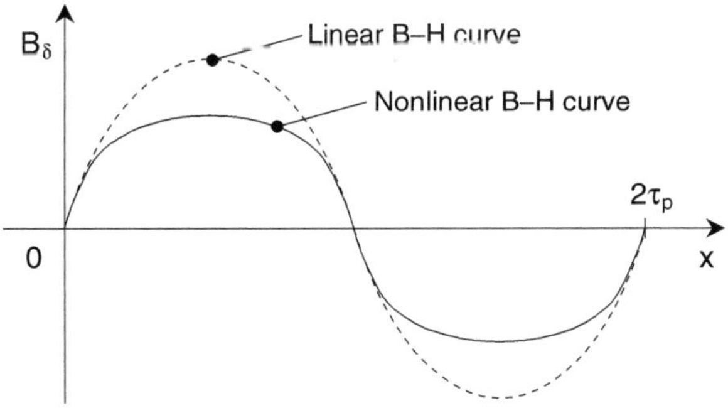

Fig. 2.54 Flattening of the air gap flux density distribution created by the fundamental harmonic of MMF and saturation in iron in a machine with constant air gap

Although there exists no analytical expression for B–H curve of iron core and, therefore, the decrease of flux density for a given MMF can only be described qualitatively, the orders of harmonics of air gap flux density due to saturation along with their influence on machine performance can be precisely quantified.

Flattening of air gap flux density distribution due to saturation in iron is a source of additional odd harmonics in its spectrum. Harmonics due to saturation act in rotating field in a different manner than harmonics created by discrete winding distribution.

Case Study 2.2: The influence of machine topology on ampere-turns demand for a given air gap flux density and on the amplitude of its pulsating component will be illustrated on five representative machine types:

- Large cylindrical rotor synchronous machine;
- Medium-size squirrel cage induction machine;
- Large low-speed surface-mounted permanent magnet machine with distributed stator windings;
- Large low-speed surface-mounted permanent magnet machine with tooth-wound stator;
- Large high-speed tooth-wound machine with embedded magnets in flux concentration geometry with data as described in Case Study 2.1.

The machines were analyzed from the point of view of excitation demand and the amplitude of the pulsating component of air gap flux density. The results of analysis are presented in Table 2.7.

The large wound rotor synchronous machine has a wide air gap and, therefore, a Carter factor close to one. The pulsating component of the air gap flux density due to stator slots is very low (4.1 % of the average). The excitation efficacy factor for the fundamental harmonic of flux density is rather high (0.893), which means that almost 90 % of a single coil ampere-turns produce the fundamental component of the air gap flux density. The value of excitation efficacy factor of 0.893 is taken as 100 % in this comparison.

The medium-size squirrel cage induction machine has high excitation efficacy, along with considerable amplitude of pulsating component of flux density due to slotting. The relatively high amplitude of B_s after Eq. 2.39 generates surface losses in the rotor, which, however, are not critical because the rotor magnetic circuit is laminated. Local effect of pulsating air gap flux density—attractive force between stator and rotor iron surfaces proportional to the square of flux density—can be large enough to generate vibrations and audible noise in this machine.

Permanent magnet machines in Table 2.7 have poorer excitation efficacy basically due to two reasons: a larger Carter factor, and, in case of tooth-wound machines, a poorer winding factor caused by large slot openings. The poorest excitation efficacy factor characterizes the tooth-wound surface-mounted PM machine, in which slightly more than 50 % of available ampere-turns can be utilized for generation of the fundamental component of air gap flux density. In other words, in order to produce the same magnitude of air gap flux density, the field winding of a wound rotor synchronous machine has to generate only 60.5 % (=0.54/0.893) of ampere-turns created by permanent magnets in a tooth-wound PM machine with equally wide air gap. This is another reason for supremacy of wound rotor over permanent magnet machines.

Table 2.7 Comparison of crucial parameters of AC machines

	Cylindrical rotor synchronous	Squirrel cage induction	PM surface mounted	PM surface mounted	PM flux concentration
Stator winding	Distributed	Distributed	Distributed	Tooth wound	Tooth wound
Air gap radius	440	155	2400	1500	600
Poles	2	4	84	70	20
Stator slots	42	48	288	72	30
Pole pitch	1382.3	243.5	179.5	134.6	188.5
Slot width	26.6	3.5	23	70	56
Stator slot height	200	40.5	117	150	156
Axial length	3420	380	770	480	700
Coil pitch	17	10	3	1	1
Slot-opening factor $f_{o,1}$	1.00	1.00	0.993	0.892	0.964
Pitch factor $f_{p,1}$	0.956	0.966	0.981	0.999	0.866
Air gap width	35	0.9	6	5	7.5
Carter factor k_C	1.07	1.082	1.236	1.65	1.364
B_s/B_O (%)	4.1	37.2	36.7	75.2	58.9
Excitation efficacy $f_{ee,1}$	**0.893** (100 %)	**0.893** (100 %)	**0.788** (88.3 %)	**0.54** (60.5 %)	**0.612** (68.5 %)

2.6 Time-Dependent Excitation, Rotating Field Generation, MMF Wave Speed, Positive and Negative Sequence Components

Rotating field in the air gap of electric machine is generated either mechanically, by rotation of current-carrying coil(s), or electrically, by supplying stationary or rotating winding(s) by alternating or constant current(s) respectively, see Table 2.8.

Only those stator and a rotor harmonics of the air gap MMF which rotate at the same speed can generate a torque with an average value different from zero and produce mechanical work. Assuming that stator winding generates a rotating field which revolves at synchronous speed n_s, mechanical work will be produced if:

- A DC-excited rotor rotates at synchronous speed, whereas the DC excitation is generated either by a constant current flowing through a coil, or by permanent magnets; and
- AC-fed windings generate rotating field with speed of rotation relative to the stator rotating field equal to zero.

Table 2.8 Modes of rotating field generation and results of their interaction

			Stator		
			At standstill		Rotating
			Polyphase excited	DC fed	DC fed
Rotor	At standstill	Polyphase excited	Rotary phase shifting transformer	–	–
	Rotating	DC fed	Synchronous machine	–	Clutch
		AC excited	Induction machine	Brake	–

One should keep in mind that spatially distributed electromagnetic quantities in a machine—its current sheet, MMF and flux density distributions—are *periodical*, as is the current, flux, induced voltage, etc.

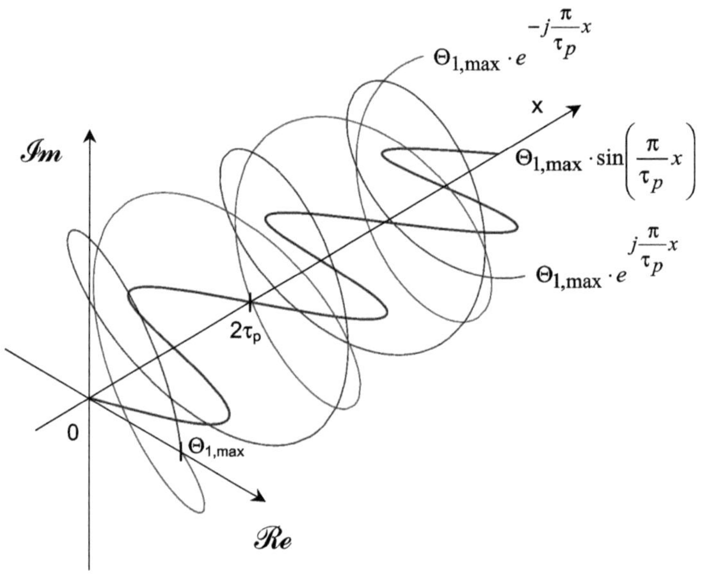

Fig. 2.55 The fundamental component of stationary air gap MMF $\Theta_1(x) = \Theta_{1,max} \sin(\pi/\tau_p \cdot x)$ represented in the complex circumferential space by means of complex conjugates $\Theta_{1,max} e^{j\frac{\pi}{\tau_p}x}$ and $\Theta_{1,max} e^{-j\frac{\pi}{\tau_p}x}$

Although the current and the MMF have the same dimension [A], these two quantities have different physical meanings and properties. Current I is a scalar, which alone cannot create physical effects in the air gap, because it does not carry information on position of conductor in the air gap. Only if the current flows through a spatially distributed winding, it can influence the electromagnetic condition of the air gap.

Whereas the alternating current can be represented in the complex plane in the manner shown in Appendix, one more dimension—the circumferential coordinate —is necessary in order to represent the air gap MMF correctly. This can be done in *the complex circumferential space*, as shown in Fig. 2.55. The complex circumferential space emerges from the complex plane, through the origin of which the machine circumferential coordinate goes perpendicularly into the plane.

The value of the fundamental component of MMF at a given circumferential coordinate x in Fig. 2.55 is proportional to the sine of angle $\pi/\tau_p \cdot x$. By applying Euler's equation, a sine function of a real argument is represented in the complex plane as a sum of complex conjugates, whereas the real argument can be time, space, or both of them

$$\Theta_1(x,t) = \Theta_{1,\max} \sin\left(\frac{\pi}{\tau_p} x - \omega t\right) = \Theta_{1,\max} \frac{e^{j\left(\frac{\pi}{\tau_p}x-\omega t\right)} - e^{-j\left(\frac{\pi}{\tau_p}x-\omega t\right)}}{2j} \tag{2.157}$$

The two functions $\Theta_{1,\max} e^{j\left(\frac{\pi}{\tau_p}x-\omega t\right)}$ and $\Theta_{1,\max} e^{-j\left(\frac{\pi}{\tau_p}x-\omega t\right)}$ in the complex circumferential space (Re, Im, x) are spirals with radius $\Theta_{1,\max}$ and axis coincident with the x-axis. Their projections to the complex plane (Re, In) are located on a circle with radius $\Theta_{1,\max}$ and angles $\pm$ arc sin ($\pi/\tau_p \cdot x - \omega \cdot t$), as shown in Fig. 2.55.

At a given time instant the coil current is constant and the MMF distribution along the x-axis is stationary. As the circumferential coordinate x increases, the positions of the corresponding points in the complex real space begin to slide along the two MMF spirals. The difference of the associated complex conjugates divided by $2j$ lies on the real axis of the complex plane and gives the amount of corresponding MMF.

The spirals $\Theta_{1,\max} e^{j\left(\frac{\pi}{\tau_p}x-\omega t\right)}$ and $\Theta_{1,\max} e^{-j\left(\frac{\pi}{\tau_p}x-\omega t\right)}$ in the complex circumferential space determine the character of the air gap MMF in the following manner:

- Stationary air gap MMF created by a constant current through coil(s): one set of stationary spirals as in Fig. 2.55, the radius $\Theta_{1,\max}$ of which is determined by the amount of coil current;
- Rotating air gap MMF created by a system of coils: one set of spirals with constant radius $\Theta_{1,\max}$ as in Fig. 2.55, rotating in the positive direction;
- Pulsating air gap MMF created by a single coil: two sets of spirals with constant radius $\Theta_{1,\max}$ as in Fig. 2.55, rotating in opposite directions.

2.6.1 MMF Waves Generated by Rotating DC-Fed Coil(s) on One Side of Air Gap

A DC-fed coil shifted for x_0 from zero point of the circumferential coordinate generates the air gap MMF distribution

$$\Theta(x, x_0) = \sum_{n=1,2,3,\ldots}^{\infty} \Theta_n \sin n \frac{\pi}{\tau_p}(x - x_0) \tag{2.158}$$

If the coil rotates at velocity $v = x_0/t$, its air gap MMF distribution rotates too:

$$\Theta(x, t) = \sum_{n=1,2,3,\ldots}^{\infty} \Theta_n \sin n \frac{\pi}{\tau_p}(x - v \cdot t) \tag{2.159}$$

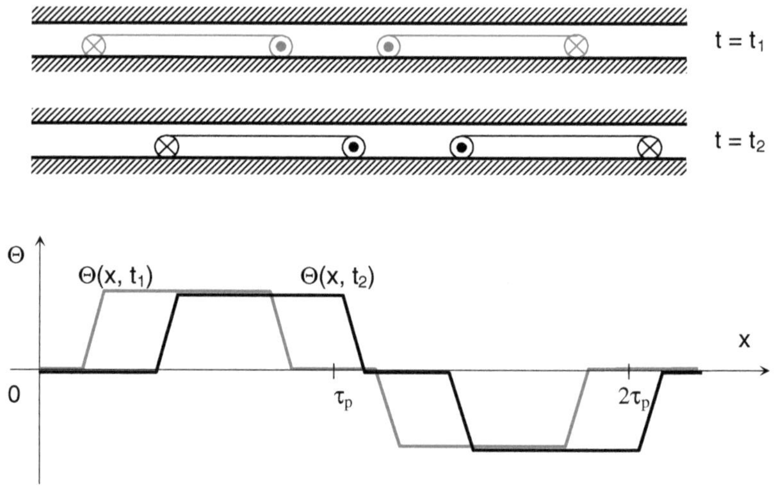

Fig. 2.56 Air gap MMF created by rotating coils fed from a DC source

The velocity of the nth harmonic of MMF is evaluated by setting the corresponding argument of sine function to a constant, the physical meaning of which is a (constant) distance between the observer and the zero crossing point of the MMF:

$$n \frac{\pi}{\tau_p}(x - v \cdot t) = \text{const} \tag{2.160}$$

the time derivative of which gives the relative velocity v_r between the observer and zero crossing point of the air gap MMF. By setting the relative velocity v_r equal to zero (the observer does not move relative to MMF), one obtains for the velocity v_n of the nth harmonic of the MMF

$$v_n = \frac{dx}{dt} = v \tag{2.161}$$

i.e., the nth harmonic of the MMF wave generated by rotating DC coil rotates at a circumferential velocity equal to the coil velocity. Since all higher harmonics travel at the same velocity as the fundamental wave of air gap MMF, they do not move relative to each other and the air gap MMF distribution retains its shape all the time, as shown in Fig. 2.56.

From the point of view of air gap MMF generation, permanent magnets have identical properties as DC-fed coils.

2.6.2 MMF Waves Generated by Symmetrically Wound Stationary Coils Carrying Symmetrical Alternating Currents on One Side of Air Gap

Consider now a single full-pitch coil per pole fed from an AC source

$$i(t) = \sum_{k=1}^{\infty} I_k \cos k(\omega t - \varphi_k) \tag{2.162}$$

If the coil axis is shifted for x_0, its MMF is equal to

$$\Theta(x, x_0, t) = \frac{4}{\pi} w \sum_{n=1,2,3,\ldots}^{\infty} \sum_{k=1}^{\infty} I_k \cos k(\omega t - \varphi_k) \frac{\sin \frac{n\pi}{2}}{n} f_{o,n} \sin n \frac{\pi}{\tau_p} (x - x_0) \tag{2.163}$$

The MMF distribution created by an alternating current does not qualitatively change its form; quantitatively, the MMF changes proportionaly to the coil current. The kth time harmonic of current creates a spatial MMF distribution component $\Theta_k(x, x_0, t)$

$$\Theta_k(x, x_0, t) = \frac{4}{\pi} w \sum_{n=1,3,5,\ldots}^{\infty} I_k \cos k(\omega t - \varphi_k) \cdot \frac{\sin \frac{n\pi}{2}}{n} f_{w,n} \sin n \frac{\pi}{\tau_p}(x - x_0) \quad (2.164)$$

The nth spatial harmonic of MMF distribution $\Theta_{k,n}(x, x_0, t)$

$$\Theta_{k,n}(x, x_0, t) = I_k \cdot \frac{4}{\pi} w \cdot \frac{\sin \frac{n\pi}{2}}{n} \cdot f_{w,n} \cdot \sin n \frac{\pi}{\tau_p}(x - x_0) \cdot \cos k(\omega t - \varphi_k) \quad (2.165)$$

is a standing wave

$$\Theta_{k,n}(x, x_0, t) = \Theta_{k,n,\max} \cdot \sin n \frac{\pi}{\tau_p}(x - x_0) \cdot \cos k(\omega t - \varphi_k) \quad (2.166)$$

which can be represented as a sum of two waves with equal amplitudes traveling in opposite directions

$$\Theta_{k,n}(x, x_0, t) = \frac{\Theta_{k,n,\max}}{2} \cdot \left\{ \sin\left[k(\omega t - \varphi_k) - n\frac{\pi}{\tau_p}(x - x_0)\right] + \sin\left[k(\omega t - \varphi_k) + n\frac{\pi}{\tau_p}(x - x_0)\right] \right\} \quad (2.167)$$

The component of wave in Eq. 2.167 traveling in the positive direction is called the *positive sequence*, and that traveling in negative direction, the *negative sequence*.

The traveling speed of the two waves is obtained by setting their arguments to a constant, and differentiating thus obtained equation with respect to time

$$\frac{\mathrm{d}}{\mathrm{d}t}\left\{ k(\omega t - \varphi_k) \pm n \frac{\pi}{\tau_p}(x - x_0) = \text{const} \right\} \Rightarrow n \frac{\pi}{\tau_p} \frac{\mathrm{d}x}{dt} = \pm k\omega \quad (2.168)$$

As a result, the circumferential velocity $v_{k,n}$ and *mechanical angular speed* $\Omega_{k,n}$ of the kth time and nth spatial harmonic are obtained

$$v_{k,n} = \pm \frac{k}{n} \cdot \frac{\tau_p}{\pi} \cdot \omega; \quad \Omega_{k,n} = \pm \frac{k}{n} \cdot \frac{\omega}{p} \quad (2.169)$$

The positive sign in the velocity equations is related to the positive sequence component of MMF, and the negative sign to the negative sequence component of MMF. The higher the order n of a spatial harmonic, the lower its speed of propagation through the air gap. This principle is a consequence of the condition that after one period T of current, i.e., at time instant $t + T$, the spatial distribution of

MMF has to have the same shape as at time instant *t*. This condition is fulfilled only if each spatial harmonic covers a distance of $2\tau_p/n$ within time *T*, *n* being the harmonic order.

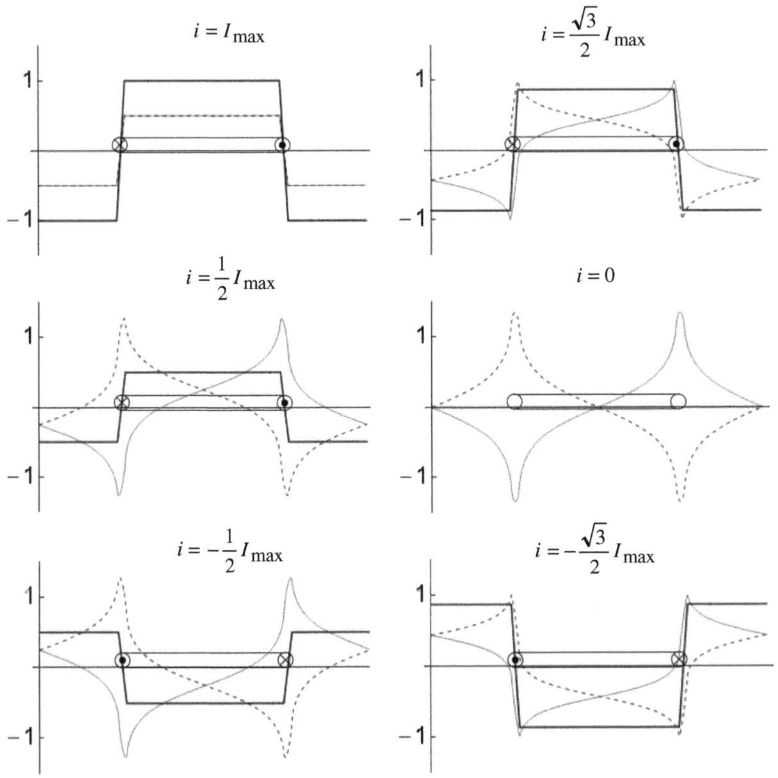

Fig. 2.57 Air gap MMF created by one stationary full-pitch coil per pole fed from an AC source. The total (pulsating) MMF is represented with a *solid black curve*, the total positive sequence component with a *solid gray*, and the total negative sequence component with *dashed gray curve*. Time step between figures is 30°

The *n*th spatial harmonic of MMF, created by the *k*th time harmonic of current, travels *k* times faster and *n* times slower than the fundamental. This means that spatial harmonics created by alternating current travel relative to each other, as opposed to spatial harmonics created by DC-carrying coil rotating at angular speed Ω, which all travel at the same, mechanical speed of rotation.

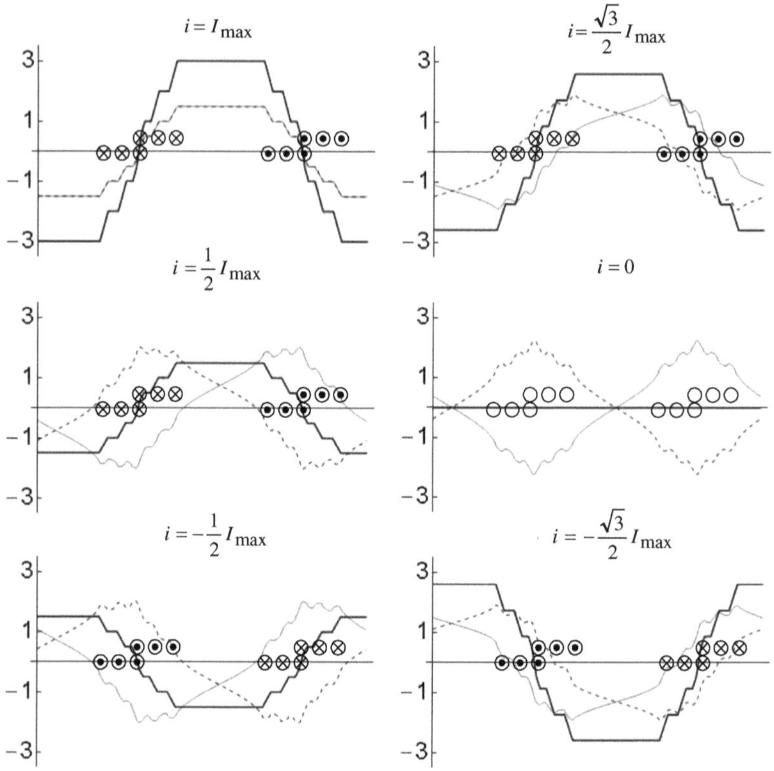

Fig. 2.58 Air gap MMF created by three stationary coils per pole with $y/\tau_p = 7/9$, carrying the same alternating current. The maximum value of resulting MMF is 3 p.u. The total (pulsating) MMF is represented with a *solid black curve*, the total positive sequence component with a *solid gray*, and the total negative sequence component with *dashed gray curve*. Time step between figures is 30°

Since the spatial harmonics created by alternating current move relative to each other, their sum—the total MMF—changes permanently its shape in time. This is illustrated in Fig. 2.57, in which the total MMF created by a full-pitch coil, along with its positive and negative sequence components within one half period of the coil current are shown.

In Fig. 2.57, the peaks of positive and negative sequence MMFs, which are fixed to the left-hand side and right-hand side of each conductor, can be recognized. At these points the MMF distribution changes abruptly its slope, i.e., its first derivative is discontinuous. Since Fourier series representation is defined only for continuously derivable functions, it cannot deliver proper results at points of discontinuity (Gibb's phenomenon).

Air gap MMF distribution created by three coils with shorted pitch 7/9 is shown in Fig. 2.58. The total MMF pulsates, as is the case with the MMF created by a single coil. In the higher harmonics spectrum of the total MMF some terms from the spectrum of a single coil are either missing, or suppressed. Therefore, both the positive and negative sequence components in Fig. 2.58 have waveforms much closer to sine function than in case of a single coil in Fig. 2.57.

Consider now a set of *2m identically wound groups of q coils* placed in adjacent slots in interval $(0,2\tau_p)$ along the circumferential coordinate x. In case of pole symmetry, oppositely wound coils are shifted for τ_p and carry the same currents as their counterparts. Therefore, only m sources (phases) with a phase shift of $2\pi/m$ are needed. The fundamental spatial harmonic ($n = 1$) of MMF created by the fundamental harmonic of current ($k = 1$) in the jth coil group ($j = 1, 2, \ldots, 2m$) is equal to

$$\begin{aligned}\Theta_{1,1,j}(x, x_{0,j}, t) \\ = \frac{\Theta_{1,1,\max}}{2} \cdot \left\{ \sin\left[(\omega t - \varphi_{1,j}) - \frac{\pi}{\tau_p}(x - x_{0,j}) \right] + \sin\left[(\omega t - \varphi_{1,j}) + \frac{\pi}{\tau_p}(x - x_{0,j}) \right] \right\}\end{aligned} \tag{2.170}$$

with $x_{0,j}$ standing for the coil group axis shift along the circumferential coordinate x, $\Theta_{1,1,\max}$ for the amplitude of the fundamental time and spatial component of the MMF created by *one coil group per pole pair*, and $\varphi_{1,j}$ for the phase shift of the fundamental current harmonic in the jth coil group. Denoting by $\gamma_{1,j,+}$ the argument of the positive, and by $\gamma_{1,j,-}$ of the negative sequence MMF

$$\gamma_{1,j,+} = (\omega t - \varphi_{1,j}) - \frac{\pi}{\tau_p}(x - x_{0,j}); \quad \gamma_{1,j,-} = (\omega t - \varphi_{1,j}) + \frac{\pi}{\tau_p}(x - x_{0,j}) \tag{2.171}$$

one can further write

$$\Theta_{1,1,j}(x, x_{0,j}, t) = \frac{\Theta_{1,1,\max}}{2} \cdot \sin\gamma_{1,j,+} + \frac{\Theta_{1,1,\max}}{2} \cdot \sin\gamma_{1,j,-} = \Theta_{1,1,j,+} + \Theta_{1,1,j,-} \tag{2.172}$$

The total positive sequence MMF is maximum if arguments $\gamma_{1,j,+}$ are equal for $j = 1, 2, \ldots, 2m$. If the condition of equal arguments is applied to arbitrary coil groups i and j, the equation is obtained

$$\gamma_{1,i,+} = \gamma_{1,j,+} \Rightarrow (\omega t - \varphi_{1,i}) - \frac{\pi}{\tau_p}(x - x_{0,i}) = (\omega t - \varphi_{1,j}) - \frac{\pi}{\tau_p}(x - x_{0,j}) \tag{2.173}$$

the solution of which is a phase angle of the current in *j*th coil group for which the positive sequence MMFs of the *i*th and *j*th coil group, $\Theta_{1,1,i,+}$ and $\Theta_{1,1,j,+}$, are in phase

$$\varphi_{1,j} = \varphi_{1,i} + \frac{\pi}{\tau_p}\left(x_{0,j} - x_{0,i}\right) \tag{2.174}$$

Positive sequence MMFs $\Theta_{1,1,i,+}$ and $\Theta_{1,1,j,+}$ created by two coil groups the axes of which are shifted along the circumferential coordinate x for $x_{0,j} - x_{0,i}$ are *in phase* if and only if *the coil group currents are out of phase*! Phase shift between coil group currents $\varphi_{1,j} - \varphi_{1,i}$ has to be equal to the electrical angle $(x_{0,j} - x_{0,i})\pi/\tau_p$ between the coil group axes. Arguments $\gamma_{1,i,+}$ and $\gamma_{1,j,+}$ of the positive sequence components $\Theta_{1,1,i,+}$ and $\Theta_{1,1,j,+}$ in the *i*th and *j*th coil group are in that case equal

$$\begin{aligned}\gamma_{1,j,+} &= \left[\omega t - \varphi_{1,i} - \frac{\pi}{\tau_p}\left(x_{0,j} - x_{0,i}\right)\right] - \frac{\pi}{\tau_p}\left(x - x_{0,j}\right) = \left(\omega t - \varphi_{1,i}\right) - \frac{\pi}{\tau_p}\left(x - x_{0,i}\right) \\ &= \gamma_{1,i,+}\end{aligned} \tag{2.175}$$

The angle between the negative sequence components $\Theta_{1,1,i,-}$ and $\Theta_{1,1,j,-}$ created by the coil group j and i can be written as

$$\begin{aligned}\gamma_{1,j,-} - \gamma_{1,i,-} &= \left(\omega t - \varphi_{1,j}\right) + \frac{\pi}{\tau_p}\left(x - x_{0,j}\right) - \left(\omega t - \varphi_{1,i}\right) - \frac{\pi}{\tau_p}\left(x - x_{0,i}\right) \\ &= 2\frac{\pi}{\tau_p}\left(x_{0,i} - x_{0,j}\right)\end{aligned} \tag{2.176}$$

Previous considerations are illustrated in Fig. 2.59, in which the MMF components created by two out of phase currents flowing through two coil groups at a given time instant t are shown. Phase shift between the currents is equal to the electrical spatial angle between the axes of two coil groups. Δ_i in Fig. 2.59 denotes the distance which the positive sequence MMF of the *i*th coil group has passed at a given time instant. At the same time, the negative sequence MMF of the *i*th coil group has passed the distance $-\Delta_i$. Similar considerations are valid for the *j*th coil group and the distance Δ_j.

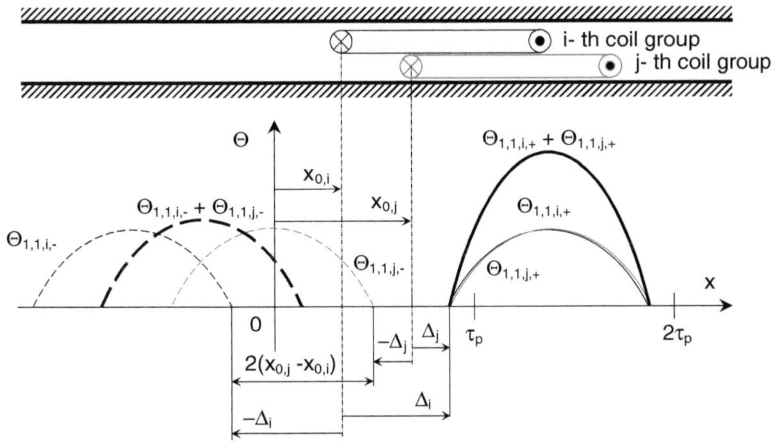

Fig. 2.59 Fundamental components of air gap MMF created by two stationary coil groups per pole fed by currents shifted for an angle equal to the electrical angle between the coil group axes. The two positive sequence components add algebraically, because the angle between them is permanently equal to zero. Phase shift between negative sequence components is equal to the twice the electrical angle between the coil axes. For the purpose of simplicity, only the positive half waves of the MMF components are shown

Whereas for a given phase shift between the coil group currents at each time instant the positive sequence components of the MMF created by the ith and jth coil group, $\Theta_{1,1,i,+}$ and $\Theta_{1,1,j,+}$, overlap along the circumferential coordinate, the negative sequence components $\Theta_{1,1,i,-}$ and $\Theta_{1,1,j,-}$ remain shifted for the amount of $2(x_{0,j} - x_{0,i})$, as shown in Fig. 2.59. If the phase shift between currents in two coil groups is equal to the electrical angle between the coil group axes, the resulting positive sequence of MMF is twice as big as its single component, and the resulting negative sequence of MMF is smaller than the algebraic sum of components $\Theta_{1,1,i,-}$ and $\Theta_{1,1,j,-}$

If the arbitrary distribution of $2m$ identical coil groups along the two poles $2\,\tau_p$ is now modified in terms of fixing the shift between adjacent coil groups to τ_p/m

$$x_{0,i+1} = x_{0,i} + \frac{\tau_p}{m} = x_{0,1} + i\frac{\tau_p}{m} \tag{2.177}$$

and the phase shift between currents in adjacent coils accordingly to π/m

$$\varphi_{1,i+1} = \varphi_{1,i} + \frac{\pi}{\tau_p}\left(x_{0,i+1} - x_{0,i}\right) = \varphi_{1,i} + \frac{\pi}{m} = \varphi_{1,1} + i\frac{\pi}{m} \tag{2.178}$$

the argument of the positive sequence MMF created by the ith coil group is

$$\gamma_{1,i,+} = \left(\omega t - \varphi_{1,i}\right) - \frac{\pi}{\tau_p}\left(x - x_{0,i}\right) = \left(\omega t - \varphi_{1,1}\right) - \frac{\pi}{\tau_p}\left(x - x_{0,1}\right) \qquad (2.179)$$

i.e., the same as for the first coil group. In other words, all positive sequence components are aligned along the same (radial) axis at each time instant.

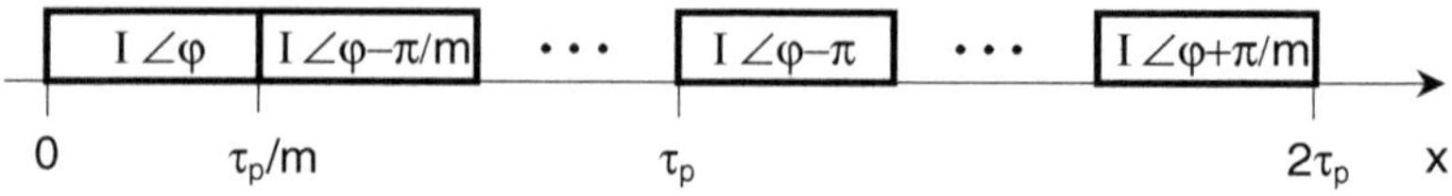

Fig. 2.60 Schematic representation of a symmetrical winding carrying symmetrical currents. As a result, the total negative sequence MMF created by fundamental components of MMF is equal to zero

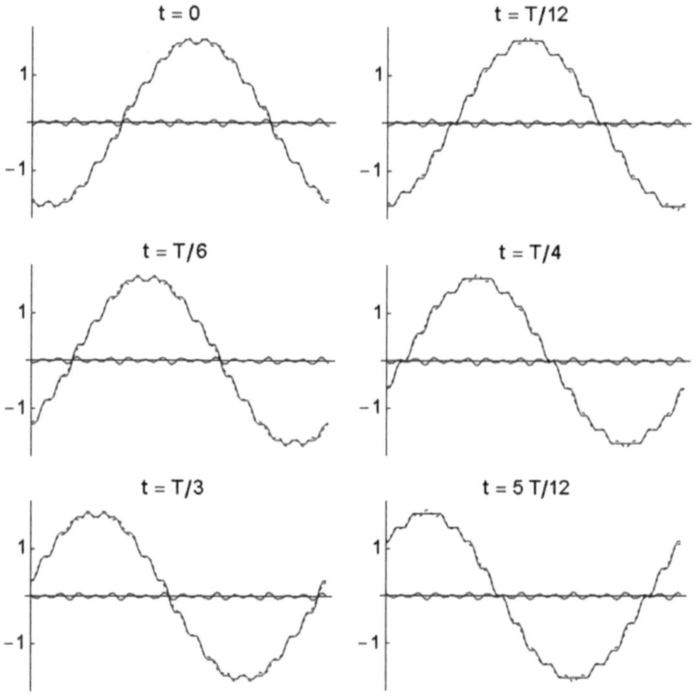

Fig. 2.61 Air gap MMF created by three groups of three stationary coils per pole with $y/\tau_p = 7/9$, carrying AC 60° out of phase in each coil group. The scale on the y-axis is in p.u.; 0.5 p.u. corresponds to the amplitude of positive sequence MMF created by a single coil group. The total MMF is represented with a *solid black curve*, the dominating resulting positive sequence component with a *dashed gray*, and the minor negative sequence component with *solid gray curve*

The total positive sequence component is equal to

$$\begin{aligned}\Theta_{1,1,+}(x,t) &= \sum_{j=1}^{2m} \frac{\Theta_{1,1,\max}}{2} \cdot \sin\left[\omega t - \varphi_{1,1} - \frac{\pi}{\tau_p}\left(x - x_{0,1}\right)\right] \\ &= m \cdot \Theta_{1,1,\max} \cdot \sin\left[\omega t - \varphi_{1,1} - \frac{\pi}{\tau_p}\left(x - x_{0,1}\right)\right]\end{aligned} \tag{2.180}$$

with $\Theta_{1,1,\max}$ denoting the amplitude of the fundamental time and spatial component of the MMF created by one out of *2m* coil groups per pole pair.

Similarly, one can express the argument of the negative sequence MMF created by the *i*th coil group as

$$\gamma_{1,i,-} = \left(\omega t - \varphi_{1,i}\right) + \frac{\pi}{\tau_p}\left(x - x_{0,i}\right) = \left(\omega t - \varphi_{1,1}\right) + \frac{\pi}{\tau_p}\left(x - x_{0,1}\right) - (i-1)\frac{2\pi}{m} \tag{2.181}$$

Consequently, if the MMF amplitudes of all *2m* coils are equal, the sum of negative sequence components of MMF over two poles is equal to zero

$$\Theta_{1,1,-}(x,t) = \sum_{j=1}^{2m} \frac{\Theta_{1,1,\max}}{2} \cdot \sin\left[\omega t - \varphi_{1,1} + \frac{\pi}{\tau_p}\left(x - x_{0,1}\right) - (j-1)\frac{2\pi}{m}\right] = 0 \tag{2.182}$$

Negative sequence components of MMFs created by *2m* coils are out of phase, with an angle of π/m between components created in adjacent coil groups. Symmetrical winding with *2m* coil groups per pole pair (Fig. 2.60) fed from a symmetrical *m*-phase source creates *rotating field* characterized by constant amplitude and constant speed of rotation. Each coil group (*zone*) of the symmetrical winding in Fig. 2.60 occupies $1/m$ of the pole pitch τ_p.

The description "*m-phase machine*" is always related to the number of supply phases; windings of conventional machines have *m* coil groups per pole. Coil groups shifted for τ_p to each other can be connected in series or parallel (the latter usually only in large turbogenerators) to the same supply phase. Electromagnetically, however, they count as separate machine phases.

Rotating field contains only one (positive or negative sequence) component of the MMF. Since the amplitude of each component of coil MMF is equal to 50 % of the total MMF, the resulting amplitude of rotating field created by *2m* coil groups per pole pair is equal to *m* times the MMF amplitude of a single coil group.

In Fig. 2.61 the air gap MMF created by fundamental components of currents flowing through 3 groups of three coils per pole at several time instants is shown. The phase shift of currents flowing through adjacent groups of coils under each pole is 60°. Fundamental components of air gap MMF distribution created by the coils create rotating field characterized by no negative sequence term. Higher harmonics of air gap MMF, however, do generate their own traveling fields with the fundamental component of the coils current. In Fig. 2.61 one recognizes different speeds of traveling waves of higher harmonics, resulting in permanently changing shape of the resulting MMF.

2.6.3 The Influence of the Number of Phases

As discussed in the previous section, a symmetrical winding consisting of *2m* coil groups per pole pair (Fig. 2.60) fed from an m-phase symmetrical source of currents with fundamental angular frequency ω generates not only the fundamental rotating field, but also a spectrum of higher harmonic fields described by equation

$$\Theta(x,t) = \sum_{j=1}^{2m}\sum_{k=1}^{\infty}\sum_{n=1}^{\infty}\Theta_{k,n,\max}\cdot\sin n\frac{\pi}{\tau_p}\left(x - x_0 - \frac{j-1}{m}\tau_p\right)\cdot\cos k\left(\omega t - \varphi_0 - \frac{j-1}{m}\pi\right) \tag{2.183}$$

the positive sequence component of which can be written as

$$\Theta_+(x,t) = \sum_{j=1}^{2m}\sum_{k=1}^{\infty}\sum_{n=1}^{\infty}\frac{\Theta_{k,n,\max}}{2}\cdot\sin\left[k(\omega t - \varphi_0) - n\frac{\pi}{\tau_p}(x - x_0) - (j-1)\frac{k-n}{m}\pi\right] \tag{2.184}$$

and the negative sequence component as

$$\Theta_-(x,t) = \sum_{j=1}^{2m}\sum_{k=1}^{\infty}\sum_{n=1}^{\infty}\frac{\Theta_{k,n,\max}}{2}\cdot\sin\left[k(\omega t - \varphi_0) + n\frac{\pi}{\tau_p}(x - x_0) - (j-1)\frac{k+n}{m}\pi\right] \tag{2.185}$$

Positive sequence components of MMF created by adjacent phases are shifted to each other for an angle of

$$\gamma_+ = \frac{k-n}{m}\pi \tag{2.186}$$

and negative sequence components for

$$\gamma_- = \frac{k+n}{m}\pi \tag{2.187}$$

If the angle γ_+ or γ_- is an integer multiple of 2π, the particular components (positive or negative) of MMF in all m phases are in phase, and the resulting sequence is $m/2$ times larger than in a single phase consisting of two coil groups over two poles. In other words, the kth time and nth spatial harmonic in an m-phase machine will generate a positive sequence MMF if

$$\frac{k-n}{m} = \text{even} \tag{2.188}$$

and a negative sequence harmonic for

$$\frac{k+n}{m} = \text{even} \tag{2.189}$$

One should note that conditions expressed in Eqs. 2.188 and 2.189 are more precise than those found in majority of textbooks on electric machines, namely that $(k \pm n)/m$ = integer. One recalls that this condition is valid for an assumption of m coil groups across two poles, which further implies the phase shift of $2\pi/m$ between currents. This result seduces, because it works perfectly for a 3-phase machine; however, in a two-phase machine, $m = 2$, it defines a phase angle of $2\pi/m = \pi$ between two currents, which is far from reality. Furthermore, fundamental space and time harmonics in a two-phase machine generate both positive and negative sequence components of MMF according to equation $(k \pm n)/m = (1 \pm 1)/2 = 0$ or 1, i.e., a pulsating MMF! In reality, however, there exists only a positive sequence MMF in a 2-phase machine. Equations 2.188 and 2.189 deliver an even number (here 0) only for a difference of harmonic orders, $(1-1)/2$, i.e., only for the positive sequence MMF!

If the angle γ_+ in Eq. 2.186 or γ_- in Eq. 2.187 is not an even multiple of π, the particular components (positive or negative) of MMF in all m phases build a symmetrical MMF star, which adds up to zero.

In Table 2.9 angular velocities of MMF waves created by the fundamental time and nth spatial harmonic as multiples of ω/p are given for various numbers of phases m. The width of the zone which belongs to a coil group under one pole is equal to 180° el./m, where m denotes the number of supply phases. This principle is illustrated in Fig. 2.62, in which axes of phase windings in symmetrically wound machines with $m = 2$–6 are shown.

One should note that in a symmetrically connected polyphase machine either pulsating, or rotating field is generated, but never elliptical. Another property of symmetrically connected coil groups in a polyphase machine is that they do not let pass air gap MMF harmonics the order n of which is an integer multiple of the number of phases. For example, in a symmetrically wound three-phase machine, the 3., 9., 15., etc., harmonic can exist in MMF of each phase, but not in the resulting air gap MMF.

When analyzing results in Table 2.9, one comes to a conclusion that in a symmetrical m-phase winding fed from a symmetrical system of sinusoidal currents

($k = 1$) the order n_ℓ of the lowest higher spatial harmonic which can create rotating air gap MMF is equal to $2m - 1$. By applying Eqs. 2.188 and 2.189, one can find the order n_ℓ of the lowest higher spatial harmonic in the air gap MMF wave created by the kth current harmonic in an m-phase machine: By inserting 2 for the lowest even number in Eqs. 2.188 and 2.189, one becomes $n_\ell = 2m - k$. The spatial harmonic of the order n_ℓ creates a negative sequence MMF.

Consider now a symmetrical winding with $q = 1$ and full-pitch coils. The winding has as many slots per pole as the number of phases, $N/(2p) = m$, and the lowest higher harmonic of rotating MMF created by the fundamental of current has the order $n_\ell = N/p - 1$, which is the order of slot harmonic, see Table 2.9. Despite the fact that the absolute value of the pitch factor for each spatial harmonic of full-pitch coils is equal to one, higher harmonics (with exception of slot harmonics) do not exist in the resulting rotating field! Obviously the winding with one slot per pole and phase, as is the case with squirrel cage, creates rotating air gap MMF the spectrum of which contains the lowest content of higher harmonics among all conventional winding types.

Table 2.9 Angular velocities of MMF waves (as multiples of ω/p) created by the fundamental harmonic of current ($k = 1$) and spatial harmonics $1 \le n \le 29$ as a function of the number of supply phases m

	Number of supply phases *m*/Zone width (° el.)								
n	$m = 1$ 180°	m = 2 90°	$m = 3$ 60°	$m = 4$ 45°	$m = 5$ 36°	$m = 6$ 30°	$m = 7$ 25.7°	$m = 8$ 22.5°	$m = 9$ 20°
1	±1	1	1	1	1	1	1	1	1
3	±1/3	−1/3	–	–	–	–	–	–	–
5	±1/5	1/5	−1/5	–	–	–	–	–	–
7	±1/7	−1/7	1/7	−1/7	–	–	–	–	–
9	±1/9	1/9	–	1/9	−1/9	–	–	–	–
11	±1/11	−1/11	−1/11	–	1/11	−1/11	–	–	–
13	±1/13	1/13	1/13	–	–	1/13	−1/13	–	–
15	±1/15	−1/15	–	−1/15	–	–	1/15	−1/15	–
17	±1/17	1/17	−1/17	1/17	–	–	–	1/17	−1/17
19	±1/19	−1/19	1/19	–	−1/19	–	–	–	1/19
21	±1/21	1/21	–	–	1/21	–	–	–	–
23	±1/23	−1/23	−1/23	−1/23	–	−1/23	–	–	–
25	±1/25	1/25	1/25	1/24	–	1/25	–	–	–
27	±1/27	−1/27	–	–	–	–	−1/27	–	–
29	±1/29	1/29	−1/29	–	−1/29	–	1/29	–	–

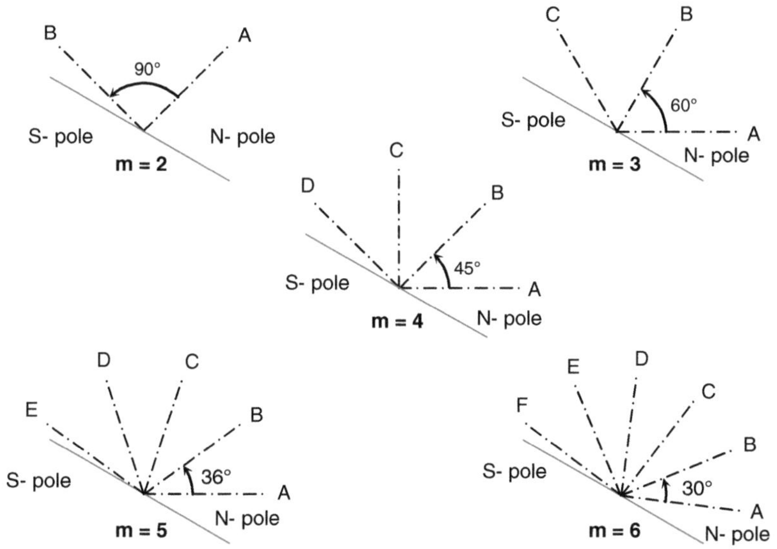

Fig. 2.62 Axes of windings in a symmetrically wound machine under one pole for different numbers of supply phases

2.6.4 MMF Waves Generated by Asymmetrically Wound Stationary Coils Carrying Asymmetrical Alternating Currents on One Side of Air Gap

If an m-phase winding is asymmetrical, or carries phase and/or amplitude asymmetrical currents, both positive and negative sequence components of air gap MMF can be generated for a given combination of spatial and time harmonics. Denoting by $\Theta_{\max,i}$ the amplitude of the MMF created by ith phase, by φ_i the phase shift of the current in the ith phase and by γ_i the electrical angle of axis of the ith phase winding, one can express the MMF created by the kth time and nth spatial harmonic in the ith phase as

$$\Theta_{i,k,n} = \Theta_{\max,i} \sin\left[n\left(\frac{\pi}{\tau_p}x - \gamma_i\right)\right] \cos[k(\omega t - \varphi_i)] \tag{2.190}$$

The amplitude $\Theta_{+,k,n}$ of the positive sequence component of the resulting MMF created by the kth time and nth spatial harmonic is equal to the geometric sum of positive sequence components created by all m phases

$$\Theta_{+,k,n} = \frac{1}{2}\sqrt{\left[\sum_{i=1}^{m} \Theta_{i,k,n} \cos(\varphi_i - \gamma_i)\right]^2 + \left[\sum_{i=1}^{m} \Theta_{i,k,n} \sin(\varphi_i - \gamma_i)\right]^2} \tag{2.191}$$

whereas the amplitude of the negative sequence component can be expressed as

$$\Theta_{-,k,n} = \frac{1}{2}\sqrt{\left[\sum_{i=1}^{m} \Theta_{i,k,n} \cos(\varphi_i + \gamma_i)\right]^2 + \left[\sum_{i=1}^{m} \Theta_{i,k,n} \sin(\varphi_i + \gamma_i)\right]^2} \tag{2.192}$$

If the amplitude of negative sequence component of MMF is different from zero and smaller than the amplitude of the positive sequence component, the air gap field is elliptical. An overview of air gap MMF forms is shown in Table 2.10, and their basic shapes are shown in Fig. 2.63. Here it is assumed that the air gap MMF is created by the fundamental time and fundamental spatial harmonics, i.e., $k = 1$ and $n = 1$.

Fundamental component of elliptical field can be represented as a sum of the fundamental positive sequence component with amplitude Θ_+, and the fundamental negative sequence component with amplitude Θ_-

Table 2.10 Forms of air gap field created by m identical coils carrying currents with equal rms values. $\Theta_{1,1}$ denotes the amplitude of the MMF created by coils in one zone

Type of MMF	Amplitude of the positive sequence component Θ_+	Amplitude of the negative sequence component Θ_-
Pulsating	$m \cdot \Theta_{1,1}$	$m \cdot \Theta_{1,1}$
Elliptical	$< m \cdot \Theta_{1,1}$	$< m \cdot \Theta_{1,1}$
Rotating	$m \cdot \Theta_{1,1}$	0

$$\Theta(x,t) = \Theta_+ \cdot \sin\left(\omega t - \frac{\pi}{\tau_p} x\right) + \Theta_- \cdot \sin\left(\omega t + \frac{\pi}{\tau_p} x\right) = \Theta_e \cdot \sin(\omega t + \gamma) \tag{2.193}$$

where

$$\Theta_e = \sqrt{\Theta_+^2 + \Theta_-^2 + 2 \cdot \Theta_+ \cdot \Theta_- \cdot \cos\frac{2\pi}{\tau_p} x} \tag{2.194}$$

and

$$\gamma = \arctan\frac{\Theta_- - \Theta_+}{\Theta_- + \Theta_+} \tan\frac{\pi}{\tau_p} x \tag{2.195}$$

The trajectory of Θ_e is an ellipse in the polar coordinate system (Θ_e, γ). The angular speed Ω of Θ_e can be written as

$$\Omega = \frac{\mathrm{d}\gamma}{\mathrm{d}t} = \frac{\frac{\Theta_- - \Theta_+}{\Theta_- + \Theta_+} \cdot \frac{\pi}{\tau_p}}{\cos^2 \frac{\pi}{\tau_p} x + \left(\frac{\Theta_- - \Theta_+}{\Theta_- + \Theta_+}\right)^2 \sin^2 \frac{\pi}{\tau_p} x} \cdot \frac{\mathrm{d}x}{\mathrm{d}t} \tag{2.196}$$

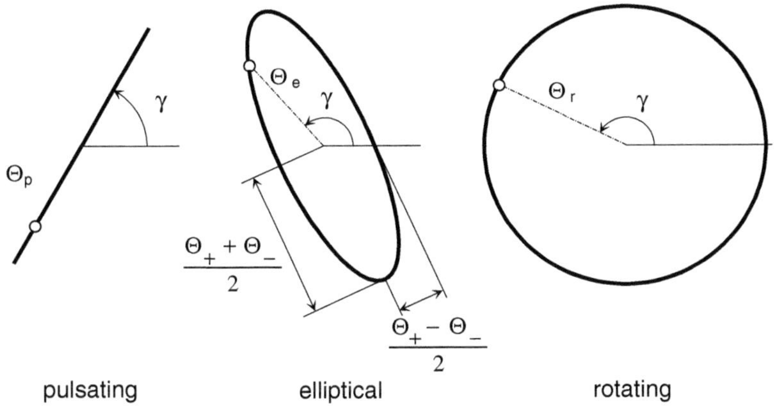

Fig. 2.63 Pulsating, elliptical, and rotating air gap MMF

Fig. 2.64 The p.u. amplitude of resulting MMF as a function of circumferential coordinate x with ratio Θ_-/Θ_+ as a parameter

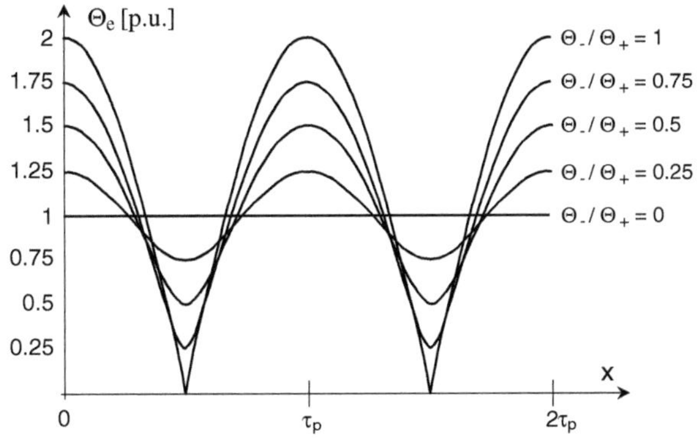

Since

$$\frac{\mathrm{d}x}{\mathrm{d}t} = v_{1,1} = \pm \frac{\tau_p}{\pi} \cdot \omega \tag{2.197}$$

one can express the angular speed Ω of the resulting MMF as

$$\Omega = \frac{d\gamma}{dt} = \pm \frac{\frac{\Theta_- - \Theta_+}{\Theta_- + \Theta_+}}{\cos^2 \frac{\pi}{\tau_p} x + \left(\frac{\Theta_- - \Theta_+}{\Theta_- + \Theta_+} \right)^2 \sin^2 \frac{\pi}{\tau_p} x} \cdot \omega \tag{2.198}$$

The variation of amplitude of elliptical MMF as a function of the grade of asymmetry and the circumferential coordinate is shown in Fig. 2.64. The largest variation of amplitude of elliptic MMF occurs for $\Theta_+ = \Theta_-$ (pulsating field), where it changes between 0 and 2. In rotating field the deviation of amplitude of resulting MMF from its average value of 1 is equal to zero.

The angular speed Ω of the resulting MMF as a function of circumferential coordinate x and the grade of asymmetry Θ_+/Θ_- as a parameter is shown in Fig. 2.65. Only in case of rotating field ($\Theta_- = 0$), the resulting MMF rotates at a constant speed; elliptic field *permanently accelerates and decelerates*. The larger the negative sequence component, the bigger the difference between maximum and minimum speed of rotation. Elliptic field rotates at highest speed at those points of circumference where the positive and negative sequence components are shifted for 180°, thus acting against each other (ellipse minor axis), and at slowest speed when they act in the same direction (ellipse major axis).

Virtual work performed by magnetic energy in the air gap can be expressed in terms of differential area swept by the magnitude Θ_e in Fig. 2.63:

$$dA = \frac{1}{2} \cdot \Theta_e \cdot (\Theta_e d\gamma) \tag{2.199}$$

and the corresponding areal velocity, which analogously has a dimension of power

$$\frac{dA}{dt} = \frac{1}{2} \cdot \Theta_e^2 \cdot \frac{d\gamma}{dt} \tag{2.200}$$

After substitution of expressions for Θ_e and $d\gamma/dt$ one can write

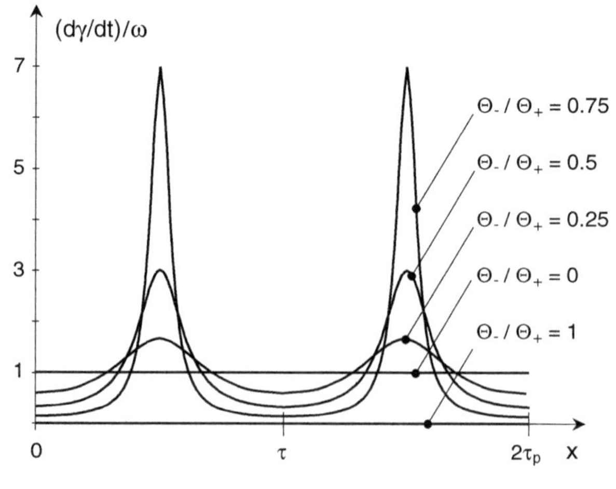

Fig. 2.65 p.u. angular speed of resulting MMF as a function of circumferential coordinate x with ratio Θ_-/Θ_+ as parameter

Fig. 2.66 Areal velocity of resulting MMF (power!) as a function of ratio Θ_-/Θ_+

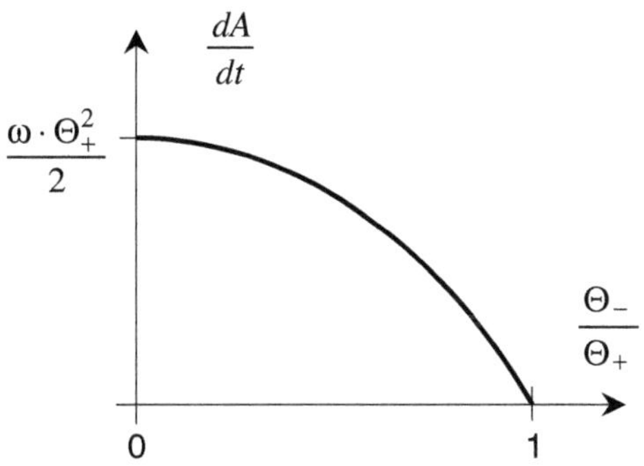

$$\frac{\mathrm{d}A}{\mathrm{d}t} = \pm \frac{\omega \cdot \Theta_+^2}{2} \cdot \frac{1 + r^2 + 2r \cos 2\frac{\pi}{\tau_p} x}{\cos^2 \frac{\pi}{\tau_p} x + \left(\frac{r-1}{r+1}\right)^2 \sin^2 \frac{\pi}{\tau_p} x} \cdot \frac{r-1}{r+1} \tag{2.201}$$

with r standing for the ratio between amplitude of negative and positive sequence component of MMF

$$r = \frac{\Theta_-}{\Theta_+} \tag{2.202}$$

Applying trigonometric identities, Eq. (2.201) for areal velocity can be simplified to

$$\frac{\mathrm{d}A}{\mathrm{d}t} = \pm \frac{\omega \cdot \Theta_+^2}{2} \cdot \left(r^2 - 1\right) = \pm \frac{\omega}{2} \cdot \left(\Theta_+^2 - \Theta_-^2\right) \tag{2.203}$$

which is constant. *The magnitude of air gap MMF sweeps equal areas during equal intervals of time*, which is analogous to **the second Kepler's law** of planetary motion!

Physical interpretation of the identical mathematical property is of course different in a rotating field machine than in celestial mechanics. In an electric machine the constant area swept by the air gap MMF means nothing but that the torque, as a partial derivative of magnetic energy accumulated in the air gap with respect to angle, is time invariant.

The dependence of areal velocity on the ratio Θ_-/Θ_+ is shown in Fig. 2.66.

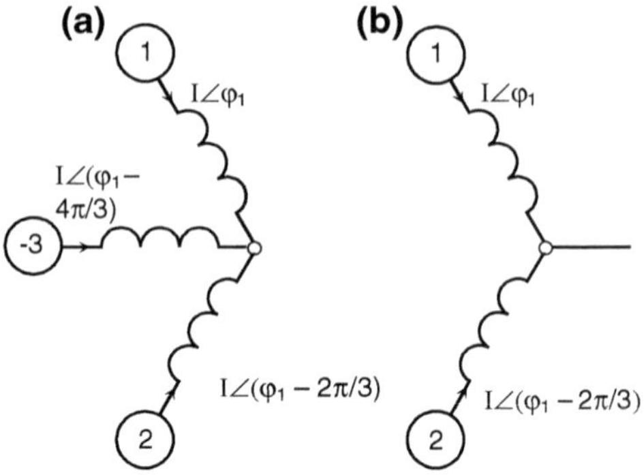

Fig. 2.67 Three-phase winding connection with one phase reversed (**a**), and with one phase disconnected (**b**)

The power generated by pulsating field, i.e. for equal amplitudes of positive and negative sequence MMF, is equal to zero.

Elliptical MMF is created in a three-phase machine when terminals of one supply phase are reversed, or if a phase is disconnected. Denoting by $\Theta_{1,1,\max}$ the amplitude of the fundamental component of MMF created by one coil per pole pair, the positive sequence MMF in a machine with reversed phase terminals as in Fig. 2.67a is equal to

$$\begin{aligned}\Theta_{1,1,,+}(x,t) &= \frac{\Theta_{1,1,\max}}{2}\cdot\sin\left[\omega t-\varphi_{1,1}-\frac{\pi}{\tau_p}(x-x_{0,1})\right]+\frac{\Theta_{1,1,\max}}{2}\cdot\sin\left[\omega t-\varphi_{1,1}-\frac{\pi}{\tau_p}(x-x_{0,1})-\pi\right]+\\ &+\sum_{j=3}^{4}\frac{\Theta_{1,1,\max}}{2}\cdot\sin\left[\omega t-\varphi_{1,1}-\frac{\pi}{\tau_p}(x-x_{0,1})\right]+\frac{\Theta_{1,1,\max}}{2}\cdot\sin\left[\omega t-\varphi_{1,1}-\frac{\pi}{\tau_p}(x-x_{0,1})-\pi\right]\\ &+\frac{\Theta_{1,1,\max}}{2}\cdot\sin\left[\omega t-\varphi_{1,1}-\frac{\pi}{\tau_p}(x-x_{0,1})\right]=\Theta_{1,1,\max}\cdot\sin\left[\omega t-\varphi_{1,1}-\frac{\pi}{\tau_p}(x-x_{0,1})\right]\end{aligned} \tag{2.204}$$

and the negative sequence

$$\begin{aligned}\Theta_{1,1,,-}(x,t) &= \frac{\Theta_{1,1,\max}}{2}\cdot\sin\left[\omega t-\varphi_{1,1}+\frac{\pi}{\tau_p}(x-x_{0,1})\right]+\frac{\Theta_{1,1,\max}}{2}\cdot\sin\left[\omega t-\varphi_{1,1}+\frac{\pi}{\tau_p}(x-x_{0,1})-\frac{5\pi}{3}\right]\\ &+\sum_{j=3}^{4}\frac{\Theta_{1,1,\max}}{2}\cdot\sin\left[\omega t-\varphi_{1,1}+\frac{\pi}{\tau_p}(x-x_{0,1})-(j-1)\frac{2\pi}{3}\right]+\frac{\Theta_{1,1,\max}}{2}\cdot\sin\left[\omega t-\varphi_{1,1}+\frac{\pi}{\tau_p}(x-x_{0,1})-\frac{5\pi}{3}\right]\\ &+\frac{\Theta_{1,1,\max}}{2}\cdot\sin\left[\omega t-\varphi_{1,1}+\frac{\pi}{\tau_p}(x-x_{0,1})-\frac{10\pi}{3}\right]=2\cdot\Theta_{1,1,\max}\cdot\sin\left[\omega t-\varphi_{1,1}+\frac{\pi}{\tau_p}(x-x_{0,1})-\frac{5\pi}{3}\right]\end{aligned} \tag{2.205}$$

The amplitude of the negative sequence component in a three-phase machine with terminals of one phase reversed is twice as large as the amplitude of the positive sequence component of the air gap MMF! Consequently, the resulting elliptic MMF rotates in negative direction.

The positive sequence MMF in a machine with one phase disconnected as shown in Fig. 2.67b is equal to

$$\begin{aligned}\Theta_{1,1,,+}(x,t) &= \sum_{j=1}^{4}\frac{\Theta_{1,1,\max}}{2}\cdot \sin\left[\omega t-\varphi_{1,1}-\frac{\pi}{\tau_p}\left(x-x_{0,1}\right)\right]\\ &= 2\cdot\Theta_{1,1,\max}\cdot\sin\left[\omega t-\varphi_{1,1}-\frac{\pi}{\tau_p}\left(x-x_{0,1}\right)\right]\end{aligned} \tag{2.206}$$

and the negative sequence

$$\Theta_{1,1,,-}(x,t) = \Theta_{1,1,\max}\cdot\sin\left[\omega t-\varphi_{1,1}+\frac{\pi}{\tau_p}\left(x-x_{0,1}\right)-\frac{4\pi}{3}\right] \tag{2.207}$$

The amplitude of the positive sequence component in a three phase machine with one phase disconnected is twice as large as the amplitude of the negative sequence component of the air gap MMF. The resulting elliptic MMF rotates in positive direction.

2.6.5 MMF Waves Generated by Rotating Coil(s) Carrying Constant Frequency Current(s)

Consider now a rotor coil fed from current source $i_r(t)$

$$i_r(t) = \sum_{k=1}^{\infty} I_{k,r}\cos k_r\left(\omega_r t-\varphi_{k,r}\right) \tag{2.208}$$

shifted for an amount of $x_{0,r}$ along the circumferential coordinate x_r and rotating at angular speed Ω. The k_rth time and n_rth spatial harmonic of air gap MMF wave created by the rotating coil is equal to

$$\begin{aligned}\Theta_{k_r,n_r}\left(x_r,x_{0,r},t\right) &= \frac{\Theta_{k_r,n_r,\max}}{2}\cdot\sin\left[k_r\left(\omega_r t-\varphi_{k,r}\right)-n_r\frac{\pi}{\tau_p}\left(x_r-x_{0,r}\right)\right]\\ &+\frac{\Theta_{k_r,n_r,\max}}{2}\cdot\sin\left[k_r\left(\omega_r t-\varphi_{k,r}\right)+n_r\frac{\pi}{\tau_p}\left(x_r-x_{0,r}\right)\right]\end{aligned} \tag{2.209}$$

Rotor circumferential coordinate x_r can be expressed in terms of stator circumferential coordinate x as

$$x_{\mathrm{r}} = x + x_0 + \Omega \cdot t \cdot R = x + x_0 + \Omega \cdot t \cdot \frac{p\tau_{\mathrm{p}}}{\pi} \tag{2.210}$$

with x_0 denoting the initial rotor shift and R the air gap mean radius. The k_{r}th time and n_{r}th spatial harmonic of the air gap MMF wave can now be expressed in the stationary system as

$$\begin{aligned}\Theta_{k_{\mathrm{r}},n_{\mathrm{r}}}(x, x_0, t) &= \frac{\Theta_{k_{\mathrm{r}},n_{\mathrm{r}},\max}}{2} \cdot \sin\left[k_{\mathrm{r}}\left(\omega_{\mathrm{r}}t - \varphi_{k,\mathrm{r}}\right) - n_{\mathrm{r}}\frac{\pi}{\tau_{\mathrm{p}}}\left(x + x_0 + \Omega \cdot t \cdot \frac{p\tau_{\mathrm{p}}}{\pi} - x_{0,\mathrm{r}}\right)\right] \\ &+ \frac{\Theta_{k_{\mathrm{r}},n_{\mathrm{r}},\max}}{2} \cdot \sin\left[k_{\mathrm{r}}\left(\omega_{\mathrm{r}}t - \varphi_{k,\mathrm{r}}\right) + n_{\mathrm{r}}\frac{\pi}{\tau_{\mathrm{p}}}\left(x + x_0 + \Omega \cdot t \cdot \frac{p\tau_{\mathrm{p}}}{\pi} - x_{0,\mathrm{r}}\right)\right]\end{aligned} \tag{2.211}$$

and

$$\begin{aligned}\Theta_{k_{\mathrm{r}},n_{\mathrm{r}}}(x, x_0, t) &= \frac{\Theta_{k_{\mathrm{r}},n_{\mathrm{r}},\max}}{2} \cdot \sin\left[(k_{\mathrm{r}}\omega_{\mathrm{r}} - n_{\mathrm{r}}p\Omega) \cdot t - n_{\mathrm{r}}\frac{\pi}{\tau_{\mathrm{p}}}x - k_{\mathrm{r}}\varphi_{k,\mathrm{r}} - n_{\mathrm{r}}\frac{\pi}{\tau_{\mathrm{p}}}(x_0 - x_{0,\mathrm{r}})\right] \\ &+ \frac{\Theta_{k_{\mathrm{r}},n_{\mathrm{r}},\max}}{2} \cdot \sin\left[(k_{\mathrm{r}}\omega_{\mathrm{r}} - n_{\mathrm{r}}p\Omega) \cdot t + n_{\mathrm{r}}\frac{\pi}{\tau_{\mathrm{p}}}x - k_{\mathrm{r}}\varphi_{k,\mathrm{r}} + n_{\mathrm{r}}\frac{\pi}{\tau_{\mathrm{p}}}(x_0 - x_{0,\mathrm{r}})\right]\end{aligned} \tag{2.212}$$

The angular speed of rotation ω of the k_{r}th time and n_{r}th spatial harmonic of the air gap MMF wave relative to the stator can be found by setting the argument of sine function equal to a constant and then differentiating it w.r.t. time

$$\begin{aligned}&\frac{\mathrm{d}}{\mathrm{d}t}\left\{(k_{\mathrm{r}}\omega_{\mathrm{r}} - n_{\mathrm{r}}p\Omega) \cdot t \mp n_{\mathrm{r}}\frac{\pi}{\tau_{\mathrm{p}}}x - k_{\mathrm{r}}\varphi_{k,\mathrm{r}} \mp n_{\mathrm{r}}\frac{\pi}{\tau_{\mathrm{p}}}(x_0 - x_{0,\mathrm{r}}) = \mathrm{const}\right\} \\ &\Rightarrow n_{\mathrm{r}}\frac{\pi}{\tau_{\mathrm{p}}}\frac{\mathrm{d}x}{\mathrm{d}t} = \pm(k_{\mathrm{r}}\omega_{\mathrm{r}} - n_{\mathrm{r}}p\Omega)\end{aligned} \tag{2.213}$$

from which the angular speed ω can be expressed as

$$\omega = \Omega \pm \frac{k_{\mathrm{r}}}{n_{\mathrm{r}}} \cdot \frac{\omega_{\mathrm{r}}}{p} \tag{2.214}$$

The air gap MMF component created by k_{r}th time harmonic of rotor current and n_{r}th spatial harmonic of rotor air gap MMF travels relative to stator at rotor mechanical speed augmented or diminished by the component speed relative to the rotor. Positive sequence component of the rotor MMF created by fundamental time and spatial harmonic is at standstill relative to the stator if the rotor mechanical speed is equal to the negative synchronous speed.

Stationary coil(s) generate the fundamental component of air gap MMF wave rotating at the same angular speed ω as in Eq. 2.214 if fed by current(s) with angular frequency ω_s

$$\omega_s = p \cdot \Omega \pm \frac{k_r}{n_r} \cdot \omega \tag{2.215}$$

Accordingly, rotor DC creates air gap MMF fundamental component ($n_r = 1$) which rotates at angular speed Ω and which is observed on the stator side as if it were generated by an AC with angular frequency $\omega_s = p\,\Omega$.

2.6.6 *MMF Waves Generated by Rotating Coil(s) Carrying Variable Frequency Currents on One Side of Air Gap*

Assume that a coil (group) rotates at mechanical angular speed Ω and that it is fed from an AC source

$$i_r(t) = \sum_{k_r=1}^{\infty} I_{k_r} \cos k_r \left(s\omega t - \varphi_{k_r}\right) \tag{2.216}$$

with s denoting the slip

$$s = \frac{\omega - p \cdot \Omega}{\omega} \tag{2.217}$$

Each coil (group) per pole creates an MMF

$$\Theta\left(x_r, x_{0,r}, t\right) = \frac{4}{\pi} w \sum_{n_r=1,3,5,\ldots}^{\infty} \sum_{k_r=1}^{\infty} I_{k_r} \cos k_r \left(s\omega t - \varphi_{k_r}\right) \cdot f_{w,n} \sin n_r \frac{\pi}{\tau_p} \left(x_r - x_{0,r}\right) \tag{2.218}$$

with x_0 denoting the shift of the coil (group) axis to the circumferential coordinate $x = 0$ and $f_{w,n}$ the winding factor for the nth spatial harmonic. The coil (group) MMF created by the kth time and nth spatial harmonic can be represented in terms of positive and negative sequence components, as given in Eq. 2.163, the speed of which *relative to the coil (group)* is

$$v_{k_r,n_r} = \pm \frac{k_r}{n_r} \cdot \frac{\tau_p}{\pi} \cdot s\omega; \quad \omega_{k_r,n_r} = \pm \frac{k_r}{n_r} \cdot \frac{s\omega}{p} \tag{2.219}$$

Since the coil (group) rotates at angular speed Ω, the angular speed of the positive and negative sequence components *relative to stator* $\omega_{k,n,\mathrm{st}}$ is equal to

$$\omega_{k,n,\mathrm{st}} = \Omega + \omega_{k_r,n_r} = \Omega \pm \frac{k_r}{n_r} \cdot \frac{s\omega}{p} = \Omega \pm \frac{k_r}{n_r} \cdot \frac{\omega - p \cdot \Omega}{p} \tag{2.220}$$

For the fundamental time (k = 1) and spatial (n = 1) harmonic one can further write

$$\omega_{1,1,\mathrm{st}} = \Omega + \omega_{1,1} = \Omega \pm \frac{s\omega}{p} = \Omega \pm \frac{\omega - p \cdot \Omega}{p} \tag{2.221}$$

The angular speed of the positive sequence is

$$\omega_{1,1,\mathrm{st},+} = \Omega + \frac{\omega - p \cdot \Omega}{p} = \frac{\omega}{p} \tag{2.222}$$

i.e., it is constant and equal to the synchronous speed of the fundamental wave of stator MMF. The angular speed of the negative sequence component of the rotor MMF is

$$\omega_{1,1,\mathrm{st},-} = \Omega - \frac{\omega - p \cdot \Omega}{p} = 2\Omega - \frac{\omega}{p} \tag{2.223}$$

For mechanical angular speed values Ω below one half of the synchronous speed $\omega/(2p)$ the negative sequence component of the rotor MMF rotates opposite to the direction of rotation of the rotor; above one half of the synchronous speed the negative sequence MMF rotates in the direction of rotation of the rotor. The negative sequence component of the rotor MMF created by rotating coil(s) is at standstill when the rotor mechanical speed is equal to one half the synchronous speed. This property of rotor pulsating field is called **Görges phenomenon**.

The slip of the negative sequence component of the rotor MMF is equal to

$$s_- = \frac{\omega - p \cdot \omega_{1,1,\mathrm{st},-}}{\omega} = \frac{\omega - 2p\Omega + \omega}{\omega} = 2s \tag{2.224}$$

2.6.7 Resulting MMF Waves Generated by Coils on Both Sides of Air Gap

Consider the fundamental component of air gap MMF wave created by a symmetrical stator winding carrying symmetrical currents, $\Theta_s(x, t)$, and the fundamental component of the air gap MMF created by rotor winding, $\Theta_r(x, t)$. The rotor MMF is created either by a symmetrical rotor winding carrying symmetrical currents with appropriate frequency, or by a DC-fed coil rotating at synchronous speed. The resulting air gap MMF $\Theta_\delta(x, t)$ can be written as

$$\Theta_\delta(x,t) = \Theta_s(x,t) + \Theta_r(x,t) \tag{2.225}$$

or

$$\Theta_\delta \sin\left(\omega t - \frac{\pi}{\tau_p}x + \gamma\right) = \Theta_s \sin\left(\omega t - \frac{\pi}{\tau_p}x + \Psi_s\right) + \Theta_r \sin\left(\omega t - \frac{\pi}{\tau_p}x + \Psi_r\right) \tag{2.226}$$

where

$$\Theta_\delta^2 = \Theta_s^2 + \Theta_r^2 + 2 \cdot \Theta_s \cdot \Theta_r \cdot \cos(\Psi_s - \Psi_r) \tag{2.227}$$

and

$$\gamma = \arctan\frac{\Theta_s \cdot \sin\Psi_s + \Theta_r \cdot \sin\Psi_r}{\Theta_s \cdot \cos\Psi_s + \Theta_r \cdot \cos\Psi_r} \tag{2.228}$$

The three MMF components $\Theta_s(x, t)$, $\Theta_r(x, t)$, and $\Theta_\delta(x, t)$ can be represented in the complex plane in the manner shown in Fig. 2.68.

By positioning the resulting air gap MMF Θ_δ to the negative real axis ($\gamma = \pi$), one can further write

$$\Theta_s = -\Theta_r \frac{\sin\Psi_r}{\sin\Psi_s} \tag{2.229}$$

Keeping resulting MMF constant, Θ_δ = const., one can determine the amplitude of the rotor MMF Θ_r necessary to generate given stator MMF Θ_s at a given angle Ψ_s. In order to do this, substitute first the equation for stator MMF into the relationship between the three MMF amplitudes

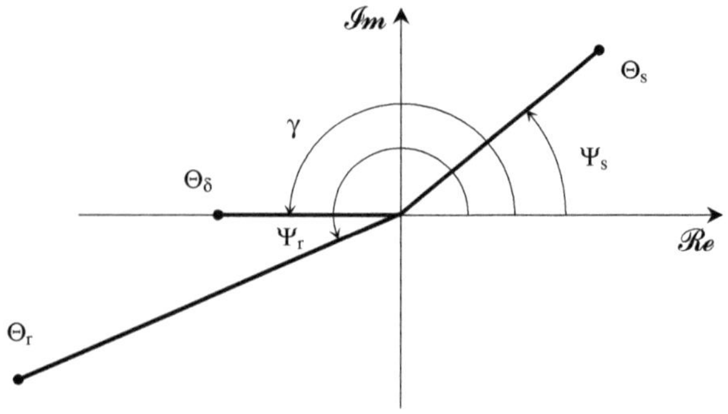

Fig. 2.68 Stator, rotor, and resulting air gap MMF

$$\Theta_\delta^2 = \left(\Theta_r \frac{\sin\Psi_r}{\sin\Psi_s}\right)^2 + \Theta_r^2 - 2\cdot\frac{\sin\Psi_r}{\sin\Psi_s}\cdot\Theta_r^2\cdot\cos(\Psi_s - \Psi_r) \tag{2.230}$$

which can be rewritten as

$$\left(\frac{\Theta_\delta}{\Theta_r}\right)^2 - 1 = \left(\frac{\sin\Psi_r}{\sin\Psi_s}\right)^2 - 2\cdot\frac{\sin\Psi_r}{\sin\Psi_s}\cdot\cos(\Psi_s - \Psi_r) \tag{2.231}$$

and solved in form of

$$\Theta_s\left(\frac{\Theta_r}{\Theta_\delta}\right)_{\Psi_s=\text{const}} \tag{2.232}$$

which is nothing but the equation of V-curves of a synchronous machine, as shown in Fig. 2.69! One should keep in mind that setting Θ_δ = const. determines the amount of air gap flux. If in addition the angle γ of the resulting MMF is kept constant, the stator-induced voltage remains unchanged, no matter how big the stator MMF and where it is relative to the resulting MMF.

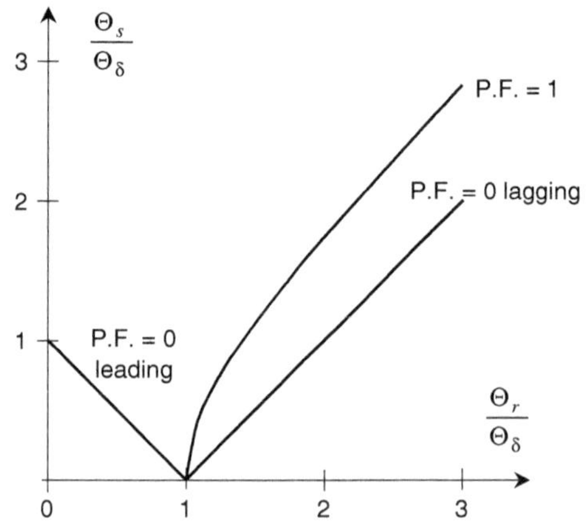

Fig. 2.69 V-curves of a synchronous machine as a solution of Eq. 2.231 for a given set of machine parameters

Neglecting voltage drops across the stator resistance and Potier reactance, the fixing of the amplitude and angle of Θ_δ means nothing but connecting the stator winding of a synchronous machine to an infinite bus.

2.6.8 Air Gap Flux Density Waves in a Single-Slotted Machine with Linear Magnetization Curve

If the B–H curve of stator and rotor irons is linear with $\mu_r >> 1$, the coil ampere-turns are equal to the MMF drop in the air gap, and the air gap flux density is proportional to the inverse of the air gap width δ. As shown previously, the amplitudes of slot harmonics with order $n = N/p \pm 1$ are neither influenced by the coil pitch, nor by the number of in series connected adjacent coils. A question can be posed whether the number of winding phases can modify the amplitude of slot harmonics in a similar manner as it eliminates the harmonics the order of which is equal to integer multiples of the number of phases. The machine is assumed to have unskewed slots.

Traveling waves created by slot harmonics of the order $n = N/p \pm 1$ and by the fundamental time component of current ($k = 1$) are always present in the spectrum of the resulting air gap MMF, because these fulfill the condition

$$\frac{1 \pm n}{m} = \text{even} \tag{2.233}$$

One can express the order n of slot harmonics by means of the *zone width* $q = N/(2 \cdot p \cdot m)$ as

$$n = \frac{N}{p} \pm 1 = 2mq \pm 1 \tag{2.234}$$

Now one can apply the condition for existence of a particular component of resulting MMF (Eqs. 2.188 and 2.189) as

$$\frac{1 \pm n}{m} = \frac{1 \pm (2mq \pm 1)}{m} = \text{even} \tag{2.235}$$

By taking altering signs inside and outside the brackets, one obtains that the condition above is fulfilled as long as $2q$ is an even number, which is always true in integer slot windings. Therefore, slot harmonics are as well present in the spectrum of resulting air gap MMF created by integer slot windings.

In fractional slot windings $2q$ is never an even number and slot harmonics of the order $n = N/p \pm 1$ do not exist in air gap distributions. Nevertheless, slot harmonics of the order $n = N/p_F \pm 1$, with p_F denoting the number of fundamental poles, are present in the air gap flux density spectrum. However, since the fundamental harmonic component of air gap MMF with period length $D \cdot \pi/p_F$ in a fractional slot machine has intentionally low amplitude, the discussed slot harmonics are of minor importance.

As shown in the previous sections, the spectrum of air gap flux density spatial components created by a coil in a machine with slotted air gap is different from the spectrum of coil MMF. If the coil is fed from an AC source, the slot harmonics $N/p - n$ and $N/p + n$ of air gap flux density $B_{n,\text{slot}}$ created by the nth harmonic of the applied MMF and the fundamental harmonic of air gap width due to slotting, $i = 1$, can be written as

$$\begin{aligned} B_{n,\text{slot}} &= 2\frac{\mu_0}{\delta_1}\frac{4}{\pi} \cdot w \cdot f_{\text{w},\frac{N}{p}-n} \sum_{k=1}^{\infty} I_k \cos(k\omega t - \varphi_k) \cos\frac{\pi}{\tau_\text{p}}\left[nx - \left(n + \frac{N}{p}\right)x_\text{c}\right] \\ &+ 2\frac{\mu_0}{\delta_1}\frac{4}{\pi} \cdot w \cdot f_{\text{w},\frac{N}{p}+n} \sum_{k=1}^{\infty} I_k \cos(k\omega t - \varphi_k) \cos\frac{\pi}{\tau_\text{p}}\left[nx - \left(n - \frac{N}{p}\right)x_\text{c}\right] \end{aligned} \tag{2.236}$$

The kth harmonic of coil current in Eq. 2.236 creates the rotating wave of air gap flux density due to slotting $B_{n-,k,\text{slot}}$ equal to

$$\begin{aligned} B_{n-,k,\text{slot}} &= \frac{\mu_0}{\delta_1}\frac{4}{\pi} I_k \cdot w \cdot f_{w,\frac{N}{p}-n} \\ &\cdot \left\{\cos\left\{k\omega t - \varphi_k + \frac{\pi}{\tau_\text{p}}\left[nx - \left(n + \frac{N}{p}\right)x_\text{c}\right]\right\} + \cos\left\{k\omega t - \varphi_k - \frac{\pi}{\tau_\text{p}}\left[nx - \left(n + \frac{N}{p}\right)x_\text{c}\right]\right\}\right\} \end{aligned} \tag{2.237}$$

as well as $B_{n+,k,\text{slot}}$, defined as

$$\begin{aligned} B_{n+,k,\text{slot}} &= \frac{\mu_0}{\delta_1}\frac{4}{\pi} I_k \cdot w \cdot f_{w,\frac{N}{p}+n} \\ &\cdot \left\{\cos\left\{k\omega t - \varphi_k + \frac{\pi}{\tau_\text{p}}\left[nx - \left(n - \frac{N}{p}\right)x_\text{c}\right]\right\} + \cos\left\{k\omega t - \varphi_k - \frac{\pi}{\tau_\text{p}}\left[nx - \left(n - \frac{N}{p}\right)x_\text{c}\right]\right\}\right\} \end{aligned} \tag{2.238}$$

As long as the coil does not move relative to the teeth, x_c = const., the speed of both components $B_{n-,k,\text{slot}}$ and $B_{n+,k,\text{slot}}$ is equal to the speed of the nth spatial and kth time harmonic of MMF which created them, namely $\pm k/n \cdot \omega/p$.

If the coil rotates at a constant mechanical angular velocity Ω relative to slots, the coil shift coordinate x_c can be represented as

$$x_\text{c} = R \cdot \Omega \cdot t \tag{2.239}$$

with R denoting the air gap radius. The angular speeds of rotation $\omega_{n-,k,\pm}$ of positive and negative sequence waves of air gap flux density due to slotting $B_{n-,k,\text{slot}}$ become in that case

$$\omega_{n-,k,\pm} = \pm\left(n - \frac{N}{p}\right) \cdot \Omega + \frac{k\,\omega}{n\,p} \tag{2.240}$$

with positive sign in front of parentheses standing for the positive, and negative sign for the negative sequence component of $B_{n-,k,\text{slot}}$.

Analogously, the angular speeds of rotation $\omega_{n+,k,\pm}$ of positive and negative sequence waves of air gap flux density due to slotting $B_{n+,k,\text{slot}}$ can be expressed as

$$\omega_{n+,k,\pm} = \pm\left(n + \frac{N}{p}\right) \cdot \Omega + \frac{k\,\omega}{n\,p} \tag{2.241}$$

The angular speeds of the four components of air gap flux density due to slotting can be expressed as

$$\omega_{n\pm,k,\pm} = \pm\left(n \pm \frac{N}{p}\right) \cdot \Omega + \frac{k\,\omega}{n\,p} \tag{2.242}$$

where every combination of signs stands for a particular higher harmonic and symmetrical component, respectively.

Coil mechanical angular velocities Ω_0 at which the positive or negative sequence wave of air gap flux density due to slotting $B_{n-,k,\text{slot}}$ and $B_{n+,k,\text{slot}}$ is at standstill relative to the teeth, i.e., $\omega_{n\pm,k,\pm} = 0$, can now be expressed as

$$\Omega_0 = \pm k \frac{\omega}{pn \pm N} \tag{2.243}$$

The positive sign in front of the expression on the right-hand side of Eq. 2.243 stands for positive sequence components, and the negative sign for negative sequence components of $B_{n-,k,\text{slot}}$ or $B_{n+,k,\text{slot}}$. The positive sign in front of the number of slots N in Eq. 2.243 is related to a slot harmonic with an order above N/p, and the negative to a slot harmonic with an order below N/p.

2.6.9 Air Gap Flux Density Waves in a Double-Slotted Machine with Linear Magnetization Curve

If the stator and rotor coils in Fig. 2.49 are fed from AC sources with angular frequencies ω_s and ω_r respectively, slot harmonics $N_s/p - n$, $N_s/p + n$, $N_r/p - n$ and $N_r/p + n$ of air gap flux density $B_{n,\text{slot}}$ created by the nth harmonic of applied MMFs and fundamental harmonics of air gap width due to slotting, $i = 1$, can be written as

$$B_{\frac{N_s}{p}-n} = 2\frac{\mu_0}{\delta_1}\frac{4}{\pi}\cdot w_s \cdot f_{w,s,\frac{N_s}{p}-n}\sum_{k=1}^{\infty} I_k \cos(k\omega_s t - \varphi_k)\cos\frac{\pi}{\tau_p}\left[nx - \left(n + \frac{N_s}{p}\right)\xi_n\right] \tag{2.244}$$

$$B_{\frac{N_s}{p}+n} = 2\frac{\mu_0}{\delta_1}\frac{4}{\pi}\cdot w_s \cdot f_{w,s,\frac{N_s}{p}+n}\sum_{k=1}^{\infty} I_k \cos(k\omega_s t - \varphi_k)\cos\frac{\pi}{\tau_p}\left[nx - \left(n - \frac{N_s}{p}\right)\xi_n\right] \tag{2.245}$$

$$\begin{aligned} B_{\frac{N_r}{p}-n} = {} & 2\frac{\mu_0}{\delta_1}\frac{4}{\pi}\cdot w_r \\ & \cdot f_{w,r,\frac{N_r}{p}-n}\sum_{j=1}^{\infty} I_j \cos\left(j\omega_r t - \varphi_j\right)\cos\frac{\pi}{\tau_p}\left[nx - \left(n + \frac{N_r}{p}\right)\left(\xi_n + \frac{N_r}{p}x_0\right)\right] \end{aligned} \tag{2.246}$$

$$\begin{aligned} B_{\frac{N_r}{p}+n} = {} & 2\frac{\mu_0}{\delta_1}\frac{4}{\pi}\cdot w_r \\ & \cdot f_{w,r,\frac{N_r}{p}+n}\sum_{j=1}^{\infty} I_j \cos\left(j\omega_r t - \varphi_j\right)\cos\frac{\pi}{\tau_p}\left[nx - \left(n - \frac{N_r}{p}\right)\left(\xi_n - \frac{N_r}{p}x_0\right)\right] \end{aligned} \tag{2.247}$$

$$B_{n,\text{slot}} = B_{\frac{N_s}{p}-n} + B_{\frac{N_s}{p}+n} + B_{\frac{N_r}{p}-n} + B_{\frac{N_r}{p}+n} \tag{2.248}$$

Each of the pulsating harmonics in Eqs. 2.244–2.248 consists of a positive and a negative sequence component, the speeds of rotation of which can be found in the manner shown previously in this chapter. When determining the rotational speed of rotor harmonics one has to consider the rotor mechanical speed of rotation Ω defined in the previous section.

Denoting by $\dot{\xi}_n$ the derivative of the position coordinate of the maximum of resulting MMF

$$\dot{\xi}_n = n\Theta_{r,n}\frac{\Theta_{r,n} + \Theta_{s,n}\cos n\frac{\pi}{\tau_p}x_0}{\Theta_{r,n}^2 + \Theta_{s,n}^2 + 2\Theta_{r,n}\Theta_{s,n}\cos n\frac{\pi}{\tau_p}x_0}\frac{dx_0}{dt} \tag{2.249}$$

one can express the angular speeds of harmonics of air gap flux density created by harmonics of stator and rotor currents with equal orders, j = k, and the fundamental harmonics due to slotting (i = 1) in the manner shown in Table 2.11.

Table 2.11 Angular speeds of air gap flux density harmonics created by nth harmonics of stator and rotor MMF and fundamental harmonics of stator and rotor slottings in a double-slotted machine

	Pos. seq.	Neg. seq.
$B_{\frac{N_s}{p}-n}$	$\frac{k\,\omega_s}{n\,p}+\frac{1}{np}\left(n+\frac{N_s}{p}\right)\frac{\pi}{\tau_p}\dot{\xi}_n$	$\frac{k\,\omega_s}{n\,p}-\frac{1}{np}\left(n+\frac{N_s}{p}\right)\frac{\pi}{\tau_p}\dot{\xi}_n$
$B_{\frac{N_s}{p}+n}$	$\frac{k\,\omega_s}{n\,p}+\frac{1}{np}\left(n-\frac{N_s}{p}\right)\frac{\pi}{\tau_p}\dot{\xi}_n$	$\frac{k\,\omega_s}{n\,p}-\frac{1}{np}\left(n-\frac{N_s}{p}\right)\frac{\pi}{\tau_p}\dot{\xi}_n$
$B_{\frac{N_r}{p}-n}$	$\frac{k\,\omega_r}{n\,p}+\frac{1}{np}\left(n+\frac{N_r}{p}\right)\left(\frac{\pi}{\tau_p}\dot{\xi}_n+\Omega\right)$	$\frac{k\,\omega_r}{n\,p}-\frac{1}{np}\left(n+\frac{N_r}{p}\right)\left(\frac{\pi}{\tau_p}\dot{\xi}_n+\Omega\right)$
$B_{\frac{N_r}{p}+n}$	$\frac{k\,\omega_r}{n\,p}+\frac{1}{np}\left(n-\frac{N_r}{p}\right)\left(\frac{\pi}{\tau_p}\dot{\xi}_n+\Omega\right)$	$\frac{k\,\omega_r}{n\,p}-\frac{1}{np}\left(n-\frac{N_r}{p}\right)\left(\frac{\pi}{\tau_p}\dot{\xi}_n+\Omega\right)$

One should note that for a given number of stator and rotor phases and slots per pole pair some harmonics of air gap flux density discussed here may have amplitudes equal to zero.

2.6.10 Air Gap Flux Density Waves in a Slotless Machine with Nonlinear Magnetization Curve

In order to increase the flux level, electric machines are built in such a manner that the operating point in the iron portion of their magnetic circuit is located more or less in saturated region.

A steep increase of MMF drop across iron core as a consequence of increasing flux density results in a slower increase of MMF drop across the air gap. The machine's magnetic circuit acts as if the air gap reluctance has increased, i.e., as if the air gap width became larger.

A comparison of influence of air gap geometry and iron B–H curve on equivalent air gap width is shown in Fig. 2.70.

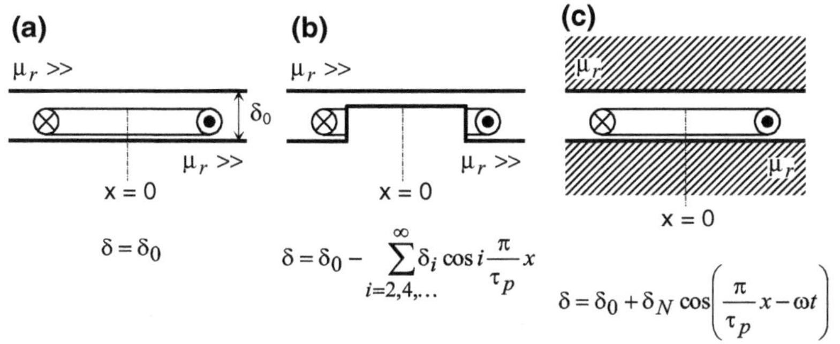

Fig. 2.70 Representation of air gap structure by means of air gap width: **a** linear iron curve, constant air gap width; **b** linear iron magnetization curve, salient poles—periodical decrease of air gap width under the poles; **c** nonlinear iron magnetization curve—periodical increase of air gap width

Whereas air gap becomes electromagnetically shorter as a consequence of saliency, saturation in iron makes it electromagnetically larger. A narrower air gap is attached to a salient pole. A wider air gap representing saturation in iron, on the other hand, travels around the periphery with angular speed of the fundamental of MMF which produced it and can be represented as

$$\delta = \delta_0 + \delta_N \cos 2\left(\frac{\pi}{\tau_\mathrm{p}}x - \omega t\right) \tag{2.250}$$

because it is assumed that only the fundamental component of applied MMF determines the level of saturation in iron.

Ampère's circuital law applied to the machine's magnetic circuit in which the increased MMF drop in iron is represented with a wider air gap can be written as

$$\left[\delta_0 + \delta_N \cos 2\left(\frac{\pi}{\tau_\mathrm{p}}x - \omega t\right)\right] \sum_{i=1,3,5,\ldots}^{\infty} B_i \cos i\left(\frac{\pi}{\tau_\mathrm{p}}x - \omega t\right) = \mu_0 \Theta_1 \cos\left(\frac{\pi}{\tau_\mathrm{p}}x - \omega t\right) \tag{2.251}$$

from which one becomes relations between amplitudes of air gap flux density harmonics as

$$\left(\delta_0 + \frac{\delta_N}{2}\right)B_1 + \frac{\delta_N}{2}B_3 = \mu_0 \Theta_1 \tag{2.252}$$

$$\frac{\delta_N}{2}B_1 + \left(\delta_0 + \frac{\delta_N}{2}\right)B_3 + \frac{\delta_N}{2}B_5 = 0 \tag{2.253}$$

$$\frac{\delta_N}{2}B_3 + \left(\delta_0 + \frac{\delta_N}{2}\right)B_5 + \frac{\delta_N}{2}B_7 = 0 \tag{2.254}$$

etc. Saturation in iron generates higher harmonics of air gap flux density which travel at the speed of the fundamental. If the MMF distribution contains higher spatial harmonics, their speed decreases with order of harmonic. This means that saturation in iron is a source of higher harmonics of air gap flux density which have the same spatial order as higher harmonics of air gap flux density created by higher harmonics of MMF, but a different time order. As a consequence, there exist two components of *n*th order of flux density harmonic which travel relative to each other at angular speeds given for a three-phase machine in Table 2.12.

Table 2.12 Angular speeds of air gap flux density harmonics due to MMF and due to saturation. For the sake of comparison, the speed differences for harmonics of the same order and different sources are shown

Order	Angular speed of MMF harmonic	Angular speed of harmonic due to saturation	Speed difference
1	ω_s	ω_s	0
3	–	ω_s	ω_s
5	$-\omega_s/5$	ω_s	6/5 ω_s
7	$\omega_s/7$	ω_s	6/7 ω_s
9	–	ω_s	ω_s

2.7 Induced Voltage

2.7.1 Rotating Air Gap Flux Density

Assume a coil with w turns and pitch y traveling in air gap at speed v_c, as shown in Fig. 2.71. The air gap is excited with a flux density wave $B(x, t)$ created by nth spatial harmonic of a symmetrical $2p$-pole winding fed by symmetrical currents with angular frequency $k\omega$. The air gap flux density wave $B(x, t)$ can be represented as

$$B(x,t) = B_{\max} \cos\left[k\omega t - \varphi_k \pm n\frac{\pi}{\tau_p}(x - x_n)\right] \tag{2.255}$$

with φ_k standing for kth time and x_n for nth spatial harmonic phase shift, respectively.

The air gap flux density wave $B(x, t)$ travels at speed $\pm k/n \cdot \tau_p/\pi \cdot \omega$ relative to the stator, with positive sign denoting the same direction of rotation as the coil, and negative sign the opposite direction of rotation to the coil velocity v_c. At time instant t the position x_c of the coil can be expressed as

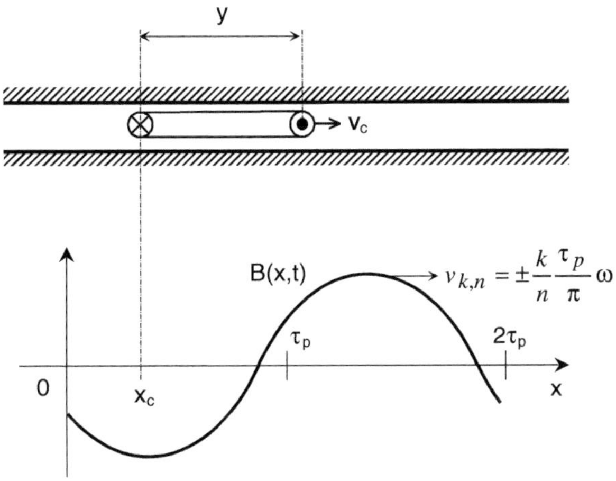

Fig. 2.71 Coil in air gap and a wave of air gap flux density

$$x_c = v_c t + x_0 \tag{2.256}$$

where x_0 denotes the coil position at t = 0. At time instant t each turn of the coil concatenates the flux $\Phi(t)$ in the amount of

$$\Phi(t) = l_{ax} \int_{x_c}^{x_c+y} B(x,t)\mathrm{d}x = l_{ax} B_{max} \int_{v_c t + x_0}^{v_c t + x_0 + y} \cos\left[k\omega t - \varphi_k \pm n\frac{\pi}{\tau_p}(x - x_n)\right]\mathrm{d}x \tag{2.257}$$

or

$$\begin{aligned}\Phi(t) = -\frac{l_{ax}\tau_p B_{max}}{n\pi}\Big\{\sin\left[k\omega t - \varphi_k \pm n\frac{\pi}{\tau_p}(v_c t + x_0 + y - x_n)\right] \\ + \sin\left[k\omega t - \varphi_k \pm n\frac{\pi}{\tau_p}(v_c t + x_0 - x_n)\right]\Big\}\end{aligned} \tag{2.258}$$

By applying addition theorem for trigonometric functions, one obtains

$$\begin{aligned}\Phi(t) = -\frac{l_{ax}\tau_p B_{max}}{n\pi}\Big\{\sin\left[\left(k\omega \pm n\frac{\pi}{\tau_p}v_c\right)t - \varphi_k \pm n\frac{\pi}{\tau_p}(x_0 - x_n)\right]\left(\cos n\frac{\pi}{\tau_p}y - 1\right) \\ - \cos\left[\left(k\omega \pm n\frac{\pi}{\tau_p}v_c\right)t - \varphi_k \pm n\frac{\pi}{\tau_p}(x_0 - x_n)\right]\sin\left(n\frac{\pi}{\tau_p}y\right)\Big\}\end{aligned} \tag{2.259}$$

or

$$\Phi(t) = -\frac{2 l_{ax}\tau_p B_{max}}{n\pi}\sin\left(n\frac{y}{\tau_p}\frac{\pi}{s}\right)\sin\left[\left(k\omega \pm n\frac{\pi}{\tau_p}v_c\right)t - \varphi_k - \varphi_{y,n} \pm n\frac{\pi}{\tau_p}(x_0 - x_n)\right] \tag{2.260}$$

where

$$\varphi_{y,n} = \arctan\frac{\cos n\frac{\pi}{\tau_p}y - 1}{\sin n\frac{\pi}{\tau_p}y} \tag{2.261}$$

Define now the maximum flux Φ_{max} concatenated by each turn of a coil with pitch y and created by sinusoidal flux density B_{max} as

$$\Phi_{max} = \frac{2}{\pi} l_{ax}\tau_p B_{max}\sin\left(n\frac{y}{\tau_p}\frac{\pi}{2}\right) \tag{2.262}$$

(recall that the factor $2/\pi$ is a consequence of the sinusoidal shape of B, the average of which over half a period is $2/\pi\ B_{\max}$).

Now one can write for the concatenated flux $\Phi(t)$:

$$\Phi(t) = -\frac{\Phi_{\max}}{n}\sin\left[\left(k\omega \pm n\frac{\pi}{\tau_p}v_c\right)t - \varphi_k - \varphi_{y,n} \pm n\frac{\pi}{\tau_p}(x_0 - x_n)\right] \tag{2.263}$$

and for the flux linkage $\Psi(t)$

$$\Psi(t) = w\Phi(t) = -\frac{w\Phi_{\max}}{n}\sin\left[\left(k\omega \pm n\frac{\pi}{\tau_p}v_c\right)t - \varphi_k - \varphi_{y,n} \pm n\frac{\pi}{\tau_p}(x_0 - x_n)\right] \tag{2.264}$$

The induced voltage $u_i(t)$ is equal to

$$u_i(t) = \left(k\omega - n\frac{\pi}{\tau_p}v_c\right)\frac{w\Phi_{\max}}{n}\cos\left[\left(k\omega \pm n\frac{\pi}{\tau_p}v_c\right)t - \varphi_k - \varphi_{y,n} \pm n\frac{\pi}{\tau_p}(x_0 - x_n)\right] \tag{2.265}$$

and its amplitude $U_{\max}$:

$$U_{\max} = \left(k\omega \pm n\frac{\pi}{\tau_p}v_c\right)\frac{w\Phi_{\max}}{n} = \left(k\omega \pm n\frac{\pi}{\tau_p}v_c\right)\frac{w}{n}\frac{2}{\pi}l_{ax}\tau_p B_{\max}\sin\left(n\frac{y}{\tau_p}\frac{\pi}{s}\right) \tag{2.266}$$

The induced voltage in Eq. 2.266 has two components, U_{wave} and U_{coil}, the amplitudes of which can be written as

$$U_{\max,\text{wave}} = \frac{k}{n}\omega w\Phi_{\max} = \frac{k}{n}\omega\frac{\tau_p}{\pi}2wl_{ax}B_{\max}\sin\left(n\frac{y}{\tau_p}\frac{\pi}{s}\right) \tag{2.267}$$

or, by utilizing expression 132 for circumferential velocity $v_{k,n}$:

$$U_{\max,\text{wave}} = 2wB_{\max}v_{k,n}l_{ax}\sin\left(n\frac{y}{\tau_p}\frac{\pi}{s}\right) \tag{2.268}$$

and

$$U_{\max,\text{coil}} = n\frac{\pi}{\tau_p}v_c\frac{w\Phi_{\max}}{n} = 2wB_{\max}v_c l_{ax}\sin\left(n\frac{y}{\tau_p}\frac{\pi}{s}\right) \tag{2.269}$$

The induced voltage in one conductor of a full-pitch coil is accordingly

$$U_{\mathrm{max,turn}} = B_{\mathrm{max}} l_{\mathrm{ax}} \left(v_{k,n} \pm v_{\mathrm{c}} \right) \tag{2.270}$$

which is nothing but $B \cdot \ell \cdot v$ equation, with v denoting the relative speed between the flux density wave and the coil!

Angular frequency of induced voltage ω_i is equal to

$$\omega_i = k\omega \pm n \frac{\pi}{\tau_{\mathrm{p}}} v_{\mathrm{c}} = k\omega \pm np\Omega \tag{2.271}$$

i.e., to a linear combination of the angular frequency $k\omega$ of the current which generated the flux density wave and an n times increased coil electrical angular speed $p\Omega$. Since the nth spatial harmonic of the air gap flux density travels n times slower than the fundamental, the apparent coil speed $np\Omega$ is n times higher than its actual electrical speed $p\Omega$.

If the coil angular speed Ω is expressed in terms of slip s and fundamental angular frequency ω of currents which generate the rotating wave of flux density $B(x, t)$, one can write for the angular frequency of induced voltage

$$\omega_i = k\omega \pm n(1 - s)\omega = \omega[k \pm n(1 - s)] \tag{2.272}$$

Angular frequencies of coil voltage induced by the fundamental time harmonic ($k = 1$) and various spatial harmonics n of air gap flux density wave $B(x, t)$ traveling in the same ($\omega_{i,+}$) and opposite ($\omega_{i,-}$) direction of the coil velocity v_{c} are given in Table 2.13. In addition, the values of slip are given in this table for which the frequencies $\omega_{i,+}$ and $\omega_{i,-}$ are equal to zero, i.e., at which the particular air gap flux density harmonic travels at the same speed as the coil. In the last two columns of Table 2.13 mechanical angular velocities are given at which a particular air gap flux density harmonic travels at the same speed as the coil (synchronism).

The nth spatial harmonic of air gap flux density travels at a speed n times smaller than the fundamental, but it has to pass an n times shorter way in order to complete a distance of two machine poles divided by its order. Therefore, the nth spatial harmonic needs the same time interval in order to complete one revolution as the fundamental, i.e., the frequency of voltage that it induces in a stationary coil is equal to the frequency of the fundamental.

As long as the coil speed is lower than the speed of flux density wave, the amplitude of induced voltage is positive. If the coil travels at a speed above the flux density wave speed, the amplitude of induced voltage formally changes its sign, which is identical to a phase angle skip of 180°. If the coil carries current with the same frequency as the induced voltage, the product of current and induced voltage —i.e., the power—changes its sign.

If the coil is at standstill ($v_{\mathrm{c}} = 0$), its position remains constant, $x_{\mathrm{c}} = x_0$, Eq. 2.256. In that case, both the amplitude and the frequency of induced voltage are a function only of the order k of time harmonic, but *not of the order n of spatial harmonic* of the air gap flux density wave:

$$u_i(t) = k\omega \frac{w\Phi_{\max}}{n} \cos\left[k\omega t - \varphi_k - \varphi_{y,n} \pm n\frac{\pi}{\tau_p}(x_0 - x_n)\right] \tag{2.273}$$

One recognizes in Eq. 2.273 that only the phase shift of induced voltage in a stationary coil is a function of coil position x_0.

Table 2.13 Frequencies of voltages induced by the fundamental time and nth spatial harmonic of air gap flux density in a coil rotating at mechanical angular speed Ω

n	$\omega_{i,+}$	$\omega_{i,-}$	s for $\omega_{i,+} = 0$	s for $\omega_{i,-} = 0$	Ω for $\omega_{i,+} = 0$	Ω for $\omega_{i,-} = 0$
1	$\mathbf{s \cdot \omega}$	$(2 - s) \cdot \omega$	**0**	2	$\boldsymbol{\omega/p}$	$-\omega/p$
3	$(-2 + 3s) \cdot \omega$	$(4 - 3s) \cdot \omega$	2/3	4/3	$\omega/(3p)$	$-\omega/(3p)$
5	$(-4 + 5s) \cdot \omega$	$\mathbf{(6 - 5s) \cdot \omega}$	4/5	**6/5**	$\omega/(5p)$	$\boldsymbol{-\omega/(5p)}$
7	$\mathbf{(-6 + 7s) \cdot \omega}$	$(8 - 7s) \cdot \omega$	**6/7**	8/7	$\boldsymbol{\omega/(7p)}$	$-\omega/(7p)$
9	$(-8 + 9s) \cdot \omega$	(10-9 s)·ω	8/9	10/9	ω/(9p)	$-\omega/(9p)$
11	$(-10 + 11-) \cdot \omega$	$\mathbf{(12 - 11s) \cdot \omega}$	10/11	**12/11**	$\omega/(11p)$	$\boldsymbol{-\omega/(11p)}$
13	$\mathbf{(-12 + 13s) \cdot \omega}$	$(14 - 13s)\cdot\omega$	**12/13**	14/13	$\boldsymbol{\omega/(13p)}$	$-\omega/(13p)$
15	$(-14 + 15s) \cdot \omega$	$(16 - 15s) \cdot \omega$	14/15	16/15	$\omega/(15p)$	$-\omega/(15p)$
17	$(-16 + 17s) \cdot \omega$	$\mathbf{(18 - 17s)\cdot\omega}$	16/17	**18/17**	$\omega/(17-)$	$\boldsymbol{-\omega/(17p)}$
19	$\mathbf{(-18 + 19s) \cdot \omega}$	$(20 - 19s) \cdot \omega$	**18/19**	20/19	$\boldsymbol{\omega/(19p)}$	$-\omega/(19p)$

In addition, slip values at which a particular harmonic travels at the same angular velocity Ω as the coil, and mechanical synchronous speeds of selected harmonics are given. Bold printed are the values related to a symmetrical 3-phase flux density distribution

2.7.2 *Elliptic Air Gap Flux Density*

Assume now that the air gap is excited with elliptic flux density consisting of a positive sequence component $B_+(x, t)$:

$$B_+(x,t) = B_{\max,+} \cos\left[k\omega t - \varphi_k - n\frac{\pi}{\tau_p}(x - x_n)\right] \tag{2.274}$$

and a negative sequence component $B_-(x, t)$:

$$B_-(x,t) = B_{\max,-} \cos\left[k\omega t - \varphi_k + n\frac{\pi}{\tau_p}(x - x_n)\right] \tag{2.275}$$

each of which induces its own voltage in a coil with coil pitch y, Eq. 2.205:

$$u_+(t) = \left(k\omega - n\frac{\pi}{\tau_p}v_c\right)\frac{w\Phi_{\max,+}}{n}\cos\left[\left(k\omega - n\frac{\pi}{\tau_p}v_c\right)t - \varphi_k - \varphi_{y,n} - n\frac{\pi}{\tau_p}(x_0 - x_n)\right] \tag{2.276}$$

$$u_-(t) = \left(k\omega - n\frac{\pi}{\tau_p}v_c\right)\frac{w\Phi_{max,-}}{n}\cos\left[\left(k\omega + n\frac{\pi}{\tau_p}v_c\right)t - \varphi_k - \varphi_{y,n} + n\frac{\pi}{\tau_p}(x_0 - x_n)\right] \tag{2.277}$$

or, in a more compact form:

$$u_+(t) = U_{max,+}\cos(\omega_+ t - \varphi_+) \tag{2.278}$$

and

$$u_-(t) = U_{max,-}\cos(\omega_- t - \varphi_-) \tag{2.279}$$

where

$$U_{max,+} = \left(k\omega - n\frac{\pi}{\tau_p}v_c\right)\frac{w}{n}\frac{2}{\pi}l_{ax}\tau_p B_{max,+}\sin\left(n\frac{y}{\tau_p}\frac{\pi}{2}\right) \tag{2.280}$$

and

$$U_{max,-} = \left(k\omega + n\frac{\pi}{\tau_p}v_c\right)\frac{w}{n}\frac{2}{\pi}l_{ax}\tau_p B_{max,-}\sin\left(n\frac{y}{\tau_p}\frac{\pi}{2}\right) \tag{2.281}$$

The total induced voltage in the coil can now be written as

$$u_{coil}(t) = U_{max,+}\cos(\omega_+ t - \varphi_+) + U_{max,-}\cos(\omega_- t - \varphi_-) \tag{2.282}$$

or

$$\begin{aligned} u_{coil}(t) = {} & U_{max,+}\left[\cos(\omega_+ t - \varphi_+) + \cos(\omega_- t - \varphi_-)\right] \\ & + \left(U_{max,-} - U_{max,+}\right)\cos(\omega_- t - \varphi_-) \end{aligned} \tag{2.283}$$

Introducing substitutions

$$\alpha - \beta = \omega_+ t - \varphi_+; \quad \alpha + \beta = \omega_- t - \varphi_- \tag{2.284}$$

or

$$\alpha = \frac{1}{2}\left[(\omega_+ + \omega_-)\cdot t - \varphi_+ - \varphi_-\right] = k\omega t - \varphi_k - n\frac{\pi}{\tau_p}x_0 \tag{2.285}$$

$$\beta = \frac{1}{2}\left[(-\omega_+ + \omega_-)\cdot t + \varphi_+ - \varphi_-\right] = n\frac{\pi}{\tau_p}(v_c t - x_n) \tag{2.286}$$

one can express the component of induced voltage with amplitude $U_{max,+}$, Eq. 2.221, as

$$
\begin{aligned}
&U_{\max,+}\left[\cos(\omega_+ t-\varphi_+)+\cos(\omega_- t-\varphi_-)\right]\\
&=2U_{\max,+}\cos\left(k\omega t-\varphi_k-n\frac{\pi}{\tau_p}x_0\right)\cos\left[n\frac{\pi}{\tau_p}(v_c t-x_n)\right]
\end{aligned}
\tag{2.287}
$$

and the total induced voltage, Eq. 2.282:

$$
\begin{aligned}
u_{\text{coil}}(t)=&\,2U_{\max,+}\cos\left(k\omega t-\varphi_k-n\frac{\pi}{\tau_p}x_0\right)\cos\left[n\frac{\pi}{\tau_p}(v_c t-x_n)\right]\\
&+\left(U_{\max,-}-U_{\max,+}\right)\cos\left[\left(k\omega+n\frac{\pi}{\tau_p}v_c\right)t-\varphi_k-\varphi_{y,n}-n\frac{\pi}{\tau_p}(x_0-x_n)\right]
\end{aligned}
\tag{2.288}
$$

In case of pulsating field, $B_{\max,+}=B_{\max,-}=B_{\max}$, one can write

$$
U_{\max,-}-U_{\max,+}=2n\frac{\pi}{\tau_p}v_c\frac{w\,2}{n\,\pi}l_{ax}\tau_p B_{\max,-}\sin\left(n\frac{y}{\tau_p}\frac{\pi}{2}\right)
\tag{2.289}
$$

Combining Eqs. 2.288 and 2.289 one can express the voltage in a stationary coil induced by pulsating field as

$$
u_{\text{coil}}(t)=2U_{\max,+}\cos\left(n\frac{\pi}{\tau_p}x_n\right)\cdot\cos\left(k\omega t-\varphi_k-\varphi_{y,n}-n\frac{\pi}{\tau_p}x_0\right)
\tag{2.290}
$$

The amplitude of voltage in a stationary coil induced by pulsating field is dependent on the spatial shift x_n of the harmonic which induced it, i.e., on the position of source of pulsating field $B_+(x,t)+B_-(x,t)$.

2.7.3 DC Flux Density Traveling at Angular Speed Ω

Field coils of conventional synchronous machines travel at synchronous speed and generate air gap MMF as described in previous equations. Assuming constant air gap width, the air gap flux density of a field coil can be expressed as

$$
B(x,t)=\sum_{n=1,2,3,\ldots}^{\infty}B_{\max,n}\sin n\frac{\pi}{\tau_p}(x-v\cdot t)
\tag{2.291}
$$

with v denoting the circumferential velocity of the field coil. The flux which nth harmonic of air gap flux density in Eq. 2.291 concatenates with an armature coil with pitch y is equal to (see Eq. 2.257)

$$\Phi(t) = l_{\mathrm{ax}} \int_{x_{\mathrm{c}}}^{x_{\mathrm{c}}+y} B(x,t)\mathrm{d}x = l_{\mathrm{ax}} B_{\max,n} \int_{v_{\mathrm{c}}t+x_0}^{v_{\mathrm{c}}t+x_0+y} \sin n\frac{\pi}{\tau_{\mathrm{p}}}(x - v\cdot t)\mathrm{d}x \tag{2.292}$$

or

$$\Phi(t) = 2\frac{\tau_{\mathrm{p}}}{n\pi}\sin\left(n\frac{y}{\tau_{\mathrm{p}}}\frac{\pi}{2}\right) l_{\mathrm{ax}} B_{\max,n} \cos n\frac{\pi}{\tau_{\mathrm{p}}}\left[(v - v_{\mathrm{c}})\cdot t - x_0 - \varphi_{y,n}\right] \tag{2.293}$$

with $\varphi_{y,n}$ already introduced in Eq. 2.202. Voltage induced in a coil with w turns is equal to

$$u_i(t) = 2w(v - v_{\mathrm{c}})\sin\left(n\frac{y}{\tau_{\mathrm{p}}}\frac{\pi}{2}\right) l_{\mathrm{ax}} B_{\max,n} \sin n\frac{\pi}{\tau_{\mathrm{p}}}\left[(v - v_{\mathrm{c}})\cdot t - x_0 - \varphi_{y,n}\right] \tag{2.294}$$

The amplitude of the induced voltage in a stationary full-pitch coil with one turn (two conductors!) is equal to

$$u_{i,1}(t) = 2vl_{\mathrm{ax}} B_{\max,n} \tag{2.295}$$

i.e., the amplitude of induced voltage in a single conductor is again $B \cdot \ell \cdot v$, as is the case with voltage induced by traveling wave of flux density, Eq. 2.269.

2.8 Fractional Slot Windings: Fundamental and Principal Poles; Single-Tooth Winding

Vast majority of rotating field electric machines is built with an integer number of slots per pole and phase q in such a manner as to create an MMF distribution the largest harmonic of which repeats at a rate of twice as large as the greatest pole pitch. As long as there is enough space along the air gap circumference, several coils carrying the same phase current can be placed next to each other into adjacent slots under one pole, resulting in appropriately large values of q. Due to manufacturing limitations, however, the slot width cannot be decreased arbitrarily in order to put high number of teeth under a pole of a high-polarity machine. An increase of number of poles, keeping all other parameters unchanged, means a decrease of q and less possibility to control the amplitudes of higher spatial harmonics by means of the zone factor. Higher harmonics of an integer slot winding with $q = 1$ can be controlled only by varying the coil pitch. In case of a three-phase winding with $q = 1$, the only reasonable chording is 2/3, which significantly deteriorates the fundamental harmonic.

Connection scheme in an integer slot winding is extremely simple and repeats each pole: the total of q adjacent coils is grouped together m times under each pole.

This means that higher spatial harmonics of MMF created by such windings are suppressed on a single-pole base, which is an inefficient approach in case of low number of phases and $q = 1$. Obviously, another method has to be applied in high-polarity machines with low numbers of phases.

It sounds reasonable to enlarge the interval along which higher harmonics are fought into r adjacent poles, instead of only one, as is the case with integer slot windings. In that case, one refers to a **fractional slot winding**. Basic difference between fractional and integer slot winding is that in an integer slot winding the poles with the largest possible pole pitch are at the same time those with the largest flux per pole, whereas in a fractional slot winding this is not the case. Accordingly, one distinguishes among *fundamental poles* and *principal poles* of a fractional slot winding. Fundamental poles are those having the largest possible pole pitch; principal poles are those having the largest flux per pole.

Denoting by p_{f} the number of fundamental pole pairs, and by p_{p} the number of principal pole pairs of a fractional pitch winding, one can write

$$\tau_{\mathrm{p,f}} = \frac{N}{2p_{\mathrm{f}}} \tag{2.296}$$

and

$$\tau_{\mathrm{p,p}} = \frac{N}{2p_{\mathrm{p}}} \tag{2.297}$$

where

$$\tau_{\mathrm{p,f}} = r \cdot \tau_{\mathrm{p,p}} \tag{2.298}$$

with $\tau_{\mathrm{p,f}}$ denoting the fundamental pole pitch, and $\tau_{\mathrm{p,p}}$ the principal pole pitch, both in number of slots, and r the (odd) number of principal poles over which given harmonic(s) are suppressed. The order n_{f} of a higher spatial harmonic over the period length of $2\tau_{\mathrm{p,f}}$ is r times larger than the order n_{p} of a higher spatial harmonic over the period length of $2\tau_{\mathrm{p,p}}$

$$n_{\mathrm{f}} = r \cdot n_{\mathrm{p}} \tag{2.299}$$

The fundamental harmonic on the principal pole basis, i.e., in interval of $2p_{\mathrm{p}}$ poles, is the rth harmonic on the fundamental pole basis, or in interval of $2p_{\mathrm{f}}$ poles.

The number of slots per phase and principal pole q_{p} of a fractional slot winding is not an integer

$$q_{\mathrm{p}} = \frac{N}{2p_{\mathrm{p}}m} = \frac{q_{\mathrm{f}}}{r} \tag{2.300}$$

as opposed to the number of slots per phase and fundamental pole q_{f}, which is always an integer

$$q_{\mathrm{f}} = \frac{N}{2p_{\mathrm{f}}m} = r \cdot q_{\mathrm{p}} \tag{2.301}$$

The mixed fraction q_{p} is equal to the ratio between the number of slots per phase and fundamental pole and the number of principal poles per fundamental pole r.

An integer slot winding is characterized by

$$p_{\mathrm{f}} = p_{\mathrm{p}} = p \tag{2.302}$$

i.e., there are as many fundamental as principal poles, $r = 1$. Whereas the fundamental and the $(N/p - 1)$st harmonics of an integer slot winding with $2p$ poles have the largest zone factor in the spectrum bandwidth of N/p spatial harmonics, the rth and $(N/p_{\mathrm{f}} - r)$th harmonics have the largest zone factor in a fractional pitch winding with $2p_{\mathrm{f}}$ fundamental poles, as shown in Fig. 2.72. Integer slot winding is, accordingly, a special case of fractional slot winding with $r = 1$ in which the fundamental component of MMF dominates as a consequence of specific winding topology—the placement of phase coils next to each other under each pole. Such connection of coils is characterized by a minimum angle between MMFs of adjacent coils and, consequently, maximum total MMF and zone factor.

Coil pitch of an integer slot winding is usually equal to, or shorter than the pole pitch, whereas the coil pitch of a fractional slot winding is approximately equal to the principal pole pitch, $y \approx \tau_{\mathrm{p,p}}$.

The pitch factor $f_{\mathrm{p,r}}$ for the rth harmonic created by a coil of a fractional slot winding with fundamental pole pitch of $\tau_{\mathrm{p,f}}$ can be expressed as

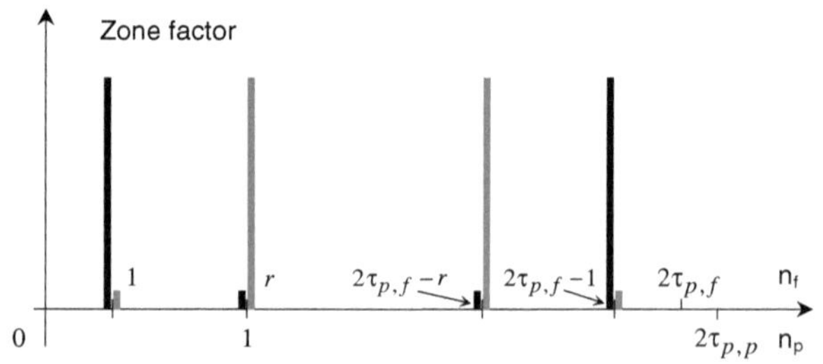

Fig. 2.72 Zone factor for selected spatial harmonics n of an integer slot winding with $p = p_{\mathrm{f}}$ (*black lines*) and a fractional slot winding (*gray lines*)

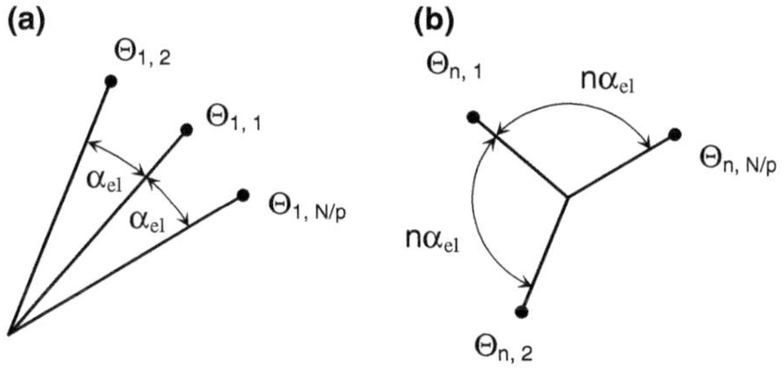

Fig. 2.73 Spatial representation of fundamental (**a**) and *n*th harmonics (**b**) of air gap MMF created by coils of an integer slot winding

$$f_{\mathrm{p,r}} = \sin r \frac{y}{\tau_{\mathrm{p,f}}} \frac{\pi}{2} \approx \sin r \frac{\tau_{\mathrm{p,p}}}{\tau_{\mathrm{p,f}}} \frac{\pi}{2} = 1 \tag{2.303}$$

with y denoting the coil pitch expressed in number of slots.

Sometimes the harmonic with period length of $2\tau_{\mathrm{p,p}}$ is described as fundamental, and one uses the construction "subharmonic" in order to describe all periodic quantities with periods longer than $2\tau_{\mathrm{p,p}}$. As opposed to higher harmonics, or simply harmonics, a "subharmonic" cannot be defined in terms of Fourier analysis and, therefore, it is not a valid quantity which could describe either spatial, or time periodic functions.

For the sake of clarity, further analysis will be performed for symmetrical fractional slot windings, i.e., for those with identical winding topologies in all m phases and with equal shift of $\tau_{\mathrm{p,f}}/m$ between adjacent phases at each fundamental pole. The number of slots per fundamental pole $N/(2p_{\mathrm{f}})$ of a symmetrical fractional slot winding is an integer multiple of the number of phases.

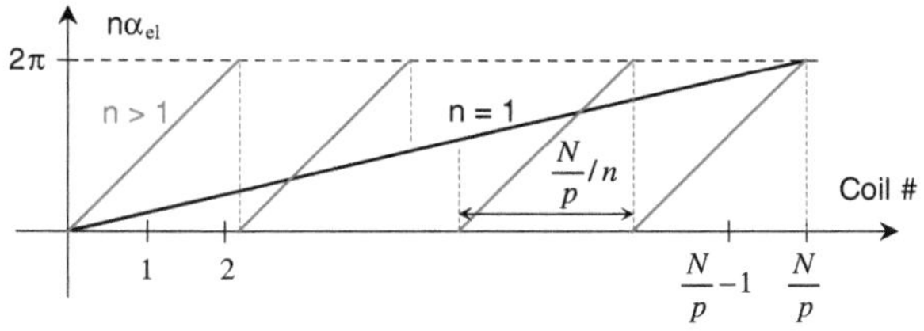

Fig. 2.74 Electrical angle between fundamental (*black line*) and *n*th harmonic components (*gray lines*) of MMF created by adjacent coils of a double-layer winding. Periodical character of angle causes the saw-tooth form for $n > 1$

If the resulting *r*th harmonic of q coils under one fundamental pole ought to be maximized, the angles between *r*th harmonic components of the coils must be minimal. This is obvious in a $2p$-pole integer slot winding ($r = 1$) placed in N slots, where the angle α_{el} between fundamental components of MMF of adjacent coils,

$\Theta_{1,N/p}$, $\Theta_{1,1}$, $\Theta_{1,2}$, etc. is equal to $\alpha_{el} = p \cdot \alpha_g = p \cdot 2\pi/N$, as shown in Fig. 2.73a. The total of N/p coils per pole pair builds the star of fundamental components of MMF with an angle α_{el} between adjacent legs. The angle between nth harmonics of MMF of adjacent coils, $\Theta_{n,N/p}$, $\Theta_{n,1}$, $\Theta_{n,2}$ etc., is $n \cdot \alpha_{el}$, as shown in Fig. 2.73b. This means that N/p coils per pole pair build n stars of nth harmonic components of MMF with an angle of $n \cdot \alpha_{el}$ between adjacent arms.

Since the angle is a periodic quantity with a period length of 2π, as shown in Fig. 2.74, the n stars of nth harmonic components of MMF overlap n times and end up in a single MMF star. The adjacent arms in the single MMF star of nth harmonic MMFs do not belong to adjacent coils, as is the case with the star of fundamental components of MMF. This key property of electrical angle of nth harmonic makes it possible to determine the sequence of coils under a fundamental pole of a fractional slot winding in such a manner as to maximize the rth spatial harmonic of MMF, as shown in Fig. 2.75.

Considering $N/(2p_f)$ slots per fundamental pole of a fractional slot winding, one groups q_f coils, the rth harmonics of which are placed next to each other in the MMF star in Fig. 2.75a, and connects them in series. The zone factor $f_{z,r}$ for q_f adjacent coils is equal to

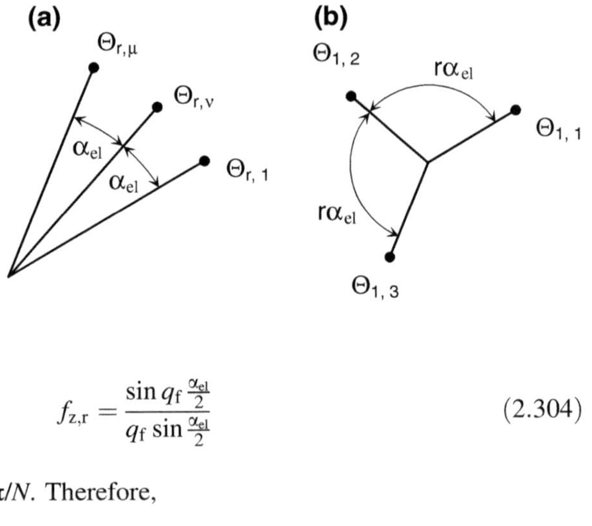

Fig. 2.75 Spatial representation of rth (**a**) and fundamental harmonics (**b**) of air gap MMF created by coils of a fractional slot winding

$$f_{z,r} = \frac{\sin q_f \frac{\alpha_{el}}{2}}{q_f \sin \frac{\alpha_{el}}{2}} \tag{2.304}$$

where $\alpha_{el} = p_f \cdot \alpha_g, = p_f \cdot 2\pi/N$. Therefore,

$$f_{z,r} = \frac{2p_f m}{N} \frac{\sin \frac{1}{m} \frac{\pi}{2}}{\sin p_f \frac{\pi}{N}} \tag{2.305}$$

If coils of a fractional slot winding have $y = 1$, one refers to **single-tooth winding**. In order to maximize the coil pitch factor, one selects the number of principal poles r close or equal to the fundamental pole pitch $\tau_{p,f}$, since in that case (Eq. 2.303)

$$f_{\mathrm{p,r}} = \sin r \frac{y}{\tau_{\mathrm{p,f}}} \frac{\pi}{2} \approx 1 \tag{2.306}$$

A single-tooth winding has short end windings. This advantage is compensated by unfavorably high zone factors for higher harmonics. Besides, in large machines large slot openings are necessary for manufacturing purposes (inserting the coils radially from the air gap), which decrease the slot-opening factor and increase the Carter factor, making ultimately the electromagnetic air gap width significantly larger than the geometric.

Since the order of the air gap MMF harmonics of a single-tooth-wound machine with the largest zone factor is close to $\tau_{\mathrm{p,f}}$, one can represent the zone factor for selected harmonics as shown in Fig. 2.76. Analogously to the zone factor distribution for selected harmonics, represented in Fig. 2.72, the zone factors for an integer slot winding are shown in Fig. 2.76 for the purpose of comparison.

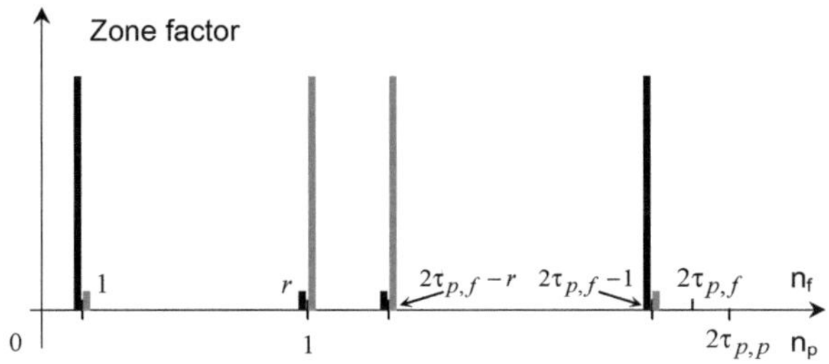

Fig. 2.76 Zone factors for selected spatial harmonics n of an integer slot winding with $p = p_{\mathrm{f}}$ (*black lines*) and a single-tooth winding (*gray lines*)

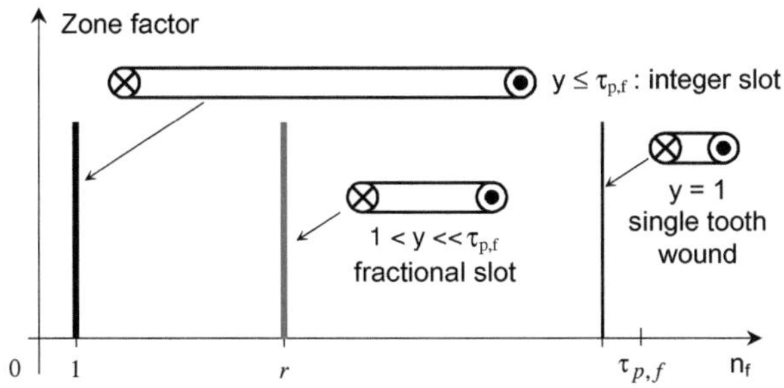

Fig. 2.77 Comparison of positions of dominating terms in air gap MMF spectra for various winding types. The pitch $\tau_{\mathrm{p,f}}$ is expressed in the number of slots, and n_{f} is the order of harmonic on the fundamental pole base

In extreme case $r = \tau_{\mathrm{p,f}} \pm 1$, and the harmonics with largest zone factor are positioned next to each other in Fig. 2.76.

One should note that groups of q_{f} adjacent coils of a single-tooth winding with number of slots close to the number of poles, i.e., with $r \approx \tau_{\mathrm{p,f}}$ belong to the same phase.

In Fig. 2.77, the strongest harmonics of air gap MMF spectra for various winding types are compared with each other.

Case Study 2.3: Stator winding of a three-phase, 84-pole electric machine is placed in 288 slots and has a coil pitch of $y = 3$. The number of slots per principal pole and phase q_{p} is equal to (Eq. 2.300)

$$q_p = \frac{N}{2p_p m} = \frac{8}{7}$$

By selecting the 7. spatial harmonic of the fundamental pole pitch to be the first harmonic of the principal pole pitch, $r = 7$, one obtains for $p_f = p_p/7 = 6$ and $\tau_{p,f} = N/(2p_f) = 24$. There are 24 slots per fundamental pole in which a symmetrical three-phase, double-layer winding ought to be placed. The number of slots per fundamental pole and phase q_f is equal to 8, the numerator of q_p. The electrical angle α_{el} between slots is equal to $\alpha_{el} = p_f \cdot 2\pi/N = \pi/24$ (7.5°), and the electrical angle between 7. harmonics of MMF created by coils in adjacent slots is $7\alpha_{el} = 52.5°$. Electrical angles of 7. harmonics of all coils, along with affiliation of a coil to a particular phase, are given in Table 2.8.

The normalized interval (0,360°) in Table 2.8 is divided into $2m = 6$ subintervals, each of which is 60°wide. In order to generate the phase sequence for a symmetrical 3-phase machine in form of A → −C → B → −A → C → −B, the coil connections have to be arranged as shown in column 4 of Table 2.14.

Table 2.14 Electrical angles of 7. harmonic components in an 84-pole, three-phase machine with 288 slots, along with phase affiliation of coils

Coil Nr. i	Electrical angle of the 7. harmonic $\alpha_{el,i}$	Electrical angle normalized to interval (0,360°)	Phase
1	0	0	+A
2	52.5°	52.5°	+A
3	105°	105°	−C
4	157.5°	157.5°	+B
5	210°	210°	−A
6	262.5°	262.5°	+C
7	315°	315°	−B
8	367.5°	7.5°	+A
9	420°	60°	−C
10	472.5°	112.5°	−C
11	525°	165°	+B
12	577.5°	217.5°	−A
13	630°	270°	+C
14	682.5°	322.5°	−B
15	735°	15°	+A
16	787.5°	67.5°	−C
17	840°	120°	+B
18	892.5°	172.5°	+B
19	945°	225°	−A
20	997.5°	277.5°	+C
21	1050°	330°	−B
22	1102.5°	22.5°	+A
23	1155°	75°	−C
24	1207.5°	127.5°	+B

In particular, coils with an electrical angle of the 7. harmonic between 0° (included) and 60° (not included) are positively oriented and belong to the phase A. Coils with an electrical angle of the 7. harmonic between 180° (included) and 240° (not included) are negatively oriented and also belong to the phase A.

The pitch factor for $r = 7$ equals to (Eq. 2.303)

$$f_{\mathrm{p},7} = \sin 7\frac{3}{24}\frac{\pi}{2} = 0.981$$

whereas the zone factors for the first 49 harmonics on interval $(0{,}2\tau_{\mathrm{p,f}})$ are

$$f_{\mathrm{z,n}} = \frac{1}{q_f}\sqrt{\left(\sum_i s_i \sin n\alpha_{\mathrm{el},i}\right)^2 + \left(\sum_i s_i \cos n\alpha_{\mathrm{el},i}\right)^2} \tag{2.307}$$

with s_i denoting the sign of a given component:

$$s_i = \frac{\sin n\alpha_{\mathrm{el},i}}{\left|\sin n\alpha_{\mathrm{el},i}\right|} \tag{2.308}$$

and $\alpha_{\mathrm{el},i}$ the angles of coils affiliated to the same phase. For the phase A one can write accordingly (Table 2.15).

Table 2.15 Angles and signs of components in Eq. 2.307

i	1	2	5	8	12	15	19	22
$\alpha_{\mathrm{el},i}$	0	52.5°	210°	367.5°	577.5°	735°	945°	1102.5°
s_i	+1	+1	−1	+1	−1	+1	−1	+1

Zone factors for the first 49 harmonics are shown in Fig. 2.78. One recognizes the dominant 7. and 48 − 7 = 41. harmonics, along with second strongest 21. and 27. harmonics on the fundamental pole basis.

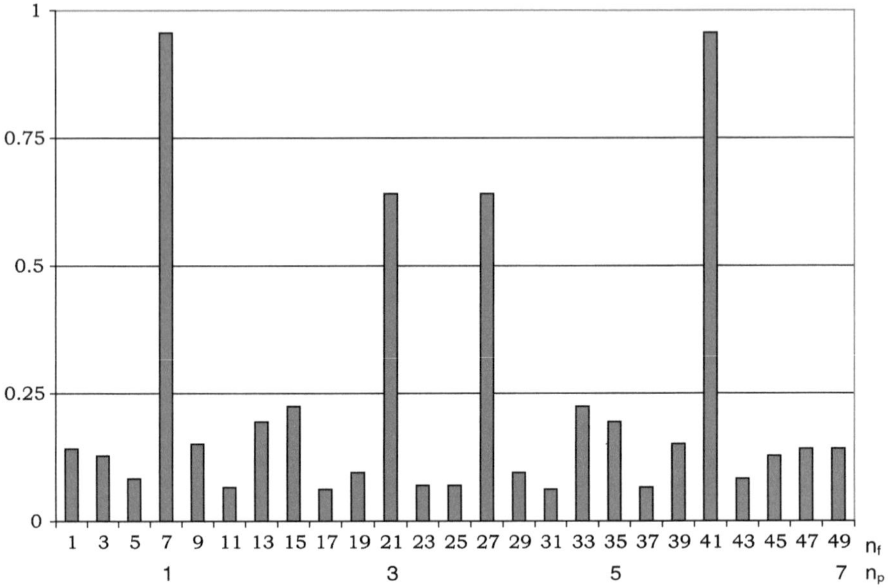

Fig. 2.78 Zone factors of the fractional pitch winding in Case Study 2.3 as a function of harmonic orders on the fundamental (n_f) and principal (n_p) pole basis

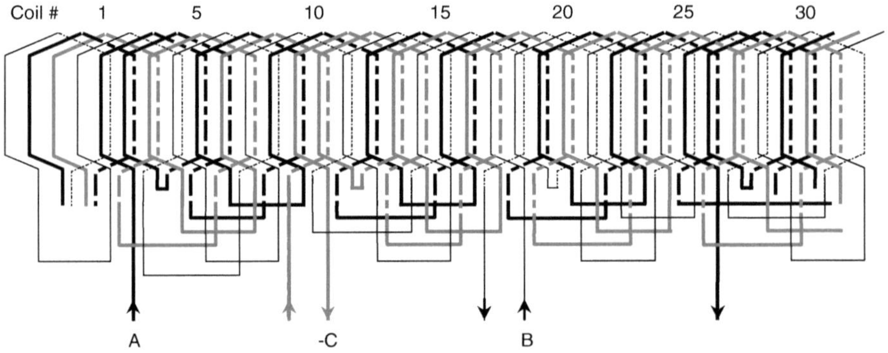

Fig. 2.79 Winding scheme for the machine in Case Study 2.3

Winding scheme for the first 30 slots of a three-phase, 84-pole generator with 288 slots and a coil pitch of $y = 3$ is given in Fig. 2.79. The winding is fully symmetrical and can be manufactured with a maximum of 288/24 = 12 parallel circuits.

Case Study 2.4: Stator winding of a three-phase, 70-pole generator is placed on 72 teeth and has a coil pitch of $y = 1$ (single-tooth winding). The number of slots per principal pole and phase q_p is equal to (Eq. 2.300)

$$q_{\mathrm{p}} = \frac{N}{2p_{\mathrm{p}}m} = \frac{12}{35}$$

and the number of slots per fundamental pole and phase q_{f} (Eq. 2.238) to

$$q_{\mathrm{f}} = \frac{N}{2p_{\mathrm{f}}m} = 12$$

By selecting the 35. spatial harmonic of the fundamental pole pitch to be the first harmonic of the principal pole pitch, $r = 35$, one obtains for the number of fundamental pole pairs $p_{\mathrm{f}} = p_{\mathrm{p}}/35 = 1$ and $\tau_{\mathrm{p,f}} = N/(2p_{\mathrm{f}}) = 36$. There are 36 teeth per fundamental pole carrying a symmetrical three-phase, double-layer winding. The two layers are placed in slots next to each other, instead of above each other, as is the case in conventional double-layer winding. The electrical angle α_{el} between slots is equal to $\alpha_{\mathrm{el}} = p_{\mathrm{f}} \cdot 2\pi/N = \pi/36$ (5°), and the electrical angle between 35. harmonics of MMF created by coils in adjacent slots is $35\alpha_{\mathrm{el}} = 175°$. Electrical angles of 35. harmonics of all coils, along with affiliation of a coil to a particular phase, are given in Table 2.16.

Table 2.16 Electrical angles of 35. harmonic components in an 70-pole, three-phase single-tooth-wound machine with 72 slots, along with phase affiliation of coils

Coil Nr. i	Electrical angle of the 35. harmonic $\alpha_{\mathrm{el},i}$	Electrical angle normalized to interval (0,360°)	Phase
1	0	0	+A
2	175°	175°	+B
3	350°	350°	−B
4	525°	165°	+B
5	700°	340°	−B
6	875°	155°	+B
7	1050°	330°	−B
8	1225°	145°	+B
9	1400°	320°	−B
10	1575°	135°	+B
11	1750°	310°	−B
12	1925°	125°	+B
13	2100°	300°	−B
14	2275°	115°	−C
15	2450°	290°	+C
16	2625°	105°	−C
17	2800°	280°	+C
18	2975°	95°	−C
19	3150°	270°	+C

(continued)

Table 2.16 (continued)

Coil Nr. i	Electrical angle of the 35. harmonic $\alpha_{el,i}$	Electrical angle normalized to interval (0,360°)	Phase
20	3325°	85°	+C
21	3500°	260°	+C
22	3675°	75°	−C
23	3850°	250°	+C
24	4025°	65°	−C
25	4200°	240°	+C
26	4375°	55°	+A
27	4550°	230°	−A
28	4725°	45°	+A
29	4900°	220°	−A
30	5075°	35°	+A
31	5250°	210°	−A
32	5425°	25°	+A
33	5600°	200°	−A
34	5775°	15°	+A
35	5950°	190°	−A
36	6125°	5°	+A

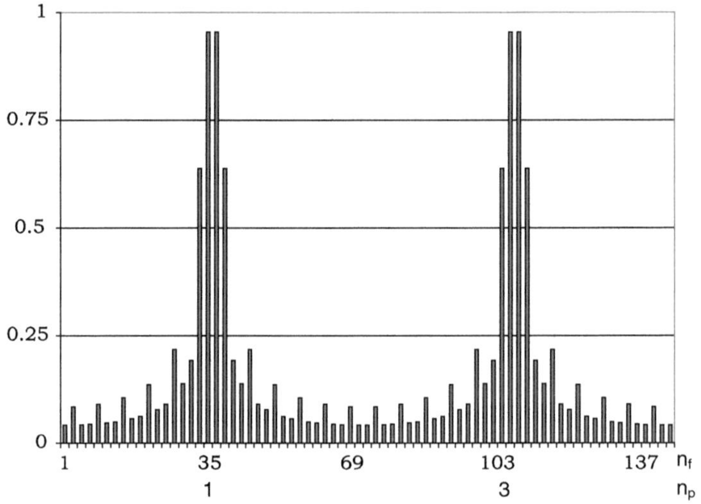

Fig. 2.80 Zone factors of the single-tooth winding in Case Study 2.4 as a function of harmonic orders on the fundamental (n_f) and principal (n_p) pole basis

In Fig. 2.80 the spectrum of air gap MMF harmonics of the tooth-wound machine in Case Study 2.4 is shown up to the range of twice the number of teeth (140 in this case). The third spatial harmonic of air gap MMF on the principal pole basis coincides with the $35 \cdot 3 = 105$. harmonic on the fundamental pole basis, the fifth harmonic on the principal pole basis with the $35 \cdot 5 = 175$. harmonic on the fundamental pole basis, etc.

Case Study 2.5: Stator winding of a three-phase, 20-pole machine is placed on $N = 30$ teeth and has a coil pitch of $y = 1$ (single-tooth winding), see also Case Study 2.1. The number of slots per principal pole and phase q_{p} is equal to (Eq. 2.300)

$$q_{\mathrm{p}} = \frac{N}{2p_{\mathrm{p}}m} = \frac{1}{2}$$

as is the number of slots per fundamental pole and phase q_{f} (Eq. 2.301), because $p_{\mathrm{p}} = p_{\mathrm{f}} = 10$. Geometric angle between stator slots is equal to $360°/N = 12°$, and electric angle $\alpha_{\mathrm{el}} = p \cdot 360°/N = 120°$.

Despite a non-integer number of slots per pole and phase, the stator winding in this case study is not a fractional slot one, because the number of fundamental poles is equal to the number of principal poles. Besides, the number of slots per pole and phase q is defined under an assumption of identical winding distribution at each pole, which is not valid in the case of a machine with one coil per pole pair.

2.9 Squirrel Cage Winding

Squirrel cage winding is unique among AC winding types because it has no electric terminals, no predefined number of phases m and no predefined number of poles $2p$. The only way a squirrel cage winding can communicate with other windings in a machine is through magnetic coupling with them, as shown in Fig. 2.81.

The number of phases of a squirrel cage winding depends on the number of rotor slots N and the number of pole pairs of stator winding. The number of phases m can be expressed as

$$m = \min\left\{N, \frac{N}{2}, \frac{N}{p}, \frac{N}{2p}\right\} \tag{2.309}$$

Since a squirrel cage winding does not have a predefined number of poles, it responds with its own MMF distribution to every spatial harmonic of stator created air gap flux density excitation of the order $n \cdot p$.

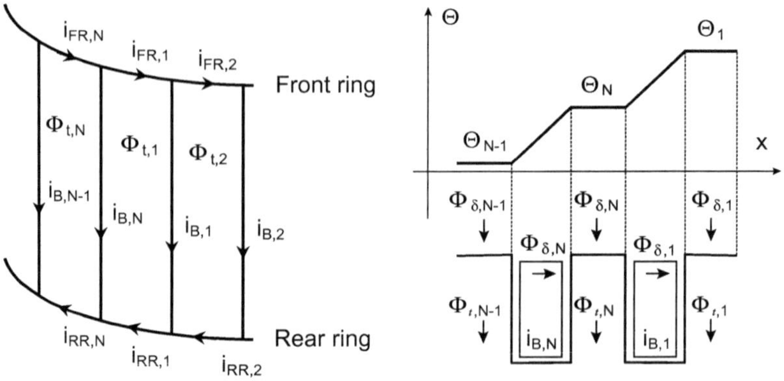

Fig. 2.81 Squirrel cage winding and the air gap MMF distribution created by bar currents

One recalls that the end winding form of a wound coil does not play any role when determining its air gap MMF distribution. Analogously, the fact that bars of a squirrel cage are short-circuited by means of rings on their ends is irrelevant for the analysis of air gap MMF created by such winding. Assume for the purpose of clarity that the number of rotor slots N is divisible by the number of poles $2p$, i.e., $m = N/2p$. In that case, the phase shift of currents in bars n, $n + N/2p$, $n + N/p$, $n + 3N/2p$, …, etc., alters for 180°, which is identical to the current distribution in slots of full-pitch coils of an m-phase machine. In other words, two bars of a squirrel cage winding placed in slots n and $n + N/2p$ create identical MMF distribution as a regular full-pitch coil. As shown in previous sections, a full-pitch coil creates a rectangularly distributed MMF characterized by unity pitch factor of all harmonics. A complete cage, however, generates only those harmonics of rotating MMF which satisfy conditions expressed in Eqs. 2.188 and 2.189. In conventional squirrel cage induction machine the number of rotor phases is large due to a large number of rotor slots. When symmetrically excited, e.g., with sinusoidal air gap flux density, a symmetrical squirrel cage winding generates an air gap MMF with fundamental component and slot harmonics only.

Taking for the number of rotor phases $m = N/2p$, one can express the total positive sequence air gap MMF Θ_+ created by the fundamental time component of squirrel cage currents as

$$\begin{aligned}\Theta_+ &= \sum_{i=0}^{\infty} \Theta_{2im+1} \cos\left[(2im+1)\frac{\pi}{\tau_p}x - \omega t\right] \\ &= \sum_{i=0}^{\infty} \Theta_{iN/p+1} \cos\left[\left(i\frac{N}{p}+1\right)\frac{\pi}{\tau_p}x - \omega t\right]\end{aligned} \tag{2.310}$$

as well as the negative sequence Θ_-

$$\Theta_{-} = \sum_{i=1}^{\infty} \Theta_{2im-1} \cos\left[(2im-1)\frac{\pi}{\tau_p}x + \omega t\right]$$
$$= \sum_{i=1}^{\infty} \Theta_{iN/p-1} \cos\left[\left(i\frac{N}{p} - 1\right)\frac{\pi}{\tau_p}x + \omega t\right] \tag{2.311}$$

The lowest order of higher harmonics of rotating MMF is $N/p - 1$. Therefore, the rotating MMF created by a squirrel cage winding is almost sinusoidal. Squirrel cage winding is a perfect example of how lower harmonics in the air gap MMF distribution can be suppressed by increasing the number of phases.

Assume now that stator winding generates an air gap flux density distribution B_δ which tries to penetrate into the squirrel cage. The corresponding fluxes $\Phi_{\delta,1} - \Phi_{\delta,N}$ are related to leakage fluxes $\Phi_{\sigma,1} - \Phi_{\sigma,\mathrm{N}}$ and tooth fluxes $\Phi_{\mathrm{t},1} - \Phi_{\mathrm{t},N}$ as

$$\Phi_{\delta,n} = \Phi_{\mathrm{t},n} + \Phi_{\sigma,n} - \Phi_{\sigma,n-1} \tag{2.312}$$

where $1 \le n \le N$. Denoting by $G_{\mathrm{s},n}$ the permeance of the nth slot for tangential flux

$$G_{\mathrm{s},n} = \frac{1}{3}\mu_0 \frac{l_{\mathrm{ax}} h_n}{b_n} \tag{2.313}$$

and by $k_{\mathrm{L},n}$ the skin effect factor for inductance, one can further write

$$\Phi_{\delta,n} = \Phi_{\mathrm{t},n} + \frac{1}{3}\mu_0 k_{\mathrm{L},n} \frac{l_{\mathrm{ax}} h_n}{b_n} i_{\mathrm{B},n} - \frac{1}{3}\mu_0 k_{\mathrm{L},n-1} \frac{l_{\mathrm{ax}} h_n}{b_n} i_{\mathrm{B},n-1} \tag{2.314}$$

and

$$\Phi_{\delta,n} = \Phi_{\mathrm{t},n} + k_{\mathrm{L},n} G_{\mathrm{s},n} i_{\mathrm{B},n} - k_{\mathrm{L},n-1} G_{\mathrm{s},n-1} i_{\mathrm{B},n-1} \tag{2.315}$$

Voltage differential equation written for the nth loop of the squirrel cage winding in Fig. 2.66, $1 \le n \le N$, yields

$$u_n = \frac{\mathrm{d}\Phi_{\mathrm{t},n}}{\mathrm{d}t} + R_{\mathrm{FR},n} \cdot i_{\mathrm{FR},n} + R_{\mathrm{B},n} \cdot i_{\mathrm{B},n} + R_{\mathrm{RR},n} \cdot i_{\mathrm{RR},n} - R_{\mathrm{B},n-1} \cdot i_{\mathrm{B},n-1} = 0 \tag{2.316}$$

with $R_{\mathrm{FR},n}$ denoting the resistance of the nth front ring segment, $R_{\mathrm{RR},n}$ the resistance of the nth rear ring segment, $R_{\mathrm{B},n}$ the AC resistance of the nth bar, and $\Phi_{\mathrm{t},n}$ the flux of the nth tooth. Considering heteropolar machine structure, one can write continuity equations

$$\sum_{j=1}^{N} \Phi_{\mathrm{t},j} = 0 \tag{2.317}$$

and

$$\sum_{j=1}^{N} i_{\mathrm{B},j} = 0 \tag{2.318}$$

as well as voltage equations for the front and rear end ring:

$$\sum_{j=1}^{N} R_{\mathrm{FR},j} \cdot i_{\mathrm{FR},j} = 0 \tag{2.319}$$

and

$$\sum_{j=1}^{N} R_{\mathrm{RR},j} \cdot i_{\mathrm{RR},j} = 0 \tag{2.320}$$

because the flux linked by front and end ring is equal to zero. Equations of the I Kirchhoff's law for bar, front and rear end ring currents can be written as

$$i_{\mathrm{B},n} = i_{\mathrm{FR},n} - i_{\mathrm{FR},n+1} \tag{2.321}$$

and

$$i_{\mathrm{B},n} = i_{\mathrm{RR},n} - i_{\mathrm{RR},n+1} \tag{2.322}$$

By combining Eqs. 2.319 and 2.321 one obtains matrix equation

$$\begin{bmatrix} 1 & -1 & 0 & \dots & 0 & 0 \\ 0 & 1 & -1 & \dots & 0 & 0 \\ 0 & 0 & 1 & \dots & 0 & 0 \\ \dots & \dots & \dots & \dots & \dots & \dots \\ 0 & 0 & 0 & \dots & 1 & -1 \\ R_{\mathrm{FR},1} & R_{\mathrm{FR},2} & R_{\mathrm{FR},3} & \dots & R_{\mathrm{FR},N-1} & R_{\mathrm{FR},N} \end{bmatrix} \cdot \begin{bmatrix} i_{\mathrm{FR},1} \\ i_{\mathrm{FR},2} \\ i_{\mathrm{FR},3} \\ \dots \\ i_{\mathrm{FR},N-1} \\ i_{\mathrm{FR},N} \end{bmatrix} = \underline{I}' \cdot \begin{bmatrix} i_{\mathrm{B},1} \\ i_{\mathrm{B},2} \\ i_{\mathrm{B},3} \\ \dots \\ i_{\mathrm{B},N-1} \\ i_{B,N} \end{bmatrix} \tag{2.323}$$

with $\underline{I}'$ denoting the modified identity matrix:

$$\underline{I}' = \begin{bmatrix} 1 & 0 & 0 & \dots & 0 & 0 \\ 0 & 1 & 0 & \dots & 0 & 0 \\ 0 & 0 & 1 & \dots & 0 & 0 \\ \dots & \dots & \dots & \dots & \dots & \dots \\ 0 & 0 & 0 & \dots & 1 & 0 \\ 0 & 0 & 0 & \dots & 0 & 0 \end{bmatrix} \tag{2.324}$$

Introducing vectors $\underline{i}_{\mathrm{B}}$, $\underline{i}_{\mathrm{FR}}$ and $\underline{i}_{\mathrm{RR}}$, defined as

$$\underline{i}_{\mathrm{B}} = \begin{bmatrix} i_{\mathrm{B},1} & i_{\mathrm{B},2} & \dots & i_{\mathrm{B},N} \end{bmatrix}^{\mathrm{T}} \tag{2.325}$$

$$\underline{i}_{\mathrm{FR}} = \begin{bmatrix} i_{\mathrm{FR},1} & i_{\mathrm{FR},2} & \dots & i_{\mathrm{FR},N} \end{bmatrix}^{\mathrm{T}} \tag{2.326}$$

$$\underline{i}_{\mathrm{RR}} = \begin{bmatrix} i_{\mathrm{RR},1} & i_{\mathrm{RR},2} & \dots & i_{\mathrm{RR},N} \end{bmatrix}^{\mathrm{T}} \tag{2.327}$$

as well as matrix $\underline{C}_{\mathrm{F}}$:

$$\underline{C}_{\mathrm{F}} = \begin{bmatrix} 1 & -1 & 0 & \dots & 0 & 0 \\ 0 & 1 & -1 & \dots & 0 & 0 \\ 0 & 0 & 1 & \dots & 0 & 0 \\ \dots & \dots & \dots & \dots & \dots & \dots \\ 0 & 0 & 0 & \dots & 1 & -1 \\ R_{\mathrm{FR},1} & R_{\mathrm{FR},2} & R_{\mathrm{FR},3} & \dots & R_{\mathrm{FR},N-1} & R_{\mathrm{FR},N} \end{bmatrix} \tag{2.328}$$

one can rewrite Eq. 2.323 as

$$\underline{C}_{\mathrm{F}} \cdot \underline{i}_{\mathrm{FR}} = \underline{I}' \cdot \underline{i}_{\mathrm{B}} \tag{2.329}$$

Similarly one can write

$$\underline{C}_{\mathrm{R}} \cdot \underline{i}_{\mathrm{RR}} = \underline{I}' \cdot \underline{i}_{\mathrm{B}} \tag{2.330}$$

where $\underline{C}_{\mathrm{R}}$ is defined as

$$\underline{C}_{\mathrm{R}} = \begin{bmatrix} 1 & -1 & 0 & \dots & 0 & 0 \\ 0 & 1 & -1 & \dots & 0 & 0 \\ 0 & 0 & 1 & \dots & 0 & 0 \\ \dots & \dots & \dots & \dots & \dots & \dots \\ 0 & 0 & 0 & \dots & 1 & -1 \\ R_{\mathrm{RR},1} & R_{\mathrm{RR},2} & R_{\mathrm{RR},3} & \dots & R_{\mathrm{RR},N-1} & R_{\mathrm{RR},N} \end{bmatrix} \tag{2.331}$$

Due to linear dependence between tooth fluxes (Eq. 2.317), only $N - 1$ linearly independent voltage equations 2.316 can be written, the solution of which are bar currents. The Nth equation for bar currents follows from 2.318. Consequently, one can write the system of equations for bar currents and tooth fluxes as

$$\underline{I}' \cdot \frac{\mathrm{d}}{\mathrm{d}t} \underline{\Phi}_{\mathrm{t}} = -\underline{I}' \cdot \underline{R}_{\mathrm{FR}} \cdot \underline{i}_{\mathrm{FR}} - \underline{I}' \cdot \underline{R}_{\mathrm{RR}} \cdot \underline{i}_{\mathrm{RR}} + \underline{R}_{\mathrm{B}} \cdot \underline{i}_{\mathrm{B}} \tag{2.332}$$

where

$$\underline{R}_{\mathrm{B}} = \begin{bmatrix} -R_{\mathrm{B},1} & 0 & 0 & \dots & 0 & 0 & R_{\mathrm{B},N} \\ R_{\mathrm{B},1} & -R_{\mathrm{B},2} & 0 & \dots & 0 & 0 & 0 \\ 0 & R_{\mathrm{B},2} & -R_{\mathrm{B},3} & \dots & 0 & 0 & 0 \\ \dots & \dots & \dots & \dots & \dots & \dots & \dots \\ 0 & 0 & 0 & \dots & -R_{\mathrm{B},N-2} & 0 & 0 \\ 0 & 0 & 0 & \dots & R_{\mathrm{B},N-2} & -R_{\mathrm{B},N-1} & 0 \\ 1 & 1 & 1 & \dots & 1 & 1 & 1 \end{bmatrix} \quad (2.333)$$

$$\underline{R}_{\mathrm{FR}} = \mathrm{diag}\left\{R_{\mathrm{FR},1}, R_{\mathrm{FR},2}, \dots, R_{\mathrm{FR},N}\right\} \quad (2.334)$$

$$\underline{R}_{\mathrm{BR}} = \mathrm{diag}\left\{R_{\mathrm{BR},1}, R_{\mathrm{BR},2}, \dots, R_{\mathrm{BR},N}\right\} \quad (2.335)$$

and

$$\underline{\Phi}_{\mathrm{t}} = \left[\Phi_{\mathrm{t},1} \quad \Phi_{\mathrm{t},2} \quad \dots \quad \Phi_{\mathrm{t},N}\right]^{\mathrm{T}} \quad (2.336)$$

Inserting Eqs. 2.326 and 2.327 in Eq. 2.332, one becomes

$$\underline{I}' \cdot \frac{\mathrm{d}}{\mathrm{d}t}\underline{\Phi}_{\mathrm{t}} = \left(-\underline{I}' \cdot \underline{R}_{\mathrm{FR}} \cdot \underline{C}_{\mathrm{F}}^{-1} \cdot \underline{I}' - \underline{I}' \cdot \underline{R}_{\mathrm{RR}} \cdot \underline{C}_{\mathrm{R}}^{-1} \cdot \underline{I}' + \underline{R}_{\mathrm{B}}\right) \cdot \underline{i}_{\mathrm{B}} \quad (2.337)$$

Introducing matrix $\underline{R}_{\mathrm{CAGE}}$, defined as

$$\underline{R}_{\mathrm{CAGE}} = \underline{R}_{\mathrm{B}} - \underline{I}' \cdot \underline{R}_{\mathrm{FR}} \cdot \underline{C}_{\mathrm{F}}^{-1} \cdot \underline{I}' - \underline{I}' \cdot \underline{R}_{\mathrm{RR}} \cdot \underline{C}_{R}^{-1} \cdot \underline{I}' \quad (2.338)$$

one can finally write for bar currents and tooth fluxes:

$$\underline{i}_{\mathrm{B}} = \underline{R}_{\mathrm{CAGE}}^{-1} \cdot \underline{I}' \cdot \frac{\mathrm{d}}{\mathrm{d}t}\underline{\Phi}_{\mathrm{t}} \quad (2.339)$$

Introducing matrix $\underline{G}_{\mathrm{mg}}$ defined as

$$\underline{G}_{\mathrm{mg}} = \begin{bmatrix} k_{\mathrm{L},1}G_{\mathrm{S},1} & 0 & \dots & 0 & 0 & 0 & \dots & -k_{\mathrm{L},N}G_{\mathrm{S},N} \\ -k_{\mathrm{L},1}G_{\mathrm{S},1} & k_{\mathrm{L},2}G_{\mathrm{S},2} & \dots & 0 & 0 & 0 & \dots & 0 \\ \dots & \dots & \dots & \dots & \dots & \dots & \dots & \dots \\ 0 & 0 & \dots & k_{\mathrm{L},n-1}G_{\mathrm{S},n-1} & 0 & 0 & \dots & 0 \\ 0 & 0 & \dots & -k_{\mathrm{L},n-1}G_{\mathrm{S},n-1} & k_{\mathrm{L},n}G_{\mathrm{S},n} & 0 & \dots & 0 \\ 0 & 0 & \dots & 0 & -k_{\mathrm{L},n}G_{\mathrm{S},n} & k_{\mathrm{L},n+1}G_{\mathrm{S},n+1} & \dots & 0 \\ \dots & \dots & \dots & \dots & \dots & \dots & \dots & \dots \\ 0 & 0 & \dots & 0 & 0 & 0 & \dots & k_{\mathrm{L},N}G_{\mathrm{S},N} \end{bmatrix} \quad (2.340)$$

and substituting time derivatives of tooth fluxes $\Phi_{\mathrm{t},1} - \Phi_{\mathrm{t},N}$ from Eq. 2.316 in Eq. 2.247, one obtains the matrix system of differential equations

$$\underline{G}_{\mathrm{mg}} \cdot \frac{\mathrm{d}}{\mathrm{d}t}\underline{i}_{\mathrm{B}} + \underline{R}_{\mathrm{CAGE}} \cdot \underline{i}_{\mathrm{B}} = \frac{\mathrm{d}}{\mathrm{d}t}\underline{\Phi}_{\delta} \tag{2.341}$$

Assuming constant parameters and sinusoidal excitation, one can rewrite Eq. 2.339 in terms of real and imaginary components of solution as

$$\omega\underline{G}_{\mathrm{mg}} \cdot \underline{I}_{\mathrm{B,Im}} - \underline{R}_{\mathrm{CAGE}} \cdot \underline{I}_{\mathrm{B,Re}} = \omega\underline{\Phi}_{\delta,\mathrm{Re}} \tag{2.342}$$

and

$$\omega\underline{G}_{\mathrm{mg}} \cdot \underline{I}_{\mathrm{B,Re}} + \underline{R}_{\mathrm{CAGE}} \cdot \underline{I}_{\mathrm{B,Im}} = \omega\underline{\Phi}_{\delta,\mathrm{Im}} \tag{2.343}$$

where

$$\underline{I}_{\mathrm{B,Re}} = \begin{bmatrix} I_{\mathrm{B},1}\cos\varphi_1 \\ I_{\mathrm{B},2}\cos\varphi_2 \\ \cdots \\ I_{\mathrm{B},N}\cos\varphi_N \end{bmatrix}; \; \underline{I}_{\mathrm{B,Im}} = \begin{bmatrix} I_{\mathrm{B},1}\sin\varphi_1 \\ I_{\mathrm{B},2}\sin\varphi_2 \\ \cdots \\ I_{\mathrm{B},N}\sin\varphi_N \end{bmatrix} \tag{2.344}$$

and

$$\underline{\Phi}_{\delta,\mathrm{Re}} = \begin{bmatrix} \Phi_{\delta,1}\sin\gamma_1 \\ \Phi_{\delta,2}\sin\gamma_2 \\ \cdots \\ \Phi_{\delta,N}\sin\gamma_N \end{bmatrix}; \quad \underline{\Phi}_{\delta,\mathrm{Im}} = \begin{bmatrix} \Phi_{\delta,1}\cos\gamma_1 \\ \Phi_{\delta,2}\cos\gamma_2 \\ \cdots \\ \Phi_{\delta,N}\cos\gamma_N \end{bmatrix} \tag{2.345}$$

The solution of the system of Eqs. 2.338 and 2.339 yields

$$\underline{I}_{\mathrm{B,Im}} = \left(\omega^2\underline{G}_{\mathrm{mg}} \cdot \underline{R}_{\mathrm{CAGE}}^{-1} \cdot \underline{G}_{\mathrm{mg}} + \underline{R}_{\mathrm{CAGE}}\right)^{-1} \omega\left(\underline{\Phi}_{\delta,\mathrm{Im}} + \omega\underline{G}_{\mathrm{mg}} \cdot \underline{R}_{\mathrm{CAGE}}^{-1} \cdot \underline{\Phi}_{\delta,\mathrm{Re}}\right) \tag{2.346}$$

and

$$\underline{I}_{\mathrm{B,Re}} = \underline{R}_{\mathrm{CAGE}}^{-1}\omega\left(\underline{G}_{\mathrm{mg}} \cdot \underline{I}_{\mathrm{B,Im}} - \underline{\Phi}_{\delta,\mathrm{Re}}\right) \tag{2.347}$$

For air gap MMFs in Fig. 2.81, one can write

$$\Theta_{n+1} = \Theta_n + i_{\mathrm{B},n+1} \tag{2.348}$$

where $1 \leq n \leq N-1$. The continuity equation for air gap MMFs can be written as

$$\sum_{j=1}^{N} \Theta_j = 0 \tag{2.349}$$

Combining $N - 1$ Eq. 2.348 with Eq. 2.329, one obtains

$$\begin{bmatrix} 1 & 0 & 0 & \dots & 0 & -1 \\ -1 & 1 & 0 & \dots & 0 & 0 \\ 0 & -1 & 1 & \dots & 0 & 0 \\ \dots & \dots & \dots & \dots & \dots & \dots \\ 0 & 0 & 0 & \dots & 1 & 0 \\ 1 & 1 & 1 & \dots & 1 & 1 \end{bmatrix} \cdot \begin{bmatrix} \Theta_1 \\ \Theta_2 \\ \Theta_3 \\ \dots \\ \Theta_{N-1} \\ \Theta_N \end{bmatrix} = \underline{I}' \cdot \underline{i}_B \tag{2.350}$$

or

$$\underline{C}_\Theta \cdot \underline{\Theta} = \underline{I}' \cdot \underline{i}_B$$

the solution of which is

$$\underline{\Theta} = \underline{C}_\Theta^{-1} \cdot \underline{I}' \cdot \underline{i}_B \tag{2.351}$$

By comparing previous equations, one concludes that the air gap MMF distribution $\underline{\Theta}$ created by a squirrel cage with identical ring segment resistances is identical to the front ($\underline{i}_{FR}$) or rear ($\underline{i}_{RR}$) ring segment current distribution.

Case Study 2.6: The rotor of a 4-pole, 100-kW, 400-V, 50-Hz squirrel cage induction machine has a total of 40 slots. Copper bars are 5.5 mm wide and 50 mm high, and ring dimensions are 50 × 10 mm. At standstill, the sinusoidally distributed tooth flux density has an amplitude of 0.35 *T*, which corresponds to some 50 % of its no-load value.

Rotor bar current density distribution in sound cage created by the flux density distribution in Fig. 2.82 is shown in Fig. 2.83, and the front ring segment current density distribution in Fig. 2.84. Ring segment currents are 90° shifted in space relative to bar currents. Bar current density at standstill equals to 32 A/mm^2, and at rated point 5.1 A/mm^2.

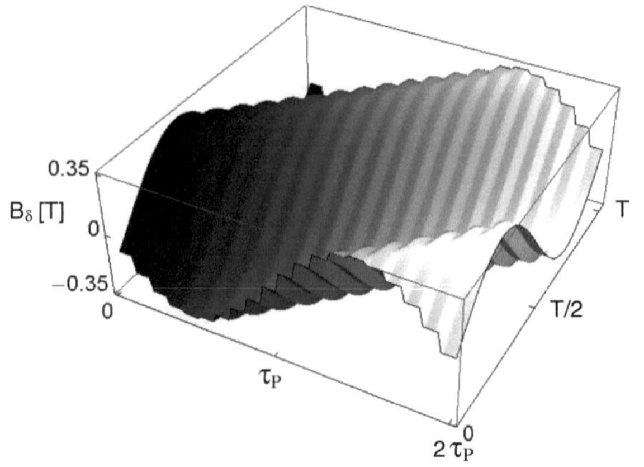

Fig. 2.82 The impressed tooth flux density distribution at short circuit

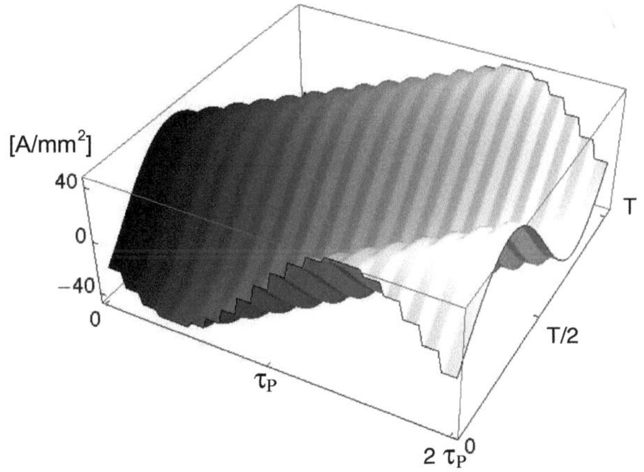

Fig. 2.83 Rotor bar current density distribution at short circuit in sound cage generated by impressed air gap fluxes in Fig. 2.82

The air gap MMF created by the rotor bar currents in Fig. 2.83 is shown in Fig. 2.85. The amplitude of air gap MMF in Fig. 2.85 is as large as the amplitude of ring current, as shown in Fig. 2.84 and equals to 40 kA.

The spectrum (in % value of the fundamental) of air gap MMF in Fig. 2.85 is shown in Fig. 2.86. The spectrum contains only the fundamental and the slot harmonics, because a symmetrical squirrel cage winding filters all harmonics besides fundamental below the order of the number of slots per pole pair decreased by 1, see Eqs. 2.188 and 2.189, and Table 2.9. Both fundamental and slot harmonics contain only positive sequence components.

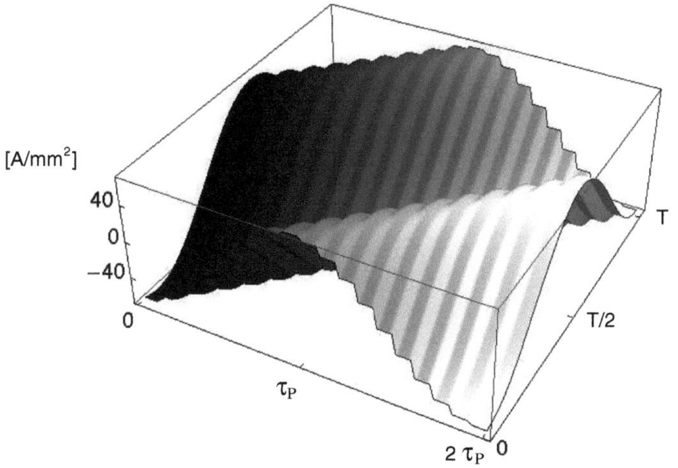

Fig. 2.84 Rotor front end ring segment current density distribution at short circuit in sound cage generated by impressed air gap fluxes in Fig. 2.67

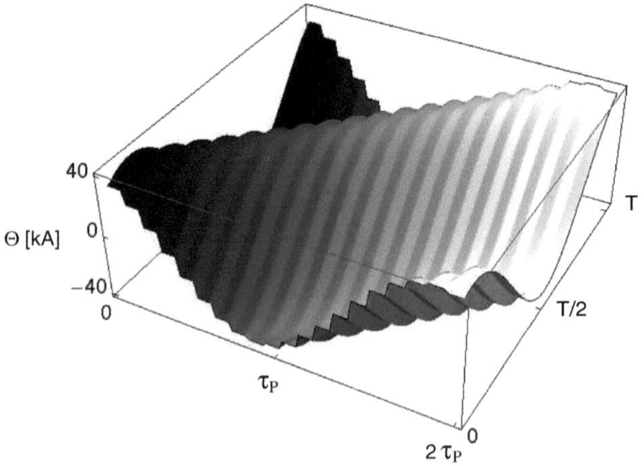

Fig. 2.85 Air gap MMF distribution created by bar currents in sound cage excited by impressed air gap fluxes in Fig. 2.82

Assume now that bars Nr. 11 and 12 are broken, i.e., without contact to the front and rear cage. The two bar currents are equal to zero, as shown in Fig. 2.87, and the front, along with rear end ring current densities in corresponding segments have equal amounts, as shown in Fig. 2.88.

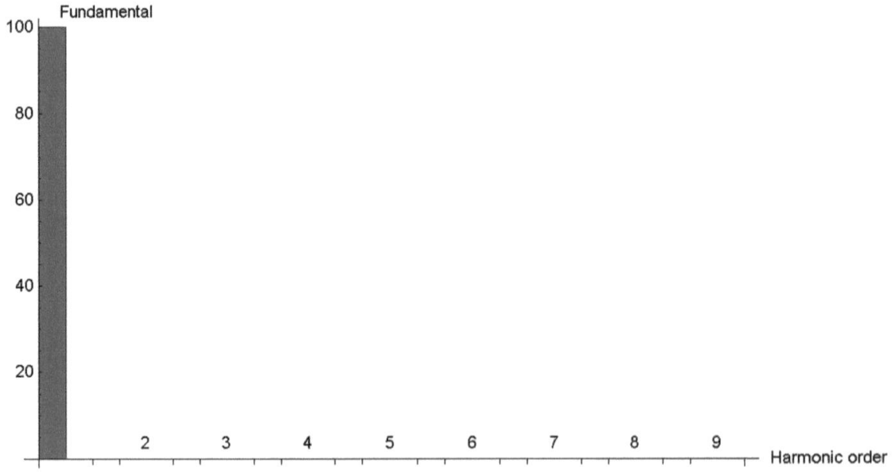

Fig. 2.86 Spectrum of air gap MMF distribution in Fig. 2.85 (in % value of the fundamental) created by bar currents in sound cage excited by impressed air gap fluxes in Fig. 2.82

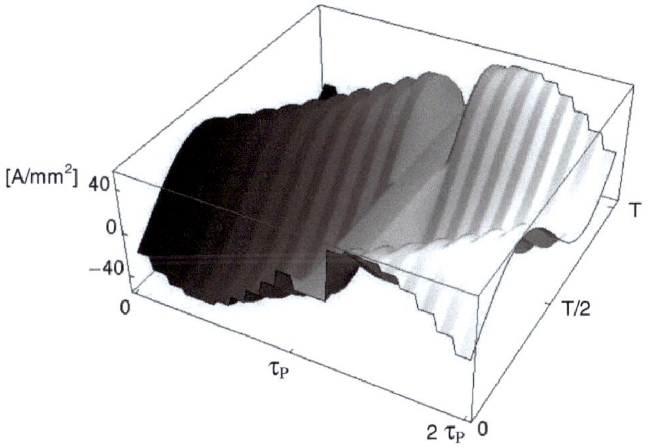

Fig. 2.87 Rotor bar current density distribution in a broken cage generated by impressed air gap flux densities in Fig. 2.82. Bars Nr. 11 and 12 are broken

The air gap MMF distribution created by currents in a broken squirrel cage is shown in Fig. 2.89. MMF of teeth surrounded by slots with broken bars is constant because there is no contribution from those slots to the MMF.

The amplitudes of bar currents are shown in Fig. 2.90. One recognizes typical increase of current amplitudes on both sides of broken bars and decrease of current amplitudes in farther bars.

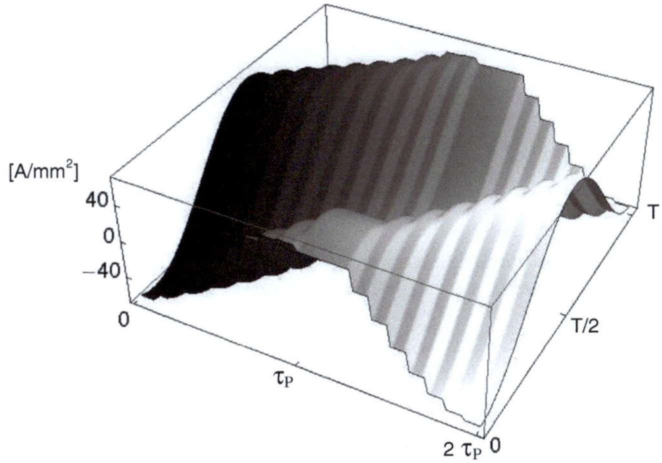

Fig. 2.88 Rotor front ring current density distribution in a broken cage generated by impressed air gap flux densities in Fig. 2.82. Since the bars Nr. 11 and 12 are broken, current densities in ring segments connected to these bars are equal to each other

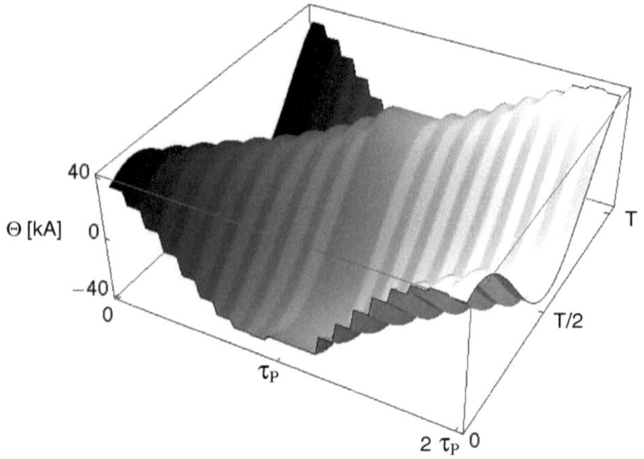

Fig. 2.89 Air gap MMF distribution created by the bar currents in a broken cage excited by impressed air gap fluxes in Fig. 2.67. Bars Nr. 11 and 12 are broken

The amplitudes of tooth MMFs are shown in Fig. 2.91. One recognizes a significant loss of MMF due to broken bars, as well as unequal amplitudes of tooth MMFs.

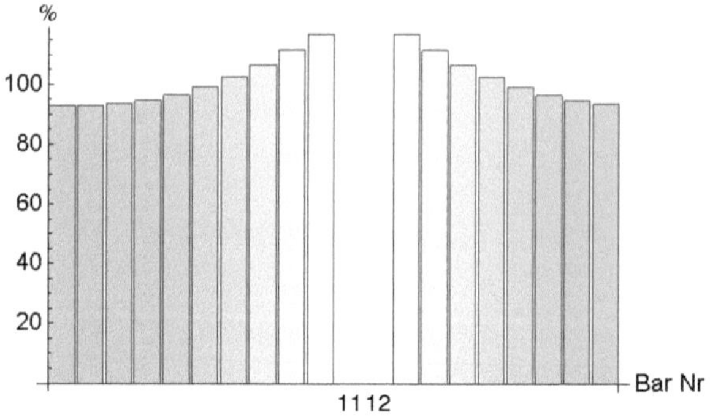

Fig. 2.90 Bar current amplitudes in a broken cage as percentage values of currents in sound cage. The cage is excited by impressed air gap flux densities shown in Fig. 2.67. Bars Nr. 11 and 12 are broken

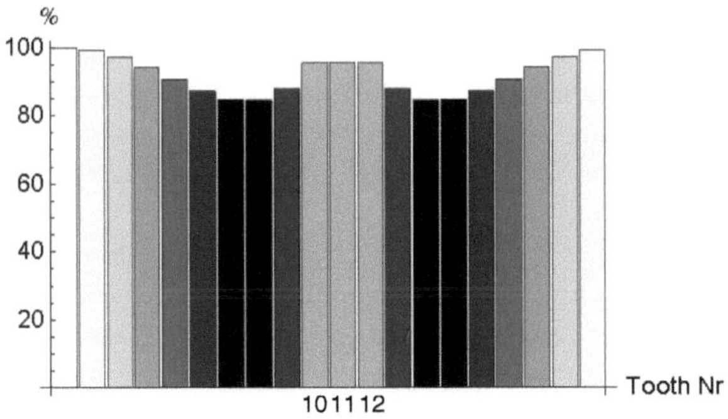

Fig. 2.91 Air gap MMF amplitudes, as percentage values of MMFs of sound cage, created by bar currents in a broken cage excited by impressed air gap flux densities in Fig. 2.82. Bars Nr. 11 and 12 are broken

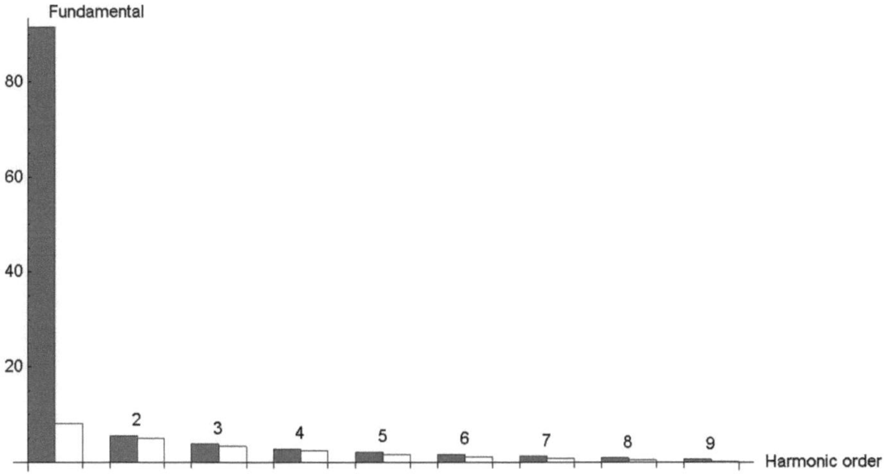

Fig. 2.92 Spectrum of the air gap MMF distribution (in % value of the fundamental of a sound cage) created by bar currents in broken cage excited by impressed air gap flux densities in Fig. 2.82. For each harmonic the maximum (*gray*) and minimum (*white*) values are given

The spectrum (in % value of the fundamental created by sound cage) of air gap MMF in Fig. 2.89 is shown in Fig. 2.92. The spectrum contains all harmonics from the interval between the fundamental and the slot harmonic, as a consequence of cage asymmetry introduced by broken bars.

The positive sequence component of the fundamental component of air gap MMF in Fig. 2.89 amounts 92.3 % of the positive sequence in a sound machine, whereas the negative sequence MMF in case of 2 broken bars increases to 2.6 % of the positive sequence in a sound machine

Case Study 2.7: The rotor of a 3.75 MVA, 50 Hz, cos φ = 0.8 lagging, salient pole synchronous machine has a damper cage with 6 slots per pole shifted 10° el. to one another. In the interpolar space, there are no bars; the front and rear end rings are continuous. The DC bar resistance equals to 82.5 μΩ and the DC ring segment resistance 5.31 μΩ. After sudden loading, the rotor oscillates for a while with a frequency of 1 Hz. During the transient, the air gap flux density is sinusoidally distributed along the periphery and has a constant amplitude of 0.75 T.

Bar currents distribution in case of *continuous ring* is shown in Fig. 2.93. One recognizes an increase of bar current amplitudes closer to pole boundaries, which compensates for a portion of missing bar currents in the interpolar space. The corresponding front end ring currents distribution is shown in Fig. 2.94, and the air gap MMF distribution in Fig. 2.95.

The amplitudes of bar currents for a given excitation are shown in Fig. 2.96. One recognizes typical increase of currents in bars closer to edges of poles.

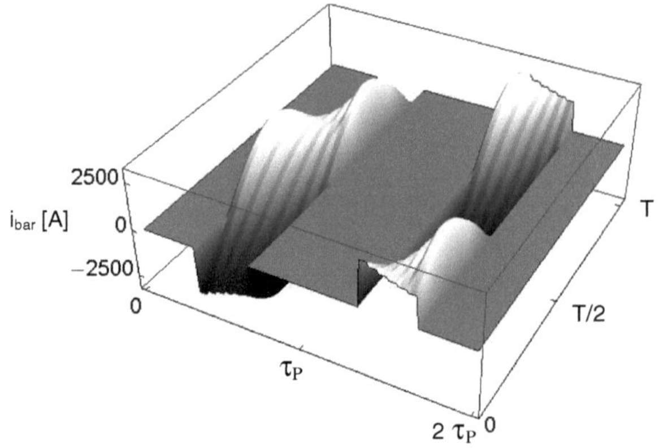

Fig. 2.93 Bar currents in a damper cage with continuous end rings, generated during load change by a 0.75 *T* air gap flux density at 1 Hz

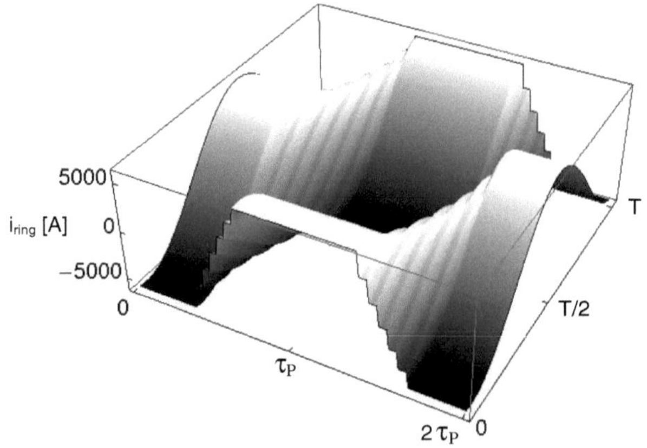

Fig. 2.94 Ring currents in a damper cage with continuous end rings, generated during load change

One of the consequences of missing bar currents in the interpolar space is constant amplitudes of tooth MMFs, as shown in Figs. 2.95 and 2.97. Denoting by 100 % the amplitudes of MMF in the interpolar space, one can represent the percentage amplitudes of spatial harmonics of air gap in the manner shown in Fig. 2.96. The amplitude of the positive sequence component of air gap in Fig. 2.97 equals to 92 %, and of the negative sequence component to 7.2 % of maximum MMF.

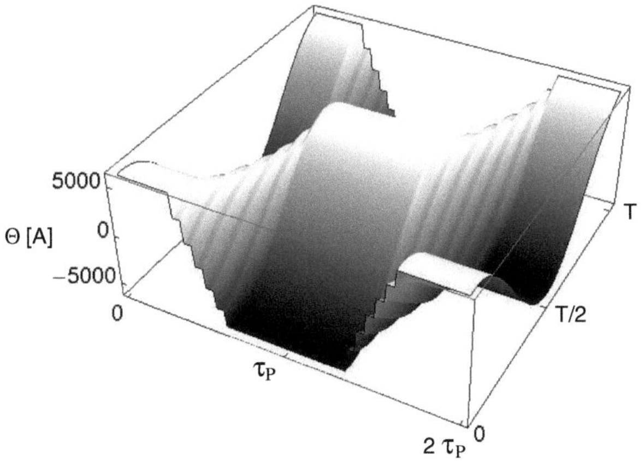

Fig. 2.95 Air gap MMF distribution created by a damper cage with continuous end rings generated during load change

Fig. 2.96 Bar current amplitudes in a damper cage with continuous end rings generated during load change

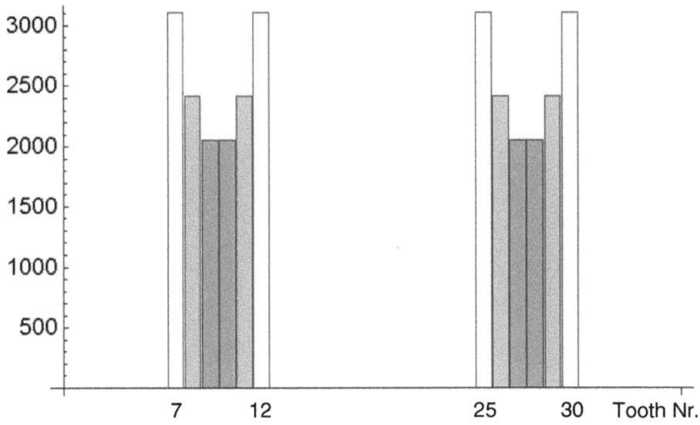

As opposed to asymmetrically broken bars in an induction machine from Case Study 2.6, which generated both odd and even spatial harmonics of MMF, the symmetrically missing bars of damper cage in this example lead to creation of odd harmonics only, as shown in Fig. 2.98.

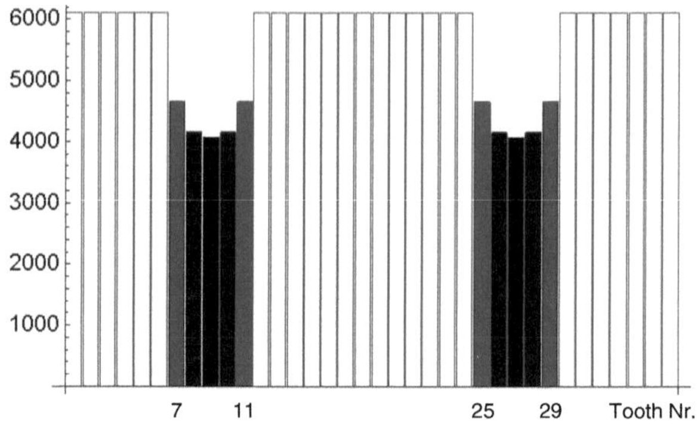

Fig. 2.97 Air gap MMF amplitudes created by currents in a damper cage with continuous end rings generated during load change

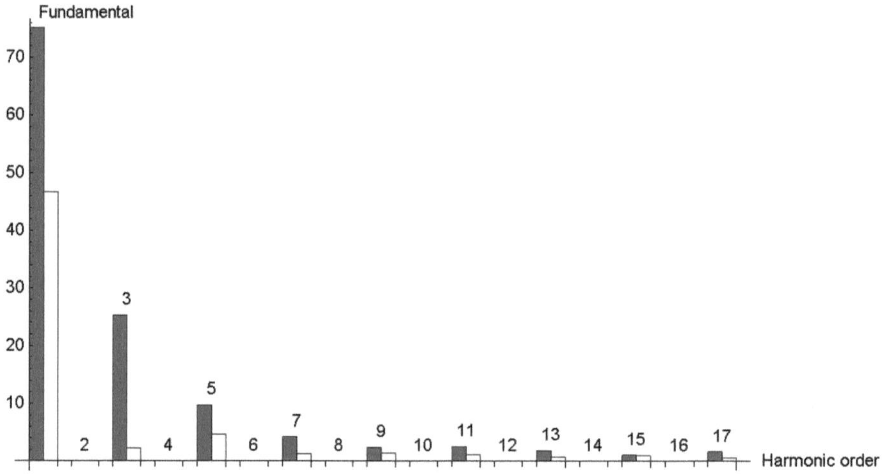

Fig. 2.98 Spectrum of the air gap MMF distribution (in % value of the largest fundamental in Fig. 2.82) created by currents in a damper cage with continuous end rings during load change. For each harmonic, the maximum (*gray*) and minimum (*white*) values are given

A continuous damper cage with bars in the interpole space creates tooth MMFs with equal amplitudes. Since in interpolar space of the analyzed machine there are no bars, the amplitude of the fundamental spatial harmonic of air gap MMF (Fig. 2.99) is obviously smaller than the amplitude of tooth harmonics in interpolar space in Fig. 2.97.

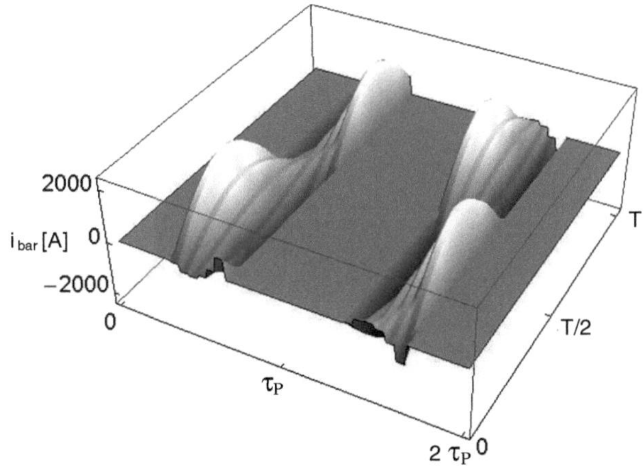

Fig. 2.99 Bar currents in a damper cage with discontinuous end rings in the interpole space during load change

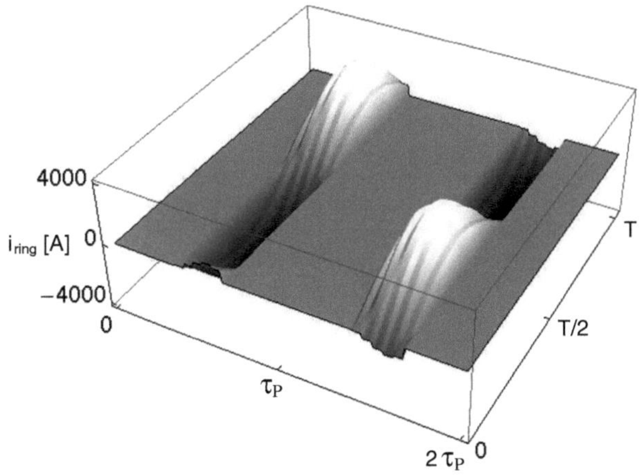

Fig. 2.100 Ring currents in a damper cage with discontinuous end rings during load change

In *discontinuous damper cage* (ring segments only on poles), the damping is not as efficient as in a continuous damper cage, since important components of ring currents are missing, Figs. 2.99 and 2.100. The missing connection between damper cage sections on salient poles results in significantly lower amplitudes of air gap MMF, as shown in Fig. 2.101. The amplitude of the positive sequence component of air gap in Fig. 2.102 equals to 14.6 %, and the amplitude of the negative sequence component to 13.2 % of the maximum MMF created by a continuous cage. The distributions of amplitudes of ring and bar segment currents of discontinuous damper cage are given in Figs. 2.102 and 2.103.

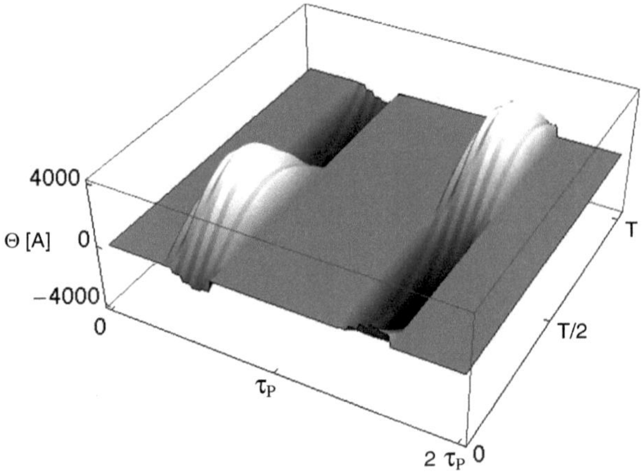

Fig. 2.101 Air gap MMF distribution created by a damper cage with discontinuous end rings during load change

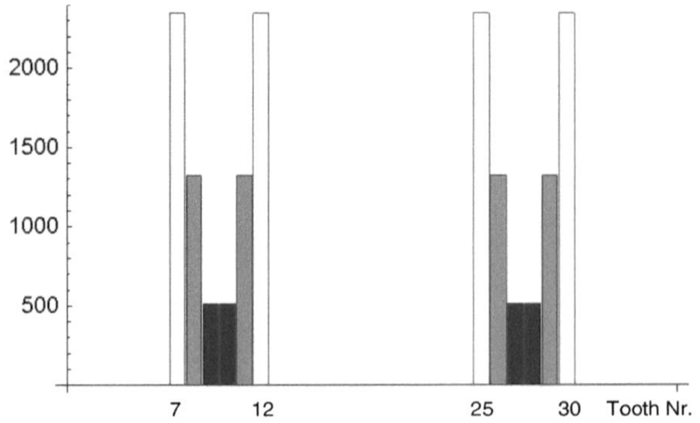

Fig. 2.102 Bar current amplitudes in a damper cage with discontinuous end rings during load change

The spectrum of air gap MMF of discontinuous cage in Fig. 2.104 shows deterioration of damper cage effects in discontinuous rings, as well as a significant increase of higher spatial harmonics of MMF created by damper cage currents.

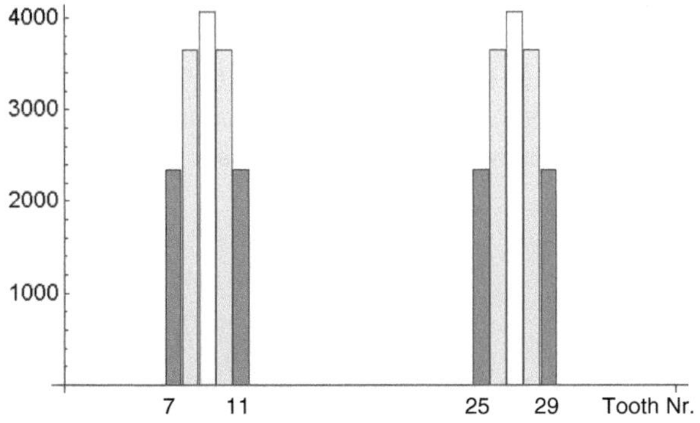

Fig. 2.103 Air gap MMF amplitudes created by currents in a damper cage with discontinuous end rings during load change. The largest MMF amplitudes are located in the center of interpole space

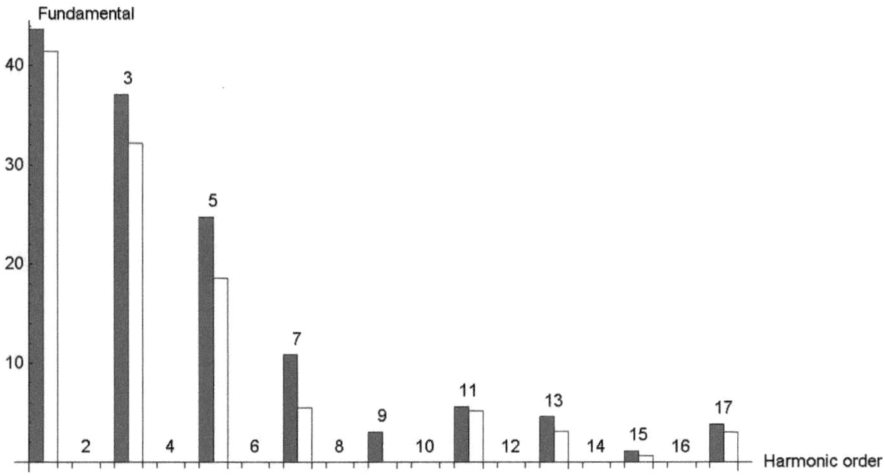

Fig. 2.104 Spectrum of the air gap MMF distribution (in % value of the largest fundamental in Fig. 2.97) created by currents in a damper cage with discontinuous end rings during load change. For each harmonic, the maximum (*gray*) and minimum (*white*) values are given

Case Study 2.8: Double-layer stator winding of a 3-phase, 281-MVA hydrogen-cooled turbogenerator 15.75 kV, 10.3 kA, two parallel circuits, cos $\varphi_r = 0.8$ is placed in 60 slots and has a pitch of 25 slots. The rotor has 14 slots per pole shifted for 9° to one another and carrying in total of 49 turns of field winding. Rated field current amounts to 4340 A and air gap is 75 mm wide. Rotor slot wedges are built of Nibrofor, a high strength copper alloy, and short-circuited on both rotor ends by means of end bells. Maximum allowed stator current unbalance is 8 %.

Negative sequence MMF caused by stator current unbalance is superimposed to the total air gap MMF. Therefore, the amount of air gap flux density created by the current unbalance of 8 % depends on the operating point of the magnetic circuit. At rated operating point the current unbalance causes a negative sequence component of the air gap flux density in the amount of 20 mT. The negative sequence component of flux density rotates at twice the synchronous speed relative to the rotor and induces in it voltages at twice the stator frequency, as shown in Fig. 2.105.

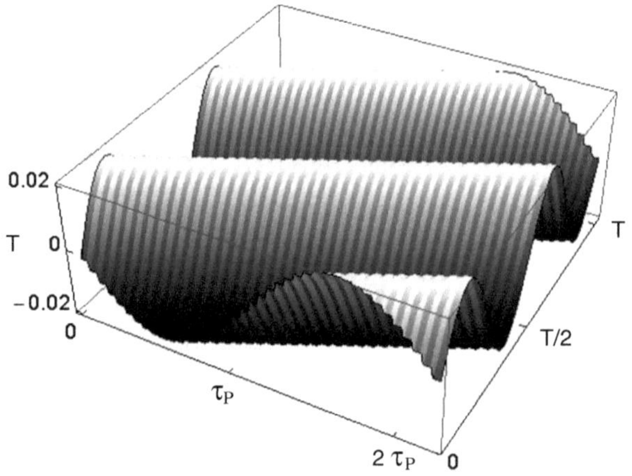

Fig. 2.105 Air gap flux density distribution at 8 % unbalanced load during one rotor revolution

The air gap flux density in Fig. 2.105 induces in damping wedges voltages which drive bar and ring currents, as shown in Figs. 2.106 and 2.107, respectively.

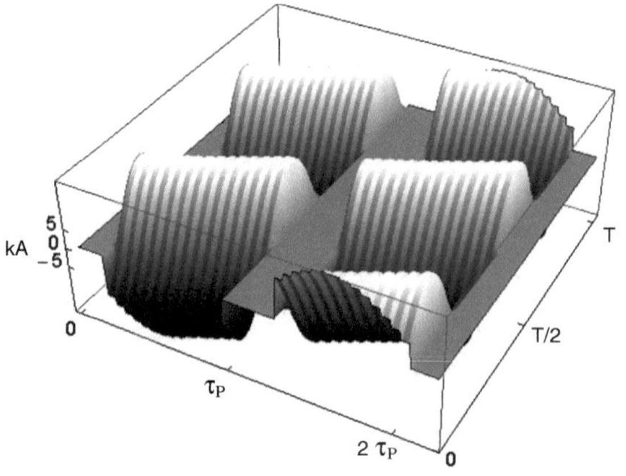

Fig. 2.106 Rotor damper bar currents at 8 % unbalanced load during one rotor revolution

The amplitudes of rotor tooth MMFs at 8 % unbalanced load are shown in Fig. 2.108. One recognizes the decreasing MMF values toward the rotor q-axis, as a consequence of the bar currents distribution in Fig. 2.109. The positive sequence component of MMF in the asymmetric distribution in Fig. 2.108 has the amount of 39.3 kA, and the negative sequence 4.7 kA. Here the positive sequence component of MMF rotates in the opposite direction to the rotor, and the negative sequence component of MMF in the direction of rotor rotation.

The spectrum of damper cage MMFs is shown in Fig. 2.110. The MMF spectrum contains only odd harmonics, since the pattern of missing bars is identical in each interpole space.

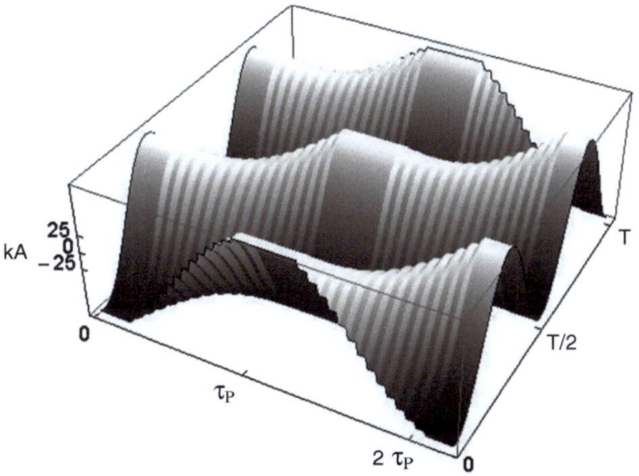

Fig. 2.107 Rotor ring currents at 8 % unbalanced load during one rotor revolution

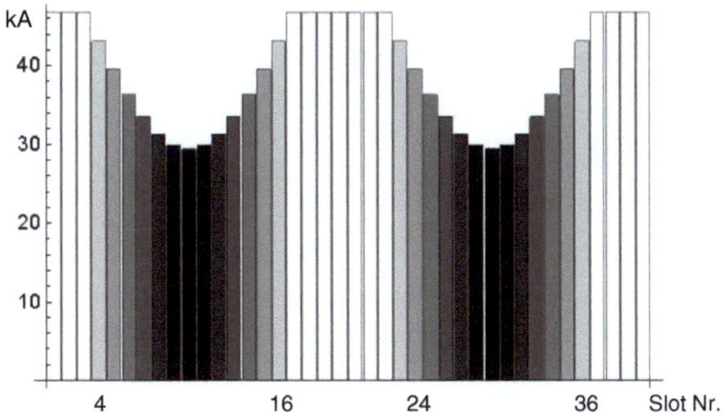

Fig. 2.108 Rotor tooth MMF amplitudes created by currents in a damper cage at 8 % unbalanced load. The largest MMF amplitudes are located in the center of poles

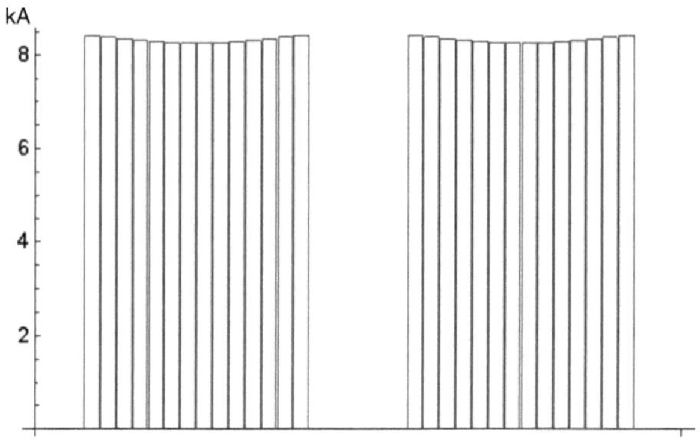

Fig. 2.109 Amplitudes of rotor bar currents corresponding to rotor tooth MMF in Fig. 2.108

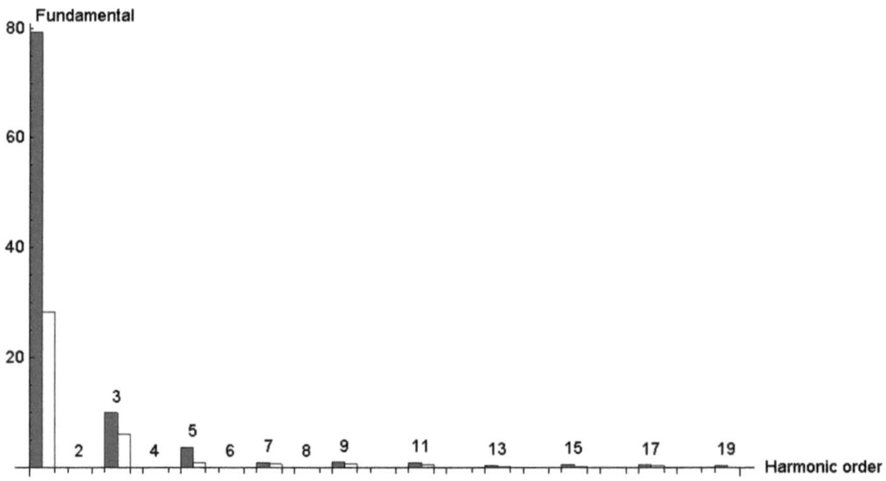

Fig. 2.110 Spectrum of the air gap MMF distribution (in % value of the largest fundamental in Fig. 2.82) created by currents in a damper cage at 8 % unbalanced load. For each harmonic, the maximum (*gray*) and minimum (*white*) values are given

2.10 Winding Failures

As illustrated in Case Study 2.6, an asymmetry of a squirrel cage winding in the form of broken bars is a source of higher harmonic components in the air gap MMF created by the squirrel cage. Very often broken bars are accompanied with pulsating torques and an increased noise level.

In case of field windings of synchronous machines, an interturn fault is a source of:

Decrease of the fundamental term of field created MMF, because short-circuited coils do not contribute to the MMF. The amount of decrease of current sheet can be quantified by means of Fig. 2.111 in which the rotor of a two-pole turbogenerator is shown. Denoting the maximum number of field coils with N_c and numbering the short-circuited coil by n, $1 \leq n \leq N_c$, one can write for the total effective number of turns w_{eff}

$$w_{eff} = \frac{1}{N_c w_c}\sqrt{\left[\sum_{i=1}^{n-1}\cos(i-1)\alpha_{el} + \sum_{i=n+1}^{N_c}\cos(i-1)\alpha_{el}\right]^2 + \left[\sum_{i=1}^{n-1}\sin(i-1)\alpha_{el} + \sum_{i=n+1}^{N_c}\sin(i-1)\alpha_{el}\right]^2} \tag{2.352}$$

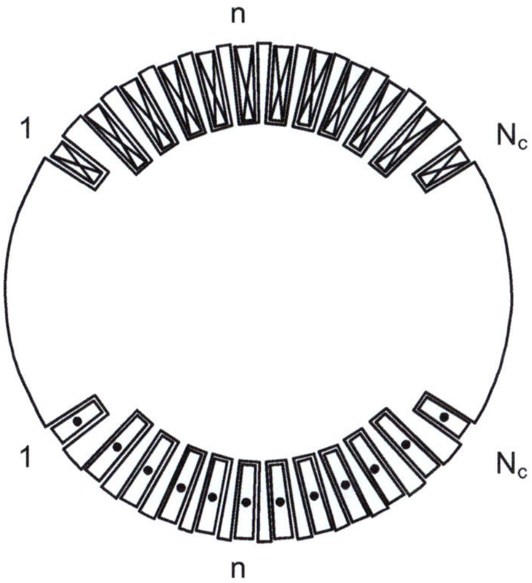

Fig. 2.111 Rotor of a two-pole turbogenerator with N_c coils. The nth coil is short-circuited

In Fig. 2.112 the winding factors of a field winding after Fig. 2.111 with 12 coils, the angle between which is 7.5° and one short-circuited coil are shown. The strongest effect on the field MMF is obtained by short-circuiting the coils in the center of the winding, i.e., slots number 6 and 7;

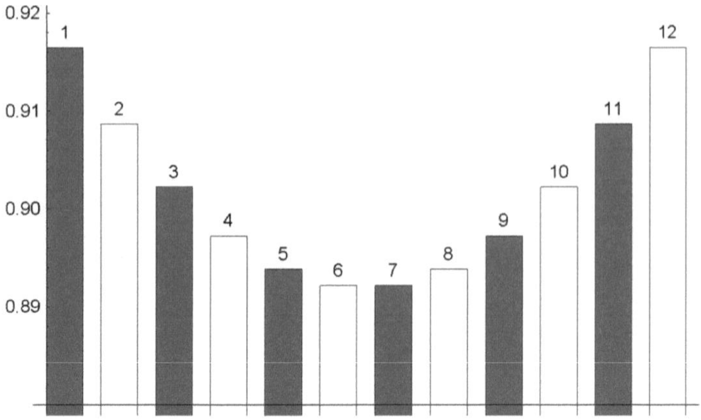

Fig. 2.112 Winding factors for a field winding with 12 coils, the angle between which is 7.5°, and one coil short-circuited

Even harmonics of MMF, which are reflected in induced voltages;

Asymmetric heating of the rotor winding, which eventually leads to rotor bowing; and

*Decrease of the field winding main and leakage inductance*s.

References

1. Ostovic V (1994) Computer-Aided Analysis of Electric Machines: a Mathematica Approach. Prentice-Hall, London
2. Richter R (1952) Lehrbuch der Wicklungen elektrischer Maschinen. Verlag G. Braun, Karlsruhe

Chapter 3
Magnetic Circuit

Contents

Some useful procedures for evaluation of field distribution in iron and air portions of magnetic circuits of electric machines are introduced in this chapter. A straightforward way for determination of flux distribution in air gap and slots based on properties of geometric series is discussed. A simple algorithm for closed form solution of magnetic circuits with nonlinear B–H curve is discussed. An accelerated procedure for solution of differential algebraic equations is applied to machine voltage equations.

The solution procedures are illustrated by an example of flux distribution in a slot and air gap of an electric machine, as well as 3-dimensional analysis of magnetic circuit of a claw pole (Lundell) generator.

3.1 A Straightforward Method for the Solution of Flux Distribution in Current-Free Air Gap and Slots

Air gap is the most important portion of an electric machine, because the air gap flux density distribution determines machine's most crucial parameters—the induced voltage and the electromagnetic torque. Both these quantities are directly dependent on the air gap flux density distribution. In the previous chapter, the harmonics of air gap flux density for a given slotting and MMF distributions were

V. Ostović, *The Art and Science of Rotating Field Machines Design*,
DOI 10.1007/978-3-319-39081-9_3

analyzed, without going into details of field distribution determined by variable geometry between stator and rotor iron surfaces, as shown in Fig. 3.1.

From the point of view of magnetic circuit representation, air gap differs from other machine portions because the flux lines in it cannot be canalized as in a tooth, or yoke segment between two teeth. Flux lines distribution in the air gap is a function of excitation and geometry for a given rotor shift, and computational tools of different art have to be employed in order to relate the air gap flux density to the MMF on its borders.

Here an analytical procedure will be presented for the determination of air gap flux density distribution based on sourcelessness of magnetic field

$$\operatorname{div}\vec{B} = 0 \tag{3.1}$$

which also can be written as

$$\frac{\partial}{\partial x}B_x + \frac{\partial}{\partial y}B_y + \frac{\partial}{\partial z}B_z = 0 \tag{3.2}$$

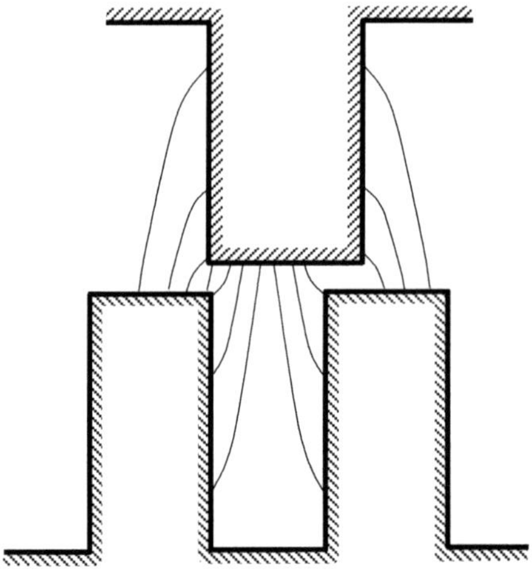

Fig. 3.1 Approximate field distribution in current-free air gap and slots

or

$$\frac{\partial}{\partial x}(\mu_x H_x) + \frac{\partial}{\partial y}(\mu_y H_y) + \frac{\partial}{\partial z}(\mu_z H_z) = 0 \tag{3.3}$$

Setting for $\mu_x = \mu_y = \mu_z = \mu_0$, one obtains

$$\frac{\partial}{\partial x}H_x + \frac{\partial}{\partial y}H_y + \frac{\partial}{\partial z}H_z = 0 \tag{3.4}$$

Being a vector quantity, the magnetic field strength $\vec{H}$ can be expressed in terms of another scalar or vector function, which formally satisfies the rules of vector analysis, not necessarily having certain physical meaning. Such scalar or vector function is called *magnetic potential* [1]. Both scalar and vector potentials have found wide area of applications in magnetism. Whereas the scalar magnetic potential φ could be interpreted in terms of MMF drop Θ between two points, i.e., $\Theta_{12} = \varphi_1 - \varphi_2$, there exists no similar physical interpretation for the vector magnetic potential $\vec{A}$.

The choice of the type of potential optimally suited for a particular problem is dependent on the character of the problem. In a current-free medium, the magnetic field strength H can be represented only as a gradient of magnetic scalar potential φ, which means further [1]

$$\frac{\partial^2 \varphi}{\partial x^2} + \frac{\partial^2 \varphi}{\partial y^2} + \frac{\partial^2 \varphi}{\partial z^2} = 0 \tag{3.5}$$

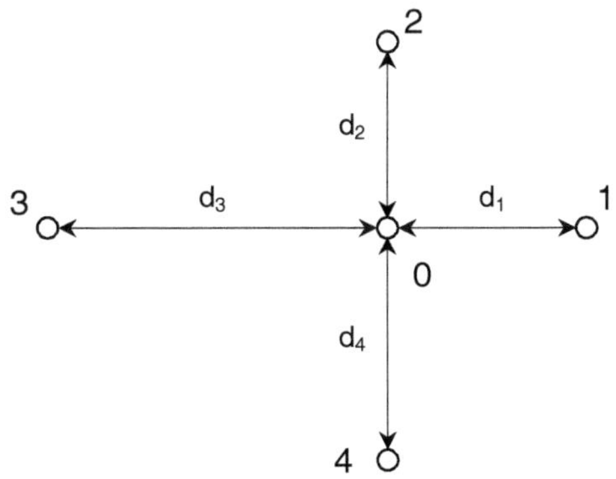

Fig. 3.2 Rectangular mesh for evaluation of magnetic scalar potential

As long as the effects in the end zone can be neglected, the field in the active part of an electric machine is two-dimensional. Therefore,

$$\frac{\partial^2 \varphi}{\partial x^2} + \frac{\partial^2 \varphi}{\partial y^2} = 0 \tag{3.6}$$

The magnetic scalar potential distribution (3.6) can be discretized in Cartesian coordinate system in the manner shown in Fig. 3.2. Along the distance d_1 in Fig. 3.2 one can discretize the first derivative of magnetic scalar potential as

$$\left(\frac{\partial\varphi}{\partial x}\right)_{d_1} = \frac{\varphi_1 - \varphi_0}{d_1} \tag{3.7}$$

and along d_3, similarly

$$\left(\frac{\partial\varphi}{\partial x}\right)_{d_3} = \frac{\varphi_0 - \varphi_3}{d_3} \tag{3.8}$$

from which one can express the second derivative of scalar magnetic potential w.r.t. coordinate x at point 0 as

$$\frac{\partial^2\varphi}{\partial x^2} = \frac{\left(\frac{\partial\varphi}{\partial x}\right)_{d_1} - \left(\frac{\partial\varphi}{\partial x}\right)_{d_3}}{\frac{d_1+d_3}{2}} = 2\frac{(\varphi_1 - \varphi_0)d_3 + (\varphi_3 - \varphi_0)d_1}{d_1 d_3 (d_1 + d_3)} \tag{3.9}$$

and analogously for the coordinate y:

$$\frac{\partial^2\varphi}{\partial y^2} = 2\frac{(\varphi_2 - \varphi_0)d_4 + (\varphi_4 - \varphi_0)d_2}{d_2 d_4 (d_2 + d_4)} \tag{3.10}$$

Now, one can write discretized Eq. 3.6 as

$$\frac{(\varphi_1 - \varphi_0)d_3 + (\varphi_3 - \varphi_0)d_1}{d_1 d_3 (d_1 + d_3)} + \frac{(\varphi_2 - \varphi_0)d_4 + (\varphi_4 - \varphi_0)d_2}{d_2 d_4 (d_2 + d_4)} = 0 \tag{3.11}$$

or

$$\varphi_0 = \frac{1}{(d_1 d_3 + d_2 d_4)}\left[\frac{d_2 d_4}{d_1 + d_3}(\varphi_1 d_3 + \varphi_3 d_1) + \frac{d_1 d_3}{d_2 + d_4}(\varphi_2 d_4 + \varphi_4 d_2)\right] \tag{3.12}$$

In case of equidistant mesh along the x-axis, $d_1 = d_3 = d_x$, and y-axis, $d_2 = d_4 = d_y$, one can write

$$\varphi_0 = \frac{(\varphi_1 + \varphi_3)d_y^2 + (\varphi_2 + \varphi_4)d_x^2}{2\left(d_x^2 + d_y^2\right)} \tag{3.13}$$

Introducing auxiliary variables c_x and c_y defined as

$$c_x = \frac{d_x^2}{2\left(d_x^2 + d_y^2\right)} \tag{3.14}$$

$$c_y = \frac{d_y^2}{2\left(d_x^2 + d_y^2\right)} \tag{3.15}$$

One can express the potential φ_0 at central point as

$$\varphi_0 = c_y\varphi_1 + c_x\varphi_2 + c_y\varphi_3 + c_x\varphi_4 \tag{3.16}$$

If $d_x = d_y$, one can write

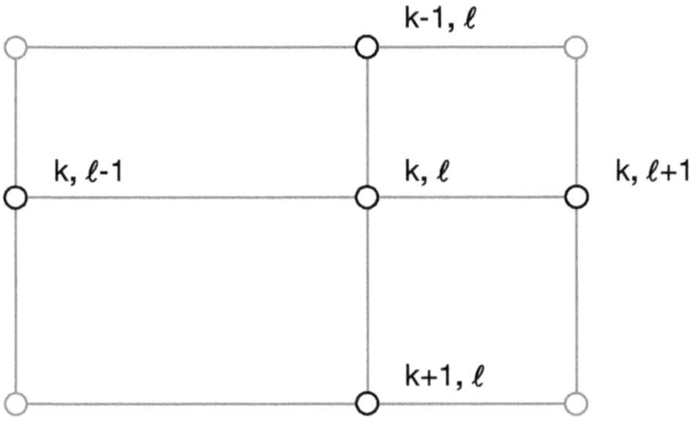

Fig. 3.3 Rectangular mesh for evaluation of scalar magnetic potential

$$\varphi_0 = \frac{\varphi_1 + \varphi_2 + \varphi_3 + \varphi_4}{4} \tag{3.17}$$

which is an intuitively comprehensible result.

In order to obtain a given accuracy, the computation of scalar potential after Eq. 3.12 is performed iteratively. The value of scalar potential at nth step of iteration at a point with coordinates k (vertical) and ℓ (horizontal), as shown in Fig. 3.3, can be expressed in terms of potential of points around it as

$$\varphi_{k,l,n} = c_y\varphi_{k,l-1,n-1} + c_x\varphi_{k-1,l,n} + c_y\varphi_{k,l-1,n} + c_x\varphi_{k-1,l,n-1} \tag{3.18}$$

when evaluating potentials from left to right and up to down. The values of scalar potentials at nth step of iteration can be expressed in terms of scalar potentials at $(n-1)$th step of iteration as

$$\varphi_{k,l,n} - c_x\varphi_{k-1,l,n} - c_y\varphi_{k,l-1,n} = c_y\varphi_{k,l-1,n-1} + c_x\varphi_{k-1,l,n-1} \tag{3.19}$$

and in matrix form as

$$\underline{C}'_1 \cdot \underline{\varphi}_n = \underline{C}'_2 \cdot \underline{\varphi}_{n-1} \tag{3.20}$$

with $\underline{C}'_1$ and $\underline{C}'_2$ denoting matrices of coefficients, $\underline{\varphi}_n$ the vector of scalar potentials at nth step of iteration, and $\underline{\varphi}_{n-1}$ the vector of scalar potentials at ($n-1$)th step of iteration. Some points in computational region can have fixed potentials, e.g., through boundary conditions. Denoting by $\underline{\varphi}_B$ the vector of scalar potentials on the region boundary, and by $\underline{B}$ the matrix of boundary conditions, one can rewrite Eq. 3.20 as

$$\underline{C}_1 \cdot \underline{\varphi}_n = \underline{C}_2 \cdot \underline{\varphi}_{n-1} + \underline{B} \cdot \underline{\varphi}_B \tag{3.21}$$

or

$$\underline{\varphi}_n = \underline{C}_1^{-1} \cdot \left(\underline{C}_2 \cdot \underline{\varphi}_{n-1} + \underline{B} \cdot \underline{\varphi}_B\right) \tag{3.22}$$

Analogously, one can write

$$\underline{\varphi}_{n-1} = \underline{C}_1^{-1} \cdot \left(\underline{C}_2 \cdot \underline{\varphi}_{n-2} + \underline{B} \cdot \underline{\varphi}_B\right) \tag{3.23}$$

$$\cdots$$

$$\underline{\varphi}_2 = \underline{C}_1^{-1} \cdot \left(\underline{C}_2 \cdot \underline{\varphi}_1 + \underline{B} \cdot \underline{\varphi}_B\right) \tag{3.24}$$

and

$$\underline{\varphi}_1 = \underline{C}_1^{-1} \cdot \left(\underline{C}_2 \cdot \underline{\varphi}_0 + \underline{B} \cdot \underline{\varphi}_B\right) \tag{3.25}$$

with $\underline{\varphi}_0$ denoting the vector of initial assumption of scalar potentials. Now, one can substitute Eq. 3.25 back in 3.24:

$$\underline{\varphi}_2 = \underline{C}_1^{-1} \cdot \left[\underline{C}_2 \cdot \underline{C}_1^{-1} \cdot \left(\underline{C}_2 \cdot \underline{\varphi}_0 + \underline{B} \cdot \underline{\varphi}_B\right) + \underline{B} \cdot \underline{\varphi}_B\right] \tag{3.26}$$

or

$$\underline{\varphi}_2 = \left(\underline{C}_1^{-1} \cdot \underline{C}_2\right)^2 \cdot \underline{\varphi}_0 + \underline{C}_1^{-1} \cdot \left(\underline{C}_2 \cdot \underline{C}_1^{-1} + \underline{I}\right) \cdot \underline{B} \cdot \underline{\varphi}_B \tag{3.27}$$

Analogously, one can write

$$\underline{\varphi}_3 = \left(\underline{C}_1^{-1} \cdot \underline{C}_2\right)^3 \cdot \underline{\varphi}_0 + \underline{C}_1^{-1} \cdot \left[\underline{C}_2 \cdot \underline{C}_1^{-1} \cdot \left(\underline{C}_2 \cdot \underline{C}_1^{-1} + \underline{I}\right) + \underline{I}\right] \cdot \underline{B} \cdot \underline{\varphi}_B \tag{3.28}$$

and

$$\underline{\varphi}_4 = (\underline{C}_1^{-1} \cdot \underline{C}_2)^4 \cdot \underline{\varphi}_0 + \underline{C}_1^{-1} \cdot \left[(\underline{C}_2 \cdot \underline{C}_1^{-1})^3 + (\underline{C}_2 \cdot \underline{C}_1^{-1})^2 + \underline{C}_2 \cdot \underline{C}_1^{-1} + \underline{I} \right] \cdot \underline{B} \cdot \underline{\varphi}_B \tag{3.29}$$

The vector $\underline{\varphi}_n$ of scalar potentials at *n*th step of iteration can now be written as

$$\underline{\varphi}_n = (\underline{C}_1^{-1} \cdot \underline{C}_2)^n \cdot \underline{\varphi}_0 + \left[\underline{C}_1^{-1} \cdot \sum_{i=1}^{n} (\underline{C}_2 \cdot \underline{C}_1^{-1})^{i-1} \right] \cdot \underline{B} \cdot \underline{\varphi}_B \tag{3.30}$$

One recognizes in square brackets of Eq. 3.30 a sum of geometric series, which also can be written as

$$\sum_{i=1}^{n} (\underline{C}_2 \cdot \underline{C}_1^{-1})^{i-1} = \left[\underline{I} - (\underline{C}_2 \cdot \underline{C}_1^{-1})^n \right] \cdot (\underline{I} - \underline{C}_2 \cdot \underline{C}_1^{-1})^{-1} \tag{3.31}$$

Now one can write the expression for direct computation of scalar magnetic potentials as

$$\underline{\varphi}_n = (\underline{C}_1^{-1} \cdot \underline{C}_2)^n \cdot \underline{\varphi}_0 + \underline{C}_1^{-1} \cdot \left[\underline{I} - (\underline{C}_2 \cdot \underline{C}_1^{-1})^n \right] \cdot (\underline{I} - \underline{C}_2 \cdot \underline{C}_1^{-1})^{-1} \cdot \underline{B} \cdot \underline{\varphi}_B \tag{3.32}$$

Since

$$|\underline{C}_1^{-1} \cdot \underline{C}_2| \leq \underline{I} \ \& |\underline{C}_2 \cdot \underline{C}_1^{-1}| \leq \underline{I} \tag{3.33}$$

one can write for n $\rightarrow \infty$:

$$\underline{\varphi}_n = \underline{C}_1^{-1} \cdot (\underline{I} - \underline{C}_2 \cdot \underline{C}_1^{-1})^{-1} \cdot \underline{B} \cdot \underline{\varphi}_B \tag{3.34}$$

As expected, the scalar magnetic potential distribution is dependent on the problem geometry and boundary conditions and independent of the assumption of initial values of $\underline{\varphi}_0$.

Case Study 3.1: Use the scalar magnetic potential distribution to calculate Carter factor of the rectangular slot in Fig. 3.4. Due to symmetry conditions, only one half of slot has to be considered.

Scalar magnetic potential distribution calculated by using Eq. 3.34 for boundary conditions $\varphi_1 = -1$ p.u. and $\varphi_2 = 1$ p.u. is shown in Fig. 3.5. One recognizes an imprint of a portion of the lower teeth from Fig. 3.4 in the potential distribution in Fig. 3.5.

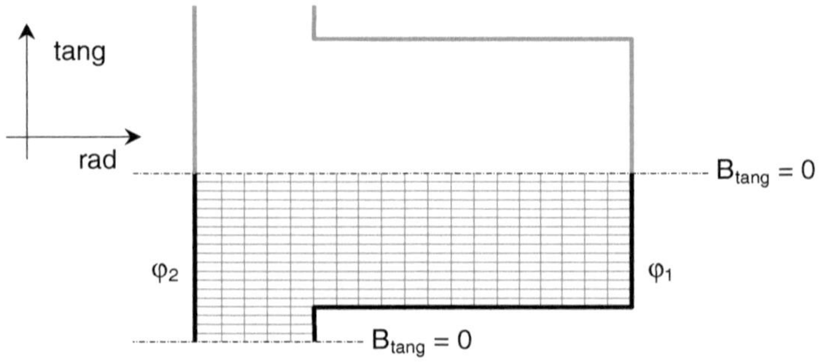

Fig. 3.4 Slot and gap geometry with computational mesh

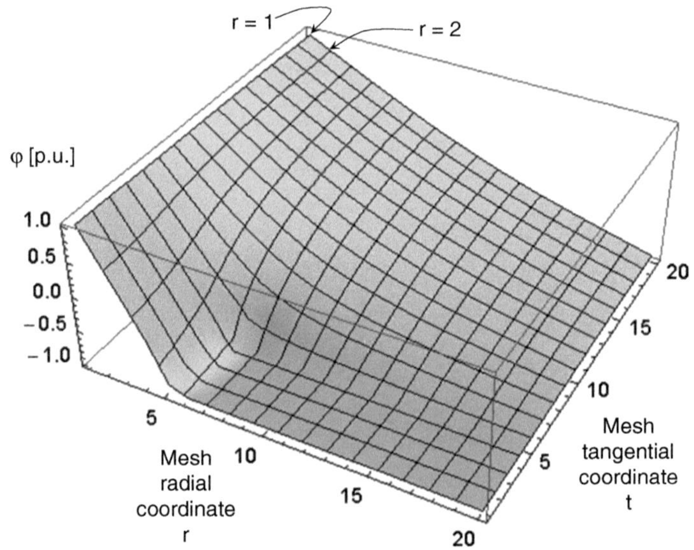

Fig. 3.5 Scalar magnetic potential distribution for geometry in Fig. 3.4 and $\varphi_1 = -1$ p.u.; $\varphi_2 = 1$ p.u

Fig. 3.6 P.u. values of the radial component of flux density at $r = 1$ as a function of tangential coordinate t in Fig. 3.5

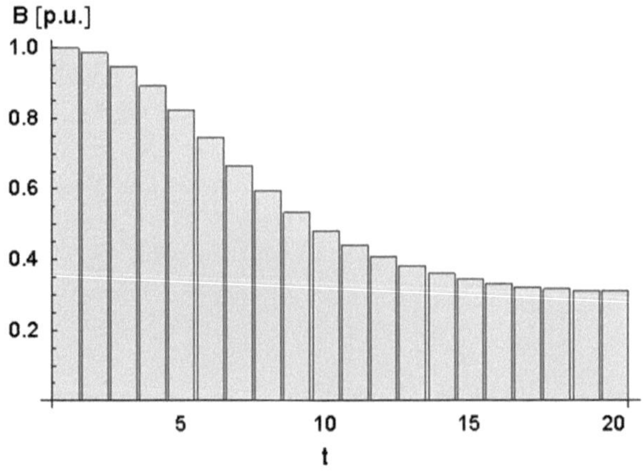

Whereas the scalar magnetic potential along the mesh radial coordinate $r = 1$ in Fig. 3.5 is set to 1 (boundary condition), its value along $r = 2$ is a function of boundary conditions and problem geometry. Since the derivative of scalar magnetic potential in air is proportional to the corresponding flux density component, one can represent the p.u. values of the radial component of flux density at radial coordinate $r = 1$ in the manner shown in Fig. 3.6.

The ratio between the maximum flux density (here 1 p.u.) and an average of 20 values along the coordinate t is equal to the Carter factor, which for the values in Fig. 3.6 equals to 1.787. When calculated by using the procedure introduced in Chap. 2, the Carter factor amounts to 1.704.

3.2 A Straightforward Method for the Solution of Flux Distribution in Air Gap and Slots with Current-Carrying Conductors

In the previous section, the scalar magnetic potential φ was introduced, which satisfies equation [1]:

$$\vec{H} = -\text{grad}\,\varphi \tag{3.35}$$

Due to the properties of vector functions, scalar magnetic potential φ can be defined only in a current-free medium. Following Ampère's circuital law

$$\text{curl}\,\vec{H} = \vec{J} \tag{3.36}$$

and inserting for magnetic field strength the substitute as defined in Eq. 3.35, one obtains

$$\text{curl}(-\text{grad}\,\varphi) = \vec{J} \tag{3.37}$$

which is a nonsense, because curl of the gradient of any scalar function is equal to zero

$$\text{curl}(-\text{grad}\,\varphi) = 0 \tag{3.38}$$

Obviously, a vector function has to be introduced for problems with current different from zero. This function is the *magnetic vector potential* $\vec{A}$, defined as

$$\vec{B} = \text{curl}\vec{A} \tag{3.39}$$

As opposed to the discretely distributed current density $\vec{J}$, magnetic vector potential $\vec{A}$ is a continuous function spread over the whole problem region, as shown in Fig. 3.7.

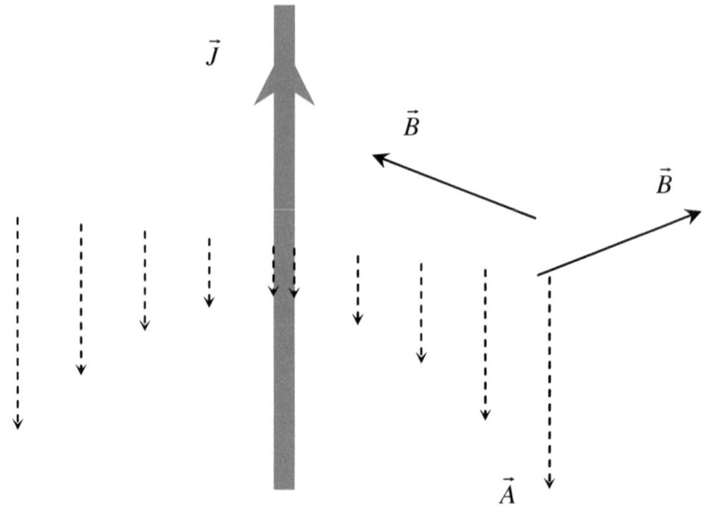

Fig. 3.7 Illustrating relationships between current density $\vec{J}$, flux density $\vec{B}$, and magnetic vector potential $\vec{A}$ affiliated to $\vec{J}$. Vectors $\vec{A}$ and $\vec{J}$ are shifted 180° to each other. The value of A at the center of conductor is set to zero ($A_0 = 0$)

Magnetic vector potential affiliated to a straight current-carrying conductor coincident with the z-axis on distance r from the conductor center can be expressed as [2]

$$A_z = -\mu_0 \frac{I}{4\pi} \cdot \frac{r^2}{r_0^2} + A_0 \tag{3.40}$$

for $r \leq r_0$, where r_0 denotes the conductor radius and I the current through it.

For $r \geq r_0$, the magnetic vector potential is equal to

$$A_z = -\mu_0 \frac{I}{2\pi} \cdot \left[\ln\left(\frac{r_0}{r}\right) - \frac{1}{2} \right] + A_0 \tag{3.41}$$

Magnetic vector potential is determined up to the constant of integration A_0, which cannot be reconstructed from physical conditions and, therefore, remains arbitrary. This does not limit the field of application of magnetic vector potential, because not its value at a point, but the difference of magnetic vector potentials at two points (which does not contain any more the constant of integration A_0) is decisive for determination of magnetic field.

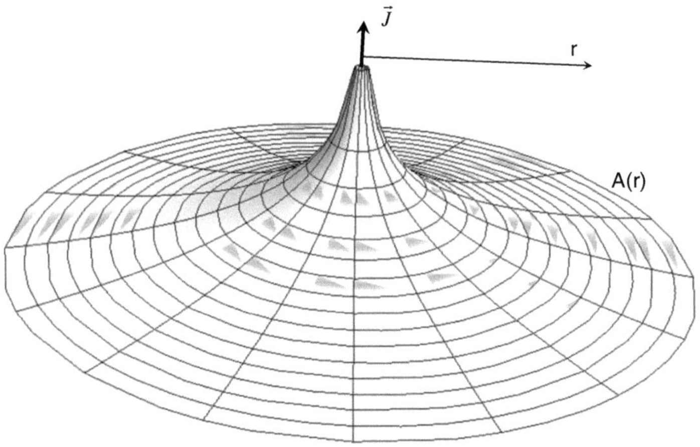

Fig. 3.8 Amplitude of magnetic vector potential affiliated to a single conductor as a function of distance r to the conductor center

The dependence of magnetic vector potential of a single current-carrying conductor on the radial distance r from the center of the conductor is shown in Fig. 3.8.

Since the amplitude of flux density B on the conductor surface is equal to

$$B_{r=r_0} = \mu_0 \frac{I}{2r_0\pi} \tag{3.42}$$

one can write for the amount of vector magnetic potential on the conductor surface

$$A_{r=r_0} = A_0 - \mu_0 \frac{I}{4\pi} = A_0 - \frac{1}{2} r_0 B_{r=r_0} \tag{3.43}$$

Magnetic vector potential is a handy quantity which helps define flux Φ through surface S

$$\Phi = \iint \vec{B} \cdot \mathrm{d}\vec{S} = \iint \left(\nabla \times \vec{A}\right) \cdot \mathrm{d}\vec{S} \tag{3.44}$$

or, by using Stoke's theorem

$$\Phi = \oint_C \vec{A} \cdot \mathrm{d}\vec{l} \tag{3.45}$$

where C is the contour bounding the surface S.

For two-dimensional problems in the $(x,\ y)$ plane, where magnetic vector potential has only the z-component, the flux Φ_{P-Q} between points P and Q is equal to

$$\Phi_{P-Q} = l_z(A_P - A_Q) \tag{3.46}$$

with ℓ_z denoting the length in z-direction, A_P the value of magnetic vector potential at point P, and A_Q at point Q. Since magnetic vector potential in case of two-dimensional problems has only the z-component, the distance between points P and Q is irrelevant for computation of flux Φ in Eq. 3.46.

Equation 3.46 helps one relate magnetic flux to magnetic vector and scalar potentials in the manner shown in Fig. 3.9. Here a two-dimensional flux tube is depicted, characterized by flux lines limited within the boundaries $A = A_P$ and $A = A_Q$. Since no flux line crosses the boundaries $A = A_P$ and $A = A_Q$, a curve defined by $A =$ const. has the properties of a flux line.

Magnetic vector potential created by two current-carrying conductors in a medium with relative permeability equal to one, both with radius r_0, placed at origin and point $(d, 0)$ parallel to the z-axis can be expressed as [2]

$$A(x,y) = \frac{\mu_0 I}{4\pi} \ln \frac{(d-x)^2 + y^2}{x^2 + y^2} \tag{3.47}$$

for a point outside both conductors and

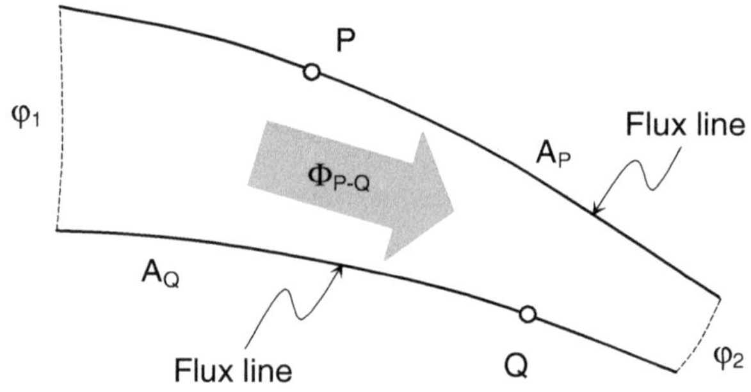

Fig. 3.9 Magnetic vector and scalar potentials of a flux tube

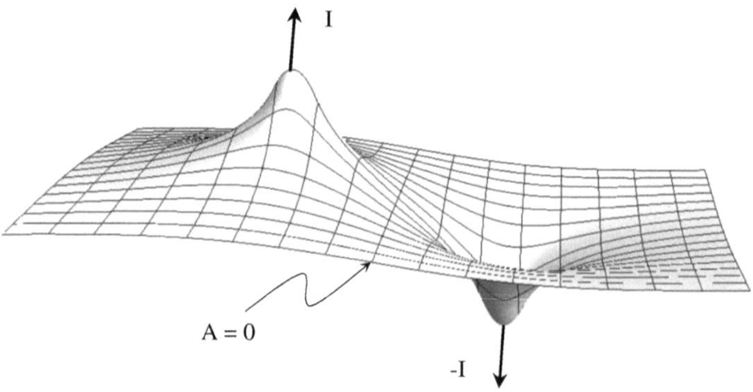

Fig. 3.10 Magnetic vector potential of two conductors carrying the same current I in opposite directions. On the centerline between the conductors, the value of magnetic vector potential is equal to zero

$$A(x,y) = \frac{\mu_0 I}{2r_0^2\pi}\left[\frac{r_0^2}{2} - \frac{x^2+y^2}{2} + r_0^2 \ln\frac{\sqrt{(d-x)^2+y^2}}{r_0}\right] \tag{3.48}$$

for a point within the conductor placed at origin and carrying current I. For a point within the second conductor having coordinates (d, 0) and carrying current $-I$ one can write

$$A(x,y) = -\frac{\mu_0 I}{2r_0^2\pi}\left[\frac{r_0^2}{2} - \frac{(d-x)^2+y^2}{2} + r_0^2 \ln\frac{\sqrt{x^2+y^2}}{r_0}\right] \tag{3.49}$$

Qualitative distribution of magnetic vector potential created by two parallel conductors carrying the same current I in opposite directions is shown in Fig. 3.10. The value of magnetic vector potential along the centerline between the conductors is equal to zero.

Curves of constant magnetic vector potential (flux lines) for two parallel conductors carrying the same current in opposite directions as in Fig. 3.10 are shown in Fig. 3.11. The value of A_{max} in Fig. 3.11 is obtained by substituting for x = 0 and y = 0 into Eq. 3.48

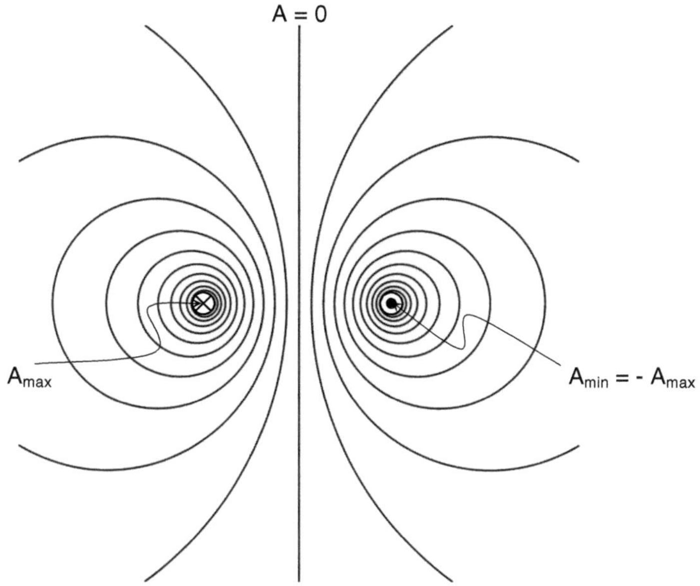

Fig. 3.11 Curves of constant magnetic vector potential—flux lines—of two parallel conductors carrying the same current in opposite directions, denoted by × and • symbols

$$A_{\max} = \frac{\mu_0 I}{2\pi}\left(\frac{1}{2} + \ln\frac{d}{r_0}\right) \tag{3.50}$$

Similarly, for $A_{\min,}$ one substitutes for x = d and y = 0 into Eq. 3.49:

$$A_{\min} = -\frac{\mu_0 I}{2\pi}\left(\frac{1}{2} + \ln\frac{d}{r_0}\right) \tag{3.51}$$

By using Eq. 3.46 one can determine the amount of flux created by the two conductors in Fig. 3.11 as

$$\Phi = l_z(A_{\max} - A_{\min}) = l_z\frac{\mu_0 I}{\pi}\left(\frac{1}{2} + \ln\frac{d}{r_0}\right) \tag{3.52}$$

and the equivalent reluctance $R_{m,\text{coil,air}}$:

$$R_{m,\text{coil,air}} = \frac{\Theta}{\Phi} = \frac{I}{l_z\frac{\mu_0 I}{\pi}\left(\frac{1}{2} + \ln\frac{d}{r_0}\right)} = \frac{1}{\mu_0 l_z}\cdot\frac{\pi}{\left(\frac{1}{2} + \ln\frac{d}{r_0}\right)} \tag{3.53}$$

Assume now that the two conductors in Fig. 3.11 are a portion of a coil with pitch d placed in slots of an electric machine with an air gap width δ and that the relative permeability of surrounding iron is very large. Coil reluctance in that case is equal to

$$R_{m,\text{coil},\delta} = \frac{1}{\mu_0 l_z}\cdot\frac{\delta}{d} \tag{3.54}$$

By placing a coil with pitch d into an air gap with width δ, the amount of created flux increases proportional to the ratio

$$\frac{\Phi_\delta}{\Phi_{air}} = \frac{R_{m,coil,air}}{R_{m,coil,\delta}} = \frac{\pi}{\left(\frac{1}{2} + \ln\frac{d}{r_0}\right)}\cdot\frac{d}{\delta} \tag{3.55}$$

which reaches two digit values in a range between 20 and 80 in practical cases.

Relationship between magnetic scalar and vector potential of a flux tube can be expressed by means of Fig. 3.9 as

$$\Phi = \Theta \cdot G_{mg} \tag{3.56}$$

with G_{mg} denoting the flux tube permeance:

$$l_z(A_P - A_Q) = (\varphi_1 - \varphi_2) \cdot \mu \cdot \frac{\overline{w} \cdot l_z}{\overline{d}} \tag{3.57}$$

By introducing the specific magnetic permeance λ_s of the flux tube [3] as a ratio between the average flux tube width $\overline{w}$ perpendicular to the direction of flux density and average length $\overline{d}$ in the direction of flux density in the flux tube:

$$\lambda_s = \frac{\overline{w}}{\overline{d}} \tag{3.58}$$

one can write

$$(A_P - A_Q) = (\varphi_1 - \varphi_2) \cdot \mu \cdot \lambda_s \tag{3.59}$$

The difference $A_P - A_Q$ between magnetic vector potentials of flux tube sides is proportional to the difference $\varphi_1 - \varphi_2$ between scalar vector potentials of flux tube bases, permeability μ of the medium, and specific magnetic permeance λ_s of the flux tube.

By combining Eqs. 3.36 and 3.39 for a two-dimensional problem, one obtains

$$\frac{\partial^2 A}{\partial x^2} + \frac{\partial^2 A}{\partial y^2} = -\mu J \tag{3.60}$$

which is analogous to Eq. 3.6 for scalar potential φ in current-free space. Following similar discretization procedure, one obtains for the mesh in Fig. 3.2

$$\frac{(A_1 - A_0)d_3 - (A_0 - A_3)d_1}{d_1 d_3 (d_1 + d_3)} + \frac{(A_2 - A_0)d_4 - (A_0 - A_4)d_2}{d_2 d_4 (d_2 + d_4)} = -\frac{\mu}{2} J_0 \tag{3.61}$$

with J_0 denoting the current density at point 0 in Fig. 3.2. For an equidistant mesh along the x-axis, $d_1 = d_3 = d_x$, and y-axis, $d_2 = d_4 = d_y$, magnetic vector potential A_0 at point 0 in Fig. 3.2 can be expressed as

$$A_0 = \frac{(A_1 + A_3)d_y^2 + (A_2 + A_4)d_x^2 + d_x^2 d_y^2 \mu J_0}{2\left(d_x^2 + d_y^2\right)} \tag{3.62}$$

For a square mesh ($d_1 = d_2 = d_3 = d_4 = d$), one can write

$$A_0 = \frac{A_1 + A_2 + A_3 + A_4 + d^2 \mu J_0}{4} \tag{3.63}$$

which is an intuitively comprehensible result. If in addition to auxiliary variables c_x and c_y, defined in Eqs. 3.14 and 3.15, one introduces c_{xy} as

$$c_{xy} = \frac{d_x^2 d_y^2}{2\left(d_x^2 + d_y^2\right)} \tag{3.64}$$

the expression for magnetic vector potential A_0 at point 0, Eq. 3.62, can be rewritten as

$$A_0 = c_y A_1 + c_x A_2 + c_y A_3 + c_x A_4 + c_{xy} \mu J_0 \tag{3.65}$$

Analogously to the procedure introduced in Eqs. 3.18 and 3.19, one can write for magnetic vector potential at node (k, ℓ) at nth step of iteration

$$A_{k,l,n} - c_x A_{k-1,l,n} - c_y A_{k,l-1,n} = c_y A_{k,l-1,n-1} + c_x A_{k-1,l,n-1} + c_{xy} \mu J_{k,l} \tag{3.66}$$

as well as

$$\underline{C}_1 \cdot \underline{A}_n = \underline{C}_2 \cdot \underline{A}_{n-1} + \underline{B} \cdot \underline{A}_B + \underline{J} \tag{3.67}$$

with $\underline{J}$ denoting the matrix of current densities of the problem and $\underline{A}_B$ the vector of boundary conditions. Finally, one can write (see Eq. 3.34)

$$\underline{A}_n = \underline{C}_1^{-1} \cdot \left(\underline{I} - \underline{C}_2 \cdot \underline{C}_1^{-1}\right)^{-1} \cdot \left(\underline{B} \cdot \underline{A}_B + \underline{J}\right) \tag{3.68}$$

Boundary conditions specified in vector $\underline{A}_B$ are defined in terms of angle between flux line and boundary, which can be either 0 (flux line parallel to the boundary) or 90° (flux line perpendicular to the boundary).

3.3 A Straightforward Method for Determination of Magnetic Parameters of a Nonlinear Permeance

A piecewise linear interpolation of B–H curve helps minimize numerical instabilities when calculating the performance of a nonlinear magnetic circuit [3]. Independently of how a B–H curve is interpolated, a nonlinear problem containing such interpolation is usually solved in iterative manner [4]. An iterative solution procedure multiplicates unnecessarily the total computational time. Here procedures will be introduced for straightforward computation of magnetic field strength H for a given flux density B and vice versa.

B given, H to be evaluated: For computations in a given interval $(0, B_m)$, the B–H curve is supplied in equidistant steps ΔB of flux density B, as shown in Fig. 3.12.

A given value of flux density B_{in} is located in the interval number i determined as

$$i = INT\left(\frac{B_{\text{in}}}{\Delta B}\right) + 1 \tag{3.69}$$

with INT denoting the integer part of expression in round brackets after it.

In the *i*th interval the output field strength Hout is found for an input flux density Bin by using a linear approximation of the *B–H* curve

$$H_{\text{out}} = H_{i-1} + \frac{H_i - H_{i-1}}{B_i - B_{i-1}}(B_{\text{in}} - B_{i-1}) \tag{3.70}$$

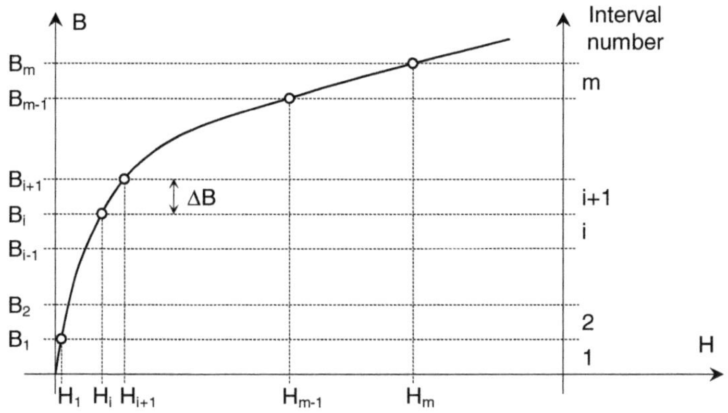

Fig. 3.12 Linear *B–H* curve approximation at m points

with $B_0 = 0$ and $H_0 = 0$.

H given, B to be evaluated: In the previous procedure, the total interval for the input variable *B* was divided into m equidistant intervals with width $\Delta B = B_m/m$, which perfectly fit the nature of saturation of *B–H* curve. An inverse approximation, where *B* has to be evaluated for a given *H*, cannot rely on equidistant strategy, because of unproportional changes of *H* for equidistantly increasing B's. Therefore, the most appropriate seems to be the procedure where the *B–H* curve is supplied at *N* points in such a manner that the magnetic field strength *H* steadily increases. The points H_i at which the magnetization curve is recorded are selected to satisfy equation

$$i = N \cdot INT\left(\frac{H_i}{H_N}\right)^{\varepsilon} + 1 \tag{3.71}$$

with ε denoting the *elasticity of approximation* (See Fig. 3.13)

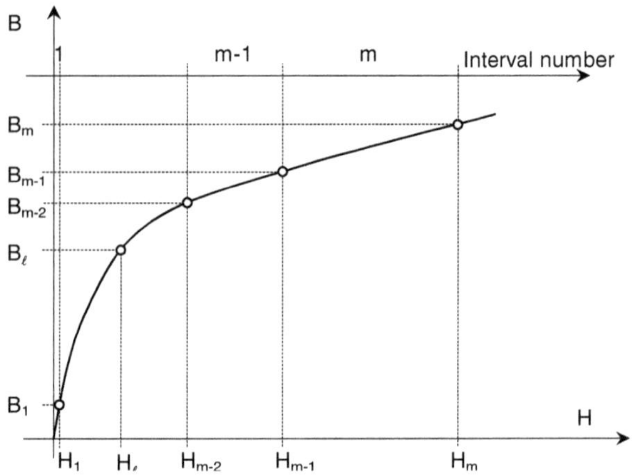

Fig. 3.13 Linear *B–H* curve approximation at m points

$$\varepsilon = \frac{\ln \frac{l-1}{N}}{\ln \frac{H_l}{H_N}} \tag{3.72}$$

Index ℓ in Eq. 3.72 is related to the point (H_ℓ, B_ℓ) of *B–H* curve after which significant effects of saturation are observed.

In the *i*th interval, the output flux density B_{out} is found for an input field strength H_{in} by using a linear approximation of the *B–H* curve

$$B_{out} = B_{i-1} + \frac{B_i - B_{i-1}}{H_i - H_{i-1}}(H_{in} - H_{i-1}) \tag{3.73}$$

with $B_0 = 0$ and $H_0 = 0$.

3.4 An Accelerated Procedure for the Solution of a System of Differential—Algebraic Equations

If n windings of a rotating field machine are connected to voltage source, the voltage differential equation for the *j*th winding, where $1 \le j \le n$, can be written as

$$u_j = \frac{d\Psi_j}{dt} + i_j R_j \tag{3.74}$$

with Ψ_j denoting the concatenated flux, i_j the current, R_j the resistance, and u_j the applied voltage in the *j*th phase. The concatenated flux in the *j*th phase is a function of currents in all phases and of the rotor angle γ:

$$\Psi_j = \Psi_j(i_1, i_2, \ldots, i_j, \ldots, i_n, \gamma) \tag{3.75}$$

The functional relationship expressed by Eq. (3.75) is nonlinear due to several reasons, out of which the saturation in iron is dominating. Therefore, the system of voltage differential equations is nonlinear, too.

The system of machine voltage equations can be solved by applying one of the procedures for numerical integration of differential equations, among which the Runge–Kutta method is one of most pronounced. As opposed to an analytical solution, which is given as a family of continuous functions, a numerical solution is a set of discrete values calculated for each state variable.

In order to solve numerically a system of differential equations, one has to write it in such a manner as to put derivatives alone on one side of the equal sign. The integration of the system starts with values of state variables as specified in initial conditions. For an extrapolated vector of state variables at the next discrete value of the independent variable, the right-hand-side vector is evaluated based on physical model, outside of the Runge–Kutta procedure. The evaluation follows by solving a system of algebraic equations, which in case of integration of voltage equations relate the vector $\underline{\Psi}$ of extrapolated fluxes to the vector $\underline{i}$ of coil currents. Coil currents stored in vector $\underline{i}$ are returned back to the Runge–Kutta procedure. If the achieved accuracy is equal to or better than requested, the computation at current step of integration is completed and the same pattern repeats for the next discrete value of the independent variable.

Any numerical solution of a system of differential equations is an iterative procedure performed step by step, where each step of integration requires certain time in order to be completed. If the system of algebraic equations which has to be solved at each step of integration is nonlinear, additional iterations have to be performed at each step of integration of the nonlinear system of differential equations.

Summarizing previous considerations, one concludes that a solution of the system of machine voltage differential equations is a time-consuming task since it requires iterations within iterations. Obviously, significant time saving could be achieved if one level of iterations is eliminated.

A way to eliminate iterative solution of the system of algebraic equations within each step of integration of the system of differential equations is to declare all relevant quantities state variables [3]. If the dependence of the vector of state variables $\underline{\Psi}$ on the vector of unknowns i can be expressed as

$$\underline{\Psi} = \underline{\Psi}(\underline{i}) \tag{3.76}$$

then

$$\frac{\mathrm{d}}{\mathrm{d}t}\underline{\Psi} = \frac{\mathrm{d}}{\mathrm{d}i}\underline{\Psi}(\underline{i}) \cdot \frac{\mathrm{d}}{\mathrm{d}t}\underline{i} \tag{3.77}$$

This way the unknowns of the system of algebraic equations became additional state variables of the system of differential equations. Instead of iterating the nonlinear system of algebraic equations at each step of integration of the system of differential equations, the right-hand sides of the extended system of differential equations are solved in straightforward way.

Case Study 3.2: Electrical engineers are used to think in terms of sources and resistances. They are trained to be able to say what happens in a circuit if certain quantity with impact on resistance(s) or source(s) changes. Therefore, not only electric, but also magnetic [5], thermal, and fluid flow problems related to electric machines have always been solved by means of lumped circuit parameters. It is interesting that largest electric machines ever built were designed by using magnetic, thermal, or fluid flow circuits, the elements of which had physically interpretable properties and reflected all peculiarities of given geometry, material properties, and boundary conditions.

Ever since advent of modern computers, another analysis method found applications in electrical machines, which was strongly related rather to mathematical, than to engineering way of thinking. Following this approach, the machine is discretized into a set of small elements (thus the name finite elements), in which a pure mathematical quantity—the magnetic vector potential—is calculated. Here, the emphasis is put on "mathematical," i.e., a non-measurable quantity, as opposed to physical, measurable quantities, such as current. Meanwhile, the finite element method has found many supporters among electrical engineers, mostly for solution of two-dimensional problems [6, 7]. The complexity of the finite element approach increases faster than exponentially when extending the model into third spatial dimension, due to the interaction between various spatial components of electromagnetic fields. Whereas a two-dimensional finite element model in the (x, y) plane can be solved by using only the z-component of the magnetic vector potential, a three-dimensional model requires all three spatial components of the magnetic vector potential to be known. Consequently, the computational time explodes when including only one additional spatial dimension.

The magnetic circuit representation of electric machines, on the other hand, does not suffer from model extension into the third spatial dimension. In order to illustrate the advantages of the magnetic circuit representation of electric machines, a three-dimensional model of a claw pole synchronous machine was built [8], as shown in Fig. 3.14.

The extension into third dimension, imposed by the rotor geometry, is handled by adding an extra set of elements (reluctances and sources) on the rotor side (Fig. 3.15), i.e., quantitatively (more elements), rather than qualitatively, as is the case with finite elements. The machine is divided in axial direction into an arbitrary number of slices, and the rotor pole trapezoidal geometry is discretized into a stepwise changing pole width. The three-dimensional model retains its straightforward nature and gives designer insight in all important machine quantities, such as time dependency of axial flux distribution in a single rotor pole at no load, as shown in Fig. 3.16.

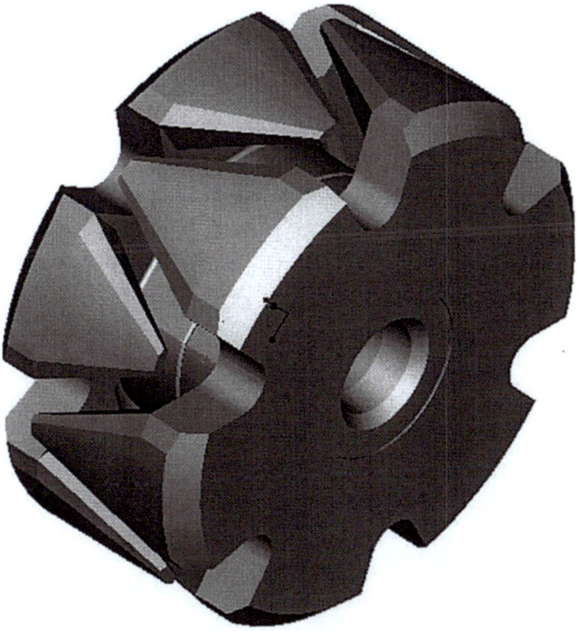

Fig. 3.14 Rotor iron of a12-pole claw pole synchronous machine [8]

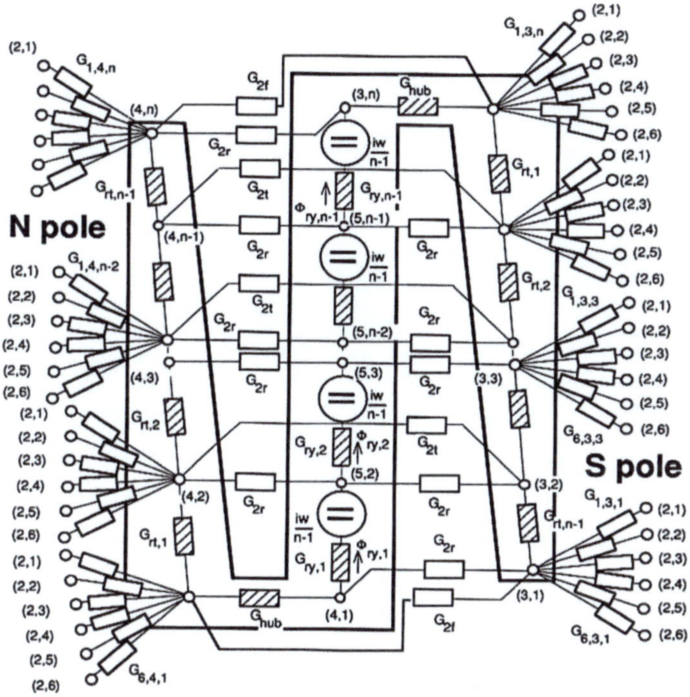

Fig. 3.15 Magnetic equivalent circuit of a rotor pole pair of a claw pole synchronous machine [8]

Rotor portion of the magnetic equivalent circuit model of the analyzed claw pole machine was divided into five axial segments. By comparing measured and calculated performance [8], the selected number of segments proved to be a good compromise between the performance and computational time.

Spatial distribution of rotor pole segment fluxes at no load is shown in Fig. 3.16. One recognizes in this figure a decrease of rotor pole flux from pole base to pole tip, primarily as a consequence of decreasing pole segment surface in the air gap.

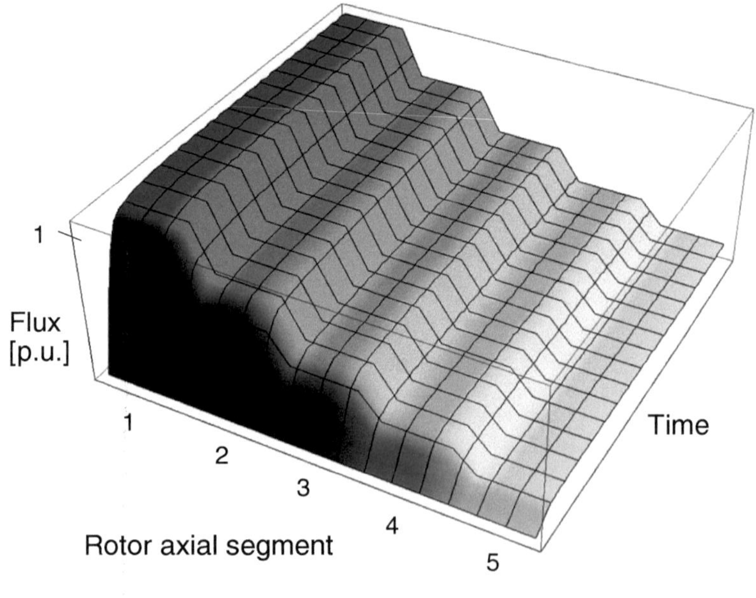

Fig. 3.16 Rotor pole flux distribution in a claw pole synchronous machine at no load [8]

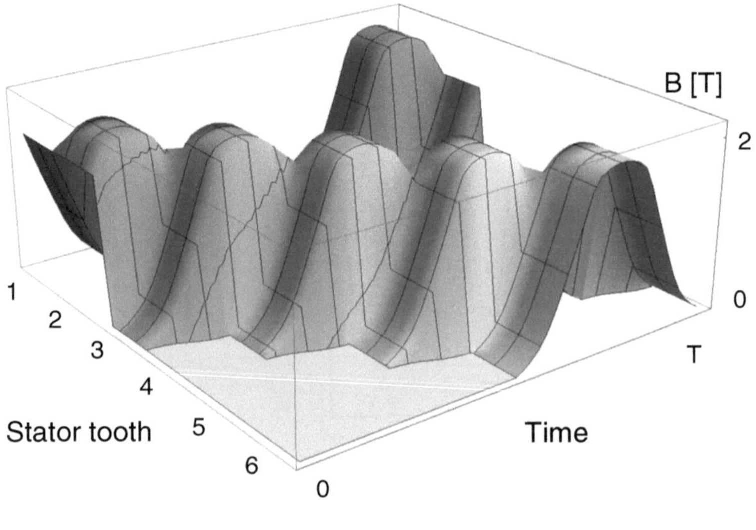

Fig. 3.17 Air gap flux density distribution in a claw pole synchronous machine at no load [8]

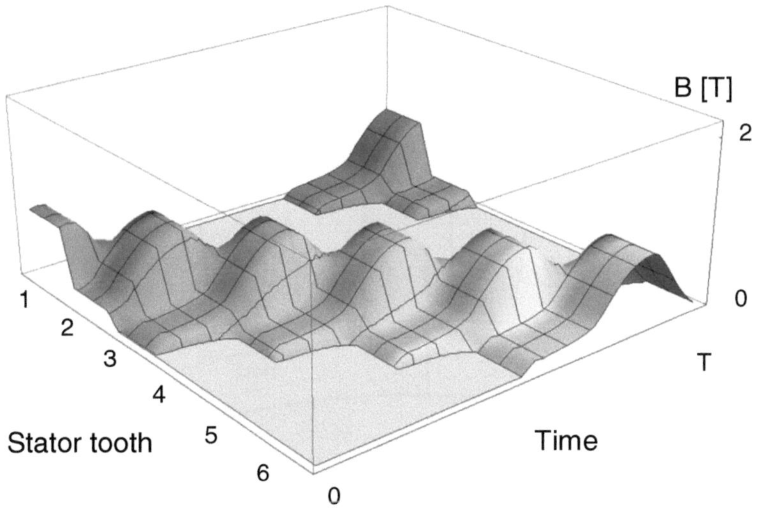

Fig. 3.18 Air gap flux density distribution in a claw pole synchronous machine at rated load [8]

In Fig. 3.17 positive half waves of air gap flux density at no load as a function of spatial coordinate, given as a number of stator tooth, and time between 0 and period length T, are shown. The machine has 6 teeth per pole pair. One recognizes in this figure a strong third spatial harmonic, as a consequence of the rotor pole form.

When the machine is loaded, the stator ampere-turns superimpose to those created by the field current in Fig. 3.18. At the given operating point, the power factor is obviously strongly lagging, since the armature reaction managed to decrease significantly the no-load air gap flux density distribution.

3.5 A Straightforward Method for the Solution of Flux Distribution in Magnets

Assume a machine with surface-mounted magnets, the geometry of which is shown in Fig. 3.19. As shown in [3], the discretization of machine's magnetic circuit is fine enough if a single tooth, or a yoke segment between two adjacent teeth, is taken as smallest elements, because of their high relative permeability and because their form coincides with flux tubes placed in the machine geometry. A magnet, on the other hand, generates flux components, the amounts of which are dependent not only on magnet parameters, but also on the geometry of the rest of the magnetic circuit [9]. Besides, the relative permeability of magnets does not differ substantially from 1, the relative permeability of air.

For this reason a magnet is subdivided into smaller units, here k in tangential and ℓ in radial direction. From the computational point of view it is interesting that

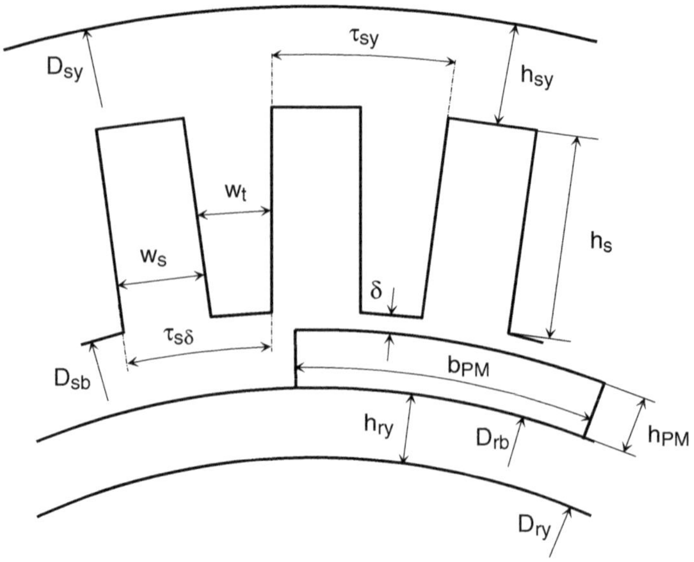

Fig. 3.19 Electric machine with surface-mounted magnets

subdividing of permanent magnet in radial direction does not necessarily increase the size of the system to be solved.

Denoting by $\mu_{r,s}$, $\mu_{r,r}$, and $\mu_{r,PM}$, relative permeabilities of stator iron and rotor iron and permanent magnets, respectively, one can define permeances

$$G_{ys} = \mu_0 \mu_{r,st} \frac{h_{sy} l_{ax}}{\tau_{sy}} \tag{3.78}$$

$$G_{ts} = \mu_0 \mu_{r,st} \frac{w_{ts} l_{ax}}{h_s} \tag{3.79}$$

$$G_{\sigma s} = \mu_0 \frac{h_s l_{ax}}{w_s} \tag{3.80}$$

$$G_{\sigma r} = \mu_0 \frac{h_{\mathrm{PM}} l_{ax}}{l(\tau_p - b_{\mathrm{PM}})} \tag{3.81}$$

$$G_d = \frac{l}{k} \mu_0 \mu_{r,\mathrm{PM}} \frac{b_{\mathrm{PM}} l_{ax}}{h_{\mathrm{PM}}} \tag{3.82}$$

$$G_q = \frac{k}{l} \mu_0 \mu_{r,\mathrm{PM}} \frac{h_{\mathrm{PM}} l_{ax}}{b_{\mathrm{PM}}} \tag{3.83}$$

$$G_{yrr} = \mu_0 \mu_{r,r} \frac{\tau_{ry} l_{ax}}{k h_{ry}} \tag{3.84}$$

$$G_{yrt} = \mu_0 \mu_{r,r} \frac{h_{ry} l_{ax}}{\tau_{ry}} k \tag{3.85}$$

$$G_{\max} = \mu_0 \frac{b_{\mathrm{PM}} l_{ax}}{k\delta} \tag{3.86}$$

which along with the amount of flux Φ_{PM} generated by each magnet segment

$$\Phi_{\mathrm{PM}} = B_r \frac{b_{\mathrm{PM}} l_{ax}}{k} \tag{3.87}$$

belong to elements outside of air gap of the machine's magnetic equivalent circuit in Fig. 3.21. Dashed gray lines outline the stator and rotor geometry in this figure.

Air gap permeance $G_{i,j}$ connecting *i*th stator tooth and *j*th magnet segment is piecewise defined in the manner shown in Fig. 3.20

Interval limits x_1–x_4 in Fig. 3.20 are equal to

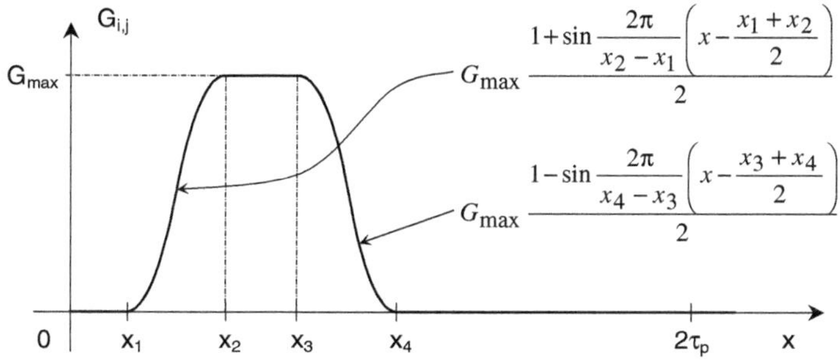

Fig. 3.20 Air gap permeance between *i*th stator tooth and *j*th rotor magnet segment as a function of rotor shift *x*

$$x_1 = \tau_{s\delta} - \frac{w_s}{2} - \left(\frac{\tau_p - b_{\mathrm{PM}}}{2} + \frac{b_{\mathrm{PM}}}{k}\right) \tag{3.88}$$

$$x_2 = \tau_{s\delta} + \frac{w_s}{2} - \frac{\tau_p - b_{\mathrm{PM}}}{2} \tag{3.89}$$

$$x_3 = 2\tau_{s\delta} - \frac{w_s}{2} - \left(\frac{\tau_p - b_{\mathrm{PM}}}{2} + \frac{b_{\mathrm{PM}}}{k}\right) \tag{3.90}$$

$$x_4 = 2\tau_{s\delta} + \frac{w_s}{2} - \frac{\tau_p - b_{\mathrm{PM}}}{2} \tag{3.91}$$

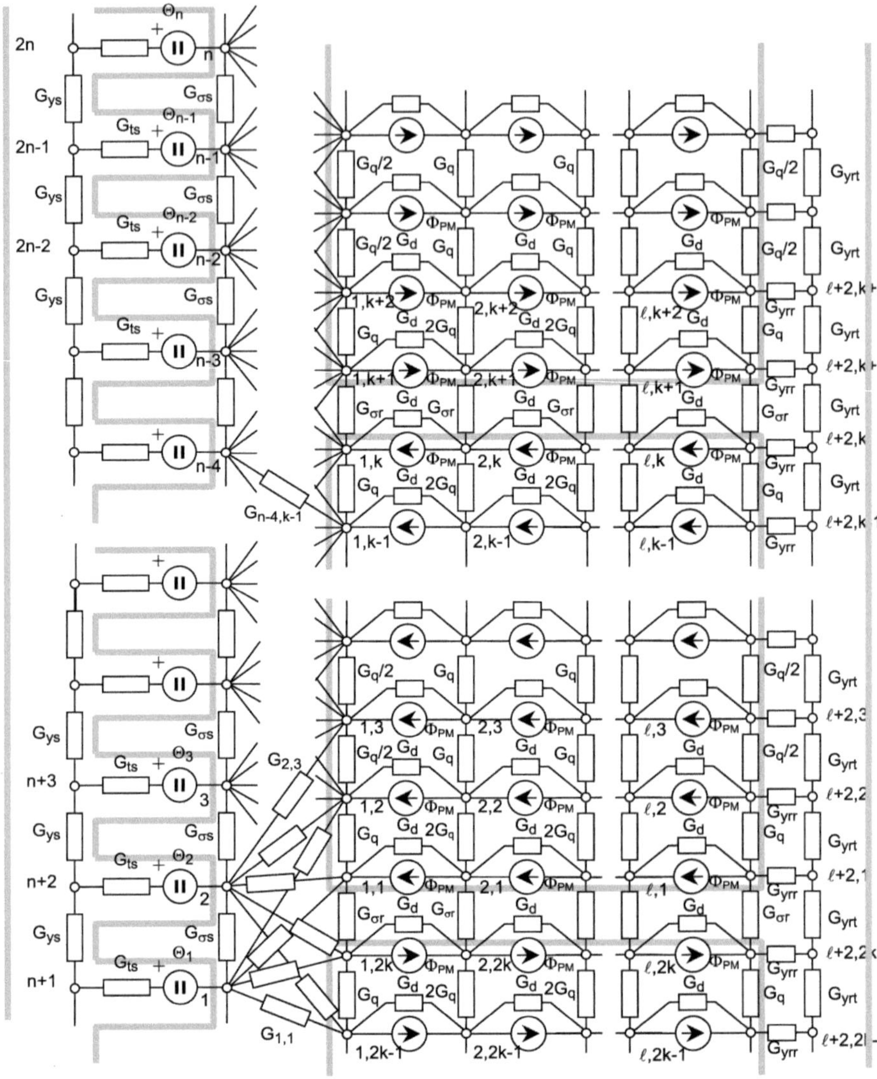

Fig. 3.21 Magnetic equivalent circuit of two poles of the machine, the geometry of which is shown in Fig. 3.19

and the rotor shift $x_{i,j}$ between ith stator tooth and jth magnet segment

$$x_{i,j} = x - (i-1)\tau_{s\delta} + (j-1)\frac{b_{PM}}{k};\ j \leq k \tag{3.92}$$

or

$$x_{i,j} = x - (i-1)\tau_{s\delta} + (j-k-1)\frac{b_{\mathrm{PM}}}{k} + \tau_p;\ k \le j \le 2k \tag{3.93}$$

Introducing

$$S_{ji} = \sum_{i=1}^{2k} G_{j,i}\ ;\ S_{ij} = \sum_{j=1}^{n} G_{j,i}\ ;\ G_{dq} = G_d + G_q \tag{3.94}$$

along with matrices

$$\underline{A}_{sy} = \begin{bmatrix} 2G_{ys}+G_{ts} & -G_{ys} & 0 & \cdots & 0 & 0 & 0 \\ -G_{ys} & 2G_{ys}+G_{ts} & -G_{ys} & \cdots & 0 & 0 & 0 \\ \cdots & \cdots & \cdots & \cdots & \cdots & \cdots & \cdots \\ 0 & 0 & 0 & \cdots & -G_{ys} & 2G_{ys}+G_{ts} & -G_{ys} \\ 0 & 0 & 0 & \cdots & 0 & -G_{ys} & 2G_{ys}+G_{ts} \end{bmatrix} \tag{3.95}$$

$$\underline{A}_{s\delta} = \begin{bmatrix} 2G_{\sigma s}+G_{ts}+S_{1,i} & -G_{\sigma s} & 0 & \cdots & 0 & 0 & -G_{\sigma s} \\ -G_{\sigma s} & 2G_{ys}+G_{ts}+S_{2,i} & -G_{\sigma s} & \cdots & 0 & 0 & 0 \\ \cdots & \cdots & \cdots & \cdots & \cdots & \cdots & \cdots \\ 0 & 0 & 0 & \cdots & -G_{\sigma s} & 2G_{ys}+G_{ts}+S_{n-1,i} & -G_{\sigma s} \\ -G_{\sigma s} & 0 & 0 & \cdots & 0 & -G_{\sigma s} & 2G_{ys}+G_{ts}+S_{n,i} \end{bmatrix} \tag{3.96}$$

$$\underline{G}_{\delta} = -\begin{bmatrix} G_{1,1} & G_{1,2} & G_{1,3} & \cdots & G_{1,2k-1} & G_{1,2k} \\ G_{2,1} & G_{2,2} & G_{2,3} & \cdots & G_{2,2k-1} & G_{2,2k} \\ G_{3,1} & G_{3,2} & G_{3,3} & \cdots & G_{3,2k-1} & G_{3,2k} \\ \cdots & \cdots & \cdots & \cdots & \cdots & \cdots \\ G_{n-1,1} & G_{n-1,2} & G_{n-1,3} & \cdots & G_{n-1,2k-1} & G_{n-1,2k} \\ G_{n,1} & G_{n,2} & G_{n,3} & \cdots & G_{n,2k-1} & G_{n,2k} \end{bmatrix} \tag{3.97}$$

$$\underline{A}_{\mathrm{PM}dq} = \begin{bmatrix} G_{\sigma r}+2G_{dq} & -2G_q & 0 & \cdots & 0 & 0 & 0 \\ -2G_q & 2G_{dq}+G_q & -G_q & \cdots & 0 & 0 & 0 \\ 0 & -G_q & 2G_{dq} & \cdots & 0 & 0 & 0 \\ \cdots & \cdots & \cdots & \cdots & \cdots & \cdots & \cdots \\ 0 & 0 & 0 & \cdots & 2G_{dq} & -G_q & 0 \\ 0 & 0 & 0 & \cdots & -G_q & 2G_{dq}+G_q & -2G_q \\ 0 & 0 & 0 & \cdots & 0 & -2G_q & G_{\sigma r}+2G_{dq} \end{bmatrix} \tag{3.98}$$

$$\underline{A}_{\mathrm{PM}\sigma} = \begin{bmatrix} 0 & 0 & \cdots & 0 & -G_{\sigma r} \\ 0 & 0 & \cdots & 0 & 0 \\ \cdots & \cdots & \cdots & \cdots & \cdots \\ 0 & 0 & \cdots & 0 & 0 \\ -G_{\sigma r} & 0 & \cdots & 0 & 0 \end{bmatrix} \tag{3.99}$$

$$\underline{A}_{\mathrm{PM}} = \begin{bmatrix} \underline{A}_{\mathrm{PM}dq} & \underline{A}_{\mathrm{PM}\sigma} \\ \underline{A}_{\mathrm{PM}\sigma} & \underline{A}_{\mathrm{PM}dq} \end{bmatrix} \tag{3.100}$$

$$\underline{A}_{r\delta 1} = \begin{bmatrix} G_{\sigma r}+G_{dq}+S_{j,1} & -G_q & 0 & \cdots & 0 & 0 & 0 \\ -G_q & \frac{G_q}{2}+G_{dq}+S_{j,2} & -\frac{G_q}{2} & \cdots & 0 & 0 & 0 \\ & -\frac{G_q}{2} & G_{dq}+S_{j,3} & \cdots & 0 & 0 & 0 \\ \cdots & \cdots & \cdots & \cdots & \cdots & \cdots & \cdots \\ 0 & 0 & 0 & \cdots & G_{dq}+S_{j,k-2} & -\frac{G_q}{2} & 0 \\ 0 & 0 & 0 & \cdots & -\frac{G_q}{2} & \frac{G_q}{2}+G_{dq}+S_{j,k-1} & -G_q \\ 0 & 0 & 0 & \cdots & 0 & -G_q & G_{\sigma r}+G_{dq}+S_{j,k} \end{bmatrix} \tag{3.101}$$

$$\underline{A}_{r\delta 2} = \begin{bmatrix} G_{\sigma r}+G_{dq}+S_{j,k+1} & -G_q & 0 & \cdots & 0 & 0 & 0 \\ -G_q & \frac{G_q}{2}+G_{dq}+S_{j,k+2} & -\frac{G_q}{2} & \cdots & 0 & 0 & 0 \\ & -\frac{G_q}{2} & G_{dq}+S_{j,k+3} & \cdots & 0 & 0 & 0 \\ \cdots & \cdots & \cdots & \cdots & \cdots & \cdots & \cdots \\ 0 & 0 & 0 & \cdots & G_{dq}+S_{j,2k-2} & -\frac{G_q}{2} & 0 \\ 0 & 0 & 0 & \cdots & -\frac{G_q}{2} & \frac{G_q}{2}+G_{dq}+S_{j,2k-1} & -G_q \\ 0 & 0 & 0 & \cdots & 0 & -G_q & G_{\sigma r}+G_{dq}+S_{j,2k} \end{bmatrix} \tag{3.102}$$

$$\underline{A}_{r\delta} = \begin{bmatrix} \underline{A}_{r\delta 1} & \underline{A}_{PM\sigma} \\ \underline{A}_{PM\sigma} & \underline{A}_{r\delta 2} \end{bmatrix} \tag{3.103}$$

$$\underline{A}_{ry} = \begin{bmatrix} 2G_{yrt}+G_{yrr} & -G_{yrt} & 0 & \cdots & 0 & 0 & 0 \\ -G_{yrt} & 2G_{yrt}+G_{yrr} & -G_{yrt} & \cdots & 0 & 0 & 0 \\ \cdots & \cdots & \cdots & \cdots & \cdots & \cdots & \cdots \\ 0 & 0 & 0 & \cdots & -G_{yrt} & 2G_{yrt}+G_{yrr} & -G_{yrt} \\ 0 & 0 & 0 & \cdots & 0 & -G_{yrt} & 2G_{yrt}+G_{yrr} \end{bmatrix} \tag{3.104}$$

$$\underline{A}_{rr1} = \begin{bmatrix} G_{\sigma r}+G_{dq}+G_{jr} & -G_q & 0 & \cdots & 0 & 0 & 0 \\ -G_q & \frac{G_q}{2}+G_{dq}+G_{jr} & -\frac{G_q}{2} & \cdots & 0 & 0 & 0 \\ & -\frac{G_q}{2} & G_{dq}+G_{jr} & \cdots & 0 & 0 & 0 \\ \cdots & \cdots & \cdots & \cdots & \cdots & \cdots & \cdots \\ 0 & 0 & 0 & \cdots & G_{dq}+G_{jr} & -\frac{G_q}{2} & 0 \\ 0 & 0 & 0 & \cdots & -\frac{G_q}{2} & \frac{G_q}{2}+G_{dq}+G_{jr} & -G_q \\ 0 & 0 & 0 & \cdots & 0 & -G_q & G_{\sigma r}+G_{dq}+G_{jr} \end{bmatrix} \tag{3.105}$$

$$\underline{A}_{rr} = \begin{bmatrix} \underline{A}_{rr1} & \underline{A}_{\mathrm{PM}\sigma} \\ \underline{A}_{\mathrm{PM}\sigma} & \underline{A}_{rr1} \end{bmatrix} \tag{3.106}$$

and vectors

$$\underline{\varphi}_{s\delta} = [\,\varphi_1 \quad \varphi_2 \quad \cdots \quad \varphi_{n-1} \quad \varphi_n\,]^T \tag{3.107}$$

$$\underline{\varphi}_{sy} = [\,\varphi_{n+1} \quad \varphi_{n+2} \quad \cdots \quad \varphi_{2n-1} \quad \varphi_{2n}\,]^T \tag{3.108}$$

$$\underline{\varphi}_1 = \left[\,\varphi_{1,1} \quad \varphi_{1,2} \quad \cdots \quad \varphi_{1,2k-1} \quad \varphi_{1,2k}\,\right]^T \tag{3.109}$$

$$\underline{\varphi}_2 = \left[\,\varphi_{2,1} \quad \varphi_{2,2} \quad \cdots \quad \varphi_{2,2k-1} \quad \varphi_{2,2k}\,\right]^T \tag{3.110}$$

$$\cdots$$

$$\underline{\varphi}_l = \left[\,\varphi_{l,1} \quad \varphi_{l,2} \quad \cdots \quad \varphi_{l,2k-1} \quad \varphi_{l,2k}\,\right]^T \tag{3.111}$$

$$\underline{\varphi}_{l+1} = \left[\,\varphi_{l+1,1} \quad \varphi_{l+1,2} \quad \cdots \quad \varphi_{l+1,2k-1} \quad \varphi_{l+1,2k}\,\right]^T \tag{3.112}$$

$$\underline{\varphi}_{ry} = \left[\,\varphi_{l+2,1} \quad \varphi_{l+2,2} \quad \cdots \quad \varphi_{l+2,2k-1} \quad \varphi_{l+2,2k}\,\right]^T \tag{3.113}$$

$$\underline{\Theta} = [\,\Theta_1 \quad \Theta_2 \quad \cdots \quad \Theta_{n-1} \quad \Theta_n\,]^T \tag{3.114}$$

$$\underline{\Phi}_{\mathrm{PM}} = \Phi_{\mathrm{PM}}[\,1 \quad 1 \quad \cdots \quad 1 \quad 1 \quad -1 \quad -1 \quad \cdots \quad -1 \quad -1\,]^T \tag{3.115}$$

one can write magnetic scalar potential equations for nodes in Fig. 3.21 as

$$\underline{A}_{sy} \cdot \underline{\varphi}_{sy} - G_{ts}\underline{\varphi}_{s\delta} = \frac{1}{G_{ts}}\underline{\Theta} \tag{3.116}$$

$$\underline{A}_{s\delta} \cdot \underline{\varphi}_{s\delta} - G_{ts}\underline{\varphi}_{sy} - \underline{G}_{\delta} \cdot \underline{\varphi}_1 = -\frac{1}{G_{ts}}\underline{\Theta} \tag{3.117}$$

$$\underline{A}_{r\delta} \cdot \underline{\varphi}_1 - \underline{G}_{\delta}^T\underline{\varphi}_{s\delta} - G_d \cdot \underline{\varphi}_2 = \underline{\Phi}_{\mathrm{PM}} \tag{3.118}$$

$$\underline{A}_{\mathrm{PM}} \cdot \underline{\varphi}_2 - \underline{G}_d\underline{\varphi}_1 - G_d \cdot \underline{\varphi}_3 = \underline{0} \tag{3.119}$$

$$\underline{A}_{\mathrm{PM}} \cdot \underline{\varphi}_3 - \underline{G}_d\underline{\varphi}_2 - G_d \cdot \underline{\varphi}_4 = \underline{0} \tag{3.120}$$

$$\cdots$$

$$\underline{A}_{\mathrm{PM}}\cdot\underline{\varphi}_l-\underline{G}_d\underline{\varphi}_{l-1}-G_d\cdot\underline{\varphi}_{l+1}=\underline{0} \tag{3.121}$$

$$\underline{A}_{rr}\cdot\underline{\varphi}_{l+1}-\underline{G}_d\underline{\varphi}_l-G_{yrr}\cdot\underline{\varphi}_{ry}=-\underline{\Phi}_{\mathrm{PM}} \tag{3.122}$$

$$\underline{A}_{ry}\cdot\underline{\varphi}_{ry}-G_{yrr}\cdot\underline{\varphi}_{l+1}=\underline{0} \tag{3.123}$$

With recursively defined matrices $\underline{\mathrm{A}}_{\mathrm{j}}, \mathrm{j}=2, \ell+1$:

$$\underline{A}_{l+1}=\underline{A}_{rr}-G_{yrr}^2\underline{A}_{ry}^{-1} \tag{3.124}$$

$$\underline{A}_l=\underline{A}_{\mathrm{PM}}-G_d^2\underline{A}_{l+1}^{-1} \tag{3.125}$$

$$\underline{A}_{l-1}=\underline{A}_{\mathrm{PM}}-G_d^2\underline{A}_l^{-1} \tag{3.126}$$

...

$$\underline{A}_3=\underline{A}_{\mathrm{PM}}-G_d^2\underline{A}_4^{-1} \tag{3.127}$$

$$\underline{A}_2=\underline{A}_{\mathrm{PM}}-G_d^2\underline{A}_3^{-1} \tag{3.128}$$

the system of algebraic equations for scalar magnetic potentials can be written as

$$\begin{bmatrix} \underline{A}_{sy} & -G_{ts}\underline{I} & \underline{0} \\ -G_{ts}\underline{I} & \underline{A}_{s\delta} & -\underline{G}_\delta \\ \underline{0} & -\underline{G}_\delta^T & \underline{A}_{r\delta}-G_d^2\underline{A}_2^{-1} \end{bmatrix}\cdot\begin{bmatrix} \underline{\varphi}_{sy} \\ \underline{\varphi}_{s\delta} \\ \underline{\varphi}_1 \end{bmatrix} = \begin{bmatrix} \frac{1}{G_{ts}}\underline{\Theta} \\ -\frac{1}{G_{ts}}\underline{\Theta} \\ -\left(G_d^l\prod_{j=2}^{l}\underline{A}_j^{-1}\cdot\underline{A}_{l+1}^{-1}-\underline{I}\right)\cdot\underline{\Phi}_{\mathrm{PM}} \end{bmatrix} \tag{3.129}$$

The core of the system of scalar magnetic potential algebraic equations has a size of [2(n + k)−1] × [2(n + k)−1], independent of the number of radial layers in magnets, which substantially decreases the computational time.

Case Study 3.3: Outer rotor synchronous generator with surface-mounted magnets has the following dimensions: axial length l_{ax} = 750 mm, 84 poles, 294 stator slots, air gap diameter 4800 mm, air gap width δ = 2 mm, rotor OD 4940 mm, rotor bore 4846 mm, stator OD 4794 mm, stator bore 4440 mm, stator slot height 118 mm, stator slot width 23 mm, magnet width per pole 136 mm, magnet height 20 mm, residual flux density 1.25 T, stator and rotor iron relative permeability 5000, and magnet relative permeability 1.05. Each magnet is divided into 50 tangential and 6 radial segments. Flux density distribution on the air gap side of one magnet at no load and x = 0 and x = 5 mm, evaluated by using Eq. 3.129 and assuming negligible MMF drop across stator iron, is shown in

Fig. 3.22. One recognizes in this figure substantial change of radial component of air gap density on the magnet surface as a function of rotor position. When the rotor is rotating, the change of air gap density becomes periodical and induces voltages in the magnet. Depending on the magnet electric conductivity, the induced voltages are accompanied by eddy currents, which create losses in magnets and which can be calculated in the manner shown in Chap. 5.

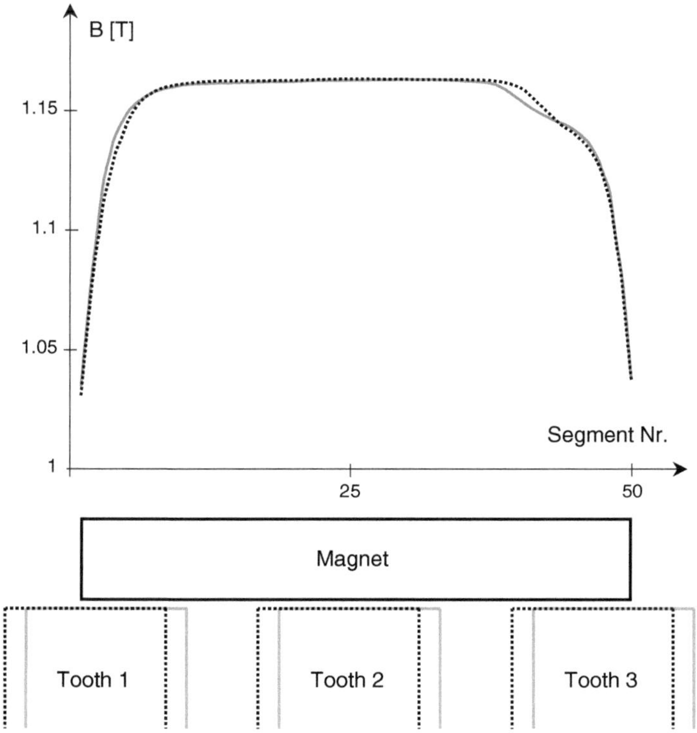

Fig. 3.22 Radial component of air gap flux density on magnet surface at two rotor positions: 0 mm (*dashed black curve*) and 5 mm (*solid gray curve*). On the right magnet edge, one recognizes substantial changes of flux density as a function of rotor position

References

1. Hayt WH (1974) Engineering electromagnetics. McGraw-Hill Kogakusha Ltd, Tokyo
2. Berberovic S (1998) Teorijska elektrotehnika-Odabrani primjeri. Graphis, Zagreb
3. Ostovic V (1989) Dynamics of saturated electric machines. Springer, New York
4. Press WH, Flannery BP, Teukolsky SA, Vetterling WT (1986) Numerical recipes- the art of scientific computing. Cambridge University Press, Cambridge
5. Roters HC (1955) Electromagnetic devices. Wiley, New York
6. Salon SJ (1995) Finite element analysis of electric machines. Kluwer Academic Publishers, Norwell

7. Bianchi N (2005) Electrical machine analysis using finite elements. Taylor & Francis, Boca Raton
8. Ostovic V, Miller J, Garg V, Schulz R, Swales S (1999) A magnetic equivalent circuit based performance computation of a Lundell alternator. IEEE Trans Ind Appl 35:825–830
9. Fouad FA, Nehl TW, Demerdash NA (1981) Permanent magnet modeling for use in vector potential finite element analysis in electric machinery. IEEE Trans Magn 17:3002–3004

Chapter 4
Machine Parameters

Contents

An electric machine interacts with its electrical environment through its terminals. Machine terminal quantities—voltages and currents—are determined by its parameters and operating point. If the machine is analyzed as a part of a power system, its terminal voltages and currents can be related to each other by means of its impedance, the components of which are resistances and inductances.

DC resistance of a coil is a function of geometry, electric properties of materials and temperature. AC resistance includes in addition the influence of skin and proximity effect.

DC inductance of a coil, as a measure of the amount of concatenated flux created by DC current(s), is significantly more complex to determine than its DC resistance. The inductance is a function of coil current; only for low current values (unsaturated iron) the inductance is constant. Besides, the total flux created by a single coil splits into main and leakage components when the coil is brought into a magnetic circuit with additional coil(s). Main flux is a measure of concatenation between

V. Ostović, *The Art and Science of Rotating Field Machines Design*,
DOI 10.1007/978-3-319-39081-9_4

magnetic fields created by two or more coils at a given number of pole pairs; the rest is referred to as leakage flux. An important component of leakage flux is the so-called *harmonic leakage*—the air gap flux created by a higher harmonic of coil MMF, which is orthogonal to the fundamental.

4.1 DC Resistance of a Coil

When determining the DC resistance of a coil, the knowledge of end winding geometry plays an important role. Whereas the end winding length of a random-wound coil is more or less a question of practical experience, for windings with a form-wound coil it can be exactly calculated.

The end winding geometry of a winding with form-wound coils is shown in Fig. 4.1.

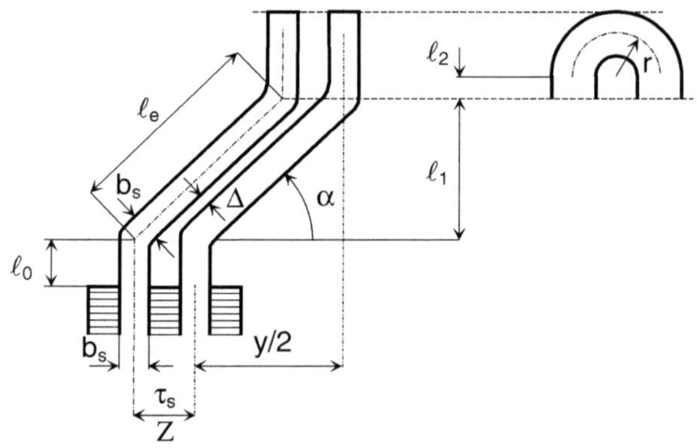

Fig. 4.1 End winding geometry of form-wound coils

For a given coil pitch y, the axial length ℓ_1 of end winding evolvent can be expressed as follows:

$$\tan\alpha = \frac{l_1}{\frac{y}{s}\tau_S} = \frac{2l_1}{y\tau_S} \Rightarrow l_1 = \frac{y}{2}\tau_S \tan\alpha, \tag{4.1}$$

where

$$\sin\alpha = \frac{b_S + \Delta}{\tau_S} \tag{4.2}$$

Since

$$\sin\alpha = \frac{\tan\alpha}{\sqrt{1 + \tan^2\alpha}},$$

one can write for axial length ℓ_1 of end winding evolvent

$$l_1 = \frac{y\tau_S(b_S + \Delta)}{2\sqrt{\tau_S^2 - (b_S + \Delta)^2}} \tag{4.3}$$

The evolvent length ℓ_e is equal to

$$l_e = \frac{y\tau_S^2}{2\sqrt{\tau_S^2 - (b_S + \Delta)^2}} \tag{4.4}$$

For a given rotor diameter, number of slots, slot width b_s, and distance Δ (Fig. 4.1), the axial length ℓ_1 of end winding evolvent is proportional to the coil pitch y or to the reciprocal of the number of pole pairs

$$l_1 \approx \frac{1}{p} \tag{4.5}$$

Low-polarity, high-speed electric machines, such as turbogenerators, have extremely long end windings, which not only increase their resistance and reactance, but also impact their mechanical strength.

4.2 Air Gap Inductance of a Coil in a Machine with Constant Air Gap Width

Self-inductance of a coil is defined as a ratio between flux concatenated by coil turns and the coil current which created it.

Consider a simple magnetic circuit in Fig. 4.2 which, to its farthest extent, represents an electric machine. Magnetic circuit in Fig. 4.2 can be split into components representing iron, air gap, and leakage paths. A significant portion of machine iron carries the air gap flux Φ_{main}; an additional (leakage) component of flux Φ_{leak} goes through the rest of the magnetic circuit. Each component of flux created by coil ampere-turns $i \cdot w$ can be characterized by an inductance L equal to the ratio between concatenated flux and current i. Accordingly:

$$\Phi_{total} = \Phi_{main} + \Phi_{leak} \Rightarrow L_{total} = L_{main} + L_{leak} \tag{4.6}$$

However, since

$$L = \frac{\Psi}{i} = w^2 \frac{\Phi}{\Theta} = \frac{w^2}{R_m} \tag{4.7}$$

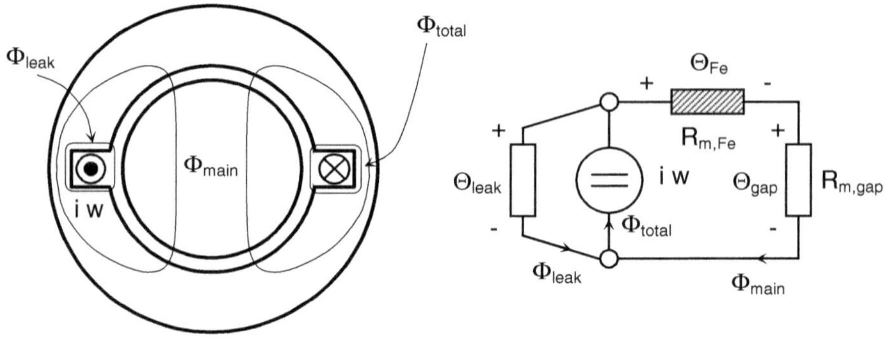

Fig. 4.2 Simplified magnetic circuit of an electric machine

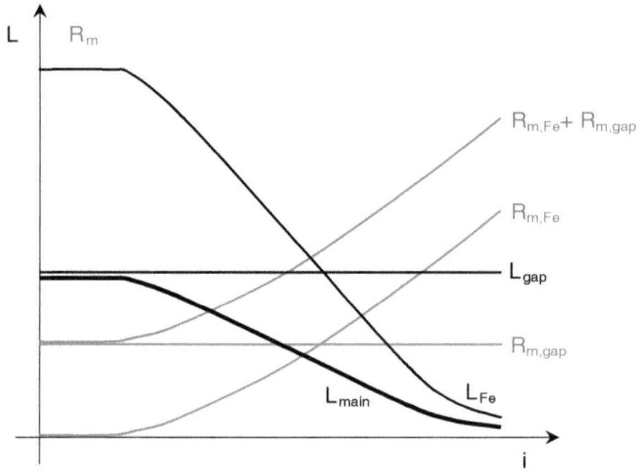

Fig. 4.3 Qualitative dependence of machine permeances and inductances on current

one can express the main inductance L_{main} as

$$L_{main} = \frac{w^2}{R_{m,Fe} + R_{m,gap}} = \frac{L_{Fe} \cdot L_{gap}}{L_{Fe} + L_{gap}} \tag{4.8}$$

where

$$L_{Fe} = \frac{w^2}{R_{m,Fe}}; L_{gap} = \frac{w^2}{R_{m,gap}} \tag{4.9}$$

The value of main inductance of a coil depends on the level of saturation in iron. At low values of flux density, the B–H curve of iron is linear; therefore, $R_{m,Fe} \ll R_{m,gap}$ and $L_{main} \approx L_{gap}$. Deep in saturation, the iron reluctance dominates, $R_{m,Fe} \gg R_{m,gap}$, and $L_{main} \approx L_{Fe}$, Fig. 4.3.

At linear portion of B–H curve (low values of flux density), the MMF drop Θ_{Fe} across iron of an electric machine is significantly lower than the MMF drop Θ_{gap} across its air gap. Accordingly, one can write for the jth harmonic of air gap flux

$$\mathrm{d}\Phi_{\mathrm{main},j} = l_{\mathrm{ax}} B_j(x)\mathrm{d}x = \frac{\mu_0 l_{\mathrm{ax}}}{\delta}\Theta_{\mathrm{gap},j}(x)\mathrm{d}x \tag{4.10}$$

considering air gap width δ constant. Furthermore,

$$\mathrm{d}\Psi_{\mathrm{main},j} = w \cdot f_{\mathrm{w},j}\mathrm{d}\Phi_{\mathrm{main},j} = w \cdot f_{\mathrm{w},j}\frac{\mu_0 l_{\mathrm{ax}}}{\delta}\Theta_{\mathrm{gap},j}(x)\mathrm{d}x \tag{4.11}$$

Assuming series connection of all coils over $2p$ poles, one can write for the flux $\Psi_{\mathrm{main},j}$ concatenated by the jth spatial harmonic

$$\Psi_{\mathrm{main},j} = 2pw \cdot f_{\mathrm{w},j}\frac{\mu_0 l_{\mathrm{ax}}}{\delta}\int_0^{\tau_\mathrm{p}} \Theta_{\mathrm{gap},j}(x)\mathrm{d}x = L_{\mathrm{gap},j} \cdot i \tag{4.12}$$

Air gap MMF distribution $\Theta_{\mathrm{gap}}(x)$ is equal to the sum of orthogonal harmonic terms $\Theta_{\mathrm{gap},j}(x)$, each of which is characterized by its own gap harmonic inductance $L_{\mathrm{gap},j}$. Accordingly, the total air gap inductance L_{gap} is equal to the sum of all gap harmonic inductances $L_{\mathrm{gap},j}$.

As shown in Chap. 2, a winding with w turns per pole-carrying current i generates an air gap MMF distribution $\Theta_{\mathrm{gap}}(x)$ described as

$$\Theta_{\mathrm{gap}}(x) = \frac{4}{\pi}\cdot i \cdot w\sum_{j=1}^{\infty}\frac{\sin\frac{j\pi}{2}}{j}f_{\mathrm{w},j}\sin j\frac{\pi}{\tau_\mathrm{p}}x \tag{4.13}$$

By substituting Eq. 4.13 in 4.12, one can express the total air gap inductance $L_{\mathrm{gap,cyl}}$ of a winding in a machine with cylindrical air gap as

$$L_{\mathrm{gap,cyl}} = \frac{8}{\pi}pw^2\frac{\mu_0 l_{\mathrm{ax}}}{\delta}\sum_{j=1}^{\infty}\left(\frac{\sin\frac{j\pi}{2}}{j}f_{\mathrm{w},j}^2\int_0^{\tau_\mathrm{p}}\sin j\frac{\pi}{\tau_\mathrm{p}}x\,\mathrm{d}x\right) \tag{4.14}$$

or

$$L_{\mathrm{gap,cyl}} = \frac{16}{\pi^2}p\frac{\mu_0 l_{\mathrm{ax}}\tau_\mathrm{p}}{\delta}w^2\sum_{j=1}^{\infty}\frac{f_{\mathrm{w},j}^2}{j^2} \tag{4.15}$$

The equivalent circuit of a coil in a cylindrical rotor machine is represented in Fig. 4.4. The total leakage flux outside the air gap is represented by the leakage inductance L_σ, whereas R stands for the coil resistance. Each harmonic of air gap MMF creates its own concatenated flux, which, divided by the coil current, yields

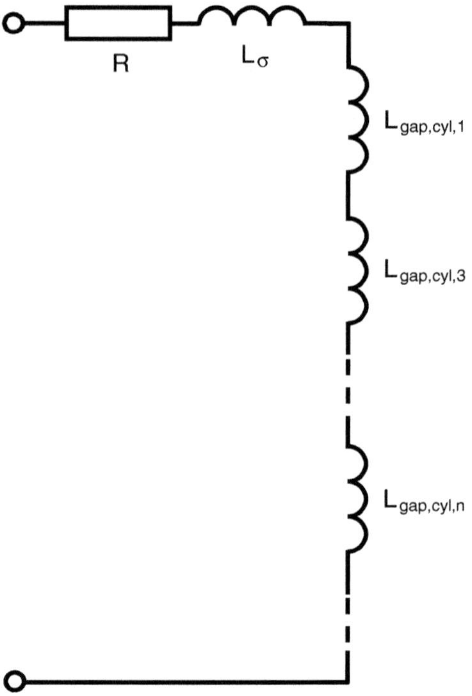

Fig. 4.4 Equivalent circuit of a coil in a machine with cylindrical rotor in Fig. 4.2

the corresponding harmonic component $L_{gap,cyl,n}$ of the gap inductance. As opposed to a simple coil of a transformer, a coil in an electric machine is characterized by an infinite number of inductances, each of which stands for a component of air gap flux created by a particular MMF harmonic.

For a given harmonic inductance $L_{gap,cyl,j}$ all harmonic inductances of the order ℓ, where $\ell \neq j$, have a meaning of leakage inductance due to orthogonality of harmonic in terms of air gap MMF. Consequently, air gap harmonic inductances with orders larger than 1 create in a machine with cylindrical air gap *the air gap leakage inductance* $L_{gap,cyl,\sigma}$ for the fundamental harmonic

$$L_{gap,cyl} = L_{gap,cyl,1} + L_{gap,cyl,\sigma} \tag{4.16}$$

where

$$L_{gap,cyl,1} = \frac{16}{\pi^2} p \frac{\mu_0 l_{ax} \tau_p}{\delta} w^2 f_{w,1}^2 \tag{4.17}$$

and

$$L_{\mathrm{gap,cyl},\sigma} = \frac{16}{\pi^2} p \frac{\mu_0 l_{\mathrm{ax}} \tau_{\mathrm{p}}}{\delta} w^2 \sum_{j>1} \frac{f_{\mathrm{w},j}^2}{j^2} \tag{4.18}$$

Maximum and minimum values of the sum in Eq. 4.15

$$S_0 = \sum_{j=1}^{\infty} \left(\frac{f_{\mathrm{w},j}}{j}\right)^2 \tag{4.19}$$

can be estimated by considering rectangular and pure sinusoidal MMF distribution. For a winding with one full-pitch coil per pole, $y = \tau_{\mathrm{p}}$, which creates rectangular MMF distribution, $f_{\mathrm{w},j} = 1$, $j = 1, 3, 5, \ldots$, and

$$S_{0,\max} = \sum_{j=1,3,5,\ldots}^{\infty} \left(\frac{1}{j}\right)^2 = \frac{\pi^2}{8} \approx 1.2337 \tag{4.20}$$

For a winding creating pure sinusoidal MMF distribution, $f_{\mathrm{w},j} = 0$, $j = 3, 5, \ldots$, one can write

$$S_{0,\min} = \sum_{j=1,3,5,\ldots}^{\infty} \left(\frac{f_{\mathrm{w},j}}{j}\right)^2 = 1 \tag{4.21}$$

Accordingly, the air gap leakage inductance $L_{\mathrm{gap,cyl},\sigma}$ of a winding with rectangular MMF distribution is equal to $\pi^2/8 - 1 = 23.4\ \%$ of its inductance due to fundamental component of air gap MMF.

For an arbitrary winding distribution, one can write

$$\frac{L_{\mathrm{gap,cyl},\sigma}}{L_{\mathrm{gap,cyl}}} = \frac{\sum\limits_{j=2,3,\ldots}^{\infty} \left(\frac{f_{\mathrm{w},j}}{j}\right)^2}{f_{\mathrm{w},1}^2 + \sum\limits_{j=2,3,\ldots}^{\infty} \left(\frac{f_{\mathrm{w},j}}{j}\right)^2} \tag{4.22}$$

Magnetic energy stored in air gap can now be calculated as (see Eqs. 4.15 and 4.18)

$$W_{\mathrm{mag,gap}} = \frac{1}{2} L_{\mathrm{gap}} i^2 = i^2 \frac{8}{\pi^2} p \frac{\mu_0 l_{\mathrm{ax}} \tau_{\mathrm{p}}}{\delta} w^2 S_0 \tag{4.23}$$

Consider again the magnetic circuit in Fig. 4.2, now connected to a voltage source u. The voltage differential equation for the coil with resistance R can be written as

$$u = iR + \frac{\mathrm{d}\Psi}{\mathrm{d}t} = iR + w \frac{\mathrm{d}\Phi}{\mathrm{d}t} \tag{4.24}$$

and the energy balance equation as

$$u i \mathrm{d}t = i^2 R \mathrm{d}t + w i \mathrm{d}\Phi \tag{4.25}$$

A portion of the differential of energy $u \cdot i \cdot \mathrm{d}t$ supplied by the source covers losses in resistor R; the rest in the amount of $w \cdot i \cdot \mathrm{d}\Phi$ is transferred to the coil and is stored in air and iron:

$$\mathrm{wid}\Phi = \mathrm{d}W_{\mathrm{mag,air}} + \mathrm{d}W_{\mathrm{mag,Fe}} \tag{4.26}$$

The differential of magnetic energy delivered to iron, $\mathrm{d}W_{\mathrm{mag,Fe}}$, splits into a component which covers eddy current and hysteresis losses in iron and a component which is stored.

The differential of magnetic energy $\mathrm{d}W_{\mathrm{mag,air}}$ goes into air of magnetic circuit (leakage paths and air gap):

$$\mathrm{d}W_{\mathrm{mag,air}} = \Theta_{\mathrm{gap}} \mathrm{d}\Phi_{\mathrm{main}} + \Theta_{\mathrm{leak}} \mathrm{d}\Phi_{\mathrm{leak}} \tag{4.27}$$

Differential of magnetic energy transferred to air gap by jth spatial harmonic of MMF is, accordingly:

$$\mathrm{d}W_{\mathrm{mag,tr,gap,2}j} = \Theta_{\mathrm{gap},j}(x) \frac{\mu_0 l_{\mathrm{ax}}}{\delta} \mathrm{d}\Theta_{\mathrm{gap},j}(x) \tag{4.28}$$

and its integral

$$W_{\mathrm{mag,tr,gap,2}j} = \frac{1}{2} \frac{\mu_0 l_{\mathrm{ax}}}{\delta} 2p \int_0^{\tau_{\mathrm{p}}} \Theta^2_{\mathrm{gap},j}(x) \mathrm{d}x \tag{4.29}$$

or

$$W_{\mathrm{mag,tr,gap,2}j} = i^2 \frac{8}{\pi^2} p \frac{\mu_0 l_{\mathrm{ax}} \tau_{\mathrm{p}}}{\delta} w^2 \frac{f^2_{\mathrm{w},j}}{j^2} \tag{4.30}$$

The total magnetic energy transferred to air gap is, accordingly

$$W_{\mathrm{mag,tr,gap}} = i^2 \frac{8}{\pi^2} p \frac{\mu_0 l_{\mathrm{ax}} \tau_{\mathrm{p}}}{\delta} w^2 S_0 \tag{4.31}$$

which is, as expected, equal to the magnetic energy stored in air gap calculated by means of winding inductance, Eq. 4.23.

Case Study 4.1: For a two-pole cylindrical rotor machine with a single full-pitch coil with $w_{2\mathrm{p}} = 2w$ turns per pole pair (Fig. 4.5), the expression for L_{gap} (Eqs. 4.15 and 4.17) becomes the following:

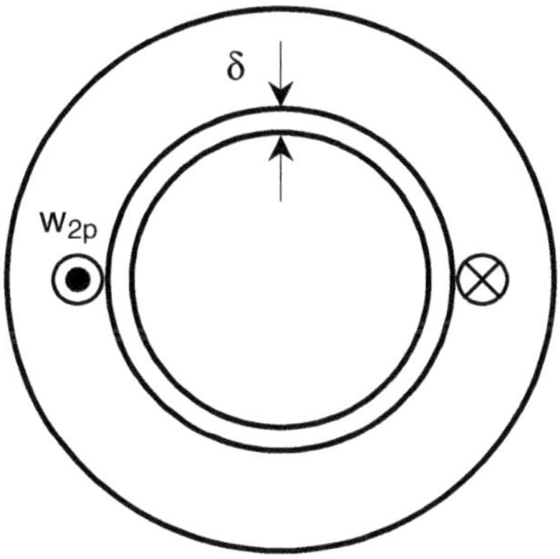

Fig. 4.5 Two-pole machine with cylindrical stator and rotor and a full-pitch coil having a total of $w_{2p} = 2w$ turns per pole pair

$$L_{\text{gap}} = \frac{16}{\pi^2} p \frac{\mu_0 l_{\text{ax}} \tau_p}{\delta} w^2 S_0 = \frac{(w_{2p})^2}{\frac{1}{\mu_0} \frac{2\delta}{\tau_p l_{\text{ax}}}} \tag{4.32}$$

which is nothing but the self-inductance of a coil with w_{2p} turns in a magnetic circuit with an air gap width 2δ and cross-sectional area $\tau_p\ \ell_{\text{ax}}$, Fig. 4.6a.

In Fig. 4.6 the magnetic circuit of a coil with w_{2p} turns containing iron portion and air gap with width 2δ (Fig. 4.6a) is compared with magnetic circuit of a two-pole machine having a full-pitch coil with w turns per pole, Fig. 4.6b.

Neglecting MMF drop across iron, flux density in the air gap of a simple magnetic circuit is constant, Fig. 4.6a. The rectangular flux density distribution in the air gap of an electric machine in Fig. 4.6b is in fact an infinite sum of sinusoidal distributions of flux density harmonics with different period lengths, each of which is characterized by its own air gap inductance. The sum of air gap inductances of all harmonics is equal to the air gap inductance of the coil.

Although the air gap inductance L_{gap} of the coil in the simple magnetic circuit in Fig. 4.6a is formally described by an identical set of parameters as the air gap inductance of a coil in an electric machine, the two have completely different physical meanings. Whereas the air gap inductance of a coil in a simple magnetic circuit as in Fig. 4.6a is a plain number describing global effects of the coil current in the magnetic circuit, the air gap inductance of a coil in an electric machine has an infinite number of components, each of which belongs to another spatial harmonic, and all of them being created by the same coil ampere-turns $i \cdot w$.

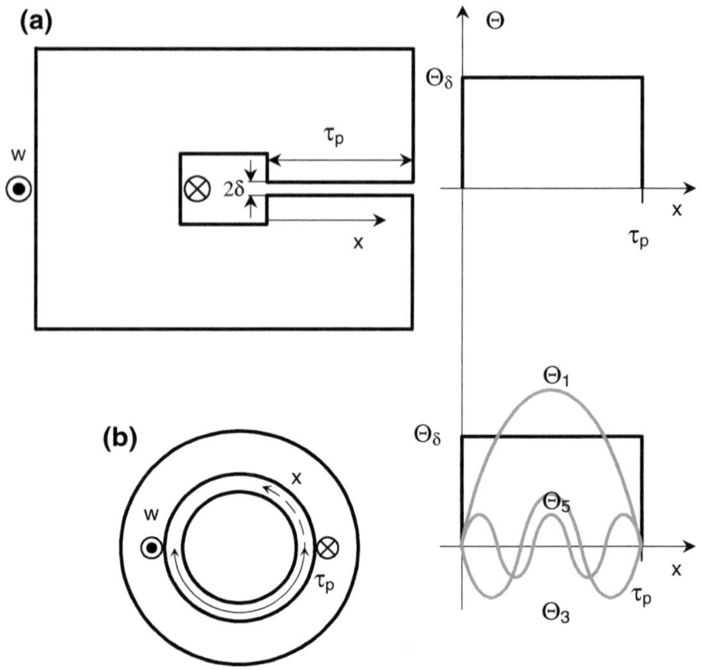

Fig. 4.6 Illustrating different physical meaning of air gap flux density distribution and air gap inductance in a simple magnetic circuit with an air gap (**a**), and in an electric machine (**b**). The horizontal coordinate x in the magnetic circuit above gives solely the direction along which one of the dimensions of the air gap is defined. The circumferential coordinate x in the machine below, on the opposite, has a more profound meaning: The rotating magnetic field spreads in a machine along this coordinate

4.3 Air Gap Inductance of a Coil in a Machine with Variable Air Gap Width—Rotational Harmonics of Concatenated Flux

Consider now a $2p$-pole electric machine with cylindrical stator and salient pole rotor, Fig. 4.7. Circumferential coordinate x in Fig. 4.7 is fixed to the stator. The coordinate x_0 stands for rotor shift relative to the point of maximum MMF denoted by $x = 0$.

Geometric air gap width $\delta(x - x_0)$ is a periodical function of circumferential coordinate x and rotor shift x_0. Since air gap width appears in denominator of the expression for air gap flux density, it is replaced with its reciprocal $\Delta(x - x_0)$ with identical periodicity properties as the air gap width $\delta(x - x_0)$, Fig. 4.8. In this way, the rules for product (instead of quotient) of trigonometric functions can be applied.

Making use of substitutions $\Delta_{min} = 1/\delta_{max}$ and $\Delta_{max} = 1/\delta_{min}$, the reciprocal of air gap width can be defined as

$$\Delta(x - x_0) = \frac{\Delta_{max} + \Delta_{min}}{2} + \\ + \frac{\Delta_{max} - \Delta_{min}}{2} \sum_{n=1,3,5,\ldots}^{\infty} \frac{\sin\frac{n\pi}{2}}{n} \cdot \sin n\frac{b_p}{\tau_p}\frac{\pi}{2} \cdot \cos 2n\frac{\pi}{\tau_p}(x - x_0) \tag{4.33}$$

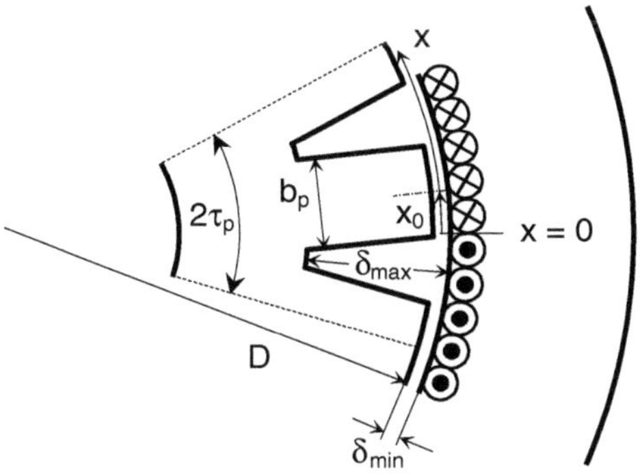

Fig. 4.7 Two poles of an electric machine with cylindrical stator and salient pole rotor. The point $x = 0$ denotes the position of the maximum of fundamental component of air gap MMF

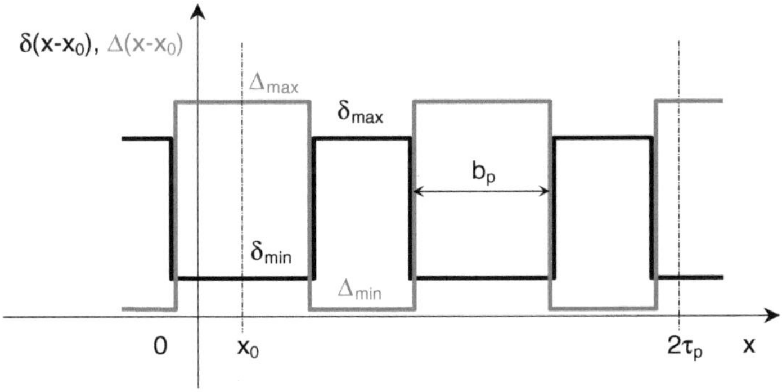

Fig. 4.8 Air gap width $\delta(x - x_0)$ and its reciprocal $\Delta(x - x_0) = 1/\delta(x - x_0)$ for the machine geometry in Fig. 4.6

Introducing

$$f_{\delta,n} = \frac{\sin\frac{n\pi}{2}}{n}\sin n\frac{b_p}{\tau_p}\frac{\pi}{2}; \ \Delta_{ave} = \frac{\Delta_{max} + \Delta_{min}}{2}; \ \Delta_{diff} = \frac{\Delta_{max} - \Delta_{min}}{2} \tag{4.34}$$

one can write for the reciprocal of air gap width

$$\Delta(x-x_0)=\Delta_{\text{ave}}+\Delta_{\text{diff}}\sum_{n=1,3,5,\ldots}^{\infty}f_{\delta,n}\cos 2n\frac{\pi}{\tau_{\text{p}}}(x-x_0) \tag{4.35}$$

The accumulated magnetic energy in the air gap, $W_{\text{mg}}(x_0)$, can be expressed for a given rotor shift x_0 by means of Eq. 4.29 as

$$W_{\text{mag}}(x_0)=p\mu_0 l_{\text{ax}}\int_0^{\tau_{\text{p}}}\sum_{j=1}^{\infty}\Theta_{\text{gap}}^2(x)\Delta(x-x_0)\text{d}x \tag{4.36}$$

As previously discussed, magnetic energy accumulated in air gap has an infinite number of components, each of which is generated by harmonics of MMF and air gap width which interact with each other at particular number of pole pairs. In particular, the jth harmonic component of the accumulated magnetic energy is generated by those harmonics of the air gap MMF and air gap width, the orders of which satisfy condition that their sum or difference is equal to j.

By using Eq. 4.13, the magnetic energy in air gap can further be expressed as

$$\begin{aligned}W_{\text{mg}}(x_0)=pl_{\text{ax}}\mu_0\left(\frac{4}{\pi}iw\right)^2\Bigg[&\Delta_{\text{ave}}\int_0^{\tau_{\text{p}}}\left(\sum_{j=1,3,5,\ldots}^{\infty}\frac{f_{\text{w},j}}{j}\sin j\frac{\pi}{\tau_{\text{p}}}x\right)^2\text{d}x+\\&+\Delta_{\text{diff}}\int_0^{\tau_{\text{p}}}\left(\sum_{j=1,3,5,\ldots}^{\infty}\frac{f_{\text{w},j}}{j}\sin j\frac{\pi}{\tau_{\text{p}}}x\right)^2\sum_{n=1,3,5,\ldots}^{\infty}f_{\delta,n}\cos 2n\frac{\pi}{\tau_{\text{p}}}(x-x_0)\text{d}x\Bigg]\end{aligned} \tag{4.37}$$

The first term in Eq. 4.37, containing Δ_{ave}, gives as a result a number representing the component of accumulated magnetic energy independent of the rotor shift x_0. The corresponding air gap inductance $L_{\text{gap,ave}}$ can be written as (Eq. 4.15):

$$L_{\text{gap,ave}}=\frac{16}{\pi^2}p\mu_0 l_{\text{ax}}\tau_{\text{p}}\Delta_{\text{ave}}w^2S_0 \tag{4.38}$$

Introduce now $L'_{\text{gap,max}}$ and $L'_{\text{gap,min}}$, as well as the saliency ratio r_δ, defined as

$$\begin{aligned}&L'_{\text{gap,max}}=\frac{16}{\pi^2}p\tau_{\text{p}}l_{\text{ax}}\mu_0 w^2\frac{1}{\delta_{\text{min}}};\,L'_{\text{gap,min}}=L'_{\text{gap,max}}\frac{\delta_{\text{min}}}{\delta_{\text{max}}};\\&r_\delta=\frac{L'_{\text{gap,min}}}{L'_{\text{gap,max}}}=\frac{\delta_{\text{min}}}{\delta_{\text{max}}}\end{aligned} \tag{4.39}$$

which help one express the component of air gap inductance corresponding to the average air gap width (as defined in Eq. 4.31) as

$$L_{\mathrm{gap,ave}} = \frac{L'_{\mathrm{gap,max}} + L'_{\mathrm{gap,min}}}{2} \sum_{j=1,3,5,\ldots}^{\infty} \left(\frac{f_{\mathrm{w},j}}{j}\right)^2 = \frac{L'_{\mathrm{gap,max}}}{2}(1 + r_\delta) \cdot S_0 \tag{4.40}$$

The second term in Eq. 4.37 is a periodical function of the rotor shift x_0. Here, the square of the infinite sum can be rewritten as

$$\begin{aligned}\left(\sum_{j=1,3,5,\ldots}^{\infty} \frac{f_{\mathrm{w},j}}{j} \sin j\frac{\pi}{\tau_{\mathrm{p}}}x\right)^2 &= \sum_{j=1,3,5,\ldots}^{\infty} \left(\frac{f_{\mathrm{w},j}}{j} \sin j\frac{\pi}{\tau_{\mathrm{p}}}x\right)^2 + \\ &+ 2\sum_{j=3,5,\ldots}^{\infty} \sum_{i=1}^{j-2} \frac{f_{\mathrm{w},j} f_{\mathrm{w},i}}{j\ i} \sin j\frac{\pi}{\tau_{\mathrm{p}}}x \cdot \sin i\frac{\pi}{\tau_{\mathrm{p}}}x\end{aligned} \tag{4.41}$$

because

$$\left(\sum_i a_i\right)^2 = \sum_i a_i^2 + 2\sum_{k<j} a_k \cdot a_j \tag{4.42}$$

The variable component of accumulated magnetic energy in the air gap can now be represented as

$$\begin{aligned}W_{\mathrm{mg}}(x_0) = p l_{\mathrm{ax}} \mu_0 \left(\frac{4}{\pi} iw\right)^2 \Delta_{\mathrm{diff}} \int_0^{\tau_{\mathrm{p}}} \sum_{n=1,3,5,\ldots}^{\infty} f_{\delta,n} \cos 2n\frac{\pi}{\tau_{\mathrm{p}}}(x - x_0) \\ \cdot \left[\sum_{j=1,3,5,\ldots}^{\infty} \left(\frac{f_{\mathrm{w},j}}{j} \sin j\frac{\pi}{\tau_{\mathrm{p}}}x\right)^2 + 2\sum_{j=3,5,\ldots}^{\infty} \sum_{i=1}^{j-2} \frac{f_{\mathrm{w},j} f_{\mathrm{w},i}}{j\ i} \sin j\frac{\pi}{\tau_{\mathrm{p}}}x \cdot \sin i\frac{\pi}{\tau_{\mathrm{p}}}x\right] \mathrm{d}x\end{aligned} \tag{4.43}$$

The expression for $W_{\mathrm{mg}}(x_0)$ in Eq. 4.43 contains two products of infinite sums, the non-orthogonal terms of which build the spatial harmonics of the periodical component of self-inductance. These products can be denoted as Π_1 and Π_2, and defined as

$$\Pi_1 = \left[\sum_{j=1,3,5,\ldots}^{\infty} \left(\frac{f_{\mathrm{w},j}}{j} \sin j\frac{\pi}{\tau_{\mathrm{p}}}x\right)^2\right] \cdot \sum_{n=1,3,5,\ldots}^{\infty} f_{\delta,n} \cos 2n\frac{\pi}{\tau_{\mathrm{p}}}(x - x_0) \tag{4.44}$$

and

$$\Pi_2 = \left[\sum_{j=3,5,\ldots}^{\infty} \sum_{i=1}^{j-2} \frac{f_{\mathrm{w},j} f_{\mathrm{w},i}}{j\ i} \sin j\frac{\pi}{\tau_{\mathrm{p}}}x \cdot \sin i\frac{\pi}{\tau_{\mathrm{p}}}x\right] \cdot \sum_{n=1,3,5,\ldots}^{\infty} f_{\delta,n} \cos 2n\frac{\pi}{\tau_{\mathrm{p}}}(x - x_0) \tag{4.45}$$

Only those factors in the product Π_1 with harmonic orders $j = n$ are non-orthogonal on interval $[0, D\pi]$. Therefore,

$$\int_0^{\tau_p} \Pi_1 dx = \frac{\tau_p}{4} \sum_{j=1,3,5,\ldots}^{\infty} \left(\frac{f_{w,j}}{j}\right)^2 f_{\delta,n} \cdot \cos 2j \frac{\pi}{\tau_p} x_0 \tag{4.46}$$

For a single coil with $f_{w,j} = 1, j = 1, 3, 5,\ldots$,the amplitude of a particular higher harmonic of air gap inductance due to rotor saliency decreases proportional to the square of reciprocal of its order, as illustrated in Fig. 4.9.

In the product Π_2 three periodic functions must fulfill simultaneously the non-orthogonality condition in order to build a particular higher harmonic of the self-inductance. The term

$$\sin j \frac{\pi}{\tau_p} x \cdot \sin i \frac{\pi}{\tau_p} x \cdot \sin 2n \frac{\pi}{\tau_p} (x - x_0) \tag{4.47}$$

in Eq. 4.39 can be represented as

$$\frac{1}{2}\left[\sin(i-j)\frac{\pi}{\tau_p}x + \sin(i+j)\frac{\pi}{\tau_p}x\right] \cdot \sin 2n \frac{\pi}{\tau_p} (x - x_0) \tag{4.48}$$

which means that the $2n$th harmonic of the air gap width can create a non-zero harmonic of self-inductance with ith and jth harmonic of the winding MMF if and only if

$$i \pm j = \pm 2n \tag{4.49}$$

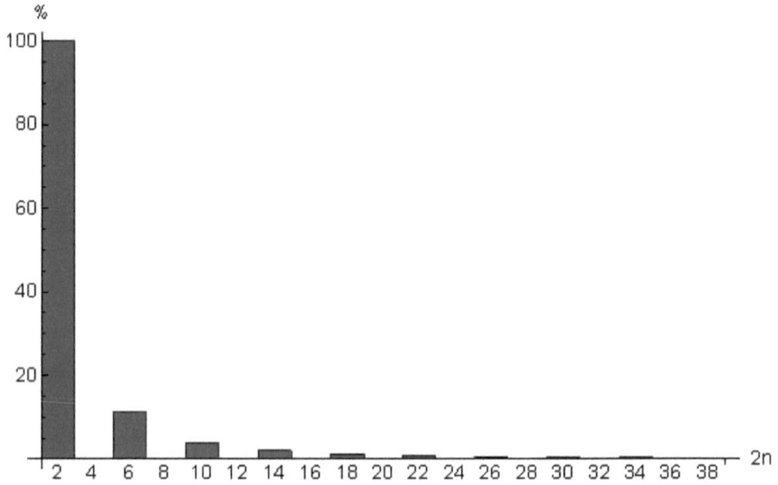

Fig. 4.9 Maximum amplitudes of components of air gap inductance created by a single harmonic of the winding MMF, n denoting the order of harmonic

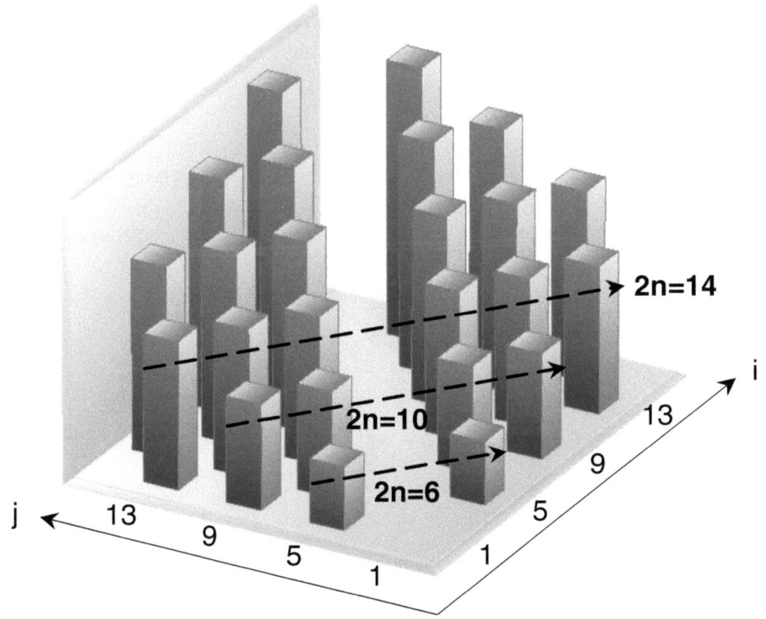

Fig. 4.10 Relationship between harmonic orders i and j of the MMF distribution and harmonic order n of the air gap width distribution satisfying $i + j = 2n$

All harmonics of orders i, j, and n which satisfy the condition 4.49 are non-orthogonal. For these harmonics, one can write

$$2\int_{0}^{\tau_p} \Pi_2 \mathrm{d}x = \pm\frac{\tau_p}{2}\cdot \sum_{n=1,3,5,\ldots}^{\infty} f_{\delta,n}\cdot\left(\sum_{j=3,5,\ldots}^{\infty}\sum_{i=1}^{j-2}\frac{f_{\mathrm{w},j}}{j}\frac{f_{\mathrm{w},i}}{i}\right)\cdot\cos 2n\frac{\pi}{\tau_p}x_0 \qquad (4.50)$$

where positive sign applies for the case $i - j = 2n$, and negative for $i + j = 2n$.

For $i + j = 2n$, one can state that the higher the order of air gap width harmonic n, the more harmonics of air gap MMF contribute to the nth harmonic of air gap inductance. The relationship between the orders i and j of the air gap MMF distribution harmonics and the order n of the air gap width distribution harmonic which together build the 2nth harmonic of the self-inductance is illustrated in Fig. 4.10.

Although an increasing number of air gap MMF harmonics contribute to a self-inductance harmonic of an increasing order, their influence on the total self-inductance decreases rapidly with increasing n.

The reason for such behavior is a decreasing amplitude of higher harmonics of the MMF, the upper limit of which $1/(i \cdot j)$ is shown in Fig. 4.11. Black drawn dashed lines in this figure denote the combinations of the MMF harmonics which built the same harmonic of self-inductance.

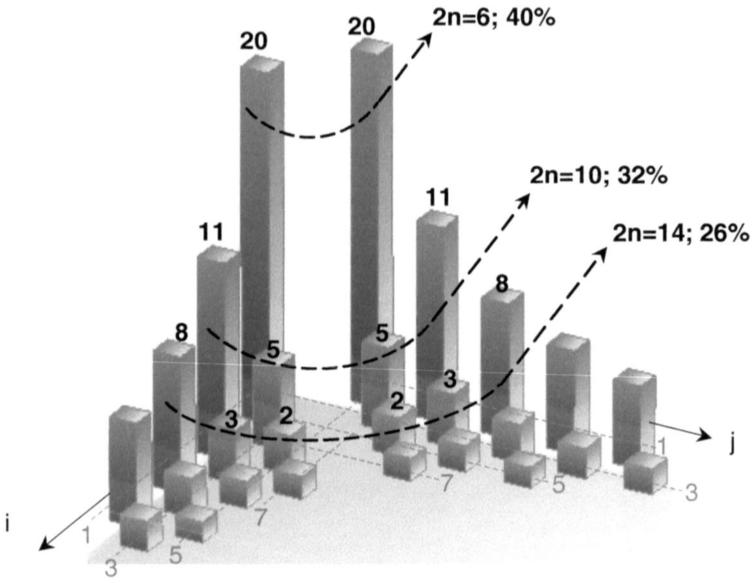

Fig. 4.11 Upper boundaries of products of amplitudes of *i*th and *j*th harmonics in % of the fundamental; only harmonics with amplitudes larger than 1 % are shown

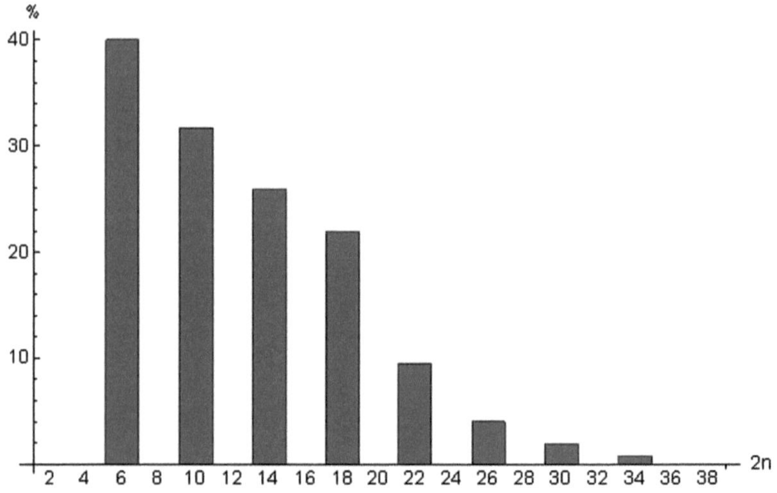

Fig. 4.12 Maximum amplitudes of components of self-inductance created by *i*th and *j*th harmonic of the winding MMF which satisfy the condition $i + j = 2n$

One can see in Fig. 4.11 that the sum of upper limits of products $1/(i \cdot j)$ decreases slowly as the order $2n$ of the self-inductance harmonic increases: The resulting amplitude of the 6. harmonic is 40 % of the fundamental, for the 10. harmonic it is equal to 32 % and for the 14. harmonic it comes to 26 %.

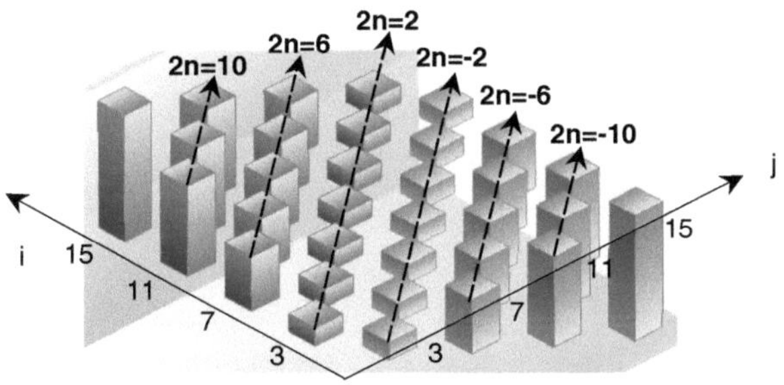

Fig. 4.13 Relationship between harmonic orders i and j of the MMF distribution and harmonic order n of the air gap width distribution, which satisfy the condition $i - j = \pm 2n$

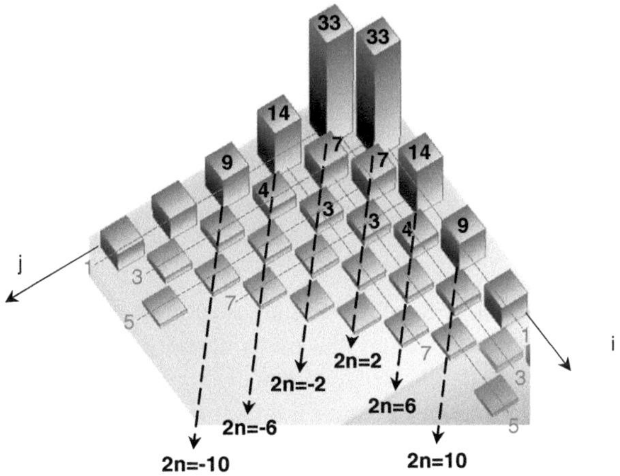

Fig. 4.14 Upper values of products of amplitudes of ith and jth harmonics in % of the fundamental; only harmonics with amplitudes larger than 1 % are shown

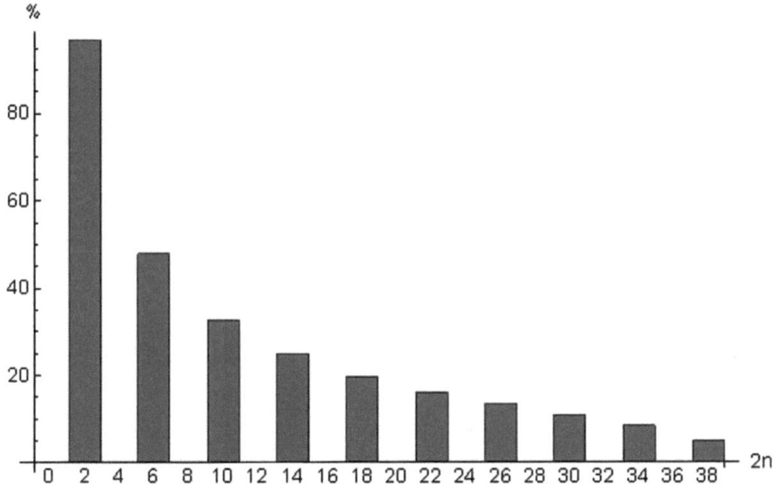

Fig. 4.15 Maximum amplitudes of components of air gap inductance created by ith and jth harmonic of the winding MMF which satisfy the condition $i - j = \pm 2n$

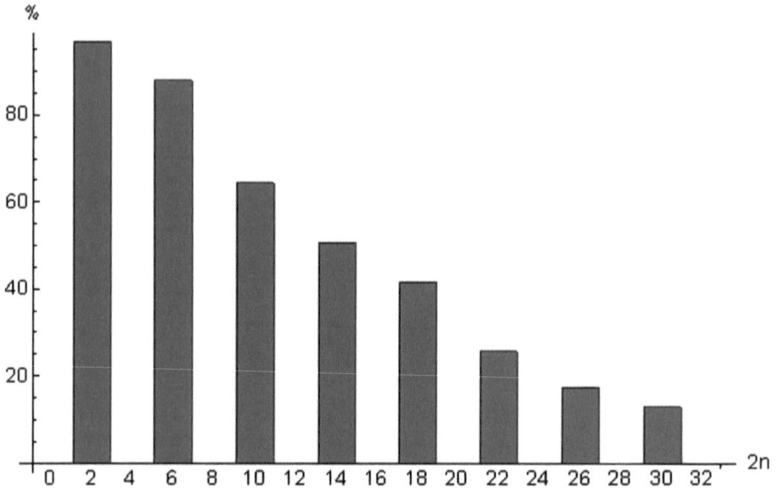

Fig. 4.16 Maximum amplitudes of components of air gap inductance created by *i*th and *j*th harmonic of winding MMF which satisfy the condition $i \pm j = \pm 2n$

Maximum amplitudes of self-inductance harmonics created by terms $i + j = 2n$ in the case of $f_{w,i} = f_{w,j} = 1$, $i = 1, 3, 5, \ldots, j = 1, 3, 5, \ldots$(single coil) are shown in Fig. 4.12. By comparing the amplitudes of higher harmonics in Figs. 4.12 and 4.9, one comes to the conclusion that the contribution of two MMF harmonics of different orders to a particular self-inductance harmonic is stronger than in the case of a single MMF harmonic—in other words, the impact of higher harmonics of the MMF must not be neglected.

The orders of harmonics which satisfy the condition $\mathrm{i} - \mathrm{j} = \pm 2\mathrm{n}$ are shown in Fig. 4.13, and the upper limit of MMF harmonics amplitudes, equal to $1/(i \cdot j)$, in Fig. 4.14. As opposed to the case $i + j = 2n$, each harmonic of self-inductance resulting from interaction of two harmonics of MMF, with a difference of orders equal to the order of harmonic of the self-inductance, has an infinite number of contributing MMF harmonics, Fig. 4.13.

Considering a much larger number of MMF harmonics contributing to a given harmonic of self-inductance through the mechanism of the difference of their order numbers than through the mechanism of sum, one comes to a conclusion that a particular harmonic of self-inductance is built predominantly as a result of interaction of two harmonics of MMF, the difference of order numbers of which is equal to the order of the self-inductance harmonic, see Figs. 4.12 and 4.15. Resulting maximum amplitudes of harmonic components of self-inductance as a function of the harmonic order n are shown in Fig. 4.16.

The component of accumulated magnetic energy dependent on the rotor shift x_0 can now be written as

$$W_{\mathrm{mg}}(x_0) = pl_{\mathrm{ax}}\mu_0\left(\frac{4}{\pi}iw\right)^2\Delta_{\mathrm{diff}}\frac{\tau_{\mathrm{p}}}{4}\sum_{n=1,3,5,\ldots}^{\infty} f_{\delta,n}\cdot\left[2\sum_{j=3,5,\ldots}^{\infty}\sum_{i=1}^{j-2}\frac{f_{\mathrm{w},j}}{j}\frac{f_{\mathrm{w},i}}{i}+ + \sum_{j=1,3,5,\ldots}^{\infty}\left(\frac{f_{\mathrm{w},j}}{j}\right)^2\right]\cdot\cos 2n\frac{\pi}{\tau_{\mathrm{p}}}x_0 \tag{4.51}$$

or by means of the variable component of air gap inductance $L_{\mathrm{var}}(x_0)$:

$$W_{\mathrm{mg}}(x_0) = \frac{1}{2}L_{\mathrm{var}}(x_0)\cdot i^2 \tag{4.52}$$

which helps one define the variable component of air gap inductance of a stator winding in a machine with cylindrical stator and salient pole rotor as

$$L_{\mathrm{gap,var}}(x_0) = \frac{8}{\pi^2}p\tau_{\mathrm{p}}l_{\mathrm{ax}}\mu_0 w^2\Delta_{\mathrm{diff}}\sum_{n=1,3,5,\ldots}^{\infty} f_{\delta,n}\cdot\left[2\sum_{j=3,5,\ldots}^{\infty}\sum_{i=1}^{j-2}\frac{f_{\mathrm{w},j}}{j}\frac{f_{\mathrm{w},i}}{i}+ + \sum_{j=1,3,5,\ldots}^{\infty}\left(\frac{f_{\mathrm{w},j}}{j}\right)^2\right]\cdot\cos 2n\frac{\pi}{\tau_{\mathrm{p}}}x_0 \tag{4.53}$$

or by means of the previously defined inductance $L'_{\mathrm{gap,max}}$ and saliency ratio r_{δ},:

$$L_{\mathrm{gap,var}}(x_0) = \frac{L'_{\mathrm{gap,max}}}{2}(1-r_{\delta})\sum_{n=1,3,5,\ldots}^{\infty} S_{2n}\cdot\cos 2n\frac{\pi}{\tau_{\mathrm{p}}}x_0 \tag{4.54}$$

where

$$S_{2n} = f_{\delta,n}\cdot\left[\sum_{j=3,5,\ldots}^{\infty}\sum_{i=1}^{j-2}\frac{f_{\mathrm{w},j}}{j}\frac{f_{\mathrm{w},i}}{i}+\sum_{j=1,3,5,\ldots}^{\infty}\left(\frac{f_{\mathrm{w},j}}{j}\right)^2\right] \tag{4.55}$$

or

$$S_{2n} = \frac{1}{n}\sin\frac{n\pi}{2}\sin n\frac{b_{\mathrm{p}}}{\tau_{\mathrm{p}}}\frac{\pi}{2}\cdot\left[\sum_{j=3,5,\ldots}^{\infty}\sum_{i=1}^{j-2}\frac{f_{\mathrm{w},j}}{j}\frac{f_{\mathrm{w},i}}{i}+S_0\right] \tag{4.56}$$

Minimum value of S_{2n} is obtained for an ideally sinusoidal MMF distribution, where $f_{\mathrm{w},j} = 0$, $j = 3, 5, \ldots$, and is equal to

$$S_{2n,\min} = \frac{1}{n}\sin\frac{n\pi}{2}\sin n\frac{b_{\mathrm{p}}}{\tau_{\mathrm{p}}}\frac{\pi}{2}\cdot S_{0,\min} = \frac{1}{n}\sin\frac{n\pi}{2}\sin n\frac{b_{\mathrm{p}}}{\tau_{\mathrm{p}}}\frac{\pi}{2} \tag{4.57}$$

Summarizing, one can state that the air gap inductance of a stator winding in a machine with salient rotor poles is a periodic function of double rotor shift x_0 and can be expressed by means of Eqs. 4.37 and 4.51 as

$$L(x_0) = L'_{\text{gap,max}} \left[\frac{S_0}{2}(1 + r_\delta) + \frac{1 - r_\delta}{2} \sum_{n=1,3,5,\ldots}^{\infty} S_{2n} \cdot \cos 2n \frac{\pi}{\tau_p} x_0 \right] \tag{4.58}$$

or simplified

$$L(x_0) = L_0 + \sum_{n=1,3,5,\ldots}^{\infty} L_{2n} \cos 2n \frac{\pi}{\tau_p} x_0 \tag{4.59}$$

Equation 4.56 is valid not only for salient pole machines ($n = 2$), but also for machines with any number of rotor teeth per pole N_{rp}, in which case one can write:

$$L(x_0) = L_0 + \sum_{n=N_{rp},3N_{rp},5N_{rp},\ldots}^{\infty} L_{2n} \cos 2n \frac{\pi}{\tau_p} x_0 \tag{4.60}$$

4.3.1 Salient Pole Rotor

If the winding with variable inductance as defined in Eq. 4.59 carries an alternating current described as

$$i(t) = \sum_{j=1,2,3,\ldots}^{\infty} I_j \cos(j\omega t - \varphi_j) \tag{4.61}$$

the concatenated flux is equal to

$$\Psi(x_0, t) = \left(L_0 + \sum_{n=1,3,5,\ldots}^{\infty} L_{2n} \cos 2n \frac{\pi}{\tau_p} x_0 \right) \cdot \sum_{j=1,2,3,\ldots}^{\infty} I_j \cos(j\omega t - \varphi_j) \tag{4.62}$$

When the rotor is at **standstill** (x_0 = const), the spectrum of concatenated flux in Eq. 4.62 is identical to the spectrum of winding current. The amplitude of a particular flux harmonic is dependent on the rotor position and on the amplitude of corresponding current harmonic.

If the rotor **rotates** at a constant angular speed Ω, its shift relative to the point of maximum MMF x_0 changes proportional to time:

$$x_0 = \Omega \frac{D}{2}(t - t_0) = p\Omega \frac{\tau_p}{\pi}(t - t_0) \tag{4.63}$$

where the rotor position at time instant $t = t_0$ is $x_0 = 0$. The air gap inductance of stator winding varies as a function of time

$$L(t_0, t) = L_0 + \sum_{n=1,3,5,\ldots}^{\infty} L_{2n} \cdot \cos[2np\Omega(t - t_0)] \tag{4.64}$$

Concatenated flux $\Psi(t_0, t)$ created by current $i(t)$ defined in Eq. 4.61 is equal to

$$\begin{aligned} \Psi(t_0, t) = L(t_0, t) \cdot i(t) = L_0 \sum_{j=1,2,3,\ldots}^{\infty} I_j \cos(j\omega t - \varphi_j) + \\ + \sum_{n=1,3,5,\ldots}^{\infty} L_{2n} \cdot \cos[2np\Omega(t - t_0)] \cdot \sum_{j=1,2,3,\ldots}^{\infty} I_j \cos(j\omega t - \varphi_j) \end{aligned} \tag{4.65}$$

The spectrum of concatenated flux can be separated in two groups. Flux harmonics in the first group, the amplitudes of which are proportional to L_0 as shown in the first indefinite sum in Eq. 4.64 have identical orders as current harmonics. The second group is composed of flux harmonics with angular frequencies

$$2np\Omega \pm j\omega \tag{4.66}$$

which depend on the rotor angular speed Ω and are in general not an integer multiple of Ω or ω.

If the electrical angular speed of rotation $\omega_{el} = p \cdot \Omega$ is equal to the angular frequency of the fundamental current harmonic ω (synchronism: $p \cdot \Omega = \omega$), the fundamental harmonic of current I_1 creates concatenated flux defined as

$$\Psi(t_0, t) = \left\{ L_0 + \sum_{n=1,3,5,\ldots}^{\infty} L_{2n} \cdot \cos[2n\omega(t - t_0)] \right\} \cdot I_1 \cos(\omega t) \tag{4.67}$$

which means that the fundamental harmonic of current produces an infinite spectrum of flux linkage, the components of which have frequencies which are odd integer multiples of the fundamental frequency of current. As a result of rotation of the salient pole rotor, spatial harmonics of self-inductance and the fundamental harmonic of current generate time harmonics of flux linkage, the order of which is equal to the sum or difference of the orders of current and inductance harmonics:

$$\begin{aligned} \Psi(t_0, t) = {} & L_0 \cdot I_1 \cos(\omega t) + I_1 \cdot \sum_{n=1,3,5,\ldots}^{\infty} L_{2n} \cdot \{\cos[(2n - 1)\omega t - 2n\omega t_0] \\ & + \cos[(2n + 1)\omega t - 2n\omega t_0]\} \end{aligned} \tag{4.68}$$

Time harmonics of concatenated flux generated by sinusoidal winding current in a salient pole machine rotating at an angular speed Ω and defined by an infinite sum in Eq. 4.68 are called **rotational flux harmonics**. At synchronous speed, the order of rotational flux harmonics is an integer multiple of the fundamental; at an arbitrary speed of rotation, the frequency of rotational harmonics of concatenated flux is not an integer multiple of the winding current frequency. Rotational harmonics of concatenated flux disappear at rotor standstill.

Time dependence of concatenated flux at standstill, synchronism, and at an arbitrary speed of rotation is shown in Fig. 4.17.

By selecting the phase shift φ_1 of the fundamental component of current equal to zero, $\varphi_1 = 0$, one has positioned the rotor centerline on the axis of stator MMF. If the corresponding rotor shift angle is denoted with $\delta_0 = \omega \cdot t_0 = -x_0\pi/\tau_p$, one can express the fundamental harmonic of concatenated flux as

$$\Psi_1(\delta_0, t) = I_1 \cdot [L_0 \cos(\omega t) + L_2 \cdot \cos(\omega t - 2\delta_0)] \quad (4.69)$$

The fundamental harmonic of concatenated flux $\Psi_1(\delta_0, t)$ in Eq. 4.69 can be represented as a sum of a component in phase with the current, and a component shifted for $\pi/2$ to the winding current, the latter being proportional to sin $2\delta_0$:

$$\Psi_1(\delta_0, t) = I_1 \cdot [(L_0 + L_2 \cos 2\delta_0) \cdot \cos(\omega t) + L_2 \sin 2\delta_0 \cdot \sin(\omega t)] \quad (4.70)$$

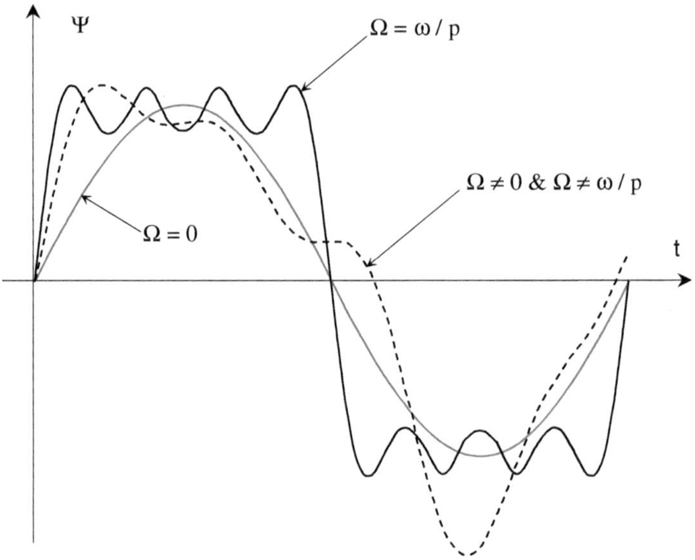

Fig. 4.17 Waveform of flux concatenated by a winding in a salient pole machine generated by sinusoidal current at different speeds of rotation

In order to shift the rotor relative to the stator MMF, mechanical force has to be applied. Mechanical work performed by force applied on the rotor is equal to electrical power dissipated on the *apparent synchronous resistance* R_{sa} of stator winding, defined as

$$R_{sa} = \omega L_2 \sin 2\delta_0 \tag{4.71}$$

In a similar manner, one can introduce the *apparent synchronous reactance* X_{sa} as

$$X_{sa} = \omega(L_0 + L_2 \cos 2\delta_0) = X_0 + X_2 \cos 2\delta_0 \tag{4.72}$$

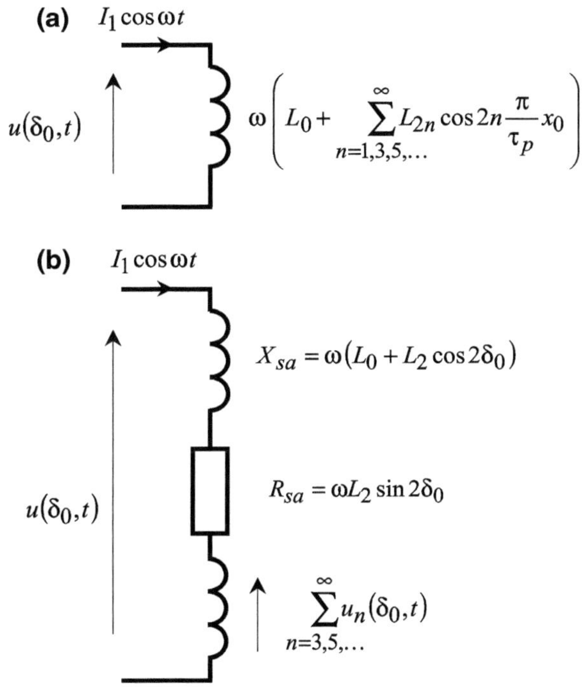

Fig. 4.18 Equivalent circuit of a winding in a machine with salient rotor poles: **a** at standstill; **b** at synchronous speed. At standstill only fundamental frequency of induced voltage exists; at synchronous speed, rotational harmonics of induced voltages are generated, the sum of which is added to the voltage drops across apparent synchronous reactance X_{sa} and resistance R_{sa}

Higher harmonics of concatenated flux in Eq. 4.68 are

$$\Psi_3(\delta_0, t) = I_1 L_2 \cos(3\omega t - 2\delta_0) \tag{4.73}$$

$$\Psi_5(\delta_0, t) = I_1 L_6 \cos(5\omega t - 6\delta_0) \tag{4.74}$$

$$\Psi_7(\delta_0, t) = I_1 L_6 \cos(7\omega t - 6\delta_0) \tag{4.75}$$

At synchronism, the fundamental component of winding current creates not only the fundamental component, but also an infinite spectrum of rotational harmonics of concatenated flux, time derivatives of which add to the voltage induced by the fundamental! The amplitudes of rotational harmonics of the order $2n \pm 1$, where $n = 1, 3, 5, \ldots$, are equal to each other, which makes them unique among other higher harmonics (e.g., current harmonics due to saturation, or MMF harmonics due to winding distribution, etc.).

Equivalent circuit of a winding in a machine with salient rotor poles is shown in Fig. 4.18a (at standstill), and 4.18b (at synchronism). Equivalent circuit in Fig. 4.18 is based on voltage differential equations:

$$u_1(\delta_0, t) = \frac{\mathrm{d}}{\mathrm{d}t}\Psi_1(\delta_0, t) = I_1 \cdot [-(X_0 + X_2\cos 2\delta_0) \cdot \sin(\omega t) + X_2 \sin 2\delta_0 \cdot \cos(\omega t)]$$
$$u_3(\delta_0, t) = \frac{\mathrm{d}}{\mathrm{d}t}\Psi_3(\delta_0, t) = -3X_2 \cdot I_1 \cdot \sin(3\omega t)$$
$$u_5(\delta_0, t) = \frac{\mathrm{d}}{\mathrm{d}t}\Psi_5(\delta_0, t) = -5X_6 \cdot I_1 \cdot \sin(5\omega t)$$
$$u_7(\delta_0, t) = \frac{\mathrm{d}}{\mathrm{d}t}\Psi_7(\delta_0, t) = -7X_6 \cdot I_1 \cdot \sin(7\omega t) \tag{4.76}$$

Voltage drop across apparent synchronous resistance R_{sa} is in phase with the coil current, whereas the voltage drop across apparent synchronous reactance X_{sa} leads to $\pi/2$.

Case Study 4.2: **Power Distribution in an Unexcited Salient Pole Synchronous Machine (Reluctance Machine) Fed by Sinusoidal Current**

Salient pole machine acts at standstill as a pure inductance, the value of which is determined by the rotor shift δ_0. At synchronism, the equivalent circuit contains three elements (Fig. 4.18b):

- Apparent synchronous reactance X_{sa}, having a component dependent on cosine of double rotor to stator MMF angle δ_0;
- Apparent synchronous resistance R_{sa}, being proportional to the sine of double rotor to stator MMF angle δ_0, and
- The reactance due to rotational harmonics, which acts as harmonic leakage.

Fundamental component of applied voltage $u_1(\delta_0, t)$ in Eq. 4.76 is a source of winding current I_1, which creates two voltage drops at the frequency of applied voltage in synchronism: $I_1 \cdot X_{sa}$ and $I_1 \cdot R_{sa}$, as in Fig. 4.19. Neglecting losses in the machine, one can state that the power dissipated on the apparent synchronous resistance R_{sa} of stator winding is equal to the mechanical power. Expressing the fundamental harmonic of winding current I_1 as

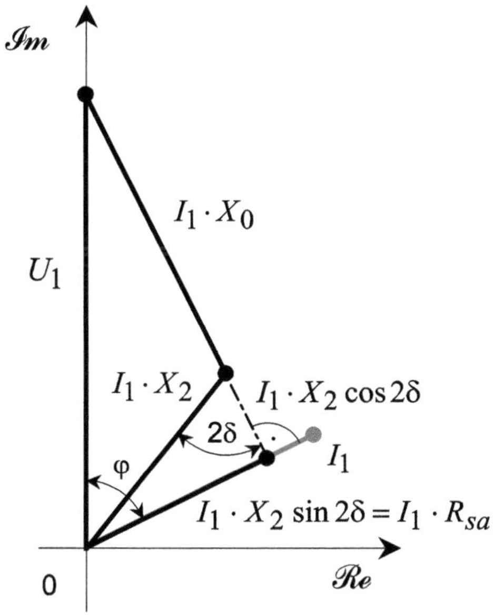

Fig. 4.19 Complex plane representation of electrical quantities (voltage diagram) in a reluctance machine at a given load

$$I_1 = \frac{U_1}{\sqrt{(X_2 \sin 2\delta)^2 + (X_0 + X_2 \cos 2\delta)^2}} \tag{4.77}$$

one can define the mechanical power P_{mech} as:

$$P_{\text{mech}} = I^2 R_{\text{sa}} = U_1^2 \frac{X_2 \sin 2\delta}{X_2^2 + X_0^2 + 2X_0 X_2 \cos 2\delta} \tag{4.78}$$

the maximum of which occurs at angle $\delta_{\text{max,P}}$ is equal to

$$\delta_{\text{max,P}} = \frac{1}{2} \arccos\left(-2\frac{X_0 X_2}{X_0^2 + X_2^2}\right) \tag{4.79}$$

Maximum mechanical power is equal to

$$P_{\text{mech,max}} = U_1^2 \frac{X_2}{X_0^2 - X_2^2} \tag{4.80}$$

or by substituting for X_0 and X_2 the maximum (X_d) and minimum (X_q) values of the coil reactance

$$X_0 = \frac{X_d + X_q}{2}; \quad X_2 = \frac{X_d - X_q}{2} \tag{4.81}$$

one obtains

$$\delta_{\max,\mathrm{P}} = \pm \arccos\left(\pm \frac{X_q}{\sqrt{X_d^2 + X_q^2}} \right) \tag{4.82}$$

and

$$P_{\mathrm{mech,max}} = \frac{U_1^2}{2}\left(\frac{1}{X_q} - \frac{1}{X_d}\right) \tag{4.83}$$

Power factor cos φ can be expressed by means of Fig. 4.19 as

$$\cos\varphi = \frac{X_2 \sin 2\delta}{\sqrt{(X_2 \sin 2\delta)^2 + (X_0 + X_2 \cos 2\delta)^2}} \tag{4.84}$$

and has an extreme at angle $\delta_{\max,\varphi}$, defined as

$$\delta_{\max,\varphi} = \frac{1}{2}\arccos\left(-\frac{X_2}{X_0}\right) \tag{4.85}$$

equal to

$$\cos\varphi_{\max} = \pm\frac{X_2}{X_0} = \pm\frac{X_d - X_q}{X_d + X_q} \tag{4.86}$$

The dependencies of winding current I, power factor cos φ, and mechanical power P_{mech} on load angle δ for $X_0 = 1$ p.u., $X_2 = 0.4$ p.u. and $U = 1$ p.u. are shown in Fig. 4.20. Here one recognizes at first sight the Achilles' heel of a reluctance machine: its poor power factor. Similarly to an induction machine, a reluctance machine consumes reactive power from the mains in order to operate, however, in a far larger extent than an induction machine.

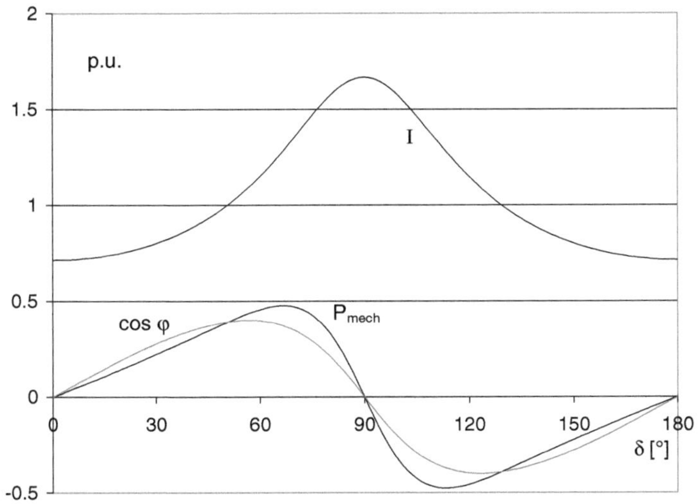

Fig. 4.20 Characteristic quantities of a reluctance machine for $X_0 = 1$ p.u., $X_2 = 0.4$ p.u. and $U = 1$ p.u

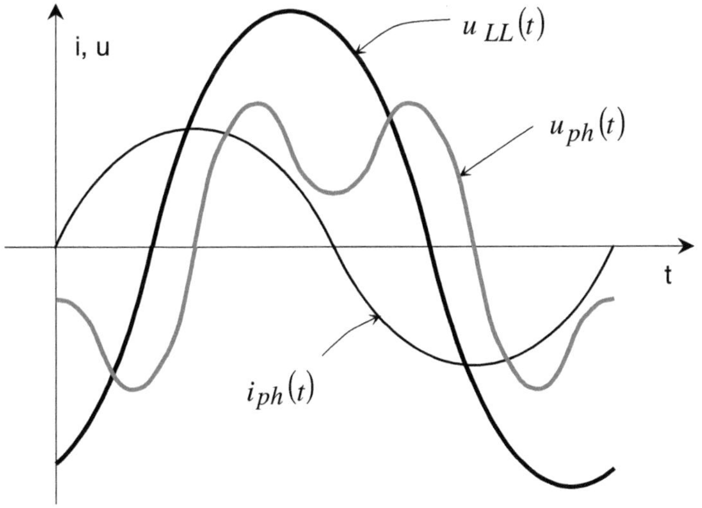

Fig. 4.21 Typical waveforms in a Y- connected stator winding of an unexcited salient pole synchronous machine (reluctance machine) at synchronism for $\delta_0 = 0$: phase current $i_{ph}(t)$, phase voltage $u_{ph}(t)$, and line-to-line voltage $u_{LL}(t)$

Previous considerations can be qualitatively illustrated with measured waveforms of stator voltages and current in an unexcited Y- connected 3-phase salient pole synchronous machine (reluctance machine) without neutral for a load angle of $\delta_0 = 0$, Fig. 4.21.

Beside fundamental component of voltage drop across the stator winding, sinusoidal phase current creates a rich spectrum of rotational harmonics of phase voltage at synchronism, see Eq. 4.76.

In a Y- connected machine without neutral, no integer multiple of the 3., 9., etc., harmonic of phase voltage due to saliency appears in the line-to-line voltage, which is almost sinusoidal, as shown in Fig. 4.21.

It is interesting that except for a very small number of references, such as [1, 2], the physics of salient pole synchronous machine, in particular of reluctance machine, in the literature on electric machines is represented in a somehow misleading manner. For example, the voltage–current diagram of a salient pole synchronous machine introduced in [3] (Fig. 4.22) is based on an assumption which suspends Kirchhoff's laws, namely that the stator winding reactances X_d and X_q are at the same time connected in series and parallel, Fig. 4.23.

Arbitrary stator current components I_d and I_q in Fig. 4.22 sum up into armature current I_a, suggesting that elements of circuit carrying the two currents are connected in parallel, as in Fig. 4.23a.

At the same time, however, voltage drops $I_d\,X_d$ and $I_q\,X_q$ are added to each other in Fig. 4.22, suggesting that the two reactances X_d and X_q are connected in series.

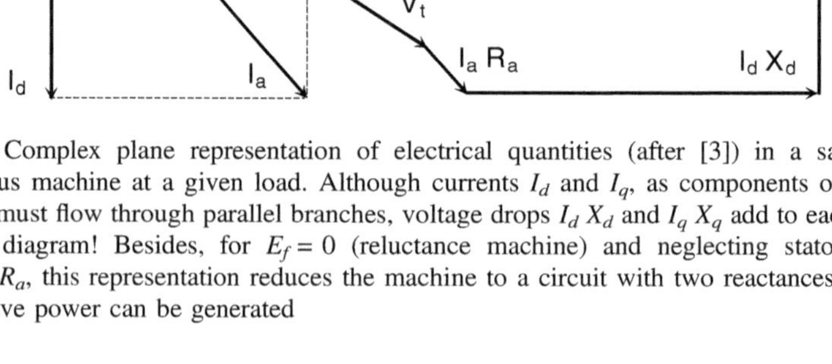

Fig. 4.22 Complex plane representation of electrical quantities (after [3]) in a salient pole synchronous machine at a given load. Although currents I_d and I_q, as components of armature current I_a must flow through parallel branches, voltage drops $I_d\,X_d$ and $I_q\,X_q$ add to each other in the same diagram! Besides, for $E_f = 0$ (reluctance machine) and neglecting stator winding resistance R_a, this representation reduces the machine to a circuit with two reactances, in which only reactive power can be generated

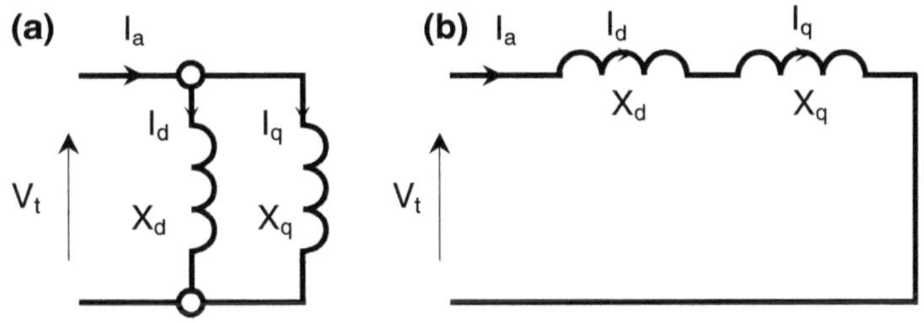

Fig. 4.23 Equivalent circuit of a reluctance machine derived from representation in Fig. 4.22 in the case of $E_f = 0$: current (**a**) and voltage (**b**) diagram. Both representations cannot obviously be valid at the same time. Furthermore, although only machine reactances are present, the angle φ between applied voltage V_t and total current I_a in Fig. 4.22 is different from 90° and active power can be generated, or absorbed! For this reason, the reluctance machine represented as suggested in Fig. 4.22 fulfills all criteria characterizing a perpetual motion machine

The applied principle, according to which the stator current separates into two components, which after that create voltage drops that add to each other, is not the only paradox in equivalent circuit in Fig. 4.22. Following this approach, a reluctance machine ($E_f = 0$) is reduced solely to reactances X_d and X_q, when armature winding resistance is neglected. Since a reactance can consume only reactive power, a machine represented exclusively by reactances X_d and X_q cannot produce mechanical power on its shaft. This simple logic seems not to work in the machine represented by the voltage and current diagram in Fig. 4.22: Despite the fact that the electric active power is equal to zero, the machine produces mechanical power! A device in which mechanical power is generated from zero active electrical power fulfills all criteria for a perpetual motion machine. This physically untenable representation is an ultimate consequence of confusing winding current with its MMF. Whereas MMF, as a spatial function, can be resolved in arbitrary spatial directions, such as d- and q-axes, winding current is a scalar, the resolving of which into components I_d and I_q is of no physical significance, because the scalar I is not dependent on spatial coordinates.

4.3.2 Slotted Rotor

If the winding with variable inductance caused by toothed rotor as defined in Eq. 4.60 carries an alternating current of the fundamental frequency ω, the concatenated flux is equal to

$$\Psi(x_0, t) = \left(L_0 + \sum_{n=N_{rp}, 3N_{rp}, \ldots}^{\infty} L_{2n} \cos 2n \frac{\pi}{\tau_p} x_0 \right) \cdot I_1 \cos(\omega t) \tag{4.87}$$

with N_{rp} denoting the number of rotor teeth per pole, $N_{rp} = N/(2p)$. By using Eq. 4.63, one becomes

$$\Psi(t_0, t) = \left\{ L_0 + \sum_{n=N_{rp},3N_{rp},\ldots}^{\infty} L_{2n} \cdot \cos[2np\omega(t - t_0)] \right\} \cdot I_1 \cos(\omega t) \tag{4.88}$$

Fundamental harmonic of air gap inductance in a $2p$-pole machine has $2p \cdot N_{rp}$ poles and is non-orthogonal with the fundamental harmonic ω of current at angular speed Ω of (see Eq. 4.63):

$$\Omega = \pm \frac{\omega}{pN_{rp}} = \pm 2\frac{\omega}{N} \tag{4.89}$$

in which case one can write

$$\begin{aligned}\Psi(t_0, t) = L_0 \cdot I_1 \cos(\omega t) + I_1 \cdot \sum_{n=N_{rp},3N_{rp},\ldots}^{\infty} L_{2n} \cdot \left\{ \cos\left[\left(\frac{2n}{N_{rp}} - 1\right)\omega t - \frac{2n}{N_{rp}}\omega t_0\right] \right. \\ \left. + \cos\left[\left(\frac{2n}{N_{rp}} + 1\right)\omega t - \frac{2n}{N_{rp}}\omega t_0\right] \right\}\end{aligned} \tag{4.90}$$

Replacing $\omega \cdot t_0$ by δ_0, one can express the fundamental harmonic of concatenated flux at angular speed Ω in a machine with N_{rp} teeth per rotor pole as

$$\Psi_1(\delta_0, t) = I_1 \cdot \left[\left(L_0 + L_{2N_{rp}} \cos 2\delta_0\right) \cdot \cos(\omega t) + L_{2N_{rp}} \sin 2\delta_0 \cdot \sin(\omega t)\right] \tag{4.91}$$

which is analogous to the expression for the fundamental harmonic of concatenated flux in a salient pole machine (Eq. 4.67). However, the load angle-dependent components L_2 in the two machine types do not have the same magnitude. Therefore, the maximum mechanical power due to rotor saliency of a toothed rotor machine is lower than in a salient pole (reluctance) machine. Nevertheless, significant torques can be developed due to change of self-inductance in a vernier motor, as well as in a squirrel cage induction motor at a low-speed rotation (synchronous torque).

4.4 Mutual Inductance Between Windings in a Machine with Cylindrical Rotor

Flux lines created by a winding in an electric machine can concatenate another winding(s), in which case one speaks of mutual inductance. Assume that the winding with index 1 generates air gap MMF after Eq. 4.13

$$\Theta_{\mathrm{gap},1}(x) = \frac{4}{\pi} \cdot i_1 \cdot w_1 \sum_{j=1}^{\infty} \frac{\sin \frac{j\pi}{2}}{j} f_{\mathrm{w1},j} \sin \frac{j\pi}{\tau_{\mathrm{p}}} x \tag{4.92}$$

and that the winding with index 2, shifted for x_0 along the air gap circumferential coordinate, concatenates flux lines of the first winding. Considering only odd harmonics in the MMF distributions of both windings, the jth harmonic of MMF created by the first winding will create a non-zero flux concatenated by the second winding equal to

$$\Psi_{12,j} = 2pw_2 \cdot f_{\mathrm{w2},j} \frac{\mu_0 l_{\mathrm{ax}}}{\delta} \int_{x_0}^{x_0+\tau_{\mathrm{p}}} \Theta_{\mathrm{gap},j}(x)\mathrm{d}x = L_{12,j} \cdot i_1 \tag{4.93}$$

Since

$$\int_{x_0}^{x_0+\tau_{\mathrm{p}}} \sin j \frac{\pi}{\tau_{\mathrm{p}}} x\mathrm{d}x = \frac{2\tau_{\mathrm{p}}}{j\pi} \sin j \frac{\pi}{2} \cos j \frac{\pi}{\tau_{\mathrm{p}}} x_0 \tag{4.94}$$

one can express the mutual inductance $L_{12,j}$ of the jth harmonic as

$$L_{12,j} = \frac{16}{(j\pi)^2} p \frac{\mu_0 l_{\mathrm{ax}} \tau_{\mathrm{p}}}{\delta} w_1 \cdot f_{\mathrm{w1},j} \cdot w_2 \cdot f_{\mathrm{w2},j} \cos j \frac{\pi}{\tau_{\mathrm{p}}} x_0 \tag{4.95}$$

The jth harmonic component of mutual inductance between two windings in a machine with cylindrical rotor is proportional to the cosine of electrical angle between coil axes.

If two coils have identical parameters, i.e., $w_1 = w_2$ and $f_{\mathrm{w1},j} = f_{\mathrm{w2},j}$, as is the case with phase windings in an m-phase machine, the mutual inductance $L_{12,j}$ can be represented as

$$L_{12,j} = L_{\mathrm{gap,cyl},j} \cos j \frac{\pi}{\tau_{\mathrm{p}}} x_0 \tag{4.96}$$

with $L_{\mathrm{gap,cyl},j}$ being defined in Eq. 4.15. The jth spatial harmonic component of flux concatenated by a winding in a symmetrically wound m-phase machine can be expressed as

$$\Psi_j = L_{\mathrm{gap,cyl},j} \sum_{n=1}^{m} i_n \cos j \frac{n-1}{m} 2\pi \tag{4.97}$$

because $x_0 = 2\,\tau_{\mathrm{p}}/m$ in a symmetrically wound m-phase machine. If the machine is in addition symmetrically fed, i.e.,

$$i_n = I_{\max} \cos\left(\alpha - \frac{n-1}{m} 2\pi\right) \tag{4.98}$$

with coefficient α denoting an arbitrary parameter, the concatenated flux Ψ_j can be written as

$$\Psi_j = L_{\mathrm{gap,cyl},j} I_{\max} \sum_{n=1}^{m} \cos\left(\alpha - \frac{n-1}{m} 2\pi\right) \cos j \frac{n-1}{m} 2\pi \tag{4.99}$$

or

$$\Psi_j = \frac{L_{\mathrm{gap,cyl},j} I_{\max}}{2} \sum_{n=1}^{m} \left\{ \cos\left[\alpha - (1-j)\frac{n-1}{m} 2\pi\right] + \cos\left[\alpha - (1+j)\frac{n-1}{m} 2\pi\right] \right\} \tag{4.100}$$

For $(1 - j)/m$ = integer the first summand yields $m \cos \alpha$; $(1 - j)/m \neq$ integer it is equal to zero. Similarly, the second summand in Eq. 4.100 yields $m \cos \alpha$ for $(1 + j)/m$ = integer and zero for $(1 + j)/m \neq$ integer. Accordingly, one can express the total flux Ψ_j of the jth spatial harmonic concatenated in one phase of a symmetrically wound, symmetrically fed m-phase winding as

$$\Psi_j = \frac{m}{2} L_{\mathrm{gap,cyl},j} I_{\max} \cos \alpha \quad \text{for} \quad \frac{1 \pm j}{m} = \text{int} \tag{4.101}$$

which means that the equivalent inductance $L_{\mathrm{eq,gap,cyl},j}$ in that case is equal to

$$L_{\mathrm{eq,gap,cyl},j} = \frac{m}{2} L_{\mathrm{gap,cyl},j} \tag{4.102}$$

All harmonics of the order j, which satisfy equation $(1 \pm j)/m$ = integer, generate resulting concatenated flux $m/2$ times larger than the concatenated flux of one phase. The effects of these harmonics can be described by an equivalent gap inductance which is $m/2$ times larger than the corresponding harmonic inductance of one phase. The equivalent inductance of all other harmonics not satisfying condition $(1 \pm j)/m$ = integer is equal to zero.

The equivalent inductance of a **squirrel cage machine** with an integer number of slots per pole, Fig. 4.24, can be found by applying Eq. 4.101 and substituting for the number of phases m the number of slots per pole $N/(2p)$.

Since the connections between conductors in the end winding zone have no influence on the air gap MMF distribution, one can replace the rings of a squirrel cage winding with separate connections between corresponding bars carrying currents in opposite directions, as in Fig. 4.25. In particular, the first bar under the first pole is connected with the first bar under the second pole, etc. This way the air gap MMF distribution created by a squirrel cage winding with $N/(2p)$ = integer is identical to the air gap MMF distribution created by an $m = N/(2p)$-phase full-pitch winding with one turn per coil.

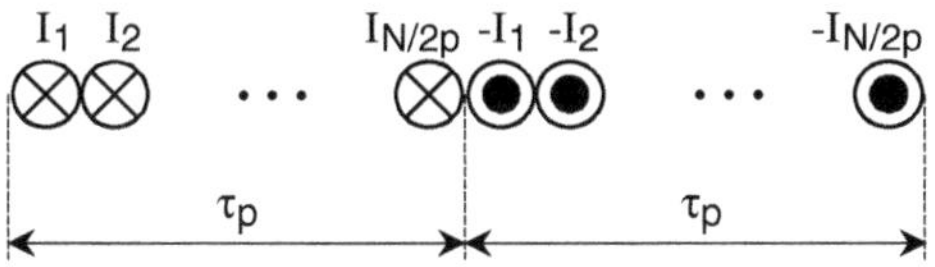

Fig. 4.24 Bar currents in adjacent poles of a squirrel cage winding with an integer number of slots per pole

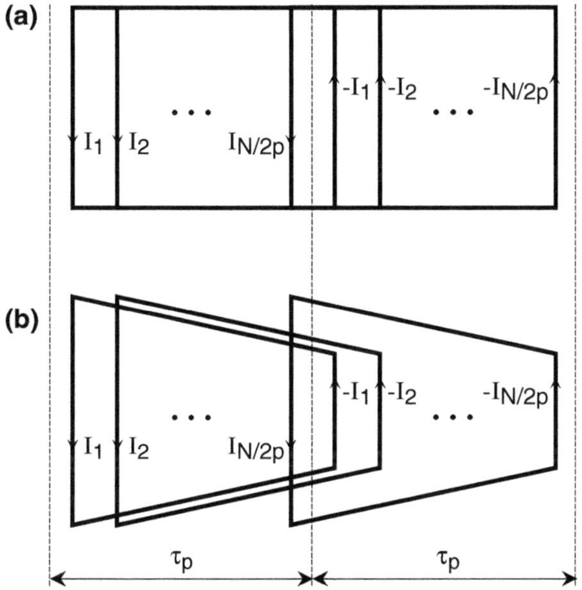

Fig. 4.25 Squirrel cage winding with an integer number of slots per pole (**a**) and its pendant—an $m = N/(2p)$—phase full-pitch winding with one turn per coil (**b**). Both winding types generate identical air gap MMF

Mutual inductance between adjacent phases for the fundamental harmonic of air gap MMF can accordingly be written as (Eq. 4.95)

$$L_{12,1} = \frac{16}{\pi^2} p \frac{\mu_0 l_{ax} \tau_p}{\delta} \cos \alpha_{el} \tag{4.103}$$

with α_{el} being equal to

$$\alpha_{el} = p \frac{2\pi}{N} \tag{4.104}$$

Corresponding equivalent inductance $L_{\text{eq,gap,cage},1}$ is now [4]

$$L_{\text{eq,gap,cage},1} = \frac{4N}{\pi^2}\frac{\mu_0 l_{\text{ax}} \tau_{\text{p}}}{\delta} \tag{4.105}$$

4.5 Slot Leakage Inductance Due to Transverse Field

4.5.1 Magnetic Energy Accumulated in One Slot

Assume a *double-layer winding* with a total of z conductors in a slot distributed uniformly over both conductors, as in Fig. 4.26. Magnetic energy stored in the slot in Fig. 4.26 can be expressed as

$$W_{\text{mg,slot}} = \frac{1}{2}\int_V BH\text{d}V = \frac{1}{2}\mu_0 l_{\text{ax}}\int_0^{h_s} H^2(r)\cdot w(r)\text{d}r \tag{4.106}$$

Denoting by H_2 and H_4 magnetic field strengths in regions 2 and 4 of the slot in Fig. 4.26

$$H_2 = \frac{z}{2}\frac{i_l}{w_s};\quad H_4 = \frac{z}{2}\frac{i_l + i_u}{w_s} \tag{4.107}$$

one can express the values of magnetic field strength in the remaining regions and the amounts of accumulated magnetic energy in all slot portions in the manner shown in Table 4.1.

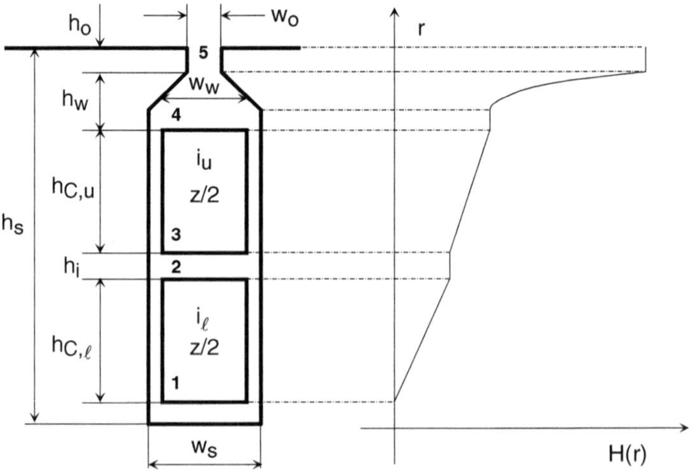

Fig. 4.26 Double-layer winding with z/2 conductors in each layer

Table 4.1 Magnetic field strength and accumulated energy components in the slot in Fig. 4.26

Region	$H(r)$	W_{mg}
1	$H_2 \frac{r}{h_{C,l}}$	$\left(\frac{z}{2}\right)^2 \frac{h_{C,l}}{3w_s} i_l^2$
2	H_2	$\left(\frac{z}{2}\right)^2 \frac{h_i}{w_s} i_l^2$
3	$H_2 + (H_4 - H_2)\frac{r - h_{C,l} - h_i}{h_{C,u}}$	$\left(\frac{z}{2}\right)^2 \frac{h_{C,u}}{w_s}\left(i_l^2 + i_l \cdot i_u + \frac{i_u^2}{3}\right)$
4	H_4	$\left(\frac{z}{2}\right)^2 \frac{h_w}{w_w}\left(i_l^2 + 2i_l \cdot i_u + i_u^2\right)$
5	$\frac{z}{2}\frac{i_l + i_u}{w_o}$	$\left(\frac{z}{2}\right)^2 \frac{h_o}{w_o}\left(i_l^2 + 2i_l \cdot i_u + i_u^2\right)$

Total accumulated magnetic energy in the slot in Fig. 4.26 can now be expressed as

$$W_{\text{mg,slot}} = \frac{1}{2}\mu_0 l_{\text{eff}} \left(\frac{z}{2}\right)^2 \left[\left(\frac{h_{C,l}}{3w_s} + \frac{h_i}{w_s} + \frac{h_{C,u}}{w_s} + \frac{h_w}{w_w} + \frac{h_o}{w_o}\right) i_l^2 + \right. \\ \left. +2 \cdot \left(\frac{h_{C,u}}{2w_s} + \frac{h_w}{w_w} + \frac{h_o}{w_o}\right) \cdot i_l \cdot i_u + \left(\frac{h_{C,u}}{3w_s} + \frac{h_w}{w_w} + \frac{h_o}{w_o}\right) \cdot i_u^2\right] \tag{4.108}$$

Introducing slot specific permeances $\lambda_{s,l}$, $\lambda_{s,\text{lu}}$, and $\lambda_{s,u}$, defined as

$$\lambda_{s,l} = \frac{h_{C,l}}{3w_s} + \frac{h_i}{w_s} + \frac{h_{C,u}}{w_s} + \frac{h_w}{w_w} + \frac{h_o}{w_o} \tag{4.109}$$

$$\lambda_{s,\text{lu}} = \frac{h_{C,u}}{2w_s} + \frac{h_w}{w_w} + \frac{h_o}{w_o} \tag{4.110}$$

$$\lambda_{s,u} = \frac{h_{C,u}}{3w_s} + \frac{h_w}{w_w} + \frac{h_o}{w_o} \tag{4.111}$$

one can express the magnetic energy accumulated in a slot of a double- layer winding in Fig. 4.26 as

$$W_{\text{mg,slot}} = \frac{1}{2}\mu_0 l_{\text{eff}} \left(\frac{z}{2}\right)^2 \left(\lambda_{s,l} \cdot i_l^2 + 2\lambda_{s,\text{lu}} \cdot i_l \cdot i_u + \lambda_{s,u} \cdot i_u^2\right) \tag{4.112}$$

By substituting $i_\ell = i_u = i$, one obtains the magnetic energy accumulated in a slot of a *single-layer* winding as

$$W_{\text{mg,slot}} = \frac{1}{2}\mu_0 l_{\text{eff}} z^2 \left(\frac{h_C}{w_s} + \frac{h_w}{w_w} + \frac{h_o}{w_o}\right) \tag{4.113}$$

4.5.2 Magnetic Energy Accumulated in All N Slots

Windings with **full-pitch**, **double-layer**, or **single-layer coils** are characterized by $i_u = i_\ell = i_{\mathrm{Ph}}$, where phase current i_{Ph} can be any of the machine's m-phase currents and is equal to I_{Ph}/a, a denoting the number of parallel circuits. Total magnetic energy $W_{\mathrm{mg,tot}}$ is equal to the sum of magnetic energies stored in all N slots, which for a symmetrically wound machine further yields the following

$$W_{\mathrm{mg,tot}} = \frac{N}{m}\frac{1}{2}\mu_0 l_{\mathrm{ax}}\left(\frac{z}{2}\right)^2\left(\lambda_{s,l} + 2\lambda_{s,lu} + \lambda_{s,u}\right)\left(i_a^2 + i_b^2 + \ldots + i_m^2\right) \tag{4.114}$$

Inserting for the number of conductors per coil $z/2$

$$\frac{z}{2} = \frac{w_{\mathrm{Ph}} \cdot m \cdot a}{N} \tag{4.115}$$

with w_{Ph} standing for the number of turns per phase, one can further write

$$W_{\mathrm{mg,tot}} = \frac{1}{2}\mu_0 l_{\mathrm{ax}} w_{\mathrm{Ph}}^2 \frac{m}{N}\left(\lambda_{s,l} + 2\lambda_{s,\mathrm{lu}} + \lambda_{s,u}\right)\left(I_a^2 + I_b^2 + \ldots + I_m^2\right) \tag{4.116}$$

Introducing the slot leakage inductance $L_{\mathrm{slot},\sigma}$, defined as

$$L_{\mathrm{slot},\sigma} = \mu_0 l_{\mathrm{ax}} w_{\mathrm{Ph}}^2 \frac{m}{N}\left(\lambda_{s,l} + 2\lambda_{s,\mathrm{lu}} + \lambda_{s,u}\right) \tag{4.117}$$

one can express the total magnetic energy accumulated in N slots of a machine with double-layer full-pitch, or single-layer winding as

$$W_{\mathrm{mg,tot}} = \frac{1}{2} L_{\mathrm{slot},\sigma}\left(I_a^2 + I_b^2 + \ldots + I_m^2\right) \tag{4.118}$$

Machines with **short pitch winding** do not have the same ampere-turns in each layer. Usually, the coil pitch y is larger than $\tau_{\mathrm{p}} - \tau_{\mathrm{p}}/m$, which means that in some slots the same current flows both in upper and in lower layers. These slots are shaded in Fig. 4.27, in which the zones of a double-layer, m-phase winding under one pole are shown. The width w_{sc} of the zone with same current in both layers of each slot (monoslots) is given as

$$w_{\mathrm{sc}} = y - \frac{m-1}{m}\tau_{\mathrm{p}} \tag{4.119}$$

whereas the width w_{dc} of the zone with different currents in slot layers (mixed slots) is obviously

$$w_{\mathrm{dc}} = \tau_{\mathrm{p}} - y \tag{4.120}$$

The total number of slots N_{sc} with the same current in both layers is accordingly

$$N_{\mathrm{sc}} = \frac{y - \frac{m-1}{m}\tau_{\mathrm{p}}}{\frac{\tau_{\mathrm{p}}}{m}} N = m\left(\frac{y}{\tau_{\mathrm{p}}} - \frac{m-1}{m}\right) N \tag{4.121}$$

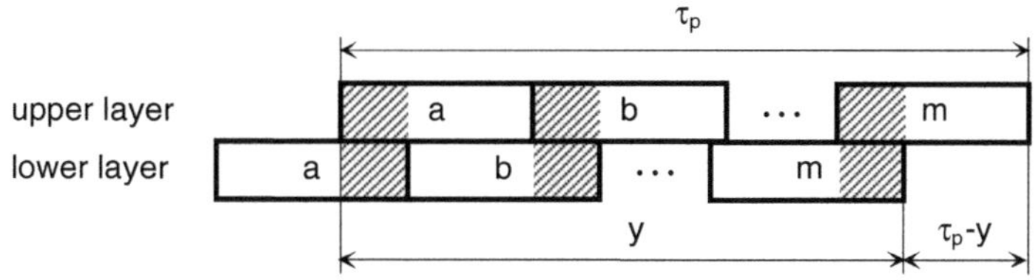

Fig. 4.27 Double-layer, m-phase winding with coil pitch $y < \tau_{\mathrm{p}}$

and with different currents

$$N_{\mathrm{dc}} = m\left(1 - \frac{y}{\tau_{\mathrm{p}}}\right) N \tag{4.122}$$

resulting in total magnetic energy accumulated in N slots (Eq. 4.116)

$$W_{\mathrm{mg,tot}} = \frac{1}{2}\mu_0 l_{\mathrm{ax}} w_{\mathrm{Ph}}^2 \frac{m}{N}\left\{\left(m\frac{y}{\tau_{\mathrm{p}}} - m + 1\right)\left(\lambda_{s,l} + 2\lambda_{s,\mathrm{lu}} + \lambda_{s,u}\right)\left(I_a^2 + I_b^2 + \ldots + I_m^2\right) + \right.$$
$$\left. + m\left(1 - \frac{y}{\tau_{\mathrm{p}}}\right)\left[\lambda_{s,l}\left(I_a^2 + I_b^2 + \ldots + I_m^2\right) + 2\lambda_{s,\mathrm{lu}}\left(I_a I_b + I_b I_c + \ldots + I_m I_a\right) + \lambda_{s,u}\left(I_a^2 + I_b^2 + \ldots + I_m^2\right)\right]\right\}$$

or

$$W_{\mathrm{mg,tot}} = \frac{1}{2}\left[L_{\sigma,\mathrm{sc}}\left(I_a^2 + I_b^2 + \cdots + I_m^2\right) + 2L_{\sigma,\mathrm{dc}}\left(I_a I_b + I_b I_c + \cdots + I_m I_a\right)\right] \tag{4.123}$$

with $L_{\sigma,\mathrm{sc}}$ denoting the leakage inductance of monoslots, and $L_{\sigma,\mathrm{dc}}$ the leakage inductance of mixed slots

$$L_{\sigma,\mathrm{sc}} = \mu_0 l_{\mathrm{ax}} w_{\mathrm{Ph}}^2 \frac{m}{N}\left[\lambda_{s,l} + 2\lambda_{s,\mathrm{lu}}\left(m\frac{y}{\tau_{\mathrm{p}}} - \frac{m-1}{m}\right) + \lambda_{s,u}\right] \tag{4.124}$$

$$L_{\sigma,\mathrm{dc}} = \mu_0 l_{\mathrm{ax}} w_{\mathrm{Ph}}^2 \frac{m^2}{N}\lambda_{s,\mathrm{lu}}\left(1 - \frac{y}{\tau_{\mathrm{p}}}\right) \tag{4.125}$$

Whereas the leakage inductance of monoslots $L_{\sigma,sc}$ has the meaning of self-inductance, the leakage inductance of mixed slots $L_{\sigma,dc}$ obviously denotes the mutual inductance between two phases.

4.6 End Winding Leakage Inductance

Currents flowing through stator and rotor end windings create magnetic fields which do not participate in electromechanical energy conversion, because they are too far from end windings on the other side of air gap. Nevertheless, magnetic field in end winding region (Fig. 4.28) can be large enough to create force on conductors and heat solid metal parts in its vicinity. In low-polarity machines with short active part the inductance belonging to the end winding field can reach significant, double digit percental values of the machine reactance. Therefore, the determination of end winding inductance is an important portion of electric machines designer's job.

None of three spatial components of the magnetic field in the end winding zone may be neglected when evaluating the end winding inductance. The procedure for the determination of end winding inductance is based on its definition as a ratio between concatenated flux and coil current. In order to find the flux, usually, Biot–Savart law is utilized with boundary conditions considering mirroring of currents on the iron lamination, shaft, and machine housing.

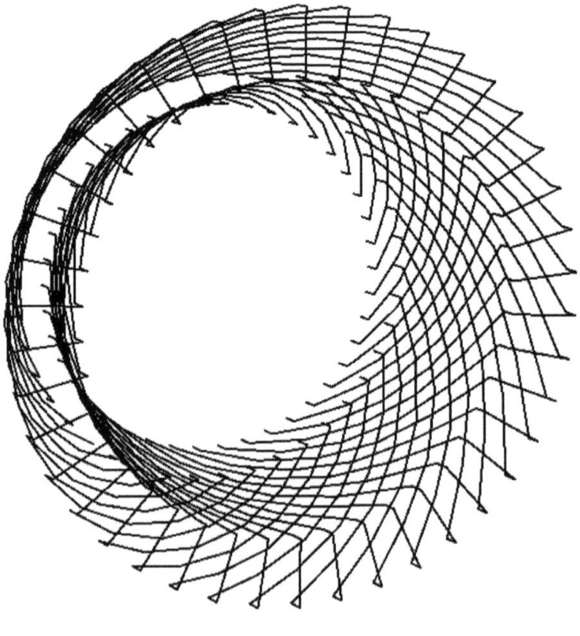

Fig. 4.28 Stator end zone of a double-layer winding with form-wound coils

This way the flux density distribution is evaluated at given points on surface S created by a single coil in Fig. 4.29. Since q is usually larger than one, the total concatenated flux belonging to one zone has to be found by adding each other all components of flux density created by currents in coils of the given zone.

Since the coils of windings on the same side of air gap are placed close to each other in the end zone, the mutual inductance between them may not be neglected. The coefficient of mutual inductance for the end winding is evaluated in the similar manner by using Biot–Savart law, as it is done for the zone inductance (Fig. 4.29).

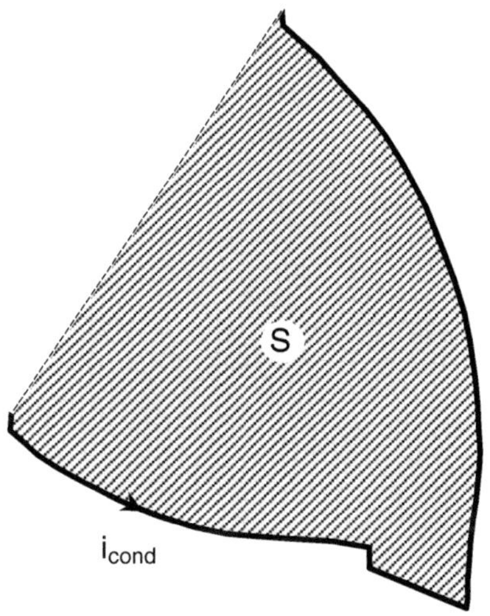

Fig. 4.29 Area determined by a single conductor in the stator end zone of a double-layer winding with form-wound coils

References

1. Yamamura S (1992) Spiral vector theory of AC circuits and machines. Clarendon Press, Oxford
2. Richter R (1930) Elektrische Maschinen- Zweiter Band: Synchronmaschinen und Einankerumformer. Verlag von Julius Springer, Berlin
3. Fitzgerald AE, Kingsley C (1952) Electric machinery—an integrated treatment of A-C and D-C machines. McGraw-Hill Book Company Inc., New York, Toronto, London
4. Schuisky W (1960) Berechnung elektrischer Maschinen. Springer- Verlag, Wien

Chapter 5
Skin and Proximity Effect

Contents

V. Ostović, *The Art and Science of Rotating Field Machines Design*,
DOI 10.1007/978-3-319-39081-9_5

Current and flux densities in electric conductors exposed to alternating magnetic fields redistribute as a consequence of Faraday's law. This fundamental observation has numerous occurrences in electric machines (skin and proximity effect), most of which lead to increase of losses. Analytical and numerical tools are developed in this chapter, which help one quantify AC losses and design a machine in such a manner as to minimize them. Among these are a method for evaluation of AC parameters of a bar with arbitrary cross section and the analytical tool for computation of proximity effect in a conductor in slot. Various procedures for minimization of AC losses are discussed. Computational procedures are presented for evaluation of skin effect in ferromagnetic media, thin plates, etc.

5.1 Analytical Solution for Current Density Redistribution in a Solid Rectangular Conductor in a Slot as a Result of Alternating Leakage Flux: One-Dimensional Skin Effect

As shown in the previous chapters, current in a conductor placed in a slot of an electric machine creates tangential leakage flux Φ_σ, Fig. 5.1. A time-dependent current creates a time-dependent leakage flux, which, following Faraday's law, induces voltages, which can drive currents in electrically conducting media. Consequently, alternating current $I_\sim$ in Fig. 5.1 creates alternating leakage flux Φ_σ, the time derivatives of which—induced voltages—are sources of eddy currents i_{e}, which superimpose to the impressed current $I_\sim$. An alternating current density within the solid conductor in Fig. 5.1 is redistributed in such a manner that its amount increases almost exponentially from the bottom to the top of the conductor. One refers to the *skin effect*.

Electromagnetic quantities creating one-dimensional sinusoidal current density redistribution in the solid conductor in Fig. 5.1 can be quantified using analytical expressions developed in [1]. Introduce first the parameter β [m^{-1}], defined as

$$\beta = \sqrt{f \pi \mu \kappa \frac{w_c}{w_s}} \tag{5.1}$$

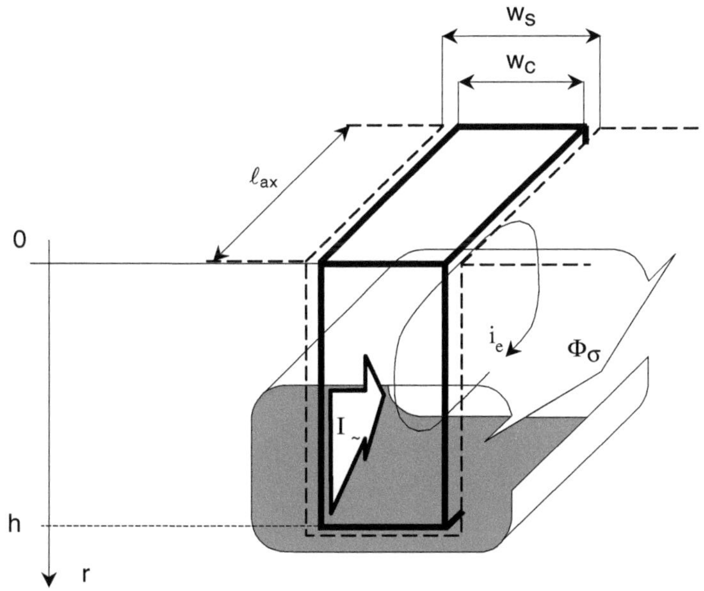

Fig. 5.1 One-dimensional current redistribution in a solid conductor placed in a slot of an electric machine. The impressed current $I_{\sim}$ is a function of time

Current density distribution created by real sinusoidal impressed current $I_{\sim}$ can be expressed [1] as:

$$\Gamma(r) = \frac{I}{w_c}\beta(1+j)\frac{\cosh[\beta(1+j)(h-r)]}{\sinh[\beta(1+j)h]} \tag{5.2}$$

The real component of current density can be written as

$$\begin{aligned}\mathrm{Re}\{\Gamma(r)\} = -\frac{2\beta I}{w_c(\cos 2\beta h - \cosh 2\beta h)}\{&\cosh[\beta(h-r)]\cos[\beta(h-r)](\cosh\beta h\cdot\sin\beta h + \cos\beta h\sinh\beta h) + \\ &+ \sinh[\beta(h-r)]\sin[\beta(h-r)](\cosh\beta h\cdot\sin\beta h - \cos\beta h\sinh\beta h)\}\end{aligned} \tag{5.3}$$

and the imaginary as

$$\begin{aligned}\mathrm{Im}\{\Gamma(r)\} = \frac{2\beta I}{w_c(\cos 2\beta h - \cosh 2\beta h)}\{&\cosh[\beta(h-r)]\cos[\beta(h-r)](\cosh\beta h\cdot\sin\beta h - \cos\beta h\sinh\beta h) - \\ &- \sinh[\beta(h-r)]\sin[\beta(h-r)](\cosh\beta h\cdot\sin\beta h - \cos\beta h\sinh\beta h)\}\end{aligned} \tag{5.4}$$

By substituting $r = 0$ (top of the conductor) in Eqs. 5.3 and 5.4, one obtains maximum values of real and imaginary components of current density as

$$\mathrm{Re}\{\Gamma_{\max}(0)\} = \frac{\beta I}{w_c} \frac{\sinh 2\beta h + \sin 2\beta h}{\cosh 2\beta h - \cos 2\beta h} \quad (5.5)$$

$$\mathrm{Im}\{\Gamma_{\max}(0)\} = \frac{\beta I}{w_c} \frac{\sinh 2\beta h - \sin 2\beta h}{\cosh 2\beta h - \cos 2\beta h} \quad (5.6)$$

For large values of argument $2\beta h$ applies $\cosh 2\beta h >> \cos 2\beta h$, $\sinh 2\beta h >> \sin 2\beta h$, and $\sinh 2\beta h \approx \cosh 2\beta h$; hence,

$$\mathrm{Re}\{\Gamma_{\max,\beta\gg}(0)\} = \mathrm{Im}\{\Gamma_{\max,\beta\gg}(0)\} = \frac{\beta I}{w_c} = \frac{I}{\delta w_c} \quad (5.7)$$

with δ denoting the *current density penetration (skin) depth* [m], defined as

$$\delta = \frac{1}{\beta} = \sqrt{\frac{2}{\omega\mu\kappa}} \quad (5.8)$$

According to Eq. 5.7, the *current density skin depth* δ is equal to the distance from the air gap side of the conductor toward the slot bottom along which the impressed current I would create *constant* current density equal to the actual current density $I/\delta w_c$ on the air gap side of the conductor.

At higher frequencies, the imaginary component of current density at air gap side of the conductor becomes as large as its real component. The rms. value of current is not dependent on the conductor height h, but on the skin depth δ.

For a DC current, one can write

$$\Gamma_{D.C.,r=0} = \lim_{\beta\to 0}\left[\mathrm{Re}\{\Gamma_{\max}(0)\}\right] = \frac{I}{hw_c} \quad (5.9)$$

since, obviously,

$$\lim_{\beta\to 0}\left[\mathrm{Im}\{\Gamma_{\max}(0)\}\right] = 0 \quad (5.10)$$

The absolute value of current density can be expressed as

$$|\Gamma(r)| = \frac{\beta I}{w_c}\sqrt{2\frac{\cosh[2\beta(h-r)] + \cos[2\beta(h-r)]}{\cosh 2\beta h - \cos 2\beta h}} \quad (5.11)$$

For large values of argument $\beta(h - r)$, the absolute value of current density decreases exponentially from the conductor air gap side toward the bottom of the slot:

$$\left|\Gamma(r)_{\beta\gg}\right| \sim \frac{I}{\delta w_c} e^{-\frac{r}{\delta}} \tag{5.12}$$

where δ, introduced in Eq. 5.8, can here be interpreted as the *exponential decay constant*. The absolute value of current density decreases exponentially at a rate of r/δ.

Case Study 5.1: A 50-mm-high copper bar is placed in an open slot of an electric machine as shown in Fig. 5.1. The dependence of current density on frequency and radial distance is illustrated in Fig. 5.2.

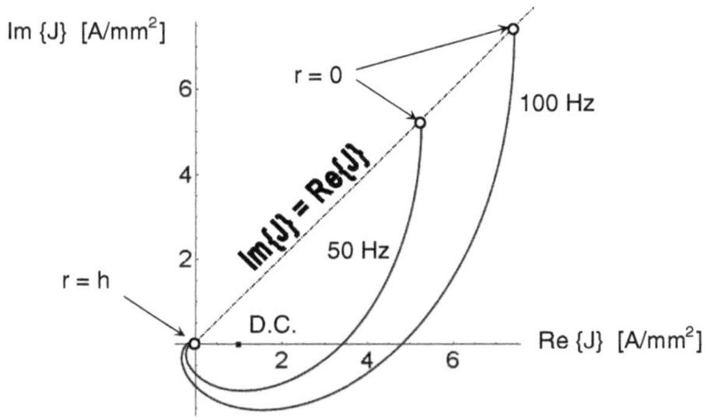

Fig. 5.2 Current density distribution in a 50-mm-high copper rectangular bar at various frequencies and radial coordinates

Real, imaginary, and absolute values of current density (in A/mm^2) in a 50-mm-high copper bar as functions of radial coordinate r are shown in Fig. 5.3a–c for various frequencies, and the dependence of the absolute value of current density on β, which is almost linear for higher values of β, is shown in Fig. 5.3d. One recalls that β is proportional to the square root of frequency. The direct current with the same value as the rms of applied current would create a current density of 1 A/mm^2.

Both real and imaginary components of current density in Fig. 5.3a, b decrease rapidly when going from the air gap surface of the conductor to the slot bottom and after a certain point change their direction. This means that total current in the conductor portion closer to the slot bottom flows in opposite direction than the impressed current. In this portion of the conductor the eddy currents overwhelm the impressed current.

In Fig. 5.4 the dependence of the absolute value of current density on the radial coordinate r at various time instants is shown both for DC and 50 Hz impressed current. The absolute value of current density in the portion of the conductor closer to the slot bottom is almost equal to zero at 50 Hz, whereas it is uniformly distributed over the whole conductor height in the case of a DC impressed current with the same amplitude as the rms value of 50 Hz current.

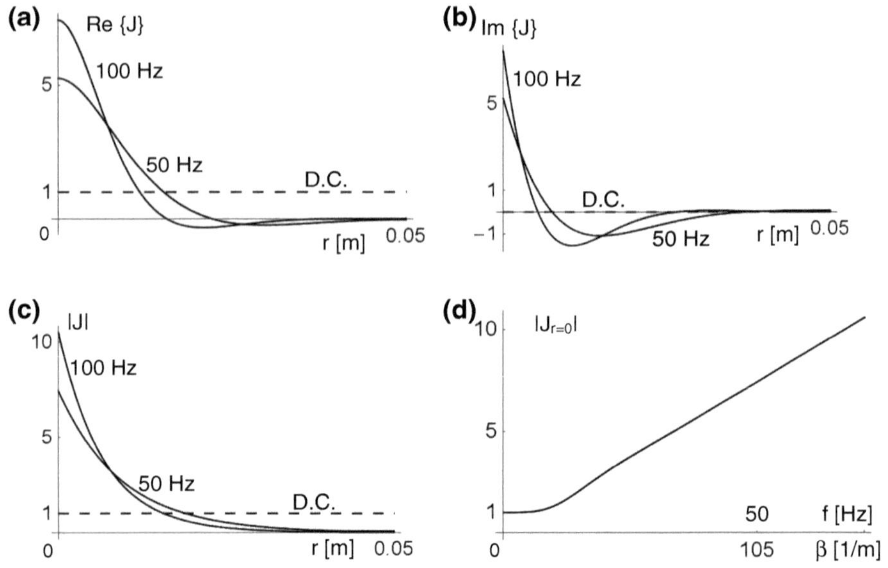

Fig. 5.3 Current density distributions in a 50-mm-high copper rectangular bar at various frequencies and radial coordinates. The impressed current creates a DC density of 1 A/mm^2. Surface integral of the real component of current density over conductor height is equal to the impressed current $\underline{\mathbf{I}}$; surface integral of the imaginary component of current density is equal to zero

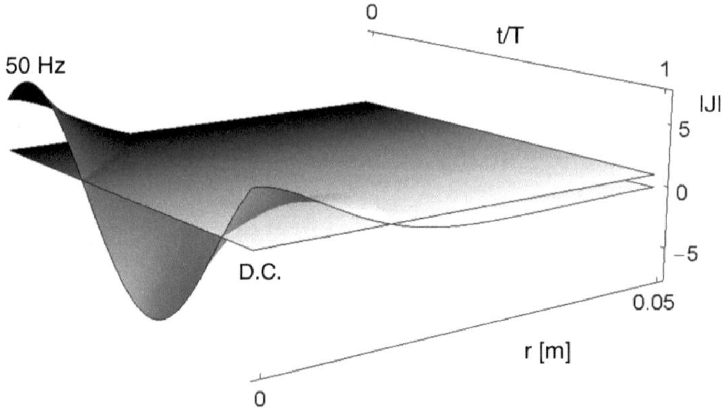

Fig. 5.4 Current density distributions in a 50-mm-high copper rectangular bar at 50 Hz and DC. The impressed current creates a DC current density of 1 A/mm^2

Total losses $P_{\sim}$ created by the impressed current $I_{\sim}$ are equal to

$$P_{\sim} = \frac{w_c l_{ax}}{\kappa} \int_0^h \Gamma \cdot \Gamma^* \mathrm{d}r = I^2 \frac{\beta l_{ax}}{\kappa w_c} \cdot \frac{\sinh 2\beta h + \sin 2\beta h}{\cosh 2\beta h - \cos 2\beta h} \tag{5.13}$$

Total losses $P_{\sim}$ are a function of conductor height h and frequency of impressed current, since β in Eq. 5.13 is proportional to $\sqrt{f}$. For direct current ($\beta = 0$) $P_{\sim}$ (h) is a hyperbola, whereas $P_{\sim}$ (h) created by alternating current ($\beta \neq 0$) has a minimum at *critical conductor height* h_c:

$$h_c = \frac{\pi}{2}\delta \tag{5.14}$$

For each frequency, represented by skin depth δ, there exists a conductor height h_c for which the I^2R losses are minimal; for example, for solid copper conductor at 50 Hz, the critical height is equal to $h_{c,50} \approx 15$ mm (see also Fig. 5.5a).

The amount of losses $P_{\sim,\,\min}$ in a solid conductor with critical height h_c is equal to

$$P_{\sim,\,\min} = I^2 \frac{\pi l_{ax}}{2\kappa w_c h_c} \cdot \frac{\sinh \pi}{1 + \cosh \pi} \approx 1.44 I^2 \frac{l_{ax}}{\kappa w_c h_c} \tag{5.15}$$

which is about 44 % higher than losses in the same conductor created by direct current with the same value. No matter how much the height of a solid conductor is increased, the AC I^2R losses in it are at least 44 % higher than the DC I^2R losses with the same amount of current. On the other hand, the thinner the conductor, the closer its AC losses to the DC value, see Fig. 5.5a.

Any increase of conductor height h above the critical height h_c increases the AC I^2R losses, which asymptotically reach the value of

$$\lim \left[P_{\sim} \atop h \to \infty \right] = I^2 \frac{\beta l_{ax}}{\kappa w_c} = I^2 \frac{l_{ax}}{\kappa \delta w_c} \tag{5.16}$$

that is, as if the conductor height h were equal to the skin depth δ. Independent of conductor dimensions, alternating current can only make use of a layer of thickness δ on the air gap side of the solid conductor with height h, see also Eq. 5.7.

The product βh in Eq. 5.13, sometimes referred to as *normalized conductor height* ξ:

$$\xi = h\beta = \frac{h}{\delta} \tag{5.17}$$

helps one express the I^2R losses in a solid conductor (Eq. 5.13) as

$$P_{\sim} = I^2 \frac{l_{ax}}{\kappa h w_c} \cdot \xi \frac{\sinh 2\xi + \sin 2\xi}{\cosh 2\xi - \cos 2\xi} \tag{5.18}$$

or

$$P_{\sim} = P_{=} \cdot \varphi(\xi) \tag{5.19}$$

where

$$P_{=} = I^2 \frac{l_{ax}}{\kappa h w_c} \tag{5.20}$$

and

$$\varphi(\xi) = \frac{P_{\sim}}{P_{=}} = \frac{R_{\sim}}{R_{=}} = \xi \frac{\sinh 2\xi + \sin 2\xi}{\cosh 2\xi - \cos 2\xi} \tag{5.21}$$

The function $\varphi(\xi)$ in Eq. 5.21 represents at the same time the ratio between conductor AC and DC resistance at a given frequency.

The **loss density**, expressed in W/m^3, is an indicator of loss distribution along the conductor height and can be expressed as

$$P'_{\sim} = \frac{dP_{\sim}}{dV} = \frac{1}{\kappa l_{ax} w_c} \Gamma^2 = 2I^2 \frac{\beta^2 l_{ax}}{\kappa w_c} \cdot \frac{\cosh[2\beta(h-r)] + \cos[2\beta(h-r)]}{\cosh 2\beta h - \cos 2\beta h} \tag{5.22}$$

Loss density is maximal on the air gap side of the conductor ($r = 0$). For medium and large values of argument βh, the maximum loss density can be expressed as

$$P'_{\sim}(r = 0) \approx 2I^2 \frac{\beta^2 l_{ax}}{\kappa w_c} \tag{5.23}$$

Analogously to the current density skin depth δ, defined in Eq. 5.8, one can introduce the *power density skin depth* δ' as a distance from the conductor air gap side at which losses generated assuming constant loss density are equal to actual losses in the conductor. Following Eq. 5.23, the power density skin depth is equal to one half of the current density skin depth:

$$\delta' = \frac{\delta}{2} \tag{5.24}$$

because losses are proportional to the square of the current amplitude.

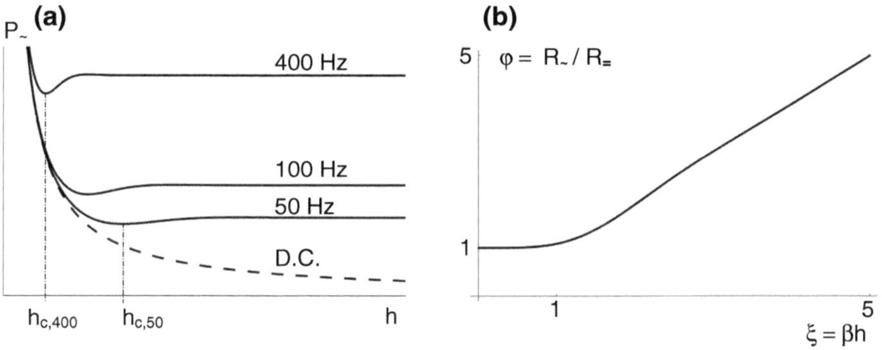

Fig. 5.5 **a** Total losses in a 50-mm-high copper rectangular bar at various frequencies; **b** the ratio between AC and DC resistance of a conductor

Case Study 5.2: Total losses in a copper conductor, introduced in Case Study 5.1, are shown in Fig. 5.5a as a function of conductor height h for various values of frequency of impressed current as parameter.

Only direct current creates less losses when conductor height increases; total losses generated by alternating current end up at the amount given in Eq. 5.16, independently of the conductor height. Total losses at a given frequency of alternating current are minimal at critical conductor height (Eq. 5.14).

The ratio between AC and DC resistance, $R_{\sim}/R_{=}$, as a function of normalized conductor height ξ (Eq. 5.17) is shown in Fig. 5.5b. The AC resistance of the conductor with height h increases proportionally to the square root of frequency. This property can also be observed in Fig. 5.6a, in which the ratio between AC and DC losses, $P_{\sim}/P_{=}$, as a function of frequency is shown. One recognizes that starting at low frequencies, the factor $\varphi(\xi)$ (represented with black curve in Fig. 5.6a) is identical with $\sqrt{f}$, represented with overlapping gray curve in the same figure. In other words, AC losses increase proportionally to the square root of frequency of impressed current.

The ratio between the loss density for alternating and direct current for various values of frequency of impressed current, as a function of the conductor radial coordinate r, is shown in Fig. 5.6b. One recognizes in Fig. 5.6b that already at 50 Hz the peak value of loss density on the air gap surface of a conductor exceeds its DC value for a factor above 50! Such extreme loss densities are a potential source of hot spots in inadequately cooled conductors.

The concatenated current at a distance r, $i(r)$, is evaluated as a surface integral of current density in Eq. 5.2, which on the other hand is proportional to the magnetic field strength H. Therefore:

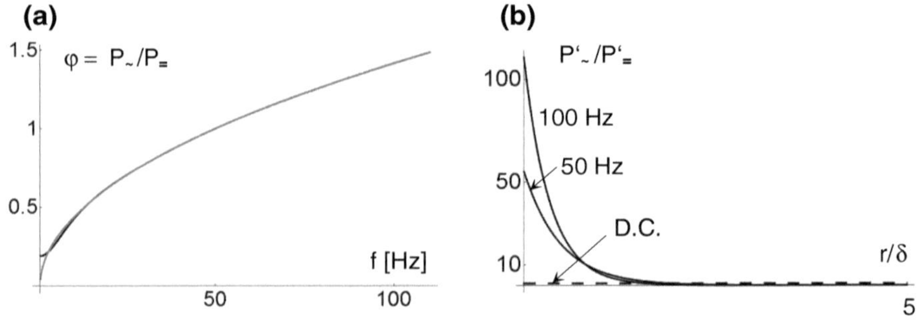

Fig. 5.6 **a** The ratio of AC to DC losses as a function of frequency; **b** loss density ratio along the conductor radial coordinate r

$$i(r) = w_c \int \Gamma(r)\mathrm{d}r_n = I\frac{\sinh[\beta(1+j)(h-r)]}{\sinh[\beta(1+j)h]} \tag{5.25}$$

and

$$H(r) = \frac{I}{w_s}\frac{\sinh[\beta(1+j)(h-r)]}{\sinh[\beta(1+j)h]} = \frac{i(r)}{w_s} \tag{5.26}$$

The real and imaginary components of the conductor current at height r can be expressed as

$$\mathrm{Re}\{i(r)\} = 2I\frac{\cosh\beta h\cosh[\beta(h-r)]\sin\beta h\sin[\beta(h-r)] + \cos\beta h\cos[\beta(h-r)]\sinh\beta h\sinh[\beta(h-r)]}{\cosh 2\beta h - \cos 2\beta h} \tag{5.27}$$

and

$$\mathrm{Im}\{i(r)\} = 2I\frac{\cos\beta h\cosh[\beta(h-r)]\sinh\beta h\sin[\beta(h-r)] - \cosh\beta h\cos[\beta(h-r)]\sin\beta h\sinh[\beta(h-r)]}{\cosh 2\beta h - \cos 2\beta h} \tag{5.28}$$

For medium and large values of β, the imaginary component of concatenated current/ magnetic field strength can be written as

$$\mathrm{Im}\left\{i(r)_{\beta\gg}\right\} \approx -Ie^{-\beta r}\sin\beta r \tag{5.29}$$

and reaches its minimum at

$$r = \frac{\pi}{4\beta} \tag{5.30}$$

The imaginary component of concatenated current/magnetic field strength has a minimum value independent of frequency equal to

$$\mathrm{Im}\left\{ i(r)_{\beta\gg} \right\}_{\min} = -Ie^{-\frac{\pi}{4}} \sin\frac{\pi}{4} \approx -0.3224I \tag{5.31}$$

The real component of concatenated current/magnetic field strength at the point of minimum of the imaginary part (Eq. 5.28) is equal to

$$Ie^{-\frac{\pi}{4}} \cos\frac{\pi}{4} \approx 0.3224I \tag{5.32}$$

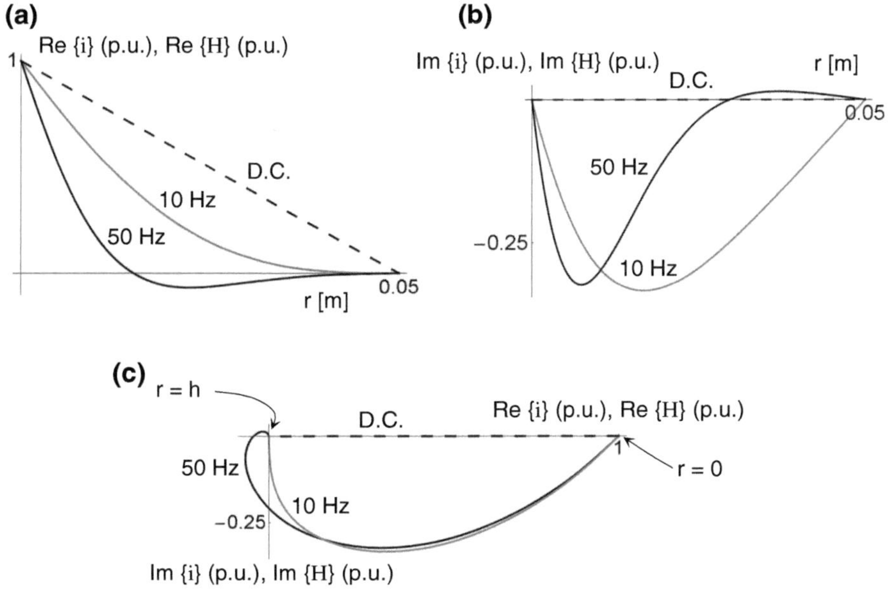

Fig. 5.7 Real (**a**) and imaginary (**b**) components of concatenated current/magnetic field strength in a 50-mm copper conductor in a slot. The parametric plot of the both components is shown in (**c**). At r = 0, the concatenated current is equal to the impressed (real) current I

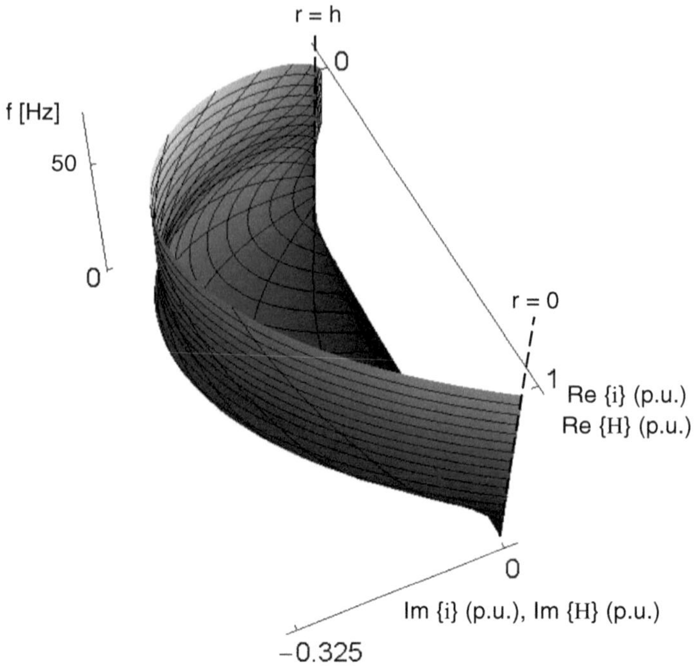

Fig. 5.8 3-dimensional parametric plot of the real and imaginary components of concatenated current/magnetic field strength

Case Study 5.3: The dependence of real and imaginary components of concatenated current/magnetic field strength (in p.u.) after Eq. 5.23 on the radial coordinate r in a 50-mm-high copper for various frequencies is shown in Fig. 5.7a, b.

The real component of concatenated direct current is a linear function of the coordinate r, and its imaginary component is equal to zero—the direct current density is constant over the whole conductor cross section. With increasing frequency, both real and imaginary components start to concentrate on the air gap side of the conductor in such a manner that the peak value of the imaginary component remains constant. This property can be recognized in Fig. 5.7c, in which the components as a function of the parameter r are shown, as well as in a 3-dimensional representation in Fig. 5.8.

Magnetic energy accumulated in the slot with a current-carrying conductor is evaluated as

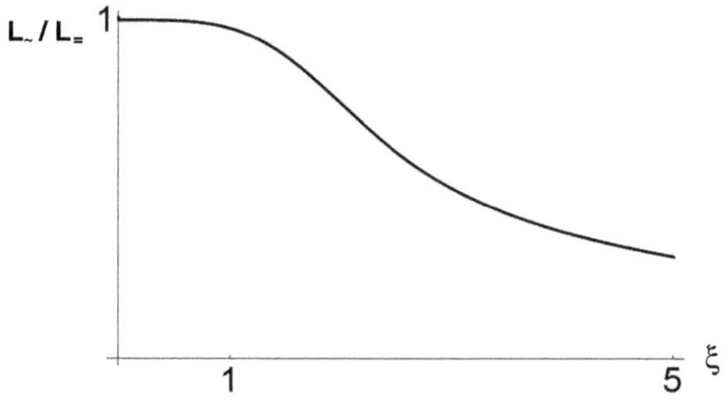

Fig. 5.9 The dependence of the ratio between AC and DC conductor inductance on its normalized height ξ

$$W_{mg} = \mu_0 l_{ax} w_s \int_0^h H(r) \cdot H^*(r) \mathrm{d}r = \frac{1}{2} i^2(t) \frac{\mu_0 l_{ax}}{\beta w_s} \frac{\sinh 2\beta h - \sin 2\beta h}{\cosh 2\beta h - \cos 2\beta h} \tag{5.33}$$

and its average value, by utilizing Eq. 5.17

$$\overline{W}_{mg} = \frac{1}{2} I^2 \frac{\mu_0 h l_{ax}}{w_s} \frac{1}{\xi} \frac{\sinh 2\xi - \sin 2\xi}{\cosh 2\xi - \cos 2\xi} \tag{5.34}$$

The accumulated magnetic energy for direct current is equal to

$$\lim \left[\underset{\xi \to 0}{\bar{W}_{mg}} \right] = \frac{1}{3} I^2 \frac{\mu_0 h l_{ax}}{w_s} \tag{5.35}$$

The ratio between AC and DC values of conductor inductance is proportional to the ratio between accumulated magnetic energies, or

$$\frac{L_{\sim}}{L_{=}} = \frac{3}{2\xi} \frac{\sinh 2\xi - \sin 2\xi}{\cosh 2\xi - \cos 2\xi} \tag{5.36}$$

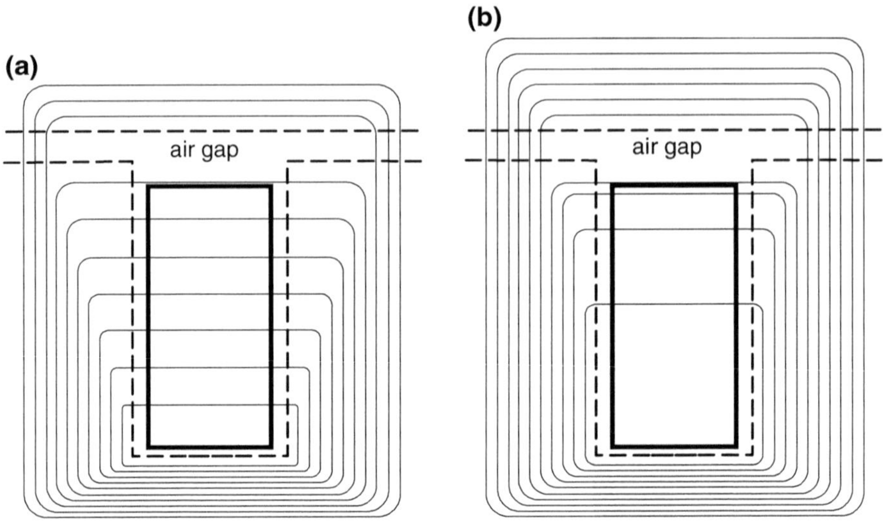

Fig. 5.10 Approximate leakage flux distribution for DC (**a**) and AC (**b**)

The dependence of the ratio $L_{\sim}/L_{=}$ on normalized conductor height ξ is shown in Fig. 5.9.

One should keep in mind that the conductor inductance evaluated here is leakage by its nature, because it is a measure of flux going through the slot, instead of through air gap. Therefore, an AC conductor current increases total air gap flux, since it forces flux out of slot into air gap, as shown in Fig. 5.10. The higher the frequency of alternating current, the stronger the effects of Faraday's law, the lower the leakage and the higher the main flux.

The decrease of leakage inductance due to skin effect is utilized in deep bar induction machines [1], where a lower rotor leakage inductance leads to a higher torque on the shaft in the low-speed range due to a higher air gap flux.

5.2 Analytical Solution for Current Density Redistribution in an Arbitrarily Shaped Solid Conductor in a Slot as a Result of Alternating Leakage Flux

Very often conductors in a slot of an electric machine do not have the simple rectangular form as shown in Fig. 5.1. As the conductor width changes along the slot height, a question is posed, how the current density, loss, field strength, etc., distributions can be found in the case of variable conductor geometry. Partial differential equations describing distributions of electromagnetic quantities remain obviously the same, independent of the conductor geometry. Their general solution in the case of sinusoidal applied current [1] yields for the magnetic field strength:

$$H = (-1+j)\frac{\beta}{\omega\mu}\left[Ce^{-\beta(1+j)(h-r)} - De^{\beta(1+j)(h-r)}\right] \tag{5.37}$$

and for current density

$$\Gamma = \kappa\left[Ce^{-\beta(j+1)(h-r)} + De^{\beta(j+1)(h-r)}\right] \tag{5.38}$$

where complex constants of integration C and D

$$C = C_r + jC_i; \quad D = D_r + jD_i \tag{5.39}$$

are determined from boundary conditions. Introducing substitutions

$$\begin{aligned} C_{\mathrm{rc}} &= C_r \cos\beta(r-h); \quad & C_{\mathrm{rs}} &= -C_r \sin\beta(r-h) \\ C_{\mathrm{ic}} &= C_i \cos\beta(r-h); \quad & C_{\mathrm{is}} &= -C_i \sin\beta(r-h) \end{aligned}$$

analogously with D_{rc}, D_{rs}, D_{ic} and D_{is} for the constant D, one can express general solutions in Eqs. 5.37 and 5.38 as

$$\begin{aligned} H = &\frac{\beta}{\omega\mu}\left[(-C_{\mathrm{rc}} + C_{\mathrm{is}} - C_{\mathrm{rs}} - C_{\mathrm{ic}})e^{-\beta(h-r)} - (-D_{\mathrm{rc}} + D_{\mathrm{is}} - D_{\mathrm{rs}} - D_{\mathrm{ic}})e^{\beta(h-r)}\right] + \\ &+ j\frac{\beta}{\omega\mu}\left[(C_{\mathrm{rc}} - C_{\mathrm{is}} - C_{\mathrm{rs}} - C_{\mathrm{ic}})e^{-\beta(h-r)} - (D_{\mathrm{rc}} - D_{\mathrm{is}} - D_{\mathrm{rs}} - D_{\mathrm{ic}})e^{\beta(h-r)}\right] \end{aligned} \tag{5.40}$$

$$\begin{aligned} \Gamma = &\kappa\left[(C_{\mathrm{rc}} - C_{\mathrm{is}})e^{-\beta(h-r)} + (D_{\mathrm{rc}} - D_{\mathrm{is}})e^{\beta(h-r)}\right] + \\ &+ j\kappa\left[(C_{\mathrm{rs}} - C_{\mathrm{ic}})e^{-\beta(h-r)} + (D_{\mathrm{rs}} - D_{\mathrm{ic}})e^{\beta(h-r)}\right] \end{aligned} \tag{5.41}$$

Particular solutions—magnetic field strength and current density distributions in each conductor segment—are obtained by applying proper boundary conditions for the uppermost and lowest conductor edge, along with continuity equations for imaginary borders between adjacent radial conductor sections. This procedure will be illustrated by the example of current redistribution due to skin effect in a hollow conductor in a slot of an electric machine.

5.2.1 *Exact Solution*

Hollow conductors, with a typical shape shown in Fig. 5.11 left, are used for direct gas or fluid cooling. The equivalent conductor, shown on the right in Fig. 5.11, can be separated into three zones (1, 2, and 3) with conductor thickness w_c, 2d, and w_c, respectively.

Complex constants of integration C and D in Eq. 5.39 have to be determined for all 3 zones, which sums up to the total of 12 unknowns: $C_{r,k}$, $k = 1, 3$; $C_{i,k}$, $k = 1, 3$; $D_{r,k}$, $k = 1, 3$; and $D_{i,k}$, $k = 1, 3$, representing real and imaginary components of C and D in all 3 zones, respectively.

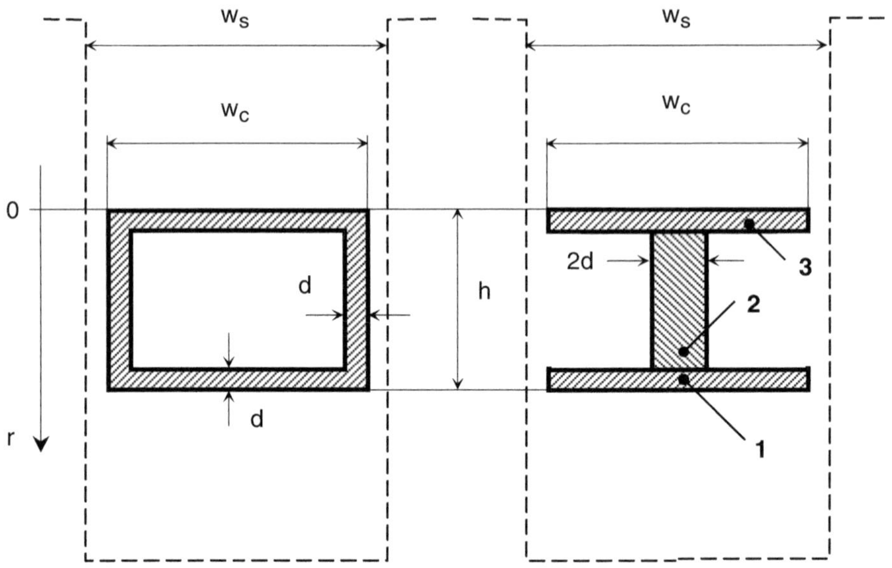

Fig. 5.11 Hollow conductor in a slot of an electric machine with wall thickness d (*left*) and equivalent conductor (*right*)

For the 12 unknowns, 12 algebraic equations can be written:

- Two boundary condition equations (one for real and one for imaginary component) state that the magnetic field strength H on the lowest bottom of the conductor is equal to zero;
- Two boundary condition equations (one for real and one for imaginary component) state that the magnetic field strength H on the uppermost edge of the conductor is equal to the applied current divided by the slot width;
- Two continuity equations (one for real and one for imaginary component) state that the magnetic field strength H on the lower side of the boundary between zones 1 and 2 is equal to the magnetic field strength H on the upper side of the boundary;
- Two continuity equations (one for real and one for imaginary component) state that the magnetic field strength H on the lower side of the boundary between zones 2 and 3 is equal to the magnetic field strength H on the upper side of the boundary;

- Two continuity equations (one for real and one for imaginary component) state that the current density Γ on the lower side of the boundary between zones 1 and 2 is equal to the current density Γ on the upper side of the boundary;
- Two continuity equations (one for real and one for imaginary component) state that the current density Γ on the lower side of the boundary between zones 2 and 3 is equal to the current density Γ on the upper side of the boundary;

The system of 12 algebraic equations determining the constants of integration can now be written as

$$\underline{A} \cdot \underline{X} = \underline{B} \tag{5.42}$$

where

$$\underline{X} = \begin{bmatrix} C_{r,1} & C_{r,2} & C_{r,3} & C_{i,1} & C_{i,2} & C_{i,3} & D_{r,1} & D_{r,2} & D_{r,3} & D_{i,1} & D_{i,2} & D_{i,3} \end{bmatrix}^{\mathrm{T}} \tag{5.43}$$

$$\underline{B} = \begin{bmatrix} 0 & 0 & 0 & 0 & 0 & 0 & \frac{I\omega\mu}{\beta_3 w_s} & 0 & 0 & 0 & 0 & 0 \end{bmatrix}^{\mathrm{T}} \tag{5.44}$$

and

$$\underline{A} = \begin{bmatrix}
a_{1,1} & 0 & 0 & a_{1,4} & 0 & 0 & a_{1,7} & 0 & 0 & a_{1,10} & 0 & 0 \\
a_{2,1} & 0 & 0 & a_{2,4} & 0 & 0 & a_{2,7} & 0 & 0 & a_{2,10} & 0 & 0 \\
a_{3,1} & a_{3,2} & 0 & a_{3,4} & a_{3,5} & 0 & a_{3,7} & a_{3,8} & 0 & a_{3,10} & a_{3,11} & 0 \\
a_{4,1} & a_{4,2} & 0 & a_{4,4} & a_{4,5} & 0 & a_{4,7} & a_{4,8} & 0 & a_{4,10} & a_{4,11} & 0 \\
0 & a_{5,2} & a_{5,3} & 0 & a_{5,5} & a_{5,6} & 0 & a_{5,8} & a_{5,9} & 0 & a_{5,11} & a_{5,12} \\
0 & a_{6,2} & a_{6,3} & 0 & a_{6,5} & a_{6,6} & 0 & a_{6,8} & a_{6,9} & 0 & a_{6,11} & a_{6,12} \\
0 & 0 & a_{7,3} & 0 & 0 & a_{7,6} & 0 & 0 & a_{7,9} & 0 & 0 & a_{7,12} \\
0 & 0 & a_{8,3} & 0 & 0 & a_{8,6} & 0 & 0 & a_{8,9} & 0 & 0 & a_{8,12} \\
a_{9,1} & a_{9,2} & 0 & a_{9,4} & a_{9,5} & 0 & a_{9,7} & a_{9,8} & 0 & a_{9,10} & a_{9,11} & 0 \\
a_{10,1} & a_{10,2} & 0 & a_{10,4} & a_{10,5} & 0 & a_{10,7} & a_{10,8} & 0 & a_{10,10} & a_{10,11} & 0 \\
0 & a_{11,2} & a_{11,3} & 0 & a_{11,5} & a_{11,6} & 0 & a_{11,8} & a_{11,9} & 0 & a_{11,11} & a_{11,12} \\
0 & a_{12,2} & a_{12,3} & 0 & a_{12,5} & a_{12,6} & 0 & a_{12,8} & a_{12,9} & 0 & a_{12,11} & a_{12,12}
\end{bmatrix} \tag{5.45}$$

Introducing

$$\begin{aligned}
c_k(r) &= \cos\beta_k(h-r); \quad s_k(r) = \sin\beta_k(h-r) \\
p_k(r) &= c_k(r) + s_k(r); \quad m_k(r) = c_k(r) - s_k(r) \\
e_k^-(r) &= e^{-\beta_k(h-r)}; \quad e_k^+(r) = e^{\beta_k(h-r)}
\end{aligned}$$

one can define the matrix coefficients in Eq. 5.45 as

$$a_{1,1} = -e_1^-(h)m_1(h); \quad a_{1,4} = -e_1^-(h)p_1(h); \quad a_{1,7} = e_1^+(h)m_1(h); \quad a_{1,10} = e_1^+(h)p_1(h)$$
$$a_{2,1} = e_1^-(h)p_1(h); \quad a_{2,4} = -e_1^-(h)m_1(h); \quad a_{2,7} = -e_1^+(h)p_1(h); \quad a_{2,10} = e_1^+(h)m_1(h)$$
$$a_{3,1} = -e_1^-(h-d)m_1(h-d); \quad a_{3,2} = e_2^-(h-d)m_2(h-d)$$
$$a_{3,4} = -e_1^-(h-d)p_1(h-d); \quad a_{3,5} = e_2^-(h-d)p_2(h-d)$$
$$a_{3,7} = e_1^+(h-d)m_1(h-d); \quad a_{3,8} = -e_2^+(h-d)m_2(h-d)$$
$$a_{3,10} = e_1^+(h-d)p_1(h-d); \quad a_{3,11} = -e_2^+(h-d)p_2(h-d)$$
$$a_{4,1} = -e_1^-(h-d)p_1(h-d); \quad a_{4,2} = e_2^-(h-d)p_2(h-d)$$
$$a_{4,4} = e_1^-(h-d)m_1(h-d); \quad a_{4,5} = -e_2^-(h-d)m_2(h-d)$$
$$a_{4,7} = e_1^+(h-d)p_1(h-d); \quad a_{4,8} = -e_2^+(h-d)p_2(h-d)$$
$$a_{4,10} = -e_1^+(h-d)m_1(h-d); \quad a_{4,11} = e_2^+(h-d)m_2(h-d)$$

$$a_{5,2} = -e_2^-(d)m_2(d); \quad a_{5,3} = e_3^-(d)m_3(d)a_{5,5} = -e_2^-(d)p_2(d);$$
$$a_{5,6} = e_3^-(d)p_3(d); \quad a_{5,8} = e_2^+(d)m_2(d); \quad a_{5,9} = -e_3^+(d)m_3(d)$$
$$a_{5,11} = e_2^+(d)p_2(d); \quad a_{5,12} = -e_3^+(d)p_3(d)$$
$$a_{6,2} = e_2^-(d)p_2(d); \quad a_{6,3} = -e_3^-(d)p_3(d); \quad a_{6,5} = -e_2^-(d)m_2(d);$$
$$a_{6,6} = e_3^-(d)m_3(d); \quad a_{6,8} = -e_2^+(d)p_2(d); \quad a_{6,9} = e_3^+(d)p_3(d)$$
$$a_{6,11} = e_2^+(d)m_2(d); \quad a_{6,12} = -e_3^+(d)m_3(d)$$

$$a_{7,3} = -e_3^-(0)m_3(0); \quad a_{7,6} = -e_3^-(0)p_3(0); \quad a_{7,9} = e_3^+(0)m_3(0); \quad a_{7,12} = e_3^+(0)m_3(0)$$
$$a_{8,3} = e_3^-(0)p_3(0); \quad a_{8,6} = -e_3^-(0)m_3(0); \quad a_{8,9} = -e_3^+(0)p_3(0); \quad a_{8,12} = e_3^+(0)m_3(0)$$
$$a_{9,1} = e_1^-(h-d)c_1(h-d); \quad a_{9,2} = -e_2^-(h-d)c_2(h-d)a_{9,4} = e_1^-(h-d)s_1(h-d);$$
$$a_{9,5} = -e_2^-(h-d)s_2(h-d); \quad a_{9,7} = e_1^+(h-d)c_1(h-d); \quad a_{9,8} = -e_2^+(h-d)c_2(h-d)$$
$$a_{9,10} = e_1^+(h-d)s_1(h-d); \quad a_{9,11} = -e_2^+(h-d)s_2(h-d)$$

$$a_{10,1} = -e_1^-(h-d)s_1(h-d); \quad a_{10,2} = e_2^-(h-d)s_2(h-d)$$
$$a_{10,4} = e_1^-(h-d)c_1(h-d); \quad a_{10,5} = -e_2^-(h-d)c_2(h-d)$$
$$a_{10,7} = -e_1^+(h-d)s_1(h-d); \quad a_{10,8} = e_2^+(h-d)s_2(h-d)$$
$$a_{10,8} = e_2^+(h-d)s_2(h-d); \quad a_{10,11} = -e_2^+(h-d)c_2(h-d)$$
$$a_{11,2} = e_2^-(d)c_2(d); \quad a_{11,3} = e_3^-(d)c_3(d); \quad a_{11,5} = e_2^-(d)s_2(d)$$
$$a_{11,6} = -e_3^-(d)s_3(d); \quad a_{11,8} = e_2^+(d)c_2(d); \quad a_{11,9} = -e_3^+(d)c_3(d)$$
$$a_{11,11} = e_2^+(h)s_2(d); \quad a_{11,12} = -e_3^+(d)s_3(d)$$

$$a_{12,2} = -e_2^-(d)s_2(d); \quad a_{12,3} = e_3^-(d)s_3(d); \quad a_{12,5} = e_2^-(d)c_2(d)$$
$$a_{12,6} = -e_3^-(d)c_3(d); \quad a_{12,8} = -e_2^+(d)s_2(d); \quad a_{12,9} = e_3^+(d)s_3(d)$$
$$a_{12,11} = e_2^+(d)c_2(d); \quad a_{12,12} = -e_3^+(d)c_3(d)$$

In order to find the distributions of electromagnetic quantities in a hollow conductor in Fig. 5.11, one has to solve the simultaneous system of 12 algebraic equations (Eq. 5.42), which is a time- and resource-consuming job. Therefore, an alternative approach seems to be more feasible, which delivers an approximate solution for an arbitrary conductor shape with less computational effort.

5.2.2 Approximate Solution

Consider a conductor with arbitrary shape placed in a slot of an electric machine as shown in Fig. 5.12. For the purpose of approximate computation of current redistribution, the conductor is separated into equidistant layers with thickness d, which is allowed since there is no radial component of current in the conductor.

Magnetic field strength H_k on the upper border of the kth layer is equal to

$$H_k w_{s,k} = \sum_{l=1}^{k} i_l + \Theta_0 \tag{5.46}$$

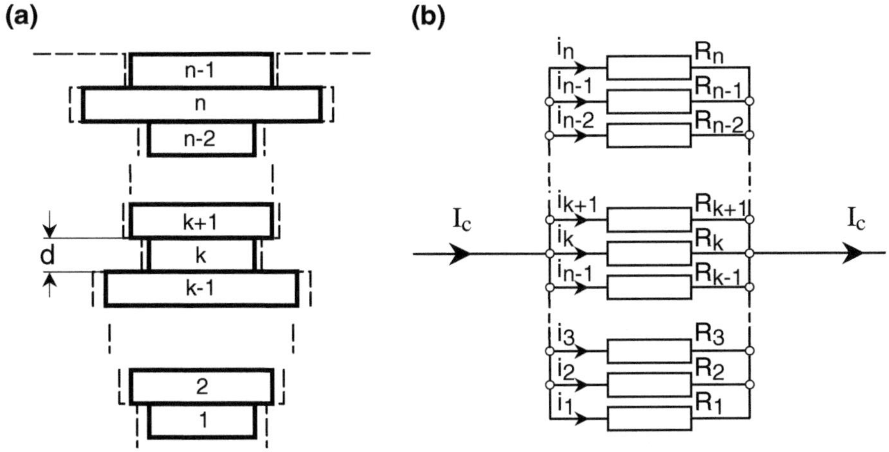

Fig. 5.12 Conductor with variable width and layer thickness d (**a**) and equivalent circuit (**b**)

with $w_{s,k}$ denoting the slot width at the position of the kth layer, i_l the unknown current in the lth layer, and Θ_0 the MMF created by current-carrying conductors placed below the conductor 1 in Fig. 5.12a.

Flux density B_k corresponding to the field strength in Eq. 5.46 is equal to

$$B_k = \frac{\mu_0}{w_{s,k}} \left(\sum_{l=1}^{k} i_l + \Theta_0 \right) \tag{5.47}$$

and the flux Φ_k through the kth layer:

$$\Phi_k = \frac{B_k + B_{k-1}}{2} l_{ax} d = \frac{l_{ax} d \mu_0}{2} \left(\frac{1}{w_{s,k}} \sum_{l=1}^{k} i_l + \frac{1}{w_{s,k-1}} \sum_{l=1}^{k-1} i_l + \frac{\Theta_0}{w_{s,k}} + \frac{\Theta_0}{w_{s,k-1}} \right) \tag{5.48}$$

or

$$\Phi_k = \frac{1}{2} \left(L_k \sum_{l=1}^{k} i_l + L_{k-1} \sum_{l=1}^{k-1} i_l + \frac{\Theta_0}{R_{m,k}} + \frac{\Theta_0}{R_{m,k-1}} \right) \tag{5.49}$$

with L_k, L_{k-1}, $R_{m,k}$, and $R_{m,k-1}$ denoting the inductances and reluctances belonging to the kth and (k-1)-th conductor layer, respectively.

For an arbitrary loop created by the kth and (k + 1)-th layer, one can write the equation of II Kirchhoff's rule

$$\frac{d}{dt} (\Phi_k + \Phi_{k+1}) + R_k i_k - R_{k+1} i_{k+1} = 0 \tag{5.50}$$

or

$$\mu_0 \frac{l_{ax} d}{2} \frac{d}{dt} \left(\frac{1}{w_{s,k-1}} \sum_{l=1}^{k-1} i_l + \frac{2}{w_{s,k}} \sum_{l=1}^{k} i_l + \frac{1}{w_{s,k+1}} \sum_{l=1}^{k+1} i_l + \frac{\Theta_0}{b_{s,k}} + \frac{\Theta_0}{b_{s,k-1}} \right) + R_k i_k - R_{k+1} i_{k+1} = 0 \tag{5.51}$$

and accordingly

$$\frac{1}{2} \frac{d}{dt} \left(L_{k-1} \sum_{l=1}^{k-1} i_l + 2 L_k \sum_{l=1}^{k} i_l + L_{k+1} \sum_{l=1}^{k+1} i_l + \frac{\Theta_0}{R_{m,k}} + \frac{\Theta_0}{R_{m,k-1}} \right) + R_k i_k - R_{k+1} i_{k+1} = 0 \tag{5.52}$$

For a sinusoidal applied total current I and linear media, the current in the ℓth layer can be expressed as a complex number by means of its real I^{Re} and imaginary I^{Im} components as

$$i_l \Rightarrow \widehat{I}_l = I_l^{Re} + j I_l^{Im} \tag{5.53}$$

whereas its derivative with respect to time is equal to

$$\frac{d}{dt} i_l \Rightarrow = j\omega \widehat{I}_l = \omega \left(-I_l^{Im} + j I_l^{Re} \right) \tag{5.54}$$

Substituting Eqs. 5.53 and 5.54 into 5.52, one obtains

$$\omega\left[(L_{k-1}+2L_k+L_{k+1})\sum_{l=1}^{k-1}\left(-I_l^{\mathrm{Im}}+jI_l^{\mathrm{Re}}\right)+(2L_k+L_{k+1})\left(-I_k^{\mathrm{Im}}+jI_k^{\mathrm{Re}}\right)+L_{k+1}\left(-I_{k+1}^{\mathrm{Im}}+jI_{k+1}^{\mathrm{Re}}\right)\right]+$$
$$+R_k\left(I_k^{\mathrm{Re}}+jI_k^{\mathrm{Im}}\right)-R_{k+1}\left(I_{k+1}^{\mathrm{Re}}+jI_{k+1}^{\mathrm{Im}}\right)=U_{0,k}^{\mathrm{Re}}+jU_{0,k}^{\mathrm{Im}};\quad k=2,3,\ldots,n \tag{5.55}$$

with $U_{0,k}^{\mathrm{Re}}$ and $U_{0,k}^{\mathrm{Im}}$ denoting the real and imaginary components of voltage induced by the MMF Θ_0 in the loop created by the kth and (k + 1)-th layers.

Complex Eq. 5.55 contains $n-1$ equations for real:

$$\begin{aligned}&-\omega(L_{k-1}+2L_k+L_{k+1})\sum_{l=1}^{k-1}I_l^{\mathrm{Im}}-\omega(2L_k+L_{k+1})I_k^{\mathrm{Im}}-\omega L_{k+1}I_{k+1}^{\mathrm{Im}}+R_kI_k^{\mathrm{Re}}\\&-R_{k+1}I_{k+1}^{\mathrm{Re}}=U_{0,k}^{\mathrm{Re}}\end{aligned} \tag{5.56}$$

and $n-1$ equations for imaginary parts

$$\begin{aligned}&\omega(L_{k-1}+2L_k+L_{k+1})\sum_{l=1}^{k-1}I_l^{\mathrm{Re}}+\omega(2L_k+L_{k+1})I_k^{\mathrm{Re}}+\omega L_{k+1}I_{k+1}^{\mathrm{Re}}+R_kI_k^{\mathrm{Im}}\\&-R_{k+1}I_{k+1}^{\mathrm{Im}}=U_{0,k}^{\mathrm{Im}}\end{aligned} \tag{5.57}$$

The nth equation for real parts yields

$$\sum_{l=1}^{n}I_l^{\mathrm{Re}}=\mathrm{Re}\{\hat{I}_c\} \tag{5.58}$$

and for imaginary

$$\sum_{l=1}^{n}I_l^{\mathrm{Im}}=\mathrm{Im}\{\hat{I}_c\} \tag{5.59}$$

The system of $2n$ algebraic equations can be written in the matrix form as

$$\begin{bmatrix}\underline{A} & \underline{B}\\ -\underline{B} & \underline{A}\end{bmatrix}\cdot\begin{bmatrix}\underline{I}^{\mathrm{Re}}\\ \underline{I}^{\mathrm{Im}}\end{bmatrix}=[\underline{Y}] \tag{5.60}$$

where the $n\times n$ submatrices $\underline{A}$ and $\underline{B}$ are defined as

$$\underline{A} = \begin{bmatrix} R_1 & -R_2 & 0 & 0 & \dots & 0 & 0 \\ 0 & R_2 & -R_3 & 0 & \dots & 0 & 0 \\ 0 & 0 & R_3 & -R_4 & \dots & 0 & 0 \\ \dots & \dots & \dots & \dots & \dots & \dots & \dots \\ 0 & 0 & 0 & 0 & \dots & -R_{n-1} & 0 \\ 0 & 0 & 0 & 0 & \dots & R_{n-1} & -R_n \\ 1 & 1 & 1 & 1 & \dots & 1 & 1 \end{bmatrix} \tag{5.61}$$

$$\underline{B} = -\omega \begin{bmatrix} 2L_1 + L_2 & L_2 & 0 & \dots & \dots & 0 & 0 \\ L_1 + 2L_2 + L_3 & 2L_2 + L_3 & L_3 & \dots & \dots & 0 & 0 \\ L_2 + 2L_3 + L_4 & L_2 + 2L_3 + L_4 & 2L_3 + L_4 & \dots & \dots & 0 & 0 \\ \dots & \dots & \dots & \dots & \dots & \dots & \dots \\ L_{n-3} + 2L_{n-2} + L_{n-1} & L_{n-3} + 2L_{n-2} + L_{n-1} & L_{n-3} + 2L_{n-2} + L_{n-1} & \dots & \dots & L_{n-1} & 0 \\ L_{n-2} + 2L_{n-1} + L_n & L_{n-2} + 2L_{n-1} + L_n & L_{n-2} + 2L_{n-1} + L_n & \dots & \dots & 2L_{n-1} + L_n & L_n \\ 0 & 0 & 0 & \dots & \dots & 0 & 0 \end{bmatrix} \tag{5.62}$$

and the vector $\underline{Y}$

$$\underline{Y} = \left[U_{0,1}^{\mathrm{Re}} \quad U_{0,2}^{\mathrm{Re}} \quad \dots \quad U_{0,n-2}^{\mathrm{Re}} \quad U_{0,n-1}^{\mathrm{Re}} \quad \mathrm{Re}\{\hat{I}_c\} \quad U_{0,1}^{\mathrm{Im}} \quad U_{0,2}^{\mathrm{Im}} \quad \dots \quad U_{0,n-2}^{\mathrm{Im}} \quad U_{0,n-1}^{\mathrm{Im}} \quad \mathrm{Im}\{\hat{I}_c\} \right]^{\mathrm{T}} \tag{5.63}$$

The solution vector of the system 5.60 can be written as

$$\begin{bmatrix} \underline{I}^{\mathrm{Re}} \\ \underline{I}^{\mathrm{Im}} \end{bmatrix} = \begin{bmatrix} I_1^{\mathrm{Re}} & I_2^{\mathrm{Re}} & \dots & I_{n-1}^{\mathrm{Re}} & I_n^{\mathrm{Re}} & I_1^{\mathrm{Im}} & I_2^{\mathrm{Im}} & \dots & I_{n-1}^{\mathrm{Im}} & I_n^{\mathrm{Im}} \end{bmatrix}^{\mathrm{T}} \tag{5.64}$$

Case Study 5.4: Assume a conductor similar to the one introduced in Case Study 5.1, the upper side of which is half as wide as its lower side, Fig. 5.13. The trapezoidal conductor form strengthens the effects of current redistribution discussed in this chapter, because of decreasing cross-sectional area in its portion closer to the air gap (Fig. 5.14).

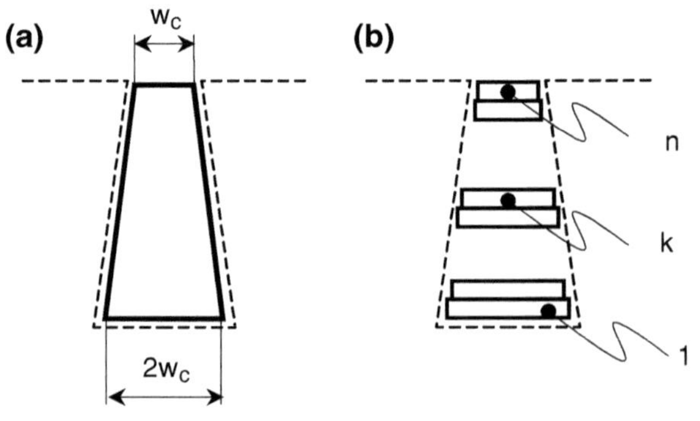

Fig. 5.13 Trapezoidal conductor in a slot (**a**) and its discretization into n layers (**b**)

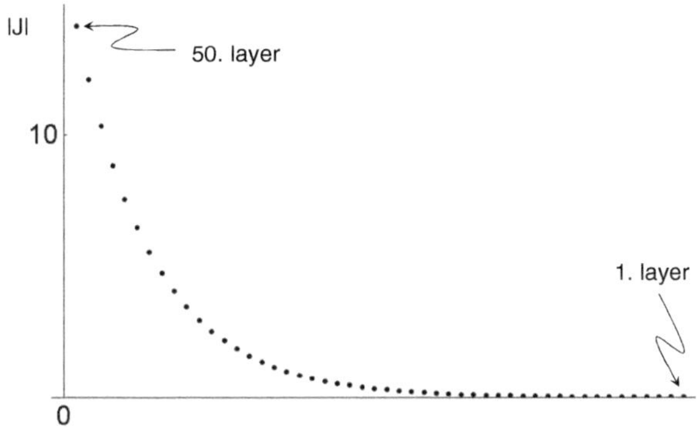

Fig. 5.14 The absolute value of current density [A/mm^2] at 50 Hz frequency of impressed current

The impressed current creates at 0 Hz the current density of 1 A/mm^2 in the lowest (k = 1), and 2 A/mm^2 in the uppermost (k = n) layer.

Alternating current density of above 14 A/mm^2 in the uppermost layer of the trapezoidal conductor, Fig. 5.14, means a more than seven-time increase as compared to the DC value, which corresponds to the analytical solution for a rectangular conductor presented in Fig. 5.3c. For this reason rotor bars of a squirrel cage induction machine often have trapezoidal, instead of rectangular form.

5.3 Analytical Solution for Current Density Redistribution in a Solid Rectangular Conductor in a Slot as a Result of Impressed Alternating Leakage Flux: One-Dimensional Proximity Effect

Consider now a conductor in a slot similar to that in Fig. 5.1, however, with an external source of MMF Θ placed on the bottom of the slot, beneath the solid conductor. The impressed ampere-turns Θ modify boundary conditions on the lower and upper edges of the conductor, which now can be written as:

(a) Magnetic field strength on the lower conductor edge:

$$H(h) = \frac{\Theta}{w_{\mathrm{s}}} \tag{5.65}$$

(b) Magnetic field strength on the upper conductor edge:

$$H(0) = \frac{\Theta + I}{w_{\mathrm{s}}} \tag{5.66}$$

whereas the impressed ampere-turns Θ have a component Θ_{Re} in phase with the conductor current I, and a component Θ_{Im}, which is 90° out phase with the conductor current I:

$$\Theta = \Theta_{\mathrm{Re}} + j\Theta_{\mathrm{Im}} \tag{5.67}$$

Magnetic field strength distribution in the solid conductor exposed to boundary conditions 5.65 and 5.66 can now be expressed as

$$H = \frac{I + \Theta_{\mathrm{Re}} + j\Theta_{\mathrm{Im}}}{w_{\mathrm{s}}} \frac{\sinh[\beta(1+j)(h-r)]}{\sinh \beta(1+j)h} + \frac{\Theta_{\mathrm{Re}} + j\Theta_{\mathrm{Im}}}{w_{\mathrm{s}}} \frac{\sinh[\beta(1+j)r]}{\sinh \beta(1+j)h} \tag{5.68}$$

and its partial derivative with respect to the radial coordinate, current density $\vec{\Gamma} = \mathrm{curl}\,\vec{H}$, is equal to

$$\Gamma = -\beta(1+j)\frac{I + \Theta_{\mathrm{Re}} + j\Theta_{\mathrm{Im}}}{w_{\mathrm{s}}} \frac{\cosh[\beta(1+j)(h-r)]}{\sinh \beta(1+j)h} + \beta(1+j)\frac{\Theta_{\mathrm{Re}} + j\Theta_{\mathrm{Im}}}{w_{\mathrm{s}}} \frac{\cosh[\beta(1+j)r]}{\sinh \beta(1+j)h} \tag{5.69}$$

Introducing parameters C_{hs} and S_{hc}, defined as

$$C_{\mathrm{hs}} = \frac{2\beta \cosh \beta h \, \sin \beta h}{w_{\mathrm{s}}(\cosh 2\beta h - \cos 2\beta h)} \tag{5.70}$$

$$S_{\mathrm{hc}} = \frac{2\beta \sinh \beta h \cos \beta h}{w_{\mathrm{s}}(\cosh 2\beta h - \cos 2\beta h)} \tag{5.71}$$

along with auxiliary functions c_{r}, s_{r}, c_{rh} and s_{rh}

$$c_{\mathrm{r}} = \cos \beta r \cosh \beta r \tag{5.72}$$

$$s_{\mathrm{r}} = \sin \beta r \sinh \beta r \tag{5.73}$$

$$c_{\mathrm{rh}} = \cos \beta(r-h) \cosh \beta(r-h) \tag{5.74}$$

$$s_{\mathrm{rh}} = \sin \beta(r-h) \sinh \beta(r-h) \tag{5.75}$$

one can define the real component of current density

$$\begin{aligned}\mathrm{Re}\{\Gamma(r)\} = {} & I[C_{\mathrm{hs}}(c_{\mathrm{r}}+s_{\mathrm{r}})+S_{\mathrm{hc}}(c_{\mathrm{r}}-s_{\mathrm{r}})]+\\ & +\Theta_{\mathrm{Re}}[C_{\mathrm{hs}}(c_{\mathrm{r}}+s_{\mathrm{r}}-c_{\mathrm{rh}}-s_{\mathrm{rh}})+S_{\mathrm{hc}}(c_{\mathrm{r}}-s_{\mathrm{r}}-c_{\mathrm{rh}}+s_{\mathrm{rh}})]+\\ & +\Theta_{\mathrm{Im}}[C_{\mathrm{hs}}(c_{\mathrm{r}}-s_{\mathrm{r}}-c_{\mathrm{rh}}+s_{\mathrm{rh}})+S_{\mathrm{hc}}(-c_{\mathrm{r}}+s_{\mathrm{r}}+c_{\mathrm{rh}}+s_{\mathrm{rh}})]\end{aligned} \tag{5.76}$$

along with its imaginary counterpart as

$$\begin{aligned}\mathrm{Im}\{\Gamma(r)\} = {} & I[C_{\mathrm{hs}}(c_{\mathrm{r}}-s_{\mathrm{r}})-S_{\mathrm{hc}}(c_{\mathrm{r}}+s_{\mathrm{r}})]+\\ & +\Theta_{\mathrm{Re}}[C_{\mathrm{hs}}(c_{\mathrm{r}}+s_{\mathrm{r}}-c_{\mathrm{rh}}+s_{\mathrm{rh}})-S_{\mathrm{hc}}(c_{\mathrm{r}}+s_{\mathrm{r}}-c_{\mathrm{rh}}-s_{\mathrm{rh}})]+\\ & +\Theta_{\mathrm{Im}}[C_{\mathrm{hs}}(-c_{\mathrm{r}}+s_{\mathrm{r}}+c_{\mathrm{rh}}+s_{\mathrm{rh}})-S_{\mathrm{hc}}(c_{\mathrm{r}}-s_{\mathrm{r}}-c_{\mathrm{rh}}+s_{\mathrm{rh}})]\end{aligned} \tag{5.77}$$

Losses in the solid conductor can be expressed as (see also Eq. 5.13)

$$P_{\sim} = \frac{w_{\mathrm{c}} l_{ax}}{\kappa} \int_0^h \Gamma \cdot \Gamma^* \mathrm{d}r = P_{\mathrm{eddy}} + P_{\mathrm{prox}} \tag{5.78}$$

I^2R losses in Eq. 5.78 consist of two components: eddy current losses, P_{eddy}:

$$P_{\mathrm{eddy}} = I^2 \frac{l_{ax}}{\kappa h w_{\mathrm{c}}} \cdot \beta h \frac{\sinh 2\beta h + \sin 2\beta h}{\cosh 2\beta h - \cos 2\beta h} \tag{5.79}$$

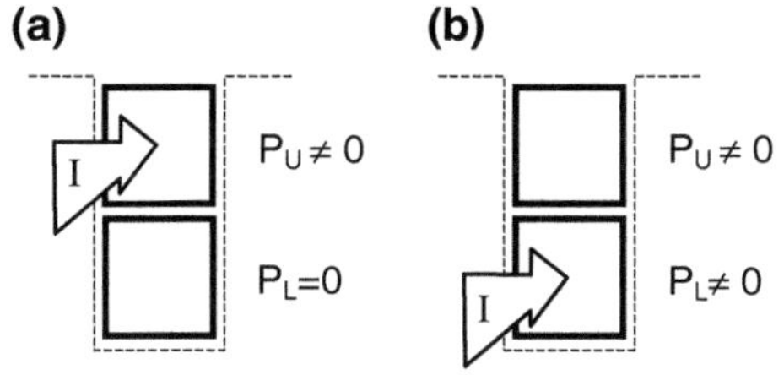

Fig. 5.15 When current *I* flows only through the conductor closer to the air gap (**a**), the losses in the conductor on the bottom of the slot are equal to zero because no flux line created by the current *I* goes through the conductor on the bottom of the slot. When current *I* flows only through the conductor on the bottom of the slot (**b**), it creates losses in both conductors

and proximity effect losses, P_{prox}:

$$P_{\mathrm{prox}} = \left(\Theta_{\mathrm{Im}}^2 + \Theta_{\mathrm{Re}}^2 + \Theta_{\mathrm{Re}} I\right) \frac{l_{ax}}{\kappa h w_{\mathrm{c}}} \cdot 2\beta h \frac{\sinh 2\beta h - \sin 2\beta h}{\cosh 2\beta h + \cos 2\beta h} \tag{5.80}$$

In addition to the function $\varphi(\xi)$ defined in Eq. 5.21 as a ratio of AC and DC losses created by current in a conductor, one can introduce the function $\psi(\xi)$ as a measure of influence of external MMF Θ on losses in a conductor as:

$$\psi(\xi) = 2\xi \frac{\sinh 2\xi - \sin 2\xi}{\cosh 2\xi + \cos 2\xi} \tag{5.81}$$

In order to get a better insight into the mechanism of proximity effect, a slot with two solid conductors as shown in Fig. 5.15 is assumed and the current density along with losses P_u in the conductor closer to the air gap are analyzed.

The conductor topology in Fig. 5.15 illustrates in the best manner the meaning of current *redistribution* due to own leakage flux, the effect which dominates in the conductor closer to the air gap in Fig. 5.15a. Eddy currents in the same conductor are completely differently distributed in the configuration described in Fig. 5.15b, when their source is an external MMF, instead of the conductor current. In the former case, there exists a strong feedback within a conductor between eddy currents and impressed current, whereas in the latter case this feedback vanishes.

The distribution of current density components in the conductor closer to the air gap is illustrated in Fig. 5.16 for the two supply modi, as defined in Fig. 5.15: Solid curves stand for current density distributions in the case of skin effect and dashed in the case of proximity effect. Current density distributions in Fig. 5.16 are evaluated for a 50-mm-high copper conductor at 50 Hz by using Eq. 5.69. Whereas the absolute value of current density due to skin effect increases in the direction of air gap, in the case of proximity effect it is symmetrical with respect to the horizontal centerline of the conductor. The same is valid for both real and imaginary components of current density, as shown in Fig. 5.16a, b.

When the conductor closer to the air gap is fed by current I, as in Fig. 5.15a, the losses P_u in it are determined by the function $\varphi(\xi)$ and can be found by means of Eq. 5.13. Losses in the conductor on the bottom of the slot are equal to zero because the conductor current is equal to zero and no flux created by current I *in the upper conductor goes through the lower conductor.*

When the conductor on the bottom of the slot is fed by current I, as in Fig. 5.15b, losses P_u in the conductor closer to the air gap are determined by the function $\psi(\xi)$ and can be found by means of Eq. 5.81.

The functions $\varphi(\xi)$ and $\psi(\xi)$, along with the ratio of $\psi(\xi)/\varphi(\xi)$, are shown in Fig. 5.17. The impact of external ampere-turns on conductor losses is obviously negligible at lower frequencies, i.e., for conductor heights significantly below the skin depth, see Fig. 5.17.

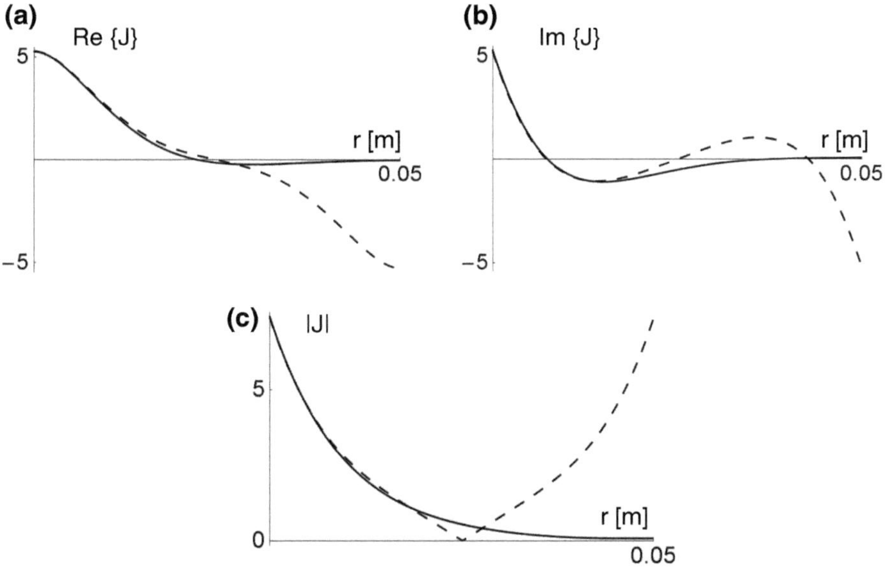

Fig. 5.16 The real (**a**), imaginary (**b**), and absolute value (**c**) of current density in the conductor closer to the air gap for two feeding modes, as defined in Fig. 5.15. The *solid curve* gives the current density distribution due to skin effect (Fig. 5.15a), and the *dashed curve* the current density due to proximity effect (Fig. 5.15b). At 0 Hz and a given impressed current, the current density is equal to 1 A/mm^2

It is interesting that losses in the current-free conductor closer to the air gap, created by ampere-turns $\Theta_{Re} = I$ and $\Theta_{Im} = 0$ in the conductor on the bottom of the slot (Fig. 5.15b), are as large as the losses created by own current I through the conductor closer to the air gap (Fig. 5.15a) already at the normalized conductor height of $\xi = 0.851$, at which $\varphi(\xi) = \psi(\xi)$.

Any further increase of parameter ξ results in higher losses due to proximity effect than due to skin effect. The maximum ratio $\psi(\xi)/\varphi(\xi)$ between the two loss components is reached at $\xi = \pi/2$, which is at the same time the conductor critical height! Above this point, the ratio $\psi(\xi)/\varphi(\xi)$ reaches asymptotically the value of 2, i.e., the same current I creates twice as much losses when it alone flows through the conductor on the bottom of the slot than through the conductor closer to the air gap.

Conductor losses defined in Eqs. 5.79 and 5.80 can further be written as

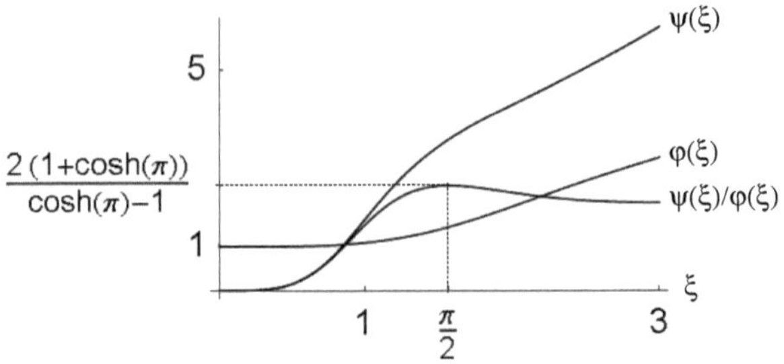

Fig. 5.17 The functions φ(ξ), ψ(ξ), and their ratio ψ(ξ)/φ(ξ)

$$P = I^2 \frac{l_{ax}}{\kappa h w_c} \cdot \left[\varphi(\xi) + \left(\frac{\Theta_{\mathrm{Im}}^2 + \Theta_{\mathrm{Re}}^2}{I^2} + \frac{\Theta_{\mathrm{Re}}}{I} \right) \psi(\xi) \right] \tag{5.82}$$

The influence of proximity effect on total losses, expressed by the function ψ (ξ), is modified by the amount and phase shift of the external MMF. If the external MMF is created by the same current I which flows through the conductor closer to the air gap, one refers to a *monoslot*.

The losses in the conductor closer to the air gap are in this case

$$P_{\mathrm{mono}} = I^2 \frac{l_{ax}}{\kappa h w_c} \cdot [\varphi(\xi) + 2\psi(\xi)] \tag{5.83}$$

because $\Theta_{\mathrm{Re}} = I$ and $\Theta_{Im} = 0$. If the external MMF is created by the current I shifted for 60° to the current flowing through the conductor closer to the air gap, which is the case in some slots of 3-phase machines with shorted pitch, one refers to a *mixed slot*. The losses in the conductor closer to the air gap are in this case

$$P_{\mathrm{mix}} = I^2 \frac{l_{ax}}{\kappa h w_c} \cdot [\varphi(\xi) + 1.5\psi(\xi)] \tag{5.84}$$

because $\Theta_{\mathrm{Re}} = I/2$ and $\Theta_{Im} = \sqrt{3/2}\mathrm{I}$.

The ratio φ(ξ) between AC and DC losses, introduced in Eq. 5.21 for a single conductor, has to be redefined when a conductor is exposed to an additional external field. Utilizing Eq. 5.82, the coefficient of resistance increase $k_{\mathrm{R}} = P_{\sim}/P_{=}$, analogous to the ratio φ(ξ), is in general case equal to

$$k_R = \frac{P_{\sim}}{P_{=}} = \varphi(\xi) + \left(\frac{\Theta_{\mathrm{Im}}^2 + \Theta_{\mathrm{Re}}^2}{I^2} + \frac{\Theta_{\mathrm{Re}}}{I} \right) \psi(\xi) \tag{5.85}$$

5.4 One-Dimensional Skin and Proximity Effect in Solid Conductors of a Coil in a Slot: Average Values of Skin Effect Factors for All Conductors in a Slot and for All Slots of a Phase

As shown in the previous chapter, the amount of losses in a conductor in a slot is dependent not only on the conductor current, but also on the amount and phase shift of ampere-turns in the slot below the conductor. Except in a single-cage winding, there is always more than one current-carrying conductor in a slot. Therefore, it is quite interesting to analyze proximity effect in turns of a coil in a slot of an electric machine. In the following analysis it will be assumed that each conductor in the coil is solid, without strands, and that all turns are connected in series and, consequently, carry the same current. The assumption that there are no strands in conductors means that eddy currents created in the slots can flow in radial direction through conductors in the end region of each turn.

Single-layer windings have half as many coils as slots. A single coil occupies two monoslots, one of them being shown in Fig. 5.18.

In a monoslot the *n*th conductor is exposed to the field of n-1 conductors below it. Therefore, the coefficient of resistance increase k_R in Eq. 5.85 can be written for the *n*th conductor as

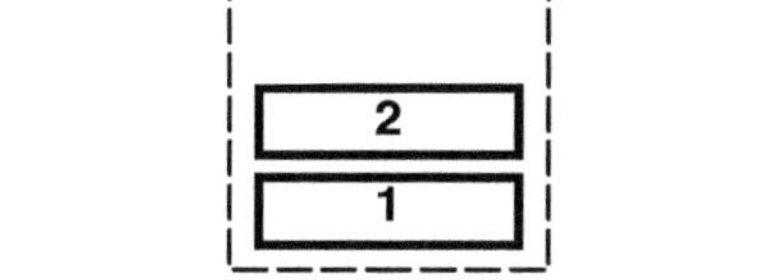

Fig. 5.18 Monoslot of a single-layer winding with w turns per coil/w conductors in slot

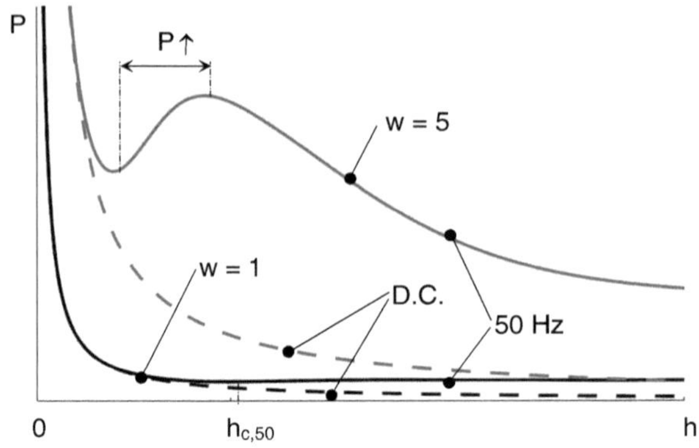

Fig. 5.19 I^2R losses in the slot portion of a coil with one turn (*black curves*) and with five turns (*gray curves*) at 50 Hz (*solid*) and for direct current (*dashed*). Whereas for direct current the coil with five turns creates five times as much losses as the coil with one turn and identical conductor dimensions, the ratio of losses at 50 Hz is a function of the conductor height h

$$k_{R,n} = \frac{P_\sim}{P_=} = \varphi(\xi) + n(n-1)\psi(\xi) \tag{5.86}$$

since $\Theta_{Re} = I$ and $\Theta_{Im} = 0$. The average coefficient of resistance increase $k_{R,av}$ for the complete coil is equal to

$$k_{R,av} = \frac{1}{w}\sum_{i=1}^{w} k_{R,i} = \varphi(\xi) + \frac{w^2-1}{3}\psi(\xi) \tag{5.87}$$

Total coil losses in the active part can now be expressed as (see also Eq. 5.74):

$$P = 2I^2\frac{l_{ax}}{\kappa\delta w_c}w\cdot\left[\frac{\sinh 2\xi+\sin 2\xi}{\cosh 2\xi-\cos 2\xi}+\frac{w^2-1}{3}2\frac{\sinh 2\xi-\sin 2\xi}{\cosh 2\xi+\cos 2\xi}\right] \tag{5.88}$$

The dependence of losses after Eq. 5.88 on conductor height in a coil with one turn and in a coil with five turns at 50 Hz and for DC are shown in Fig. 5.19, black and gray, respectively. Loss curves for a single-turn coil correspond to those shown in Fig. 5.5a.

One recognizes in Fig. 5.19 the influence of proximity effect on total losses, which substantially increase as the external MMF increases proportionally to the number of conductors in slot. Starting already at a low number of turns, proximity effect is strong enough to reverse the natural tendency of losses to decrease as the conductor height increases. For each number of turns losses increase in a certain range of conductor height h, as depicted with “P ↑” in Fig. 5.19 for $w = 5$.

Double-layer windings have as many coils as slots. Windings with shorted coils ($y < \tau_p$) have both mono- and mixed slots. Since the proximity effect in a mixed slot is not as strong as in a monoslots (compare Eqs. 5.83 and 5.84), total losses in conductors in the layer of a mixed slot closer to air gap are lower than in a monoslot, as shown in Fig. 5.20.

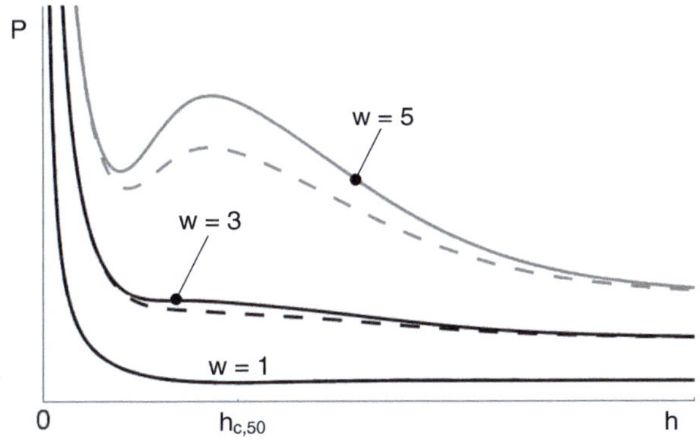

Fig. 5.20 I^2R losses in the side of a coil closer to air gap of a mono- (*solid curves*) and mixed (*dashed curves*) slot for various numbers of turns and at 50 Hz

5.5 Coil Manufacturing Techniques for Suppression of Current Redistribution Due to Skin Effect: Roebel Bar, Ringland Bar, Willyoung Bar, Strand Transposition

It has been shown in the previous chapters that a solid conductor is vulnerable against skin and proximity effect as long as it offers a relatively large area to the slot leakage flux. Therefore, the basic idea of mitigating the negative consequences of alternating current redistribution is to cut the conductor along its radial height into strands isolated to each other, Fig. 5.21.

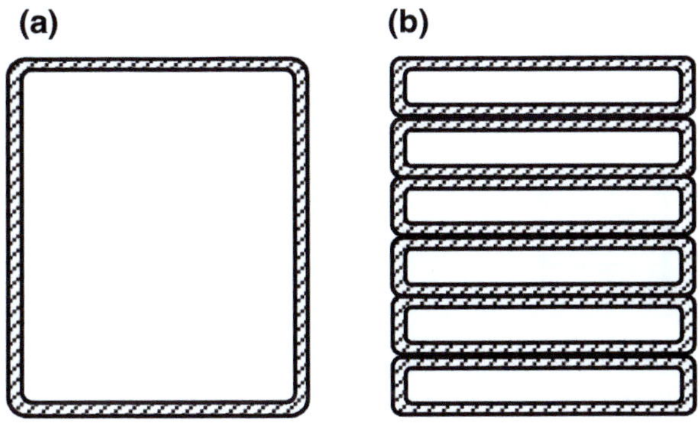

Fig. 5.21 Solid (**a**) and stranded (**b**) conductor

5.5.1 *Multi-turn Coil with Straight Strands*

Assume that each conductor of a multi-turn coil is built out of ℓ insulated strands connected with strands on the other side of the coil in a straight manner, without crossing in the end winding zone, as shown in Fig. 5.22. Strands denoted by "1" in conductors in Fig. 5.22 occupy in both coil sides the lowermost positions of all conductors and strands denoted by "ℓ" the uppermost conductor positions, consequently. The strands are connected to each other at the beginning of the first turn and at the end of the last (wth) turn. The position of strands within a conductor in end windings remains the same as in slots.

Real and imaginary components of the voltage induced at radial coordinate r by slot leakage flux in the nth conductor, $1 < n \leq w$, can be expressed by means of Eqs. 5.76 and 5.77 as

$$\mathrm{Re}\{U_n(r)\} = \frac{l_{ax}}{\kappa}\mathrm{Re}\{\Gamma(r)\} \tag{5.89}$$

$$\mathrm{Im}\{U_n(r)\} = \frac{l_{ax}}{\kappa}\mathrm{Im}\{\Gamma(r)\} \tag{5.90}$$

One recalls that the radial coordinate r is local for each conductor. For a given local coordinate r, the functions c_{r}, s_{r}, c_{rh} and s_{rh} (Eqs. 5.72–5.75) have the same value in each of w conductors in both slots. Since all w conductors carry the same real current I, Θ_{Im} in Eqs. 5.76 and 5.77 is equal to zero, and the real and imaginary components of induced voltage in the nth conductor can be expressed as

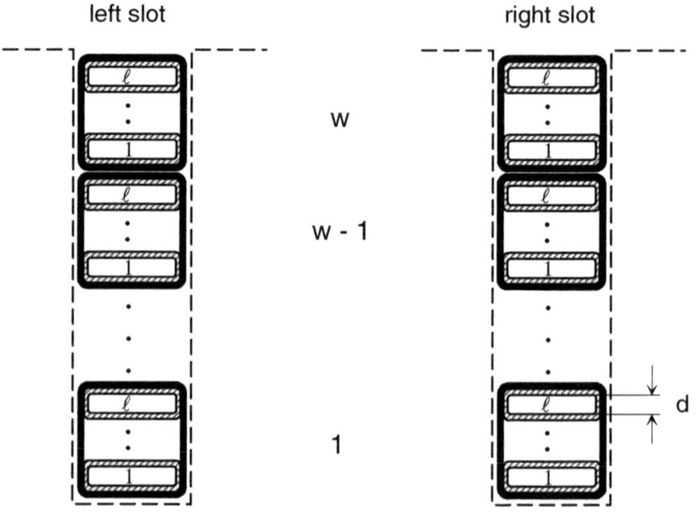

Fig. 5.22 Multi-turn coil with ℓ strands in each conductor connected to strands in another slot in a straight manner, without crossing in the end windings

$$\mathrm{Re}\{U_n(r)\} = \frac{l_{ax}}{\kappa}\mathrm{Re}\{\Gamma(r)\} = \frac{l_{ax}}{\kappa}I[a_r + (n-1)b_r] \tag{5.91}$$

$$\mathrm{Im}\{U_n(r)\} = \frac{l_{ax}}{\kappa}\mathrm{Re}\{\Gamma(r)\} = \frac{l_{ax}}{\kappa}I[f_r + (n-1)e_r] \tag{5.92}$$

where

$$a_r = C_{hs}(c_r + s_r) + S_{hc}(c_r - s_r) \tag{5.93}$$

$$b_r = C_{hs}(c_r + s_r - c_{rh} - s_{rh}) + S_{hc}(c_r - s_r - c_{rh} + s_{rh}) \tag{5.94}$$

$$e_r = C_{hs}(c_r + s_r - c_{rh} + s_{rh}) - S_{hc}(c_r + s_r - c_{rh} - s_{rh}) \tag{5.95}$$

$$f_r = C_{hs}(c_r - s_r) - S_{hc}(c_r + s_r) \tag{5.96}$$

Combining Eqs. 5.91 and 5.92, one can state that the induced voltage U_n at radial coordinate r in the nth conductor has two components: U_o, created by conductor own current I, and U_e, created by the same current I flowing through $n-1$ conductors below:

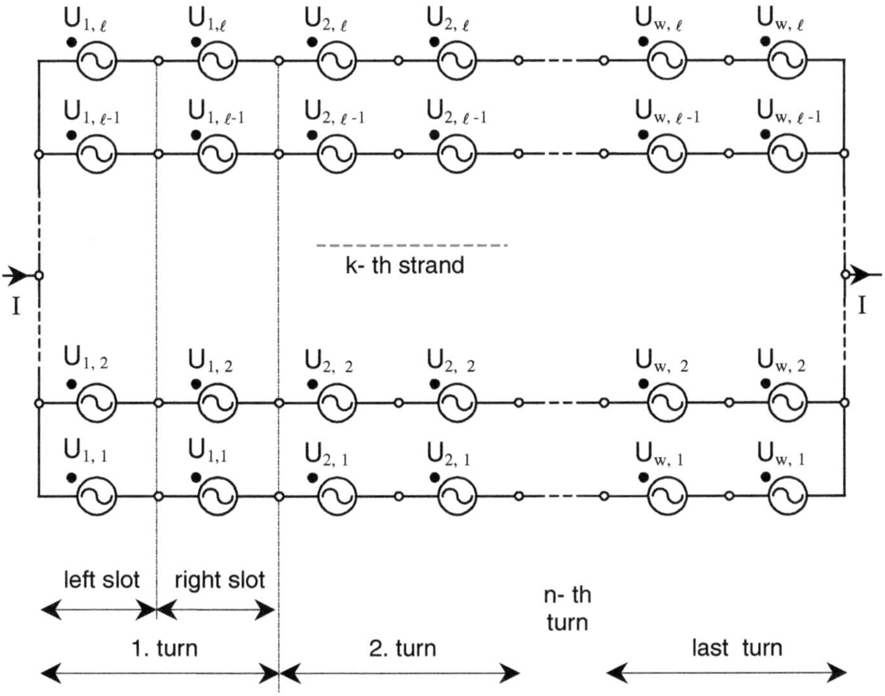

Fig. 5.23 Equivalent circuit for induced voltages in a coil with ℓ strands per conductor connected in straight manner in end windings

$$U_{\mathrm{o}}(r) = \frac{l_{ax}}{\kappa} I(a_{\mathrm{r}} + jf_{\mathrm{r}}) \tag{5.97}$$

$$U_{\mathrm{e}}(r) = \frac{l_{ax}}{\kappa} I(b_{\mathrm{r}} + je_{\mathrm{r}}) \tag{5.98}$$

resulting in

$$U_{n,\mathrm{r}} = U_{\mathrm{o}}(r) + (n-1)U_{\mathrm{e}}(r) \tag{5.99}$$

If the radial strand height is equal to d, the induced voltage in the center of the kth strand, $1 < k \le \ell$ of the nth conductor, $U_{n,k}$, can be written as

$$U_{n,k} = U_{\mathrm{o}}[(2k-1)d/2] + (n-1) \cdot U_{\mathrm{e}}[(2k-1)d/2] \tag{5.100}$$

The equivalent circuit for induced voltages in ℓ strands and w turns is shown in Fig. 5.23.

Circulating currents between two arbitrary parallel branches in Fig. 5.23 are as larger as higher the difference between induced voltages in the branches. Driving force for circulating current in a loop containing ith and jth strand is the voltage difference ΔU defined as

$$\begin{aligned}\Delta U = 2\sum_{n=1}^{w}\left(U_{n,i} - U_{n,j}\right) &= 2\sum_{n=1}^{w}\{U_{\mathrm{o}}[(2i-1)\mathrm{d}/2] - U_{\mathrm{o}}[(2j-1)\mathrm{d}/2]\} \\ &+ 2\sum_{n=1}^{w}\{(n-1)\cdot U_{\mathrm{e}}[(2i-1)\mathrm{d}/2] - (n-1)\cdot U_{\mathrm{e}}[(2j-1)\mathrm{d}/2]\}\end{aligned} \tag{5.101}$$

Introducing

$$\Delta U_{\mathrm{o},i,j} = U_{\mathrm{o}}[(2i-1)\mathrm{d}/2] - U_{\mathrm{o}}[(2j-1)\mathrm{d}/2] \tag{5.102}$$

and

$$\Delta U_{\mathrm{e},i,j} = U_{\mathrm{e}}[(2i-1)\mathrm{d}/2] - U_{\mathrm{e}}[(2j-1)\mathrm{d}/2] \tag{5.103}$$

one can write

$$\Delta U = 2w\Delta U_{\mathrm{o},i,j} + w(w-1)\Delta U_{\mathrm{e},i,j} \tag{5.104}$$

The voltage difference ΔU in Eq. 5.104 contains a sum of w induced voltages, each of which is proportional to $w - 1$ times the conductor current. Therefore, the voltage difference ΔU has to be proportional to the square of the number of turns.

5.5.2 Multi-turn Coil with All Strands Twisted

Formed coils are manufactured in such a manner as to loop the strands in the form shown in Fig. 5.24. After looping, the coils are spread to the coil pitch. The spreading of looped coils leads to swapping of the relative position of a strand within a conductor—the uppermost strand in the left side of the coil changes its position in the end winding zone and becomes the lowermost strand in the right–side coil. In other words, conductor sides on one coil side are upside down to the conductor sides on the other. All strands get twisted by passing through front- and back-end winding zone, which leads to the induced voltages equivalent circuit as shown in Fig. 5.25. Circulating current in a loop containing *i*th and *j*th strand is determined by the voltage difference ΔU defined as

$$\begin{aligned}\Delta U = {} & U_{1,i} + U_{1,j} + U_{2,i} + U_{2,j} + \cdots + U_{w-1,i} + U_{w-1,j} + U_{w,i} + U_{w,j} \\ & - U_{w,j} - U_{w,i} - U_{w-1,j} - U_{w-1,i} - \cdots - U_{2,j} - U_{2,i} - U_{1,j} - U_{1,i} = 0\end{aligned} \quad (5.105)$$

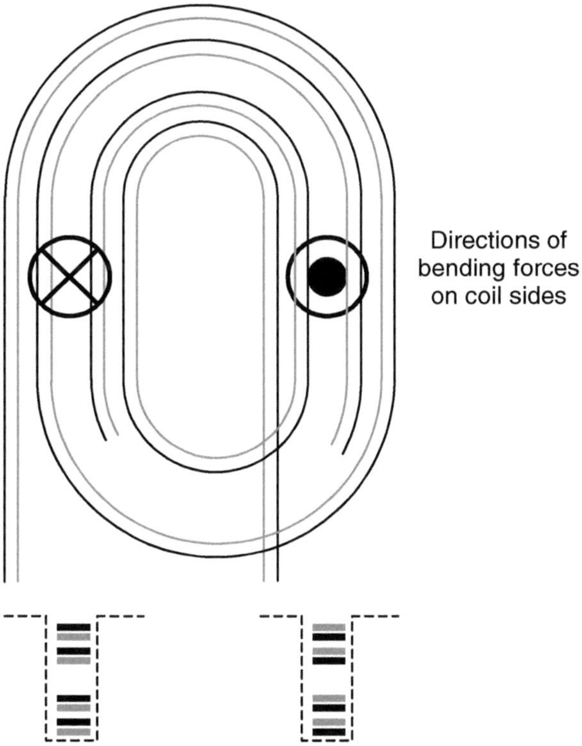

Fig. 5.24 Formed coil with two strands per conductor. Due to looping and spreading process, the position of strands in two coil sides is opposite

Twisting of strands in the end winding zone substantially reduces the total induced voltage in an arbitrary loop created by *i*th and *j*th strand. Therefore, formed coils with twisted strands in each end winding are free from eddy current losses within conductors. Strand twisting does not influence eddy current losses within a single strand.

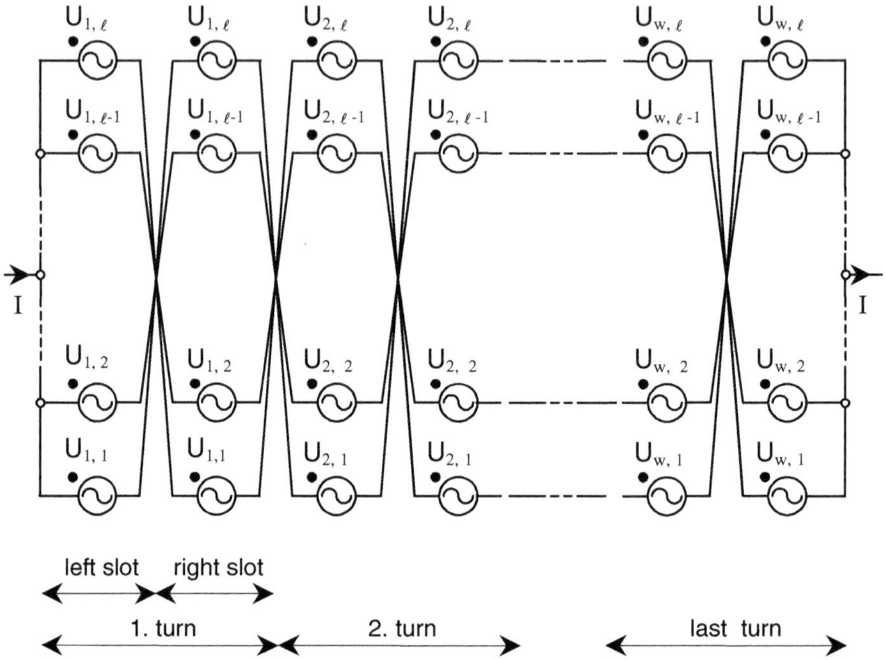

Fig. 5.25 Equivalent circuit for induced voltages in a coil with ℓ strands per conductor twisted in each end winding

5.5.3 *Multi-turn Coil with Arbitrarily Twisted Strands*

Coils of single-tooth-wound machines are usually manufactured in the manner shown in Fig. 5.22, without natural twisting of strands as in the case of formed coil in Fig. 5.24. In order to minimize losses due to proximity effect, strands in some conductors are twisted. Considering quadratic increase of induced voltage due to proximity effect (Eq. 5.104), it is enough to twist only the turn(s) closest to the air gap. The voltage difference ΔU_1 (Eq. 5.104) in a coil in which only the turn closest to the air gap is twisted can be written as

$$\Delta U = 2(w-2)\Delta U_{o,i,j} + (w-1)(w-4)\Delta U_{e,i,j} \tag{5.106}$$

Strands positioned in conductors next to each other in tangential, instead above each other in radial direction as shown in Fig. 5.26, have to be twisted in such a manner as to compensate for induced voltage due to radial slot flux. This is illustrated in Fig. 5.26b, in which strands in the last turn are twisted in the end windings. Twisting of selected, instead of all, strands disturbs the end winding geometry and requires additional space for realization. Since there exists no closed form solution for radial distribution of slot flux density, the effects of strand twisting on radial distribution of slot flux density cannot be expressed in the form of analytical function of geometry and excitation, as is the case with tangential distribution. The only way to quantify the impact of twisting on proximity effect losses in a tooth-wound coil is by means of numerical analysis.

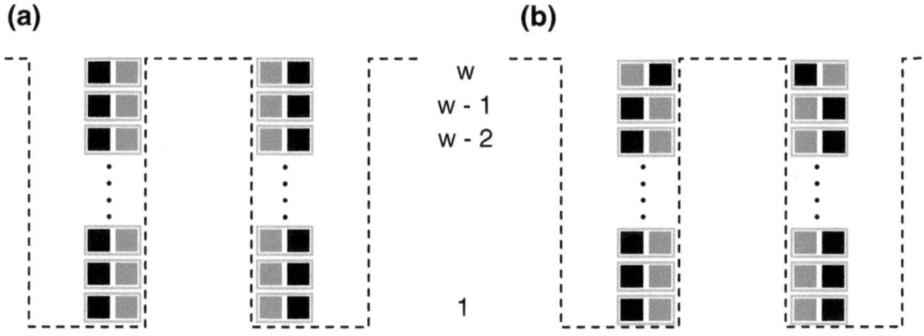

Fig. 5.26 Single-tooth winding with two strands in tangential direction: straight (**a**) and two times twisted (**b**)

Case Study 5.5: Minimization of losses due to tangential field in a slot by twisting strands in some conductors Assume a coil with 4 conductors and 6 strands per conductor wound in such a manner that the strands in the conductor closest to the gap are twisted, Fig. 5.27 [2]. The voltage difference after Eq. 5.106 for $w = 4$ is equal to

$$\Delta U_{\text{twisted}} = 4\Delta U_{\text{o},i,j}$$

which is significantly less than for untwisted, straight strands (Eq. 5.104):

$$\Delta U_{\text{straight}} = 8\Delta U_{\text{o},i,j} + 12\Delta U_{\text{e},i,j}$$

As a consequence, the circulating current is significantly reduced when only the strands in conductor closest to the air gap are twisted.

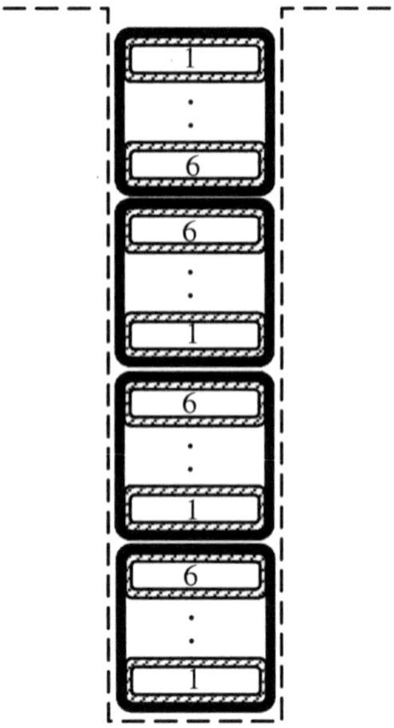

Fig. 5.27 Single-layer winding with four turns and six strands per turn

Case Study 5.6: **Minimization of losses due to radial field in a slot by twisting strands in some conductors** Large slot opening in machines with single wound teeth not only increases the Carter factor, but also strengthens the radial component of slot flux density. Whereas tangential component of slot leakage flux can be expressed analytically, its radial component can be found only numerically, e.g., by applying some of methods for field computation introduced in Chap. 3.

Coils of the machine introduced in Case Study 2.1 have 8 turns, each of which is manufactured of two strands next to each other in tangential direction, Fig. 5.28. Strand currents at rated load are shown in Fig. 5.29.

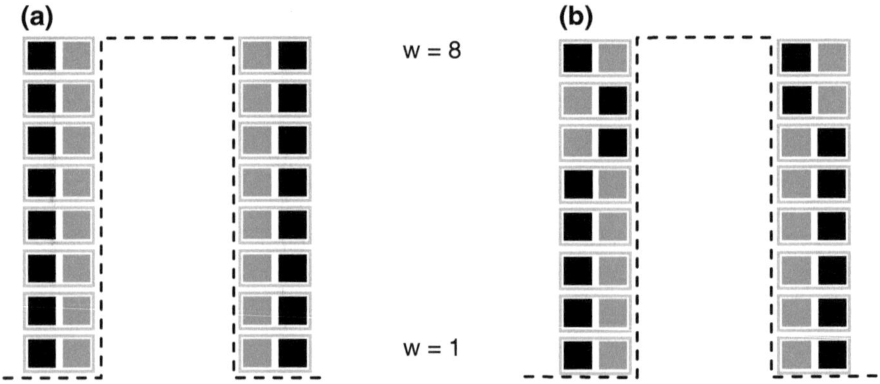

Fig. 5.28 Single tooth winding with eight turns and two strands per turn: straight (a) and once twisted in the 5. and 8. turn (b)

If the machine is wound with straight strands as in Fig. 5.28a, the circulating current between the strands at rated load exceeds the rated current, as in Fig. 5.29. When added to the load current, the circulating current exceeds the current in the inner strand for a factor larger than 2.

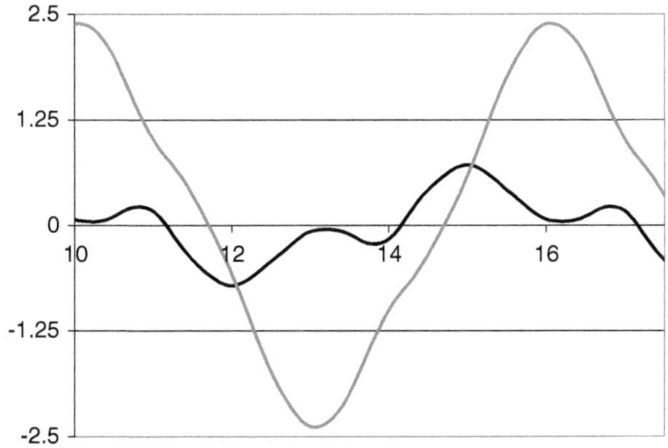

Fig. 5.29 Strand currents in coil with straight strands after Fig. 5.27a in p.u. of the rated current (*black curve* outer strand; *gray curve* inner strand). Time in ms

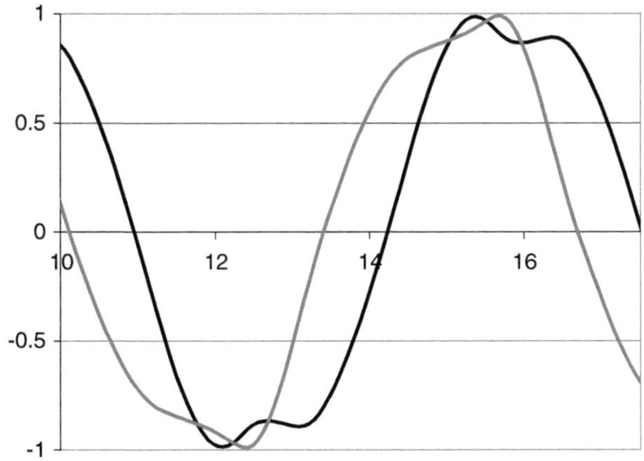

Fig. 5.30 Strand currents in coil with straight strands after Fig. 5.27a in p.u. of the rated current (*black curve* outer strand; *gray curve* inner strand). Time in ms

If the strands are twisted once in the 5 and once in the 8 turn as shown in Fig. 5.28b, the circulating current is significantly decreased. Current waveforms in both strands in this case are shown in Fig. 5.30.

5.5.4 *Transposed Strands Within a Slot: Roebel Bar, Ringland Bar, Willyoung Bar*

When a machine is designed to operate at such conditions that only one turn per coil is necessary, as is the case with large 50 and 60 Hz alternators, or medium size 400 Hz aircraft generators, the conductors are made out of a number of strands insulated to each other. The strands are usually connected in parallel (welded) at the beginning and at the end of each single conductor and transposed either along the complete length, or only in the machine active part.

By transposing the strands within the machine active part (i.e., in slots), the condition is fulfilled that each strand acts on another arbitrary strand in identical manner. The transposition of conductors, introduced by Roebel [1], is known as 360° one, since each strand returns at the end of active part to its initial position in the slot as at the beginning.

The 360° transposition of strands in Roebel bar compensates for effects of leakage flux only in slots. If the induced voltages in end windings in addition have to be compensated, a phase shift between them has to be created equal or close to 180°. The exact phase shift between induced voltages in front- and rear-end winding of 180° is achieved by using transposition proposed in [3] (Ringland bar), in which case one talks of a 540° bar. Compensation of effects of front- and rear-end winding fields with a 540° bar is complete only if these are absolutely symmetrical; otherwise, the transposition proposed in [4] (Willyoung bar) has to be applied.

The active part portion of a bar built after [4] is separated into three zones (Fig. 5.31). Each strand in the central zone along with one end zone occupies all positions within a conductor. This, however, is not identical to a 360° transposition because the slope of strand transposition in the end zones is different from the slope in the central zone. Position of a particular strand within a conductor in one end zone is identical to its position in the other end zone. If the positions of a strand within a conductor at the beginning and at the end of front-end zone are denoted by p_1 and p_2, respectively, the strand in the central zone is transposed in such a manner as to restore the identical strand position in the rear-end zone.

Denoting by n the total number of strands in the conductor, and by S the difference between p_2 and p_1, $S = p_2 - p_1$, one can introduce the *strand shift ratio* K [4] as

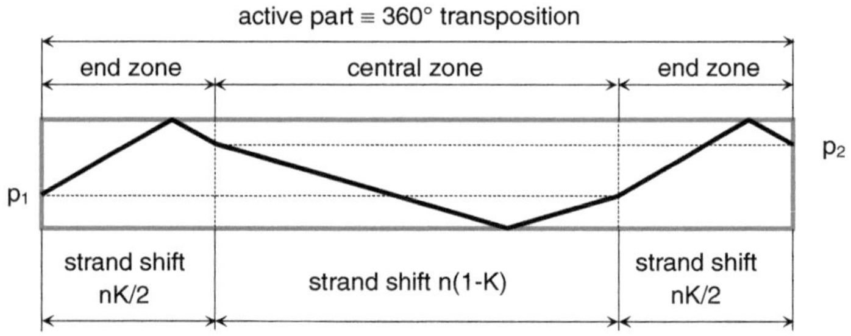

Fig. 5.31 Transposition of strands within a conductor after [4]

$$K = \frac{S}{n} \tag{5.107}$$

$K = 0$ corresponds to Roebel transposition, and $K = 0.5$ to the 540° transposition introduced in [3].

5.6 Analytical Method for the Determination of Three-Dimensional Proximity Effect in Strands in the End Winding Zone; Circulating Currents

It was shown in the previous chapter that the flux density in the end winding zone of an electrical machine can be evaluated by applying Biot–Savart law. The results obtained give the three-dimensional distribution of flux density, which can generate losses in all closed loops of strands through the mechanism of proximity effect. A closed loop is formed by two arbitrary strands connected in parallel at the beginning and at the end of each conductor, i.e., in the front- and rear-end winding zone. In order to determine the amount of circulating current in a given loop, the induced voltage in it is evaluated by differentiating the flux concatenated with the loop and added to other induced voltages in it. The sum of all induced voltages in a particular loop is equal to the voltage drop on the total strand resistance, since the applied voltage in each such loop is equal to zero (Eq. 5.50).

5.7 Skin Effect in a Ferromagnetic, Conducting Half-Space

Assume a current-carrying conductor placed parallel to a ferromagnetic, electrically conducting half-space with relative permeability $\mu_{r,Fe}$ on a distance d from it, as in Fig. 5.32. The tangential component of magnetic field strength $H(t)$ on the

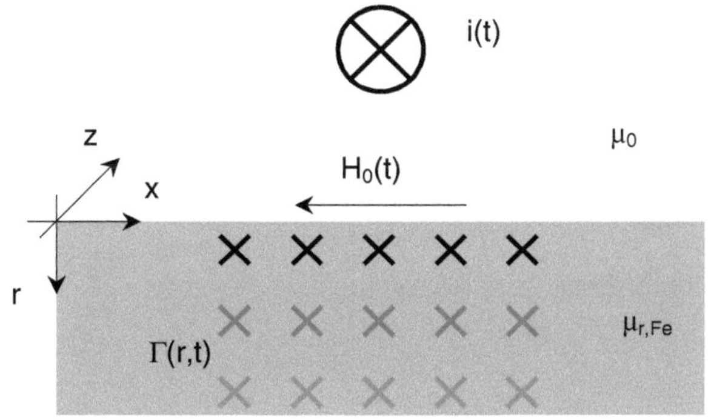

Fig. 5.32 Current-carrying conductor in front of ferromagnetic, electrically conducting half-space with relative permeability $\mu_{r,Fe}$

ferromagnetic half-space surface in case of DC excitation is given by Eq. 1.15. The field strength $H(t)$ is an impressed quantity for the electrically conducting, ferromagnetic half-space, in which eddy currents with current density Γ flow following the Faraday's law.

By applying Ampère's circuital law for an arbitrary loop in Fig. 5.32, one obtains

$$\oint \vec{H} \cdot \mathrm{d}\vec{s} = H(r,t)\mathrm{d}x - H_0(t)\mathrm{d}x = \int_0^r \Gamma(r,t)\mathrm{d}r\mathrm{d}x \tag{5.108}$$

from which one can write

$$\frac{\partial}{\partial t}H(r,t) = \Gamma(r,t) \tag{5.109}$$

Faraday's law for induced voltage u_i in an arbitrary loop with dimensions $\mathrm{d}r \times \mathrm{d}x$ can be written as

$$u_i = -\frac{\partial}{\partial t}\mu H(r,t)\mathrm{d}r\mathrm{d}x = [\Gamma(r+\mathrm{d}r,t) - \Gamma(r,t)]\rho\mathrm{d}x \tag{5.110}$$

or by replacing ρ through 1/κ:

$$\frac{\partial}{\partial x}\Gamma(r,t) = -\mu_0\mu_r\kappa\frac{\partial}{\partial t}H(r,t) \tag{5.111}$$

where μ_r = const. By combining Eqs. 5.109 and 5.11, one obtains the partial differential equation

$$\mu_0\mu_r\kappa\frac{\partial}{\partial t}\Gamma(r,t) = -\frac{\partial^2\Gamma(r,t)}{\partial r^2} \tag{5.112}$$

For sinusoidal time dependence of current density

$$\Gamma(r,t) = \sqrt{2}\Gamma(r)e^{j\omega t} \tag{5.113}$$

one obtains further

$$\frac{\partial}{\partial t}\Gamma(r,t) = \sqrt{2}\Gamma(r)j\omega e^{j\omega t} \tag{5.114}$$

which leads to the ordinary differential equation for current density distribution

$$\frac{d^2\Gamma(r)}{dr^2} = -j\omega\mu\kappa\Gamma(r) \tag{5.115}$$

Defining parameter α [m^{-1}] as

$$\alpha^2 = j\omega\mu\kappa \tag{5.116}$$

one can write the ordinary differential equation of the second order (Eq. 5.115) as

$$\frac{d^2\Gamma(r)}{dr^2} = -\alpha^2\Gamma(r) \tag{5.117}$$

The solution of Eq. 5.117 is a linear combination of exponential functions of complex argument

$$\Gamma(r) = D_1 e^{\alpha r} + D_2 e^{-\alpha r} \tag{5.118}$$

where the values of integration constants can be found from boundary conditions of the problem. Since boundary conditions are known both for the magnetic field strength and current density, it sounds reasonable to express the solution in Eq. 5.118 in terms of magnetic field strength by using Ampère's circuital law:

$$\frac{d}{dr}\Gamma(r) = -\alpha^2 H(r) \tag{5.119}$$

from which

$$-\alpha H(r) = D_1 e^{\alpha r} - D_2 e^{-\alpha r} \tag{5.120}$$

The amounts D_1 and D_2 of constants of integration are now found by inserting the value of magnetic field strength at $r = 0$, $H(0)$:

$$H(0) = H_0 \Rightarrow D_1 - D_2 = -\alpha H_0 \tag{5.121}$$

and current density at $r \to \infty$, $\Gamma(\infty)$:

$$\Gamma(\infty) = 0 \Rightarrow D_1 = 0 \tag{5.122}$$

Thus:

$$D_2 = \alpha H_0 \tag{5.123}$$

and, finally,

$$\Gamma(r) = \alpha H_0 e^{-\alpha r} \tag{5.124}$$

Substituting for α (see Eqs. 5.116 and 5.8)

$$\alpha = (1+j)\sqrt{\frac{\omega\mu\kappa}{2}} = \frac{1+j}{\delta} \tag{5.125}$$

and assuming H_0 to be real, one can write for eddy current density (Eq. 5.124)

$$\Gamma(r) = \frac{1+j}{\delta} H_0 e^{-\frac{1+j}{\delta}r} \tag{5.126}$$

the real, $\text{Re}\{\Gamma(r)\}$, and imaginary, $\text{Im}\{\Gamma(r)\}$, components of which can be expressed as

$$\text{Re}\{\Gamma(r)\} = \frac{H_0}{\delta} e^{-\frac{r}{\delta}}\left(\cos\frac{r}{\delta} + \sin\frac{r}{\delta}\right) \tag{5.127}$$

$$\text{Im}\{\Gamma(r)\} = \frac{H_0}{\delta} e^{-\frac{r}{\delta}}\left(\cos\frac{r}{\delta} - \sin\frac{r}{\delta}\right) \tag{5.128}$$

Current density is maximal on the surface ($r = 0$), where its phase shift to the impressed field strength H_0 is equal to 45°, see also Fig. 5.2.

The absolute value of current density is, accordingly:

$$|\Gamma(r)| = \sqrt{2}\frac{H_0}{\delta} e^{-\frac{r}{\delta}} \tag{5.129}$$

The amplitude of eddy current density decreases exponentially from its value on the surface $\Gamma(0) = \sqrt{2}\ \alpha\ H_0$ down to zero in the inner portion of half-space with a decay constant α. This helps visualize the physical meaning of the skin depth δ: The total current, which in reality flows in the whole ferromagnetic, conducting half-space, would flow in the layer with thickness δ in the case of fictitious constant current density equal to $\Gamma(0)$:

$$\int_0^\infty |\Gamma(r)| \mathrm{d}r = \Gamma(0)\delta \tag{5.130}$$

Eddy current losses generated in a differential of volume $\mathrm{d}r \cdot \mathrm{d}x \cdot \mathrm{d}z$ are equal to

$$P_\mathrm{e} = \frac{\mathrm{d}x\mathrm{d}z}{\kappa} \int_0^\infty \Gamma(r)\Gamma^*(r)\mathrm{d}r = \frac{\mathrm{d}x\mathrm{d}z}{\kappa}\frac{H_0^2}{\delta} \tag{5.131}$$

and the surface loss density

$$P_\mathrm{e}' = \frac{P_\mathrm{e}}{\mathrm{d}x\mathrm{d}z} = \frac{H_0^2}{\kappa\delta} \tag{5.132}$$

5.8 The Influence of Saturation on Skin Effect in Iron

Saturation changes relationships between field quantities in iron substantially and makes analytical computations impossible. Nevertheless, based on the character of B–H curve, reliable quantitative estimations of current density, losses, etc., can be made for saturated iron, which correlate well with experimental data [4–7].

Steep increase of flux density in the interval of low-field strengths allows for the simple approximation of B–H curve, as shown in Fig. 5.33. One can estimate that in the layer with thickness δ the flux density remains constant, as a consequence of high values of magnetic field strength.

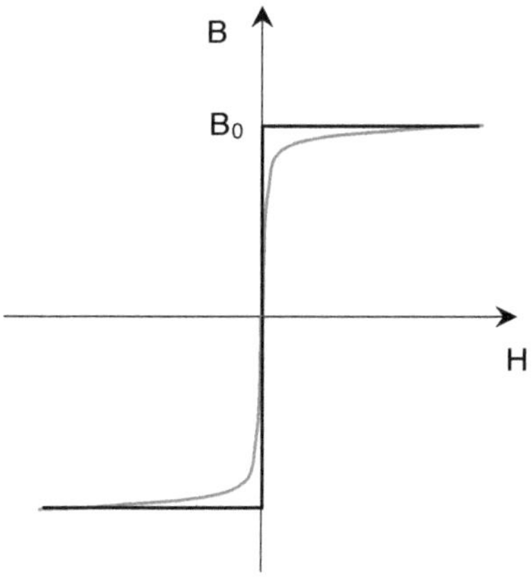

Fig. 5.33 The actual B–H curve (*gray*) and its approximation (*black*)

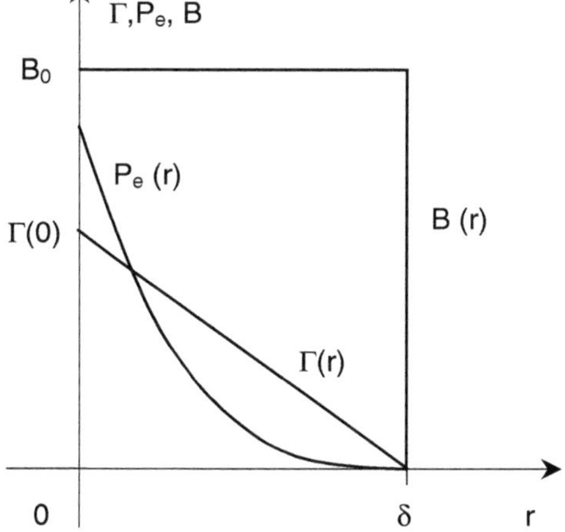

Fig. 5.34 Approximations of current density and loss curves

Skin depth δ can now be defined as

$$\delta = \sqrt{\frac{2}{\omega\mu\kappa}} = \sqrt{\frac{2}{\omega\kappa}\frac{H_0}{B_0}} \tag{5.133}$$

The actual exponential decay of magnetic field strength within skin depth δ is replaced by linear decrease, which means that losses decrease quadratically, Fig. 5.34.

With approximations introduced in Fig. 5.34, one can write for the total losses

$$P_e = \frac{1}{3}\Gamma^2(0)\frac{hC}{\kappa\delta} \tag{5.134}$$

with C denoting the axial and h tangential distance in which the skin effect takes place. Replacing the value of surface current density $\Gamma(0)$ by field strength (Eq. 5.129), one can express the surface loss density as

$$P'_e = \frac{2}{3}\sqrt{\frac{\omega}{2\kappa}}H_0^3 B_0 \tag{5.135}$$

5.9 Skin Effect in a Thin Plate

Assume now a plate with thickness Δ and height b through which an alternating magnetic field with flux density B spreads in the direction shown in Fig. 5.35.

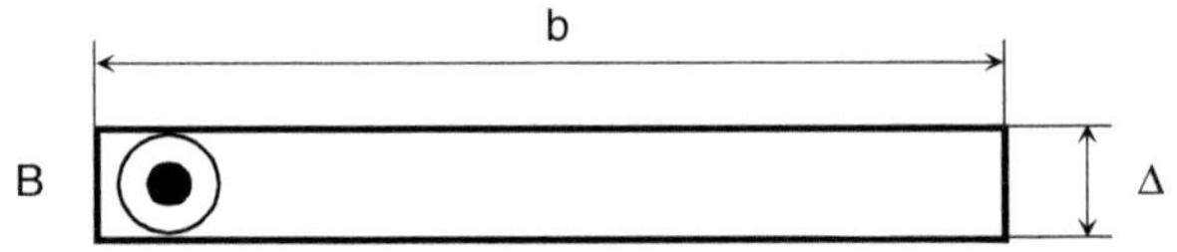

Fig. 5.35 Illustrating skin effect in a thin ferromagnetic plate

The induced voltage on the plate circumference is equal to

$$U_i = \frac{2\pi}{\sqrt{2}}f \cdot B \cdot b \cdot \Delta \tag{5.136}$$

and the current density on the plate margin

$$\Gamma = \kappa \cdot \frac{U_i}{2(b+\Delta)} \approx \kappa \cdot \frac{U_i}{2b} = \frac{\pi}{\sqrt{2}}f \cdot \kappa \cdot B \cdot \Delta \tag{5.137}$$

Eddy current losses are now (see Eq. 5.134):

$$P_e = \frac{1}{3}\Gamma^2 \frac{b \cdot \Delta \cdot l}{\kappa} = \frac{\pi^2}{6}\kappa \cdot f^2 B^2 b \cdot l \cdot \Delta^3 \tag{5.138}$$

with ℓ denoting the length of the plate in the direction of flux density B. Since the MMF created by eddy currents is significantly smaller than the impressed MMF, the flux density distribution over the plate area remains constant.

5.10 Skin Effect in a Solid Ferromagnetic Cylinder

Assume a ferromagnetic cylinder with radius R which carries an alternating current I in axial direction, Fig. 5.36.

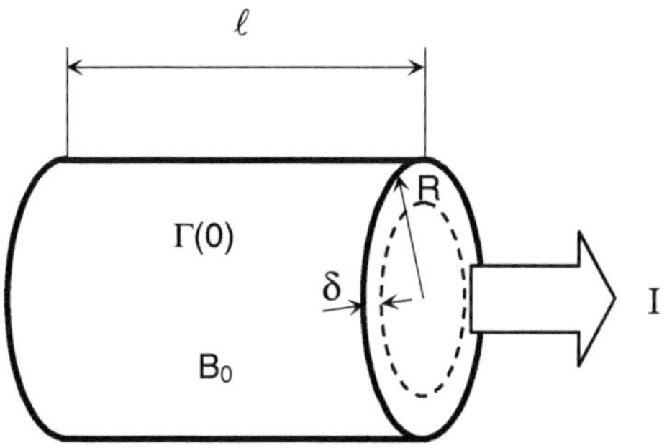

Fig. 5.36 Illustrating skin effect in a ferromagnetic cylinder

Considering the current density distribution as shown in Fig. 5.34, one can express the maximum current density $\Gamma(0)$ on the cylinder surface as

$$\Gamma(0) = \frac{I}{R\pi\delta} \tag{5.139}$$

and the flux through cylinder as

$$\Phi = B_0 \delta l \tag{5.140}$$

The induced voltage on the cylinder circumference is now

$$U_i = \frac{2\pi}{\sqrt{2}} f B_0 \delta l = \Gamma(0)\frac{l}{\kappa} = \frac{I}{R\pi\delta}\frac{l}{\kappa} \tag{5.141}$$

from which

$$\delta = \sqrt{\frac{I}{B_0}\frac{\sqrt{2}}{\omega\kappa R\pi}} \tag{5.142}$$

and the eddy current losses

$$P_e = \frac{2}{3}\Gamma^2(0)\frac{R\pi l\delta}{\kappa} = \frac{2}{3}l\sqrt{B_0 I^3 \frac{\omega}{\kappa R\pi\sqrt{2}}} \tag{5.143}$$

If a ferromagnetic cylinder with diameter D rotates at angular velocity Ω in heteropolar field created by DC, Fig. 5.37, eddy current losses in it can be evaluated in the following manner:

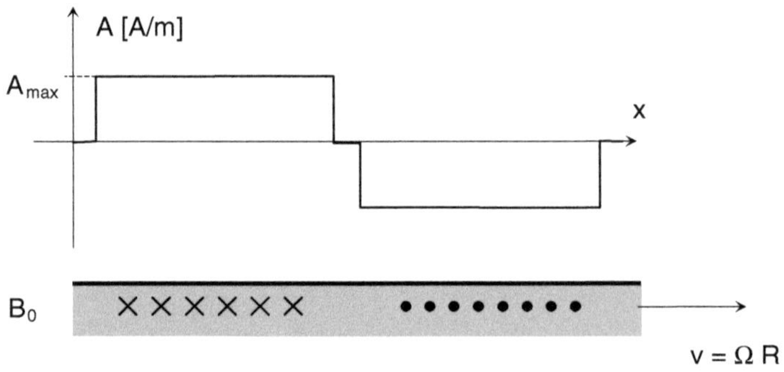

Fig. 5.37 Skin effect in a ferromagnetic cylinder rotating at angular velocity Ω relative to a rectangularly distributed current sheet with amplitude A_{max}

The equivalent skin depth is (see Eq. 5.133)

$$\delta = \sqrt{\frac{A_{max}}{B_0}\frac{1}{p\Omega\kappa'}} \tag{5.144}$$

where

$$\kappa' = \frac{\kappa}{1 + \frac{D\pi}{2pl}} \tag{5.145}$$

In equation above the electrical conductivity of material κ is replaced by a term taking into account the finite length ℓ of the analyzed cylinder with diameter D and number of poles 2p.

The rms value of current density on the cylinder surface is now

$$\Gamma(0) = \sqrt{p\Omega\kappa' A_{\max} B_0} \tag{5.146}$$

and the total eddy current losses

$$P_e = \frac{2}{3} D\pi l \sqrt{\frac{p\Omega}{2\kappa'}} A_{\max}^3 B_0 \tag{5.147}$$

If the current sheet is sinusoidally distributed and rotates at angular velocity ω relative to the ferromagnetic cylinder, the losses are reduced to

$$P_e = \frac{1}{3} D\pi l \sqrt{\frac{p\Omega}{2\kappa'}} A_{\max}^3 B_0 \tag{5.148}$$

5.11 Losses in Surface-Mounted Permanent Magnets

It has been shown in Sect. 3.5 how the magnetic scalar potential distribution is evaluated in a surface-mounted permanent magnet. For known MMF drops between points in a permanent magnet one can evaluate proximity effect losses during one period by utilizing Eq. 5.80.

References

1. Ostovic V (1994) Computer-Aided Analysis of Electric Machines: a Mathematica Approach. Prentice-Hall, London
2. Punga F (1948) Elektromaschinenbau. Demmig Verlag, Darmstadt
3. Ringland WL (1958) US Patent 2,821,641
4. Willyoung DM (1964) US Patent 3,118,015
5. Rosenberg E (1923) Wirbelströme in massivem Eisen. Elektrotechnik und Maschinenbau 22:317–325
6. MacLean W (1954) Theory of strong electromagnetic waves in massive iron. J Appl Phys 10:1267–1270
7. Lammeraner J, Štafl M (1964) Vířivé proudy. Státni nakl. Technické lit, Praha

Chapter 6
Force and Torque

Contents

Magnetic energy density is the basic quantity used to determine force and torque in electric machines, because it also can be interpreted as shear force. Magnetic energy is used to evaluate force on the border between two media with different relative permeabilities. Torque components are calculated for various machine types and modes of operation, also taking into account the influence of slotting, saliency, and rotor eccentricity. Radial air gap force is analyzed, and means to suppress it are

V. Ostović, *The Art and Science of Rotating Field Machines Design*,
DOI 10.1007/978-3-319-39081-9_6

discussed. Forces acting on conductors in slots and in the end winding are evaluated. A method for measurement of torque is proposed and compared with back-to-back procedure.

6.1 Magnetic Field as a Medium in Which Electromechanical Energy Conversion Takes Place, the Role of Accumulated Magnetic Energy

One of the things which make electrical engineering and especially electric machines so exciting is extremely simple equations which describe them and which, however, not always work. There seems to be a reasonable explanation for such situation—there are only a few general laws governing operation of electric machines, which result in a huge variety of system equations under different boundary conditions. Very often, the conditions under which a particular equation was derived fall into oblivion and one gets an impression that it is generally valid, which is not always necessarily true.

A perfect example which illustrates possible confusion caused by inconsequent use of formulas in electrical engineering is the well-known equation for the force acting on a current-carrying conductor $F = B \cdot \ell \cdot I$, B denoting the flux density in the conductor, I the conductor current, and ℓ the length of the conductor, as in Fig. 6.1a.

If the conductor is placed into an iron tube, which shields it from external field B, the flux density in it goes down to zero, as in Fig. 6.1b. However, despite zero flux density in the conductor, the force on it remains unchanged, i.e., $F = B \cdot \ell \cdot I$.

Therefore, it seems to be more appropriate to define the force as a result of general state in electromagnetic field, in particular as a function of magnetic energy stored in it [1, 2]. This way one comes to a conclusion that energy accumulated in the space around a current-carrying conductor only slightly changes (decreases) when the conductor is shielded with iron tube as shown in Fig. 6.1 [2]. Consequently, the force acting on a current-carrying conductor in magnetic field practically does not differ in the two cases illustrated in Fig. 6.1.

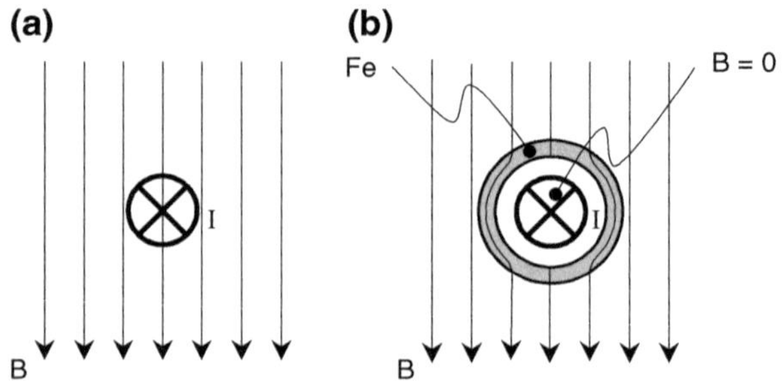

Fig. 6.1 Illustrating the force acting on a current-carrying conductor in magnetic field. The force acting on the conductor is the same in both configurations (**a**) and (**b**), although the external flux density in the conductor in case (**a**) is equal to B and in case (**b**) to zero

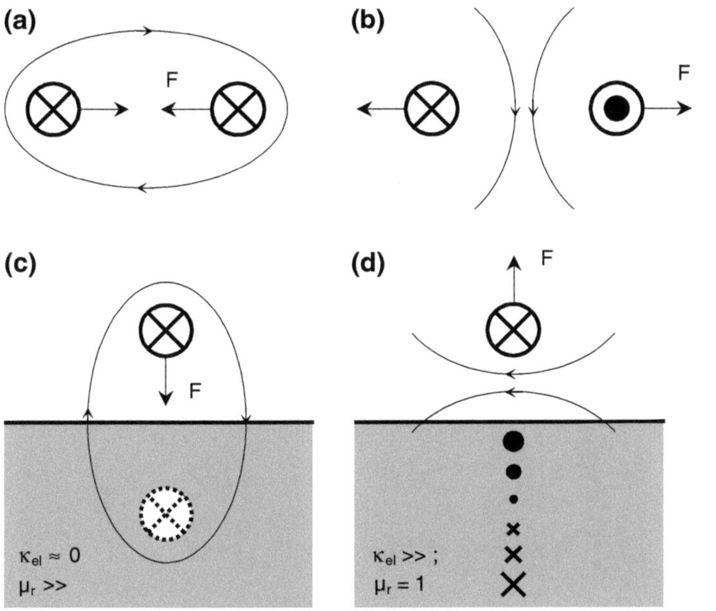

Fig. 6.2 Force in electromagnetic field: **a** between two conductors carrying currents in the same direction; **b** between two conductors carrying currents in opposite directions; **c** on a current-carrying conductor in front of a high-permeance, non-conducting semispace; **d** on a current-carrying conductor in front of a low permeance, conducting semispace

In Fig. 6.2 typical cases of forces acting on conductor(s) in magnetic field are illustrated.

Force acting between two conductors carrying currents in the same direction is always attractive (Fig. 6.2a). If the currents flow in opposite directions, the force is repulsive (Fig. 6.2b). Current imaging in Fig. 6.2c in front of magnetic half-space results in attractive force acting on the conductor, whereas the conductor carrying alternating current in front of an electrically conducting half-space (Fig. 6.2d) gets repelled from it as a result of action of eddy currents.

One should note that force and torque in magnetic field depend only on instantaneous state in magnetic field.

6.2 Shear Force on Contact Surfaces Between Media with Different Permeabilities

In linear media with μ = const, the magnetic energy volume density $W_{mg,V}$ can be expressed as

$$W_{mg,V} = \frac{H \cdot B}{2} = \frac{\mu H^2}{2} = \frac{B^2}{2\mu} = \sigma \tag{6.1}$$

and has a dimension of Ws/m^3 = N/m^2, i.e., the same as pressure σ. Consequently, one can express the magnetic energy density in linear media as the amount of pressure on a given surface, defined by means of vector of magnetic field strength and unit surface vector, as in Fig. 6.3. Note that the vector of the pressure $\vec{\sigma}$ is perpendicular to the surface in two cases: when the field strength $\vec{H}$ is also perpendicular to the surface and when $\vec{H}$ is parallel to the surface.

Vector $\vec{\sigma}$ can be resolved into a component $\vec{\sigma}_c$ collinear with $\vec{H}$ and a component $\vec{\sigma}_p$ perpendicular to $\vec{H}$

$$\vec{\sigma} = \vec{\sigma}_c + \vec{\sigma}_p \tag{6.2}$$

where

$$\sigma_c = \sigma \cos \alpha; \quad \sigma_p = \sigma \sin \alpha \tag{6.3}$$

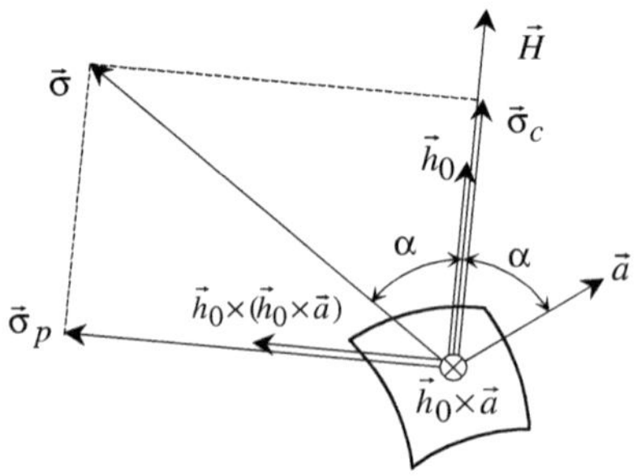

Fig. 6.3 Pressure $\vec{\sigma}$ acting on unit surface $\vec{a}$ in magnetic field with intensity $\vec{H}$. In linear media, the amount σ of pressure is numerically equal to volume density of stored magnetic energy if the vectors $\vec{\sigma}$, $\vec{a}$, and $\vec{H}$ lie in the same plane and if the vector $\vec{H}$ halves the angle between vectors $\vec{\sigma}$ and $\vec{a}$

Introducing the unit vector $\vec{h}_0$ of magnetic field strength, defined as

$$\vec{h}_0 = \frac{\vec{H}}{H} \tag{6.4}$$

one can write by using unity surface vector $\vec{a}$

$$\sin\alpha = \left|\vec{h}_0 \times \vec{a}\right|; \quad \cos\alpha = \vec{h}_0 \bullet \vec{a} \tag{6.5}$$

as well as

$$\vec{\sigma}_c = \sigma \cdot \left(\vec{h}_0 \bullet \vec{a}\right) \cdot \vec{h}_0 \tag{6.6}$$

and

$$\vec{\sigma}_p = \sigma \cdot \vec{h}_0 \times \left(\vec{h}_0 \times \vec{a}\right) \tag{6.7}$$

where symbol "•" denotes scalar product of two vectors and "·" a regular product of two scalars, or of a scalar and a vector.

Now, one can rewrite Eq. 6.2 as

$$\vec{\sigma} = \sigma \cdot \left[\left(\vec{h}_0 \bullet \vec{a}\right) \cdot \vec{h}_0 + \vec{h}_0 \times \left(\vec{h}_0 \times \vec{a}\right)\right] \tag{6.8}$$

Substituting for value of $\sigma = \mu H^2/2$ (Eq. 6.1), one can further write

$$\vec{\sigma} = \frac{\mu H^2}{2}\left[\left(\frac{\vec{H}}{H} \bullet \vec{a}\right) \cdot \frac{\vec{H}}{H} + \frac{\vec{H}}{H} \times \left(\frac{\vec{H}}{H} \times \vec{a}\right)\right] = \frac{\mu}{2}\left[(\vec{H} \bullet \vec{a}) \cdot \vec{H} + \vec{H} \times (\vec{H} \times \vec{a})\right] \tag{6.9}$$

On a boundary surface between media with equal permeability on both sides, the pressure σ remains unchanged. If the permeability on one side of the boundary is different than on the other, a pressure difference is generated. If there is no current sheet on the boundary, the corresponding force acts in the direction from the medium with bigger to the medium with smaller permeability, as in Fig. 6.4.

Boundary conditions for magnetic field vectors are [2]

$$H_{1,t} = H_{2,t} \tag{6.10}$$

and

$$\mu_1 H_{1,n} = \mu_2 H_{2,n} \tag{6.11}$$

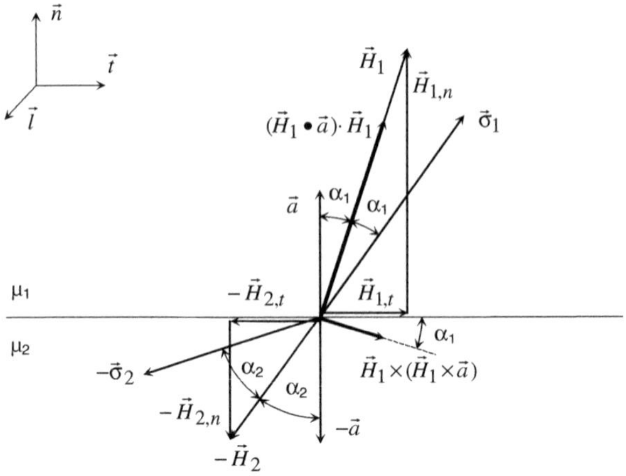

Fig. 6.4 Pressure $\vec{\sigma}$ acting on unit surface $\vec{a}$ in magnetic field with intensity $\vec{H}$ in the absence of current sheet on the boundary. $\vec{t}$, $\vec{n}$, and $\vec{l}$ are unit vectors in tangential, normal, and lateral directions, respectively. Tangential components of magnetic field strength are equal on both sides of the boundary

resulting in

$$\frac{\tan \alpha_1}{\tan \alpha_2} = \frac{\mu_1}{\mu_2} \tag{6.12}$$

One can further write for both sides of the boundary

$$\vec{H} \bullet \vec{a} = H_n \tag{6.13}$$

$$\vec{H} \times \left(\vec{H} \times \vec{a}\right) = \left(\vec{t}H_t + \vec{n}H_n\right) \times \left[\left(\vec{t}H_t + \vec{n}H_n\right) \times \vec{n}\right] = H_t\left(\vec{t}H_n - \vec{n}H_t\right) \tag{6.14}$$

The pressure in the medium with μ_1 is now

$$\begin{aligned}\vec{\sigma}_1 &= \frac{\mu_1}{2}\left[H_{1,n} \cdot \left(\vec{t}H_{1,t} + \vec{n}H_{1,n}\right) + H_{1,t}\left(\vec{t}H_{1,n} - \vec{n}H_{1,t}\right)\right] \\ &= \frac{\mu_1}{2}\left[2\vec{t}H_{1,t}H_{1,n} + \vec{n}\left(H_{1,n}^2 - H_{1,t}^2\right)\right]\end{aligned} \tag{6.15}$$

and in the medium with μ_2 on the other side of the boundary, analogously,

$$\vec{\sigma}_2 = \frac{\mu_2}{2}\left[2\vec{t}H_{2,t}H_{2,n} + \vec{n}\left(H_{2,n}^2 - H_{2,t}^2\right)\right] \tag{6.16}$$

The total pressure is equal to

$$\vec{\sigma} = \vec{\sigma}_1 - \vec{\sigma}_2$$
$$= \frac{1}{2}\left\{\mu_1\left[2\vec{t}H_{1,t}H_{1,n} + \vec{n}\left(H_{1,n}^2 - H_{1,t}^2\right)\right] - \mu_2\left[2\vec{t}H_{2,t}H_{2,n} + \vec{n}\left(H_{2,n}^2 - H_{2,t}^2\right)\right]\right\} \tag{6.17}$$

and has a normal component only. By applying boundary conditions (Eqs. 6.10–6.11), one obtains

$$\vec{\sigma} = \frac{\mu_1}{2}\left[H_{1,n}^2\left(1 - \frac{\mu_1}{\mu_2}\right) - H_{1,t}^2\left(1 - \frac{\mu_2}{\mu_1}\right)\right]\vec{n} \tag{6.18}$$

or, by means of normal and tangential components of flux density

$$\vec{\sigma} = \frac{1}{2\mu_1}\left[B_{1,n}^2\left(1 - \frac{\mu_1}{\mu_2}\right) - B_{1,t}^2\left(1 - \frac{\mu_2}{\mu_1}\right)\right]\vec{n} \tag{6.19}$$

The force F_n on the boundary between two media with different permeabilities without current excitation has only one component, which is perpendicular to the boundary surface. This force acts in the direction from the medium with a higher permeability to the medium with a lower permeability.

In case of iron-to-air boundary, with $\mu_1 = \mu_r\mu_0$ and $\mu_2 = \mu_0$, one can write for the amplitude of normal pressure

$$\sigma_n = \frac{1}{2\mu_r\mu_0}\left[B_{1,n}^2(1 - \mu_r) - B_{1,t}^2\left(1 - \frac{1}{\mu_r}\right)\right] \tag{6.20}$$

Consider now two extreme cases:

(a) $B_{1,n} = B$; $B_{1,t} = 0$; $\mu_r \gg 1$ (flux lines perpendicular to the boundary surface on air side, as in Fig. 6.5a):

$$\sigma_n \approx \frac{B^2}{2\mu_0} \tag{6.21}$$

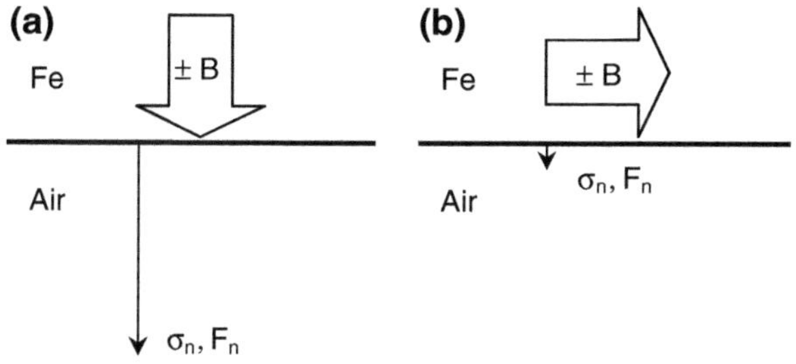

Fig. 6.5 Pressure acting on air–iron boundary for a given flux density *B* in iron in absence of current sheet. The force acts always from iron to air, and the amount of pressure in case of tangential flux density (**b**) is μ_r times smaller than it is for normal flux density of the same amount *B*

(b) $B_{1,n} = 0$; $B_{1,t} = B$; $\mu_r \gg 1$ (flux lines parallel to the boundary surface on air side, as in Fig. 6.5b):

$$\sigma_n \approx \frac{B^2}{2\mu_r\mu_0} \tag{6.22}$$

The ratio of normal pressure for the same value of normal (Eq. 6.21) to tangential (Eq. 6.22) flux density is equal to μ_r.

Normal force on a boundary surface acts from the medium with a higher to the medium with a lower permeability. Equations 6.21 and 6.22 reflect the physical reality that a flux tube, here represented by the half-space with a relative permeability μ_r, tends to shorten and widen in magnetic field. Equation 6.21 can be used to find attractive force of an electromagnet, whereas the pressure after Eq. 6.22 is the reason why the lamination in electric machines has to be fastened in axial direction by means of screws/bolts and press fingers. Another application of the principle that pressure on the iron–air border always acts from iron to air is magnetic separation of iron sheets, as in Fig. 6.6. When vertically stacked single iron sheets have to be separated from each other, permanent magnets are placed around the stack, which create magnetic flux density parallel to the surfaces of each sheet, as in Fig. 6.6. Vertical distances between single sheets in Fig. 6.6 b are largest on the top of the stack and smallest on its bottom, as a consequence of interaction between the gravitational and magnetic forces acting on the sheets. The uppermost sheet levitates and exerts the repelling magnetic force and the (attracting) gravitational force on the sheet below it. The sheet below it carries the uppermost sheet and exerts the repelling force against the third sheet, etc. This way, several upper sheets levitate above each other.

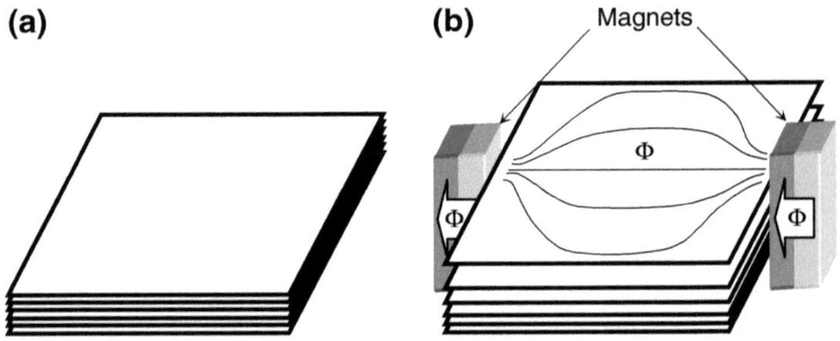

Fig. 6.6 Magnetic sheet separation principle based on repulsion due to tangential field in the sheets: **a** sheets in field-free space; **b** sheets between two permanent magnets generating flux Φ

If there exists current sheet A [A/m] on the boundary surface, Fig. 6.7, the boundary condition 6.10 changes to

$$H_{2,t} = H_{1,t} \pm A \tag{6.23}$$

and the expression for pressure (Eq. 6.17) into:

$$\vec{\sigma} = \pm\vec{t}B_{1,n}A + \vec{n}\left[\frac{B_{1,n}^2}{2\mu_1}\left(1 - \frac{\mu_1}{\mu_2}\right) + \frac{B_{1,t}^2}{2\mu_1}\left(\frac{\mu_2}{\mu_1} - 1\right) + \frac{\mu_2}{2}A^2 \pm \frac{\mu_2}{\mu_1}AB_{1,t}\right] \quad (6.24)$$

For $A \neq 0$, the pressure difference $\vec{\sigma}$ has both normal and tangential components

$$\vec{\sigma} = \vec{t}\sigma_t + \vec{n}\sigma_n \quad (6.25)$$

where

$$\sigma_t = \pm B_{1,n}A \quad (6.26)$$

and

$$\sigma_n = \frac{B_{1,n}^2}{2\mu_1}\left(1 - \frac{\mu_1}{\mu_2}\right) + \frac{B_{1,t}^2}{2\mu_1}\left(\frac{\mu_2}{\mu_1} - 1\right) + \frac{\mu_2}{2}A^2 \pm \frac{\mu_2}{\mu_1}AB_{1,t} \quad (6.27)$$

Tangential component of pressure/force is different from zero only if there is a component of flux density perpendicular to the boundary, as in Fig. 6.8.

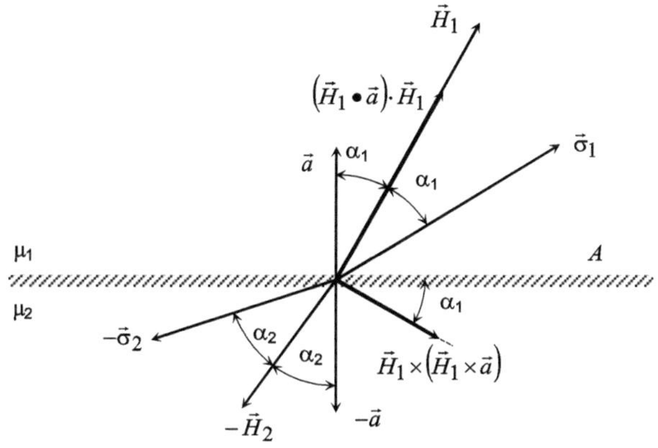

Fig. 6.7 Pressure $\vec{\sigma}$ acting on unit surface $\vec{a}$ in magnetic field with intensity $\vec{H}$ and current sheet A on the boundary. Tangential components of magnetic field strength are not equal on the two sides of the boundary

For iron-to-air boundary, with $\mu_1 = \mu_r\mu_0$ and $\mu_2 = \mu_0$, the amplitude of normal pressure is equal to

$$\sigma_n = \frac{B_{1,n}^2}{2\mu_r\mu_0}(1 - \mu_r) - \frac{B_{1,t}^2}{2\mu_r\mu_0}\left(1 - \frac{1}{\mu_r}\right) + \frac{\mu_0}{2}A^2 \pm \frac{AB_{1,t}}{\mu_r} \quad (6.28)$$

Consider again the two extreme cases:

(a) $B_{1,n} = B$; $B_{1,t} = 0$; $\mu_r \gg 1$ (flux lines perpendicular to the boundary surface on air side, as in Fig. 6.8a):

$$\sigma_t = \pm B \cdot A \tag{6.29}$$

$$\sigma_n \approx \frac{1}{2}\left(-\frac{B^2}{\mu_0} \pm \mu_0 A^2\right) \tag{6.30}$$

(b) $B_{1,n} = 0$; $B_{1,t} = B$; $\mu_r \gg 1$ (flux lines parallel to the boundary surface on air side, as in Fig. 6.8b):

$$\sigma_t = 0 \tag{6.31}$$

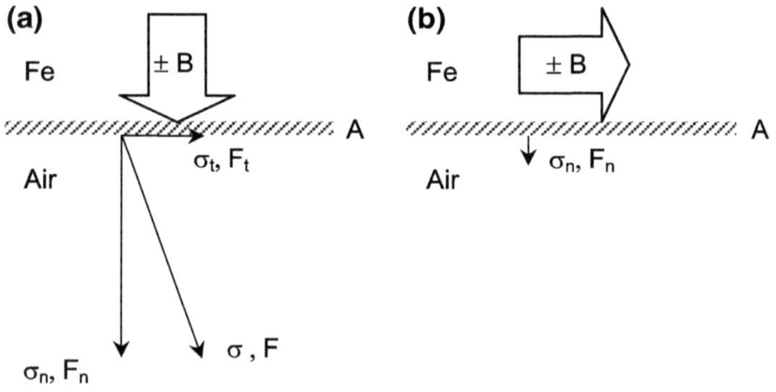

Fig. 6.8 Pressure acting on air–iron boundary for a given flux density *B* in iron carrying current sheet *A*. Current sheet generates a tangential component of pressure/force

$$\sigma_n \approx -\frac{B^2}{2\mu_r\mu_0} + \frac{\mu_0}{2}A^2 \pm \frac{AB}{\mu_r} \tag{6.32}$$

For current sheet of 100 kA/m and flux density in iron of 1.5 T, one obtains in case (a) above $\sigma_t = \pm 150$ kN/m^2 and $\sigma_n \approx 895 \pm 6$ kN/m^2, whereas in case (b) $\sigma_n \approx 7.2 \pm 0.23$ kN/m^2.

6.3 Force Due to External Field Acting on Current-Carrying Conductors in Slots of Electric Machines

In conventional electric machines conductors are placed in slots, as in Fig. 6.9, where the flux density B_s is significantly smaller than that in the air gap. Following equation $F = B_s \cdot \ell \cdot I$, the force acting on conductor in a slot should also be significantly smaller than that in the air gap.

The pressure acting along the slot circumference in Fig. 6.9 has both normal and tangential components after Eqs. 6.29–6.32. Since the direction of normal and tangential components of pressure changes when passing the slot along tooth flanks and its bottom, the resulting normal pressure acting on tooth flanks after Eq. 6.17 is

$$\sigma_n = 2\mu_{\mathrm{Fe}} H_t A \tag{6.33}$$

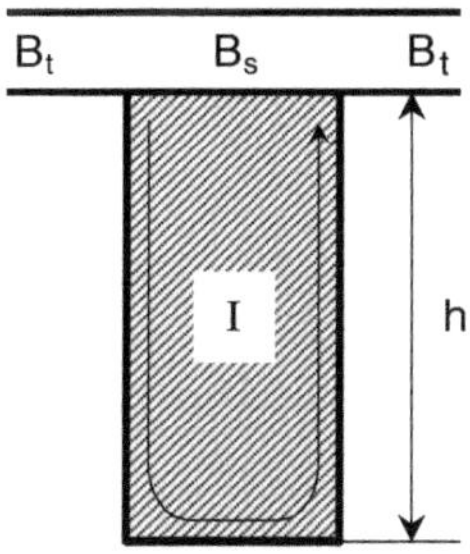

Fig. 6.9 Illustrating the force due to external field on a current-carrying conductor in a slot of an electric machine

Considering constant current density in the conductor, the current is proportional to the conductor area. Since one half of the conductor borders with the tooth on the left side of the conductor, and the other half on the right side, the current sheet A is equal to $I/(2h)$, and one can write for the normal force F_n acting on tooth flanks

$$F_n = B_t l I \tag{6.34}$$

with ℓ denoting the machine axial length. Total force acting on tooth flanks and conductor is consequently $F = B \cdot \ell \cdot I$.

6.4 Torque as a Function of Air Gap Quantities

In Eqs. 6.29 and 6.30 the tangential and normal components of pressure in case of current sheet on iron surface and normal component of flux density are given. The two components of pressure are illustrated in Fig. 6.8. Applied to the machine geometry, the meaning of tangential direction remains unchanged, whereas normal direction in Fig. 6.8 is identical with machine radial direction. The ratio between the two components can be expressed as

$$\frac{\sigma_n}{\sigma_t} \approx \frac{-\frac{B^2}{\mu_0} + \mu_0 A^2}{2B \cdot A} = -\frac{B}{2\mu_0 A} + \frac{\mu_0 A}{2B} \tag{6.35}$$

and is typically a double-digit number. In other words, radial force in an electric machine is typically an order of magnitude larger than the useful, tangential force.

Torque on the shaft is a result of common action of all tangential forces in the air gap. Thus, torque is an integral, global quantity, whereas radial force is a local quantity, since it changes from one point to another in the air gap.

For a given continuously distributed force along the air gap circumference the torque can be evaluated as an integral of tangential force F_t. Denoting by R the air gap radius, one can write for the torque

$$M = \int_0^{2R\pi} F_t(x)\mathrm{d}x = R\int_0^{2R\pi}\int_0^{l_{ax}} \sigma_t(x,z)\mathrm{d}z\mathrm{d}x = R\int_0^{2R\pi}\int_0^{l_{ax}} A(x)B(x)dzdx \qquad (6.36)$$

with x standing for air gap circumferential coordinate and z for the machine axial coordinate.

In the analysis of torque after Eq. 6.36, the air gap geometry plays a decisive role.

6.4.1 Constant Air Gap Width

A single winding in a machine with constant air gap width cannot create torque on itself, because in that case none of the components of magnetic energy stored in the air gap is a function of the rotor to stator angle. In other words, a current-carrying conductor does not exert any force on itself. In a machine with constant air gap width at least two independent windings are necessary to create a torque after Eq. 6.36, which in that case is named the *pure electromagnetic torque*.

The mechanism of torque creation in a constant air gap width rotating field electric machine can be visualized by using a simplified machine model in which the current sheet A and the flux density B in Eq. 6.36 originate from separate sources, here permanent magnets on each side of the air gap. In the model, B belongs to permanent magnets on the inner side of air gap, and A is created by a separate set of permanent magnets on its outer side, as in Fig. 6.10. Since the relative permeability of magnets is close to 1, the machine acts as if it had a constant air gap width.

If the orders of harmonics of flux density and current sheet are different, which is identical to a different number of pole pairs on the stator and rotor sides, as in Fig. 6.10a, the torque for each and every rotor to stator angle γ is equal to zero. If the orders of two harmonics are equal, which is identical to an equal number of poles of stator and rotor, as in Fig. 6.10b, the torque different from zero is generated as a function of the rotor to stator angle γ.

Obviously, the orders of the harmonics of air gap flux density and current sheet have to be equal if a torque different from zero in a machine with constant air gap width ought to be produced.

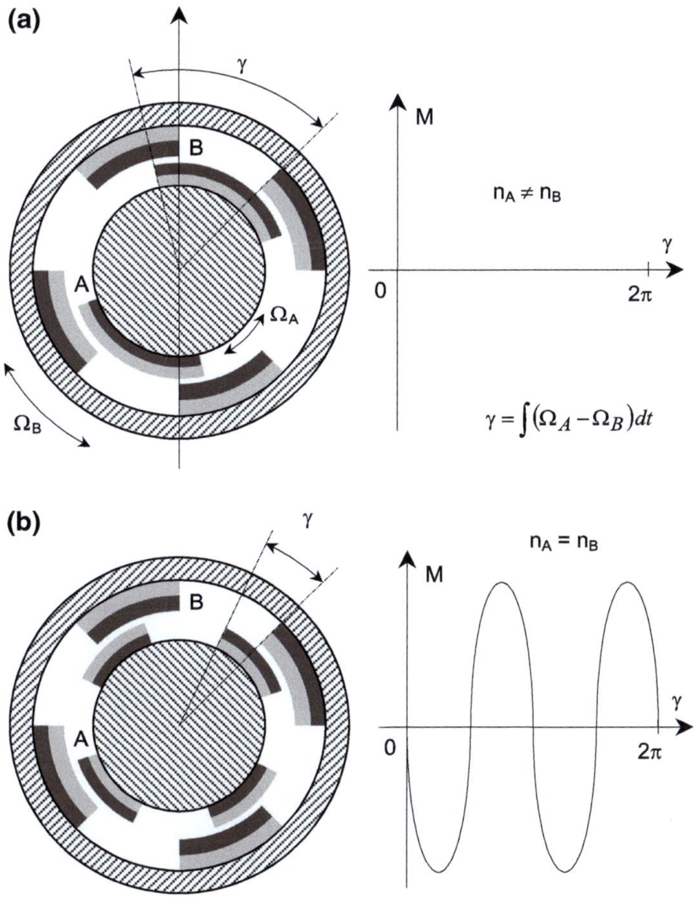

Fig. 6.10 The principle of generation of torque between a current sheet and flux density harmonic of different orders (**a**) and of identical orders (**b**) in a machine with electromagnetically constant air gap width

Another conclusion that can be drawn from Fig. 6.10b is that the two harmonics must be at standstill to each other if the generated torque is used to generate mechanical work, i.e., $\Omega_A = \Omega_B$, as in Fig. 6.10. If this condition is not fulfilled, i.e., if the two harmonics do not rotate at the same speed, the torque permanently changes as a function of angle γ between them (Fig. 6.10b) and the mechanical work performed by the two harmonics is equal to zero.

Assume now a 2p-pole machine with windings on one side of air gap creating current sheet $A(x)$, and windings on the other side of air gap creating flux density $B(x)$. The n_Bth spatial and k_Bth time harmonic of air gap flux density with amplitude $B_{\max}$ generates with n_Ath spatial and k_Ath time harmonic of current sheet with amplitude $A_{\max}$ the torque equal to

$$M = RA_{\max}B_{\max}$$
$$\int_0^{2R\pi}\int_0^{l_{ax}} \cos\left(n_A\frac{\pi}{\tau_p}x \pm k_A\omega_s t - \varphi_A\right)\cos\left[n_B\frac{\pi}{\tau_p}\left(x - \frac{R\beta}{l_{ax}}z\right) \pm k_B\omega_r t \pm p\gamma_g - \varphi_B\right]\mathrm{d}z\mathrm{d}x \tag{6.37}$$

or

$$M = Rl_{ax}f_{sk,n}A_{\max}B_{\max}$$
$$\int_0^{2R\pi} \cos\left(n_A\frac{\pi}{\tau_p}x \pm k_A\omega_s t - \varphi_A\right)\cos\left(n_B\frac{\pi}{\tau_p}x - \frac{np\beta}{2} \pm k_B\omega_r t \pm p\gamma_g - \varphi_B\right)\mathrm{d}x \tag{6.38}$$

with γ_g denoting the rotor to stator geometric angle, β the geometric skewing angle, and $f_{sk,n}$ the skewing factor for the nth harmonic. The amplitude of torque is directly proportional to the skewing factor. The orders n_A and n_B of spatial harmonics are defined on interval length of $D \cdot \pi/p$, i.e., the order $n_A = n_B = 1$ is identical to the order of fundamental spatial term over 2p poles. The altering sign of the product $p \cdot \gamma_g$ allows for both positive and negative directions of rotation, whereas the altering signs in front of terms multiplying time variable t stand for positive and negative sequence components of current sheet and flux density.

The character of torque after Eq. 6.38 depends on several parameters:

Current sheet and air gap flux density harmonics rotate in the same direction: The sign of $k_B \cdot p \cdot \gamma_g$ is negative, and the torque is equal to zero as long as the spatial orders of harmonics n_A and n_B are not equal, no matter how big k_A and k_B are.

<u>Current sheet and air gap flux harmonics have the same spatial order $n_A = n_B = n$ and different time orders k_A and k_B</u>: The expression for torque reduces to

$$M = VA_{\max}B_{\max}f_{sk,n}\cos[(k_A\omega_s - k_B\omega_r)t + \varphi_A - \varphi_B] \tag{6.39}$$

with V standing for the machine air gap volume. The two harmonics have the same number of pole pairs $n \cdot p$, but they travel at different speeds of rotation in the same direction around the circumference. The torque has only a pulsating component, the amplitude of which is dependent on the spatial order n of harmonics.

<u>Current sheet and air gap flux harmonics have the same spatial order $n_A = n_B = n$ and the same time order $k_A = k_B = k$</u>; angular frequencies of stator and rotor currents are equal, $\omega_s = \omega_r$: The torque is independent of the order numbers n and k and equal to:

$$M = VA_{\max}B_{\max}f_{\mathrm{sk},n}\cos(\varphi_A - \varphi_B) \tag{6.40}$$

When the harmonics of current sheet and flux density have the same time and spatial order and travel at the same speed along the circumference, they can create a constant component of torque if they are not perpendicular to each other, i.e., if $\varphi_A - \varphi_B \neq \pi/2$. A constant component of pure electromagnetic torque in an electric machine create only those harmonics in traveling waves of flux density and current sheet which stand still relative to each other.

Current sheet and air gap flux density harmonics rotate in opposite directions: The sign of $k_B \cdot \mathrm{p} \cdot \gamma$ is positive, and the torque is again equal to zero as long as the spatial orders of harmonics n_A and n_B are not equal, no matter how big k_A and k_B are.

Current sheet and air gap flux harmonics have the same spatial order n and different time orders k_A and k_B: The expression for torque yields

$$M = VA_{\max}B_{\max}f_{\mathrm{sk},n}\cos[(k_A\omega_s + k_B\omega_r)t + \varphi_A - \varphi_B] \tag{6.41}$$

The two harmonics have the same number of pole pairs $n \cdot p$, but they travel at different speeds of rotation at opposite directions around the circumference. The torque has only a pulsating component which, again, is independent of the spatial order n of harmonics.

Current sheet and air gap flux harmonics have the same spatial order n and the same time order k: The torque is equal to:

$$M = VA_{\max}B_{\max}f_{\mathrm{sk},n}\cos(2k\omega t + \varphi_A - \varphi_B) \tag{6.42}$$

When two harmonics have the same time and spatial order and travel at the same speed in opposite directions along the circumference, the torque they create pulsates with twice the angular speed of their rotation.

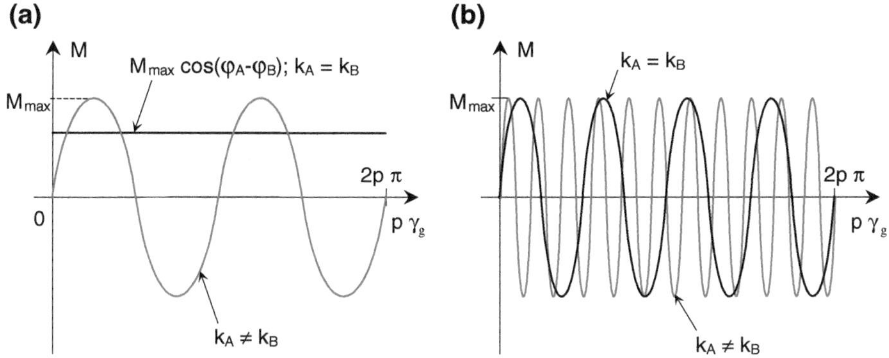

Fig. 6.11 Torque created by current sheet and air gap flux density harmonics of the same spatial order n and time orders k_A and k_B: rotating in the same direction (**a**) and in opposite directions (**b**); $M_{\max} = \mathrm{V} \cdot A_{\max} \cdot B_{\max}$

The dependence of torque on rotor electrical angle $p \cdot \gamma_g$ after Eqs. 6.39–6.42 is shown in Fig. 6.11. In Fig. 6.11a the torques created by harmonics of current sheet and air gap flux density with the same spatial order and rotating in the same direction are shown, whereas in Fig. 6.11b the torques created by harmonics of current sheet and air gap flux density with the same spatial order and rotating in opposite directions.

Mechanical work W_{mech} performed in one rotor revolution can be expressed as

$$W_{\mathrm{mech}} = \int_0^{2\pi} M d\gamma_g = l_{\mathrm{ax}} r \int_0^{2\pi} \int_0^{2r\pi} A(x)B(x)\mathrm{d}x\mathrm{d}\gamma_g \tag{6.43}$$

Out of four cases analyzed here, characterized by expressions for torque in Eqs. 6.39–6.42, obviously only the combination *equal orders of time and spatial harmonics, the same direction of rotation, and the same frequency of stator and rotor currents* (Eq. 6.40) can produce mechanical energy different from zero. Accordingly, one can formulate:

- The **trivial** condition for generation of pure electromagnetic torque: The orders of spatial harmonics n_A and n_B of current sheet and air gap flux density harmonics must be equal in order to generate an electromagnetic torque different from zero;
- The **necessary** condition for generation of pure electromagnetic torque: Besides equal orders of spatial harmonics, the orders of time harmonics k_A and k_B of current sheet and air gap flux density along with frequencies of stator and rotor currents must be equal in order to generate an electromagnetic torque with an average value different from zero;
- The **sufficient** condition for continuous electromechanical energy conversion: The two harmonics of current sheet and air gap flux density with identical spatial and time orders must rotate in the same direction in order to be able to generate mechanical energy different from zero.

6.4.2 Variable Air Gap Width

As shown in Chap. 2, in Table 2.4, the air gap flux density distribution generated by the fundamental spatial harmonic of MMF and fundamental spatial harmonic of air gap permeance with order *N/p* contains a spectrum of higher harmonics, the lowest orders of which are equal to $N/p \pm 1$. Since the fundamental and the harmonics of air gap flux density are orthogonal to each other, one can contemplate them as if they were generated by separate sources and, as such, being capable of generating a torque.

MMF harmonics with orders $N/p \pm 1$ (slot harmonics!), generated by an m-phase winding carrying alternating current with fundamental frequency ω, fulfill the conditions expressed in Eqs. 2.184–2.185 for positive and negative sequence components, in particular:

The slot harmonic $N/p - 1$ creates *negative* sequence component of MMF, because

$$\frac{1 + \left(\frac{N}{p} - 1\right)}{m} = \frac{1 + (2qm - 1)}{m} = 2q \tag{6.44}$$

is an even number.

The slot harmonic $N/p + 1$ creates *positive* sequence component of MMF, because

$$\frac{1 - \left(\frac{N}{p} + 1\right)}{m} = \frac{1 - (2qm + 1)}{m} = -2q \tag{6.45}$$

is an even number, too. The corresponding rotating harmonics of MMF can be defined as

$$\Theta_{\frac{N}{p}-1} = \Theta_{\frac{N}{p}-1,\max} \cos\left[\left(\frac{N}{\mathrm{p}} - 1\right)\frac{\pi}{\tau_p}x + \omega t\right] \tag{6.46}$$

and

$$\Theta_{\frac{N}{p}+1} = \Theta_{\frac{N}{p}+1,\max} \cos\left[\left(\frac{N}{p} + 1\right)\frac{\pi}{\tau_\mathrm{p}}x - \omega t\right] \tag{6.47}$$

The positive sign of the term ωt in Eq. 6.46 denotes the negative direction of rotation of the MMF harmonic of the order $N/p - 1$.

However, the harmonics of the orders $N/p \pm 1$ in the air gap flux density distribution of a slotted machine are generated by the fundamental harmonic of the MMF, which rotates in the positive direction, and by the variable air gap geometry. Since only the fundamental component of MMF has a time dependent term, all harmonics of air gap flux density created by the fundamental component of air gap MMF must also rotate in the positive direction, i.e., the sign of the term ωt has to be negative for them. In particular,

$$\Theta_{\frac{N}{p}-1} = \Theta_{\frac{N}{p}-1,\max} \cos\left[\left(\frac{N}{p} - 1\right)\frac{\pi}{\tau_\mathrm{p}}x - \omega t\right] \tag{6.48}$$

Variable parts of the product of the fundamental of current sheet and ($N/p - 1$)st harmonic of MMF can now be described as

$$\cos\left(\frac{\pi}{\tau_p}x-\omega t\right)\cos\left[\left(\frac{N}{p}-1\right)\frac{\pi}{\tau_p}x-\omega t\right]=\frac{1}{2}\cos\left(\frac{N}{p}\frac{\pi}{\tau_p}x-2\omega t\right)+\frac{1}{2}\cos\left[\left(\frac{N}{p}-2\right)\frac{\pi}{\tau_p}x\right] \tag{6.49}$$

The first term on the right-hand side of Eq. 6.49 rotates at a mechanical speed Ω equal to

$$\Omega = 2\frac{\omega}{N} \tag{6.50}$$

whereas the second term of the order $N/p - 2$ is stationary.

Similarly, one can write for the product of the fundamental and $(N/p + 1)$st harmonic

$$\begin{aligned}&\cos\left(\frac{\pi}{\tau_p}x-\omega t\right)\cos\left[\left(\frac{N}{p}+1\right)\frac{\pi}{\tau_p}x-\omega t\right]\\&=\frac{1}{2}\cos\left[\left(\frac{N}{p}+2\right)\frac{\pi}{\tau_p}x-2\omega t\right]+\frac{1}{2}\cos\left(\frac{N}{p}\frac{\pi}{\tau_p}x\right)\end{aligned} \tag{6.51}$$

The first term on the right-hand side of Eq. 6.51 rotates at a mechanical speed Ω equal to

$$\Omega = 2\frac{\omega}{N+2p} \tag{6.52}$$

and the second term of the order N/p is stationary.

Following Eq. 6.36, the torque is proportional to the integral of the product of current sheet and air gap flux density. Therefore, the products introduced in Eqs. 6.49 and 6.51 have to be multiplied by the fundamental harmonic of air gap width distribution, which has the order of N/p, as in Eq. 2.77. Following the criteria for torque generation introduced in the previous subsection, one can claim that only those terms in Eqs. 6.49 and 6.51, which have the same number of poles as the air gap width in Eq. 2.77, can generate torque. Since only the first summand on the right-hand side of Eq. 6.49 fulfills this criterion, one can further write for the *reluctance torque*

$$M_{\mathrm{rel}} \sim \int_0^{2p\tau_p} \cos\left(\frac{N}{p}\frac{\pi}{\tau_p}x-2\omega t\right)\cos\left[\frac{N}{p}\frac{\pi}{\tau_p}(x-\Omega Rt-\gamma_0)\right]\mathrm{d}x \approx \sin(2p\gamma_0) \tag{6.53}$$

with γ_0 denoting the geometric angle between the centerline of one tooth and the axis of the resultant field.

Based upon previous results, one can formulate

- The **trivial** condition for generation of reluctance torque: The air gap has to be slotted and not skewed for a slot pitch, with windings placed on unslotted side;

- The **necessary** condition for generation of reluctance torque with an average value different from zero: Current(s) in winding(s) has/have to be time-dependent;
- The **sufficient** condition for continuous electromechanical energy conversion by means of reluctance torque: The rotor has to rotate at mechanical speed equal to the angular frequency *f* the winding current(s), multiplied by the factor of 2/*N*, with *N* denoting the number of teeth.

Since the pure electromagnetic torque is created by two fundamental harmonics of air gap quantities, and the reluctance torque by the fundamental and a slot harmonic, it is realistic to expect that the amplitude of the reluctance torque is smaller than the amplitude of the pure electromagnetic torque for a given volume and electromagnetic and thermal loading of the machine.

6.5 Spectral Components of Torque in a Constant Width Air Gap

Rotating field electric machines create torque spectra which can vary from one operating point to another. Pulsating torque components do not contribute to the electromechanical energy conversion. Instead, they can cause torsional vibrations in the machine and mechanical load. If a particular torque component is a result of interaction of a current sheet and an air gap flux harmonic, its properties can be analyzed by means of Eqs. 6.39–6.42.

As illustrated in Chap. 2, the source of spatial harmonics in current sheet distribution are discretely distributed ampere-turns along the air gap circumference. Higher spatial harmonics in conventional rotating field machines interact with the fundamental and with each other and, as a result, a torque is created.

6.5.1 Symmetrically Wound Polyphase Machine Fed Symmetrically with Sinusoidal Currents

Time harmonics in current spectrum are either a consequence of nonlinearities in magnetic circuit, or they are injected from the source. Stator currents in some types of rotating field machines are perfectly sinusoidal. Symmetrical stator winding in a rotating field machine carrying sinusoidal symmetrical currents generates armature current sheet A_A with fundamental time harmonic and spatial harmonics of the order n:

$$A_A = \sum_{n}^{\infty} A_{\max,n} \cos\left(n\frac{\pi}{\tau_p}x - \omega t - \varphi_A\right) \tag{6.49}$$

The nth harmonic of current sheet rotates at an angular velocity which is n times smaller than the angular velocity of the fundamental.

Symmetrically wound polyphase synchronous machine at steady state is characterized by the rotor flux density distribution B_F the harmonics of which all rotate at the same mechanical synchronous speed $\Omega = \omega/p$

$$B_F = \sum_{n}^{\infty} B_{\max,n} \cos\left(n\frac{\pi}{\tau_p}x - np\Omega t - \varphi_B\right) \tag{6.50}$$

In Fig. 6.12 spatial distributions of stator current sheet and rotor air gap flux density containing fundamental terms and the 5th and 7th spatial harmonics at two time instants, denoted by t_0 and $t_0 + T/4$, are shown, with T standing for the period length of the stator current. After $T/4$, the rotor along with the air gap flux density distribution shifts for 90° electrically. Since all harmonics of the air gap flux density distribution rotate at the same angular (synchronous) speed, i.e., they do not move relative to each other, the resulting air gap flux density distribution does not change its shape during rotation (compare Fig. 6.12c, d).

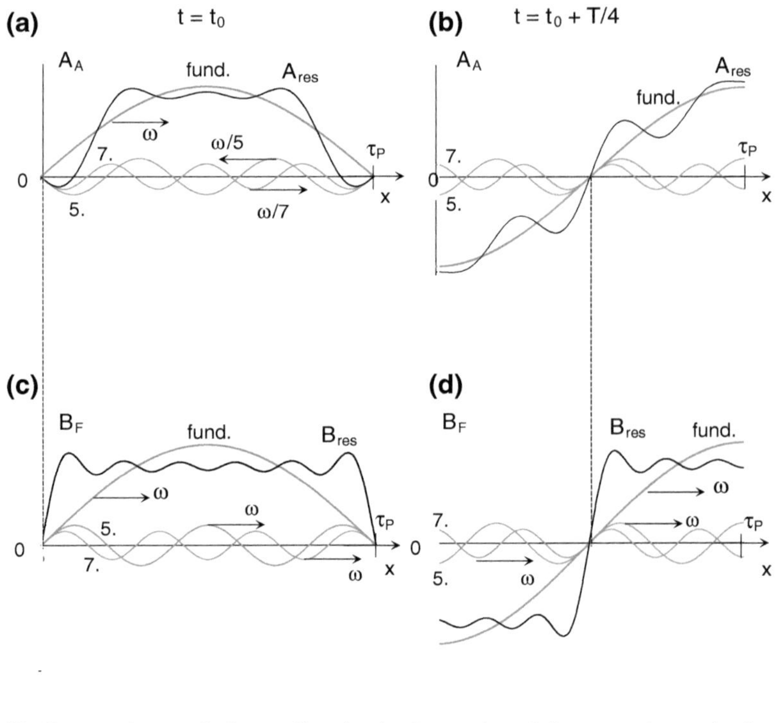

Fig. 6.12 Current sheet and air gap flux density harmonics of the same time order $k = 1$ and different spatial orders $n = 1$, 5, and $n = 7$ in a symmetrically wound 3-phase synchronous machine at two time instants shifted for one quarter of the period of stator current. *Dashed vertical lines* denote zero crossing points of the fundamental components of current sheet and flux density. Only the fundamental spatial harmonics traveling at synchronous speed ω can create a constant torque. The 5th spatial harmonic of stator current sheet travels at $-\omega/5$ and the 5th spatial harmonic of rotor flux density at ω; therefore, they create only a pulsating torque. The 7th spatial harmonic of stator current sheet travels at $\omega/7$ and the 7th spatial harmonic of rotor flux density at ω; therefore, they again create only a pulsating torque

The rotation of the stator current sheet, on the other hand, is caused by alternating currents flowing through stationary windings. Therefore, each harmonic of the rotating current sheet travels at its own angular speed. Since current sheet harmonics travel relative to each other, the form of resulting current sheet varies from one time instant to another, as illustrated in Fig. 6.12a, b.

Whereas the 5th spatial harmonic of current sheet rotates at angular speed of $\omega_{1,5} = -\Omega/5$, Ω being the mechanical synchronous speed of the fundamental, the 5th spatial harmonic of air gap flux density created on the rotor side rotates at mechanical angular speed of the rotor Ω. Relative angular speed between the two is $\Omega - (-\Omega/5) = 6/5\ \Omega$, which is six times the synchronous speed of the fifth harmonic. Therefore, the torque created by the fifth spatial harmonic of stator current sheet and fifth spatial harmonic of rotor air gap flux density pulsates at six times the stator angular frequency. Frequencies of pulsation of torque components created by other higher spatial harmonics are determined in similar manner, as illustrated in Table 6.1, in which the pure electromagnetic torque components at steady state of a 3-phase synchronous machine created by current sheet and air gap flux density distributions after Eqs. 6.39 and 6.40 are evaluated for the first 19 spatial harmonics. One recognizes in Table 6.1 that besides the dominating constant torque, components of electromagnetic torque are generated which pulsate at $2 \cdot k \cdot m$ multiple frequencies of the stator angular synchronous speed, m being the number of phases, and $k = 1, 2, \ldots$

Table 6.1 Pure electromagnetic torque components in a 3-phase synchronous machine at steady state

Spatial order n	Current sheet		Air gap flux density		Torque
	k_A	$\omega_{k,n}/\omega_{1,1}$	k_B^a	$\omega_{k,n}/\omega_{1,1}$	$M/(V \cdot A_{\max,n} \cdot B_{\max,n})$
1	1	1	1	1	$\cos(\varphi_{A,1} - \varphi_{B,1})$
3	1	N.A.	3	1	N.A.
5	1	−1/5	5	1	$\cos(6p\Omega t + \varphi_{A,5} - \varphi_{B,5})$
7	1	1/7	7	1	$\cos(6p\Omega t + \varphi_{A,7} - \varphi_{B,7})$
9	1	N.A.	9	1	N.A.
11	1	−1/11	11	1	$\cos(12p\Omega t + \varphi_{A,11} - \varphi_{B,11})$
13	1	1/13	13	1	$\cos(12p\Omega t + \varphi_{A,13} - \varphi_{B,13})$
15	1	N.A.	15	1	N.A.
17	1	−1/17	17	1	$\cos(18p\Omega t + \varphi_{A,17} - \varphi_{B,17})$
19	1	1/19	19	1	$\cos(18p\Omega t + \varphi_{A,19} - \varphi_{B,19})$

[a]Equivalent order of the time harmonic, which would result in the same angular velocity as in case of rotor rotating at synchronous speed

Since all higher harmonics of stator current sheet travel relative to the rotor, they induce voltages in rotor components. The induced voltages cause currents in electrically conducting parts of the rotor, which dissipate losses and increase temperature in them.

Proper synchronization and/or hunting of a symmetrically wound poly-phase synchronous machine: The rotor rotates at an angular velocity Ω which swings periodically around the synchronous speed ω/p, i.e.,

$$\Omega = \frac{\omega}{p} + D\cos(\Delta\omega t) \tag{6.51}$$

with D denoting the amplitude and $\Delta\omega$ the angular frequency of the deviation component of angular velocity Ω. Along with the rotor, the fundamental and all harmonics of the rotor flux density distribution B_F rotate at the same angular velocity Ω. Formally, the varying rotor angular velocity can be replaced by variable factor k_B in Eq. 6.39

$$k_B = 1 + \frac{D}{\Omega}\cos(\Delta\omega t) \tag{6.52}$$

Table 6.2 Pure electromagnetic torque components in a 3-phase synchronous machine during proper synchronization and/or hunting

Spatial order n	Current sheet		Air gap flux density		Torque
	k_A	$\omega_{k,n}/\omega_{1,1}$	k_B^a	$\omega_{k,n}/\omega_{1,1}$	$M/(V \cdot A_{\max,n} \cdot B_{\max,n})$
1	1	1	1[a]	Eq. 6.52	See Eq. 6.48
3	1	N.A.	3[a]	Eq. 6.52	N.A.
5	1	−1/5	5[a]	Eq. 6.52	$\cos[p(6\Omega + D\cos\Delta\omega t)t + \varphi_{A,5} - \varphi_{B,5}]$
7	1	1/7	7[a]	Eq. 6.52	$\cos[p(6\Omega + D\cos\Delta\omega t)t + \varphi_{A,7} - \varphi_{B,7}]$
9	1	N.A.	9[a]	Eq. 6.52	N.A.
11	1	−1/11	11[a]	Eq. 6.52	$\cos[p(12\Omega + D\cos\Delta\omega t)t + \varphi_{A,11} - \varphi_{B,11}]$
13	1	1/13	13[a]	Eq. 6.52	$\cos[p(12\Omega + D\cos\Delta\omega t)t + \varphi_{A,13} - \varphi_{B,13}]$
15	1	N.A.	15[a]	Eq. 6.52	N.A.
17	1	−1/17	17[a]	Eq. 6.52	$\cos[p(18\Omega + D\cos\Delta\omega t)t + \varphi_{A,17} - \varphi_{B,17}]$
19	1	1/19	19[a]	Eq. 6.52	$\cos[p(18\Omega + D\cos\Delta\omega t)t + \varphi_{A,19} - \varphi_{B,19}]$

[a]Equivalent order of the time harmonic, which would result in the same angular velocity as in case of rotor rotating at synchronous speed

resulting in

$$M_1 = VA_{\max,1}B_{\max,1}\cos[pD\cos(\Delta\omega t)t + \varphi_A - \varphi_B] \tag{6.53}$$

for the fundamental current sheet time harmonic, i.e., $k_A = 1$.

Relationship between current sheet and air gap flux density harmonics during proper synchronization and/or hunting can be illustrated by means of modified Fig. 6.12. Since the rotor does not rotate at synchronous speed, the dashed line between Fig. 6.12b, d would not go through the zero crossing point of the fundamental in Fig. 6.12d. Instead, the rotor flux density distribution is shifted to the left or to the right of the position in Fig. 6.12d.

In Table 6.2 the pure electromagnetic torque components created during proper synchronization and/or hunting of a 3-phase synchronous machine by current sheet and air gap flux density distributions after Eqs. 6.54–6.55 are evaluated for the first 19 spatial harmonics.

Faulty synchronization of a symmetrically wound polyphase synchronous machine is a consequence of opposite directions of rotation of stator current sheet and rotor. Consequently, all harmonics of the rotor flux density distribution B_F rotate at the same, negative synchronous speed $\omega_s = -p \cdot \Omega$, as in Fig. 6.13.

Table 6.3 Pure electromagnetic torque components in a 3-phase synchronous machine during faulty synchronization

Spatial order n	Current sheet		Air gap flux density		Torque
	k_A	$\omega_{k,n}/\omega_{1,1}$	k_B^a	$\omega_{k,n}/\omega_{1,1}$	$M/(V \cdot A_{\max,n} \cdot B_{\max,n})$
1	1	1	1	−1	$\cos(2p\Omega t + \varphi_{A,1} - \varphi_{B,1})$
3	1	N.A.	3	−1	N.A.
5	1	−1/5	5	−1	$\cos(4p\Omega t + \varphi_{A,5} - \varphi_{B,5})$
7	1	1/7	7	−1	$\cos(8p\Omega t + \varphi_{A,7} - \varphi_{B,7})$
9	1	N.A.	9	−1	N.A.
11	1	−1/11	11	−1	$\cos(10p\Omega t + \varphi_{A,11} - \varphi_{B,11})$
13	1	1/13	13	−1	$\cos(14p\Omega t + \varphi_{A,13} - \varphi_{B,13})$
15	1	N.A.	15	−1	N.A.
17	1	−1/17	17	−1	$\cos(16p\Omega t + \varphi_{A,17} - \varphi_{B,17})$
19	1	1/19	19	−1	$\cos(20p\Omega t + \varphi_{A,19} - \varphi_{B,19})$

[a]Equivalent order of the time harmonic, which would result in the same angular velocity as in case of rotor rotating at synchronous speed

Fundamental component of stator current sheet rotates in positive direction, whereas the fundamental component of air gap flux density rotates along with the rotor in negative direction. As a result, the two fundamental components create a pure electromagnetic torque which pulsates at twice the mains frequency, as in Eq. 6.42.

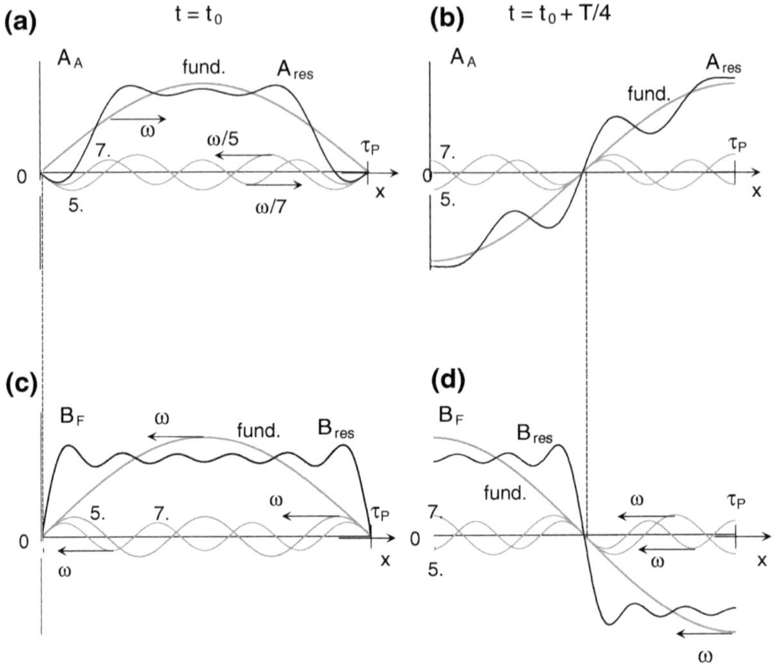

Fig. 6.13 Current sheet and air gap flux density harmonics of the same time order $k = 1$ and different spatial orders $n = 1$, 5, and $n = 7$ in a symmetrically wound synchronous machine during faulty synchronization at two time instants shifted for one quarter of period of the stator current. The fundamental spatial harmonics traveling at synchronous speed ω in opposite directions create torque which pulsates at twice the mains frequency. The 5th spatial harmonic of stator current sheet travels at $-\omega/5$ and the 5th spatial harmonic of rotor flux density at $-\omega$; therefore, they create a pure electromagnetic torque which pulsates with the difference of two angular speeds, i.e., with 4 times the stator angular frequency. The 7th spatial harmonic of stator current sheet travels at $\omega/7$ and the 7th spatial harmonic of rotor flux density at $-\omega$; therefore, the frequency of pulsations of the pure electromagnetic torque created by these two components is 8 times the stator angular frequency, etc.

Since the 5th harmonic of stator current sheet rotates in the same (negative) direction as the 5th harmonic of the air gap flux density, the relative speed between them equals 4 Ω. The difference between the positive speed of rotation of the 7th harmonic of stator current sheet and the negative speed of rotation of the 7th harmonic of the air gap flux density is equal to 8 Ω.

In the same manner, one comes to the conclusion that the 11th and 13th harmonics generate pulsating torques at 10 and 14 times the mains frequency, respectively, whereas the 17th and 19th harmonics create pulsating torques at 16 and 20 times the mains frequency.

In Table 6.3 the pure electromagnetic torque components created during faulty synchronization of a symmetrically wound 3-phase synchronous machine are evaluated.

Doubly fed polyphase machine generates at steady state a stator current sheet equal to

$$A_{\mathrm{s}} = \sum_{n}^{\infty} A_{\max,n} \cos\left(n\frac{\pi}{\tau_{\mathrm{p}}}x - k\omega_{\mathrm{s}}t - \varphi_A\right) \tag{6.54}$$

and the rotor flux density relative to stator

$$B_r = \sum_{n}^{\infty} B_{\max,n} \cos\left[n\frac{\pi}{\tau_{\mathrm{p}}}(x - x_0) - k\omega_{\mathrm{r}}t - \varphi_B\right] \tag{6.55}$$

with $\omega_{\mathrm{s}} = 2 \cdot \pi \cdot f_{\mathrm{s}}$ and $\omega_{\mathrm{r}} = 2 \cdot \pi \cdot f_{\mathrm{r}}$ standing for stator and rotor angular frequencies, respectively, k denoting the order of time harmonic of currents and x_0 the rotor shift

$$x_0 = \frac{D}{2}\Omega t \tag{6.56}$$

whereas Ω stands for the rotor angular speed. With substitutions above, one can express the rotor flux density distribution as

$$B_{\mathrm{r}} = \sum_{n}^{\infty} B_{\max,n} \cos\left[n\frac{\pi}{\tau_{\mathrm{p}}}x - (np\Omega \pm k\omega_{\mathrm{r}})\,t - k\varphi_B\right] \tag{6.57}$$

where positive sign in front of ω_{r} denotes the positive and negative sign the negative sequence harmonics. Introducing slip s for the fundamental stator time harmonic

$$\Omega = (1 - s)\frac{\omega_s}{p} \tag{6.58}$$

and taking into account the connection between the angular frequencies of stator and rotor fundamental time harmonics

$$\omega_r = s\omega_s \tag{6.59}$$

one can express the rotor flux density as

$$B_r = \sum_n^{\infty} B_{\max,n} \cos\left\{n\frac{\pi}{\tau_p}x - \omega_s[n - s(n \mp k)]\,t - \varphi_B\right\} \tag{6.60}$$

The nth spatial harmonic of rotor flux density created by the kth time harmonic of rotor currents rotates at slip s relative to the stator with an angular speed of

$$\omega^r_{1,n} = \frac{n - s(n \mp k)}{n}\frac{\omega_s}{p} = \left[1 - s\left(1 \mp \frac{k}{n}\right)\right]\frac{\omega_s}{p} \tag{6.61}$$

Negative sign of the ratio k/n denotes positive and positive sign the negative sequence component. Accordingly, one can write

$$\omega^r_{1,1} = \frac{\omega_s}{p};\quad \omega^r_{1,5} = \left(1 - \frac{6}{5}s\right)\frac{\omega_s}{p};\quad \omega^r_{1,7} = \left[1 - \frac{6}{7}s\right]\frac{\omega_s}{p}\ \text{etc.} \tag{6.62}$$

Relative angular speed between the stator current sheet and rotor flux density component created by kth time harmonic of current and nth spatial harmonic of winding distribution can now be expressed as

$$\omega^{s,r}_{k,n} = \pm\frac{k}{n}\frac{\omega_s}{p} - \left[1 - s\left(1 \mp \frac{k}{n}\right)\right]\frac{\omega_s}{p} = \frac{\omega_s}{p}(1 - s)\left(\pm\frac{k}{n} - 1\right) \tag{6.63}$$

and in particular

$$\omega^{s,r}_{1,1} = 0;\ \omega^{s,r}_{1,5} = -\frac{6}{5}\frac{\omega_s}{p}(1 - s);\quad \omega^{s,r}_{1,7} = -\frac{6}{7}\frac{\omega_s}{p}(1 - s)\ \text{etc.} \tag{6.64}$$

(see Fig. 6.14).

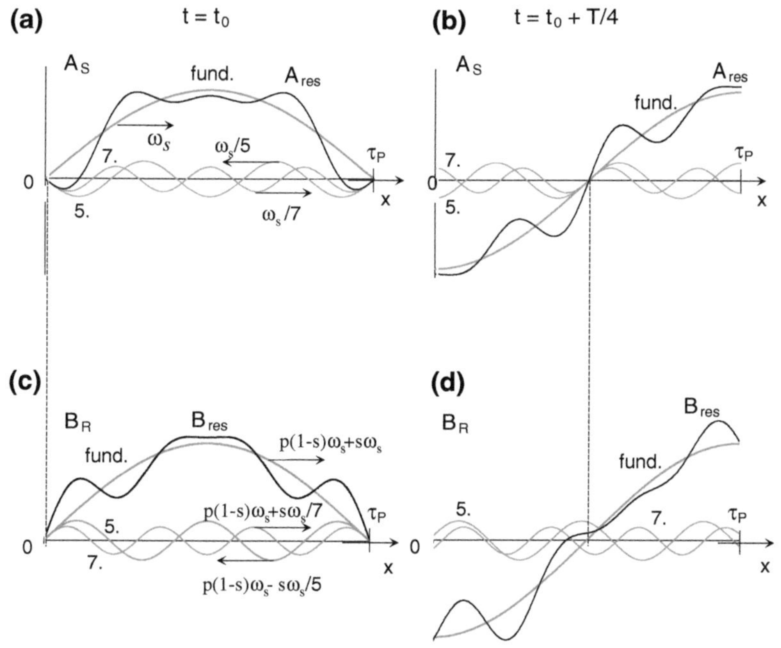

Fig. 6.14 Current sheet and air gap flux density harmonics of the same time order $k = 1$ and different spatial orders $n = 1$, 5, and $n = 7$ in a symmetrically wound induction machine with slip rings at slip $s = 0.7$ and two time instants shifted for one quarter of period of the stator current. *Dashed lines* denote zero crossing points of the fundamental components of current sheet and flux density. Only the fundamental spatial harmonics traveling at synchronous speed ω_s can create a constant torque. The 5th spatial harmonic of stator current sheet travels at $-\omega_s/5$ and the 5th spatial harmonic of rotor flux density at $p(1 - s)\omega_s - s\omega_s/5$; therefore, they create only a pulsating torque. The 7th spatial harmonic of stator current sheet travels at $\omega_s/7$ and the 7th spatial harmonic of rotor flux density at $p(1 - s)\omega_s + s\omega_s/7$; therefore, they again create only a pulsating torque

The value of slip s_0 at which stator kth time and nth spatial harmonic rotates at the same angular speed as the rotor harmonic of the same order can be found by setting the two angular speeds equal to each other, i.e.,

$$\pm\frac{k}{n}\frac{\omega_s}{p} = \frac{\omega_s}{p}\left[1 - s_0\left(1 \mp \frac{k}{n}\right)\right] \tag{6.65}$$

which has a solution for $s_0 = 1$. In a wound rotor induction machine, a stator harmonic and a rotor harmonic of arbitrary order are in synchronism only at standstill. Only the fundamental stator and rotor time and spatial harmonics travel at the same angular speed ω_s/p at every rotor mechanical speed of rotation Ω.

In Table 6.4 the components of pure electromagnetic torque generated by the fundamental stator and rotor time harmonics and first 19 spatial harmonics in a 3-phase induction machine are given. The frequency of pulsations of torque components is a function of slip; at zero speed, all components of pure electromagnetic torque become constant.

In Fig. 6.14 the stator current sheet and rotor air gap density distribution created by windings of a wound rotor induction machine carrying pure sinusoidal currents at two time instants and the slip of 0.7 are shown. Whereas the fundamental harmonic of rotor air gap flux density travels at the same angular speed as the fundamental harmonic of current sheet, each higher spatial harmonic in both distributions travels at its own speed.

Table 6.4 Pure electromagnetic torque components at steady state of a 3-phase wound rotor induction machine fed with sinusoidal currents

Spatial order n	Current sheet		Air gap flux density		Torque
	k_A	$\omega_{k,n}/\omega_{1,1,s}$	k_B	$\omega_{k,n}/\omega_{1,1,r}$	$M/(V \cdot A_{\max,n} \cdot B_{\max,n})$
1	1	1	1	1	$\cos(\varphi_{A,1} - \varphi_{B,1})$
5	1	−1/5	1	−1/5	$\cos[6/5\omega_s(1 - s)t + \varphi_{A,5} - \varphi_{B,5}]$
7	1	1/7	1	1/7	$\cos[6/7\omega_s(1 - s)t + \varphi_{A,7} - \varphi_{B,7}]$
11	1	−1/11	1	−1/11	$\cos[12/11\omega_s(1 - s)t + \varphi_{A,11} - \varphi_{B,11}]$
13	1	1/13	1	1/13	$\cos[12/13\omega_s(1 - s)t + \varphi_{A,13} - \varphi_{B,13}]$
17	1	−1/17	1	−1/17	$\cos[18/17\omega_s(1 - s)t + \varphi_{A,17} - \varphi_{B,17}]$
19	1	1/19	1	1/19	$\cos[18/19\omega_s(1 - s)t + \varphi_{A,19} - \varphi_{B,19}]$

Symmetrically wound squirrel cage induction machine is characterized by stator current sheet distribution at steady state

$$A_s = \sum_{n}^{\infty} A_{\max,n} \cos\left(n\frac{\pi}{\tau_p}x - k\omega_s t - \varphi_A\right) \tag{6.66}$$

The nth spatial harmonic of corresponding air gap flux density induces in rotor squirrel cage a 2n·p-pole voltage distribution and, as a consequence, a 2n·p-pole MMF distribution. Since both the stator current sheet and rotor MMF have the same number of poles 2n·p, they create a pure electromagnetic torque with a constant component since they rotate at the same angular speed (Fig. 6.15).

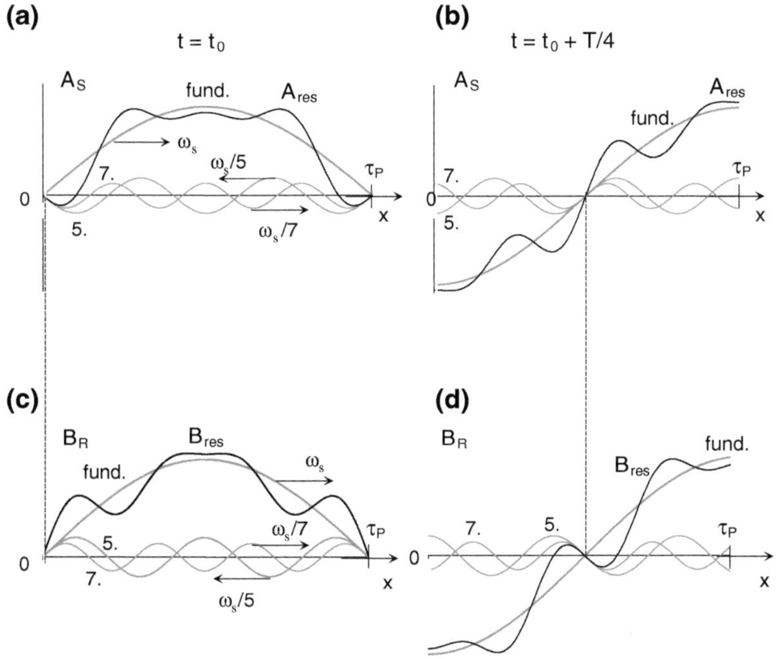

Fig. 6.15 Current sheet and air gap flux density harmonics of the same time order $k = 1$ and different spatial orders $n = 1$, 5, and $n = 7$ in a symmetrically wound squirrel cage induction machine at two time instants shifted for one quarter of period of the stator current. *Dashed lines* denote zero crossing points of the fundamental components of current sheet and flux density. Not only the fundamental spatial harmonics traveling at synchronous speed ω_s create a constant torque: The 5th spatial harmonic of stator current sheet travels at $-\omega_s/5$ and the 5th spatial harmonic of rotor flux density also at $-\omega_s/5$; therefore, they create a true electromagnetic torque with constant component. The 7th spatial harmonic of stator current sheet travels at $\omega_s/7$ and the 7th spatial harmonic of rotor flux density also at $\omega_s/7$—they both create a constant component of torque as well

In Table 6.5 components of pure electromagnetic torque generated by the fundamental stator and rotor time harmonics and first 19 spatial harmonics in a 3-phase squirrel cage induction machine are given. In this table the pure electromagnetic torque components created by slot harmonics are not included because their order is out of range of interest.

Table 6.5 Pure electromagnetic torque components at steady state of a 3-phase squirrel induction machine fed with sinusoidal currents

Spatial order n	Current sheet		Air gap flux density		Torque
	k_A	$\omega_{k,n}/\omega_{1,1,s}$	k_B	$\omega_{k,n}/\omega_{1,1,r}$	$M/(V \cdot A_{\max,n} \cdot B_{\max,n})$
1	1	1	1	1	$\cos(\varphi_{A,1} - \varphi_{B,1})$
5	1	−1/5	1	−1/5	$\cos(\varphi_{A,5} - \varphi_{B,5})$
7	1	1/7	1	1/7	$\cos(\varphi_{A,7} - \varphi_{B,7})$
11	1	−1/11	1	−1/11	$\cos(\varphi_{A,11} - \varphi_{B,11})$
13	1	1/13	1	1/13	$\cos(\varphi_{A,13} - \varphi_{B,13})$
17	1	−1/17	1	−1/17	$\cos(\varphi_{A,17} - \varphi_{B,17})$
19	1	1/19	1	1/19	$\cos(\varphi_{A,19} - \varphi_{B,19})$

6.5.2 Symmetrically Wound Machine Fed with Sinusoidal Unbalanced Currents

Unbalanced operation of a symmetrically wound synchronous machine at steady state is caused by amplitude and/or phase asymmetry of stator currents. As a consequence, both positive and negative sequence components of stator air gap current sheet and MMF are generated. Since the rotor is at synchronism, the relative speed between the fundamental term of negative sequence component of stator current sheet and rotor is twice the synchronous speed. In addition to torques created by positive sequence component of stator current sheet, as in Table 6.1, the negative sequence of stator current sheet generates with rotor air gap flux density a spectrum of pulsating torque components with frequencies identical to those at faulty synchronization represented in Table 6.3.

Pure electromagnetic torque components created by negative sequence components of stator current sheet and rotor air gap flux density terms with corresponding numbers of poles are represented in Table 6.6.

In Fig. 6.16, negative sequence harmonics of current sheet created by stator unbalanced currents along with rotor air gap flux density harmonics are shown.

The fundamental component of stator current sheet rotates in opposite direction to the rotor and induces in it voltages with twice the stator frequency. The induced voltages generate eddy currents in rotor solid parts, which dissipate losses in them. In order to minimize rotor eddy current losses due to unbalanced load, the stator negative sequence current is limited to typically 5–12 % of rated current, the value depending on the machine size and the rating of the damper cage.

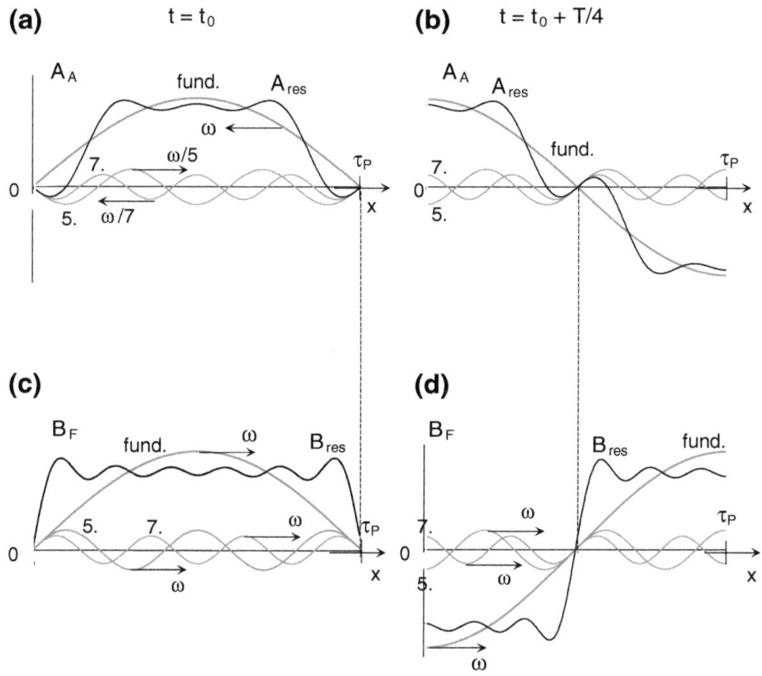

Fig. 6.16 Negative sequence current sheet and air gap flux density harmonics of the same time order $k = 1$ and different spatial orders $n = 1, 5$, and $n = 7$ in a symmetrically wound synchronous machine fed with sinusoidal unbalanced currents at two time instants shifted for one quarter of period of the stator current. The fundamental spatial harmonics traveling at synchronous speed ω in opposite directions create a pulsating torque. The 5th spatial harmonic of stator current sheet travels at ω/5 and the 5th spatial harmonic of rotor flux density at ω; therefore, they create a pure electromagnetic torque which pulsates with the difference of two angular speeds, i.e., with 4 times the stator angular frequency. The 7th spatial harmonic of stator current sheet travels at −ω/7 and the 7th spatial harmonic of rotor flux density at ω; therefore, the frequency of pulsations of the pure electromagnetic torque created by these two components is 8 times the stator angular frequency

Table 6.6 Pure electromagnetic torque components in a 3-phase synchronous machine at steady state created by negative sequence component of stator current sheet

Spatial order n	Current sheet		Air gap flux density		Torque
	k_A	$\omega_{k,n}/\omega_{1,1}$	k_B^{a}	$\omega_{k,n}/\omega_{1,1}$	$M/(V \cdot A_{\max,n} \cdot B_{\max,n})$
1	1	−1	1	1	$\cos(2p\Omega t + \varphi_{A,1} - \varphi_{B,1})$
3	1	N.A.	3	1	N.A.
5	1	1/5	5	1	$\cos(4p\Omega t + \varphi_{A,5} - \varphi_{B,5})$
7	1	−1/7	7	1	$\cos(8p\Omega t + \varphi_{A,7} - \varphi_{B,7})$

(continued)

Table 6.6 (continued)

Spatial order n	Current sheet		Air gap flux density		Torque
	k_A	$\omega_{k,n}/\omega_{1,1}$	k_B^{a}	$\omega_{k,n}/\omega_{1,1}$	$M/(V \cdot A_{\max,n} \cdot B_{\max,n})$
9	1	N.A.	9	1	N.A.
11	1	1/11	11	1	$\cos(10p\Omega t + \varphi_{A,11} - \varphi_{B,11})$
13	1	−1/13	13	1	$\cos(14p\Omega t + \varphi_{A,13} - \varphi_{B,13})$
15	1	N.A.	15	1	N.A.
17	1	1/17	17	1	$\cos(16p\Omega t + \varphi_{A,17} - \varphi_{B,17})$
19	1	−1/19	19	1	$\cos(20p\Omega t + \varphi_{A,19} - \varphi_{B,19})$

[a]Equivalent order of the time harmonic, which would result in the same angular velocity as in case of rotor rotating at synchronous speed

Polyphase machine with symmetrically wound, symmetrically fed stator and unbalanced rotor is a mode of operation characteristic for both synchronous and induction machines, e.g., during:

- Start-up transient of a synchronous machine with short-circuited field winding;
- Start-up transient of a wound rotor induction machine with asymmetrically connected rotor winding;
- Start-up transient of a squirrel cage induction machine with broken bar(s) and/or ring segment(s).

The impressed stator currents generate rotating current sheet distribution at steady state described as

$$A_s = \sum_{n}^{\infty} A_{\max,n} \cos\left(n\frac{\pi}{\tau_p}x - k\omega_s t - \varphi_A\right) \tag{6.67}$$

Independent of the machine type, rotor asymmetric currents create besides a positive B_+ also a negative sequence air gap flux density B_-, the speed of rotation of which depends on slip (Eq. 6.64)

$$B_r = \sum_{n}^{\infty} \begin{Bmatrix} B_{+,n} \cos\left\{n\frac{\pi}{\tau_p}x - \omega_s[n - s(n \mp k)]t - \varphi_B\right\} \\ +B_{-,n} \cos\left\{n\frac{\pi}{\tau_p}x + \omega_s[n - s(n \mp k)]t - \varphi_B\right\} \end{Bmatrix} \tag{6.68}$$

The order n of harmonic in Eq. 6.68 can be 1, 3, 5, 7, … in an asymmetrical 3-phase and single-phase rotor, or 1, $N/p \pm 1$, $2N/p \pm 1$, … in an asymmetrical squirrel cage with $N/2p$ phases. Since in the current sheet spectrum of symmetrically wound, symmetrically fed stator there exist no harmonics with order equal to odd multiples of the number of phases, pure electromagnetic torque components created by these harmonics are equal to zero.

The effects of positive sequence harmonics in the spectrum of rotor air gap flux density B_r were discussed previously. Here torque components created by negative sequence harmonics $B_{\_,n}$ will be found in a similar manner as for positive sequence harmonics. If in Eq. 6.72 the positive sign of the ratio k/n is applied for positive and negative sign for negative sequence components, one obtains for angular speed of the first 3 harmonics of the negative sequence component of the air gap flux density

$$\omega_{1,1}^{\mathrm{r}} = (1 - 2s)\frac{\omega_{\mathrm{s}}}{p}; \quad \omega_{1,5}^{\mathrm{r}} = \left(1 - \frac{4}{5}s\right)\frac{\omega_{\mathrm{s}}}{p}; \; \omega_{1,7}^{\mathrm{r}} = \left(1 - \frac{8}{7}s\right)\frac{\omega_{\mathrm{s}}}{p} \; \text{etc.} \qquad (6.69)$$

The fundamental term of the negative sequence component of the rotor air gap flux density caused by rotor winding asymmetry is at standstill for slip $s = 1/2$, the fifth spatial harmonic at slip $s = 5/4$, the seventh at $s = 7/8$, etc.

Table 6.7 Relative angular speeds between harmonics of the rotor negative sequence flux density and current sheet after Eq. 6.67 along with slip values at which the corresponding harmonics are synchronized

Spatial order n	Relative angular speed $/(\omega_s/p)$	Harmonics synchronized at slip
1	$-2s$	0
5	$2(3 - 2s)/5$	3/2
7	$2(3 - 4s)/7$	3/4
11	$2(6 - 5s)/11$	6/5
13	$2(6 - 7s)/13$	6/7
17	$2(9 - 8s)/17$	9/8
19	$2(9 - 10s)/19$	9/10

Relative angular speed between the stator current sheet in Eq. 6.67 and negative sequence components of the rotor flux density created by kth time harmonic of current and nth spatial harmonic of winding distribution can now be expressed as

$$\omega_{k,n}^{\mathrm{s,r}} = \frac{\omega_{\mathrm{s}}}{p}\left[1 \pm \frac{k}{n} + s\left(-1 \pm \frac{k}{n}\right)\right] \qquad (6.70)$$

In Table 6.7 relative angular speeds after Eq. 6.70 are calculated for the first 19 spatial harmonics in a machine with 3-phase stator winding.

The fifth spatial harmonic of the rotor negative sequence flux density travels at one-fifth of the rotor synchronous speed plus one-fifth of the stator synchronous speed relative to the fifth spatial harmonic of the stator positive sequence current sheet. At slip $s = 3/2$, the rotor rotates with one half of the negative synchronous speed, whereas the fifth harmonic of the rotor flux density rotates relative to rotor at 1/5 of the rotor synchronous speed. Since the rotor synchronous speed at that operating point is 3/2 of the stator synchronous speed, the relative speed of the fifth harmonic of the rotor flux density to the rotor is $1/5 \cdot 3/2 = 3/10$ of the stator synchronous speed. Added to the rotor mechanical speed of one half of the negative stator synchronous speed, the speed of the fifth rotor spatial harmonic of flux density is equal to the speed of the fifth spatial harmonic of stator current sheet, i.e., –2/10 of the stator synchronous speed. Therefore, at slip $s = 3/2$, the fifth harmonic of rotor flux density in a machine with unbalanced rotor is synchronized with the fifth harmonic of stator current sheet and the torque they create can have a constant component. At all other speeds of the rotor, the two harmonics create only a pulsating torque.

In Table 6.8 components of pure electromagnetic torque generated by the fundamental time harmonics and first 19 spatial harmonics of stator current sheet after Eq. 6.71 and negative sequence rotor air gap flux density are presented. The pure electromagnetic torque components created by slot harmonics are not included in this table because their order is out of scope of interest.

Table 6.8 Pure electromagnetic torque components at steady state of a 3-phase machine with unbalanced rotor created by stator currents with angular frequency·ω_s and negative sequence air gap flux density harmonics

Spatial order n	Current sheet		Air gap flux density		Torque
	k_A	$\omega_{k,n}/\omega_{1,1,s}$	k_B	$\omega_{k,n}/\omega_{1,1,r}$	$M/(V \cdot A_{max,n} \cdot B_{max,n})$
1	1	1	1	1	$\cos(2s\omega_s t + \varphi_{A,1} - \varphi_{B,1})$
5	1	−1/5	1	−1/5	$\cos[2/5\omega_s(3 - 2s)t + \varphi_{A,5} - \varphi_{B,5}]$
7	1	1/7	1	1/7	$\cos[2/7\omega_s(3 - 4s)t + \varphi_{A,7} - \varphi_{B,7}]$
11	1	−1/11	1	−1/11	$\cos[2/11\omega_s(6 - 5s)t + \varphi_{A,11} - \varphi_{B,11}]$
13	1	1/13	1	1/13	$\cos[2/13\omega_s(6 - 7s)t + \varphi_{A,13} - \varphi_{B,13}]$
17	1	−1/17	1	−1/17	$\cos[2/17\omega_s(9 - 8s)t + \varphi_{A,17} - \varphi_{B,17}]$
19	1	1/19	1	1/19	$\cos[2/19\omega_s(9 - 10s)t + \varphi_{A,19} - \varphi_{B,19}]$

In addition to the currents which create current sheet after Eq. 6.67, stator winding carries currents the source of which are voltages induced by the negative sequence component of rotor air gap flux density, as in Eq. 6.68. Since the stator windings do not move, the frequency of induced voltage in them is independent of the order of spatial harmonic of flux density wave which induced it.

Stator currents with angular frequency ω_r driven by voltages induced by the negative sequence component of rotor flux density create current sheet $A\_$ equal to

$$A_- = \sum_n^{\infty} A_{max,n,-} \cos\left(n\frac{\pi}{\tau_p}x + \omega_- t - \varphi_{A,-}\right) \tag{6.71}$$

where the frequency $\omega\_$ is equal to

$$\omega_- = p(\Omega - \omega_r) = (1 - 2s)\omega_s \tag{6.72}$$

because the fundamental time component of the rotor negative sequence flux density travels at electrical angular speed—$p \cdot \omega_r = -s \cdot p \cdot \omega_r$ relative to the rotor mechanical speed $p \cdot \Omega$.

The fundamental ($n = k = 1$) negative sequence component of air gap flux density in Eq. 6.68 can be expressed as

$$B_{r,1} = B_{-,1} \cos\left[\frac{\pi}{\tau_p}x + (1 - 2s)\omega_s t - \varphi_B\right] \tag{6.73}$$

The fundamental negative sequence component of air gap flux density $B_{r,1}$ generates with the fundamental component of current sheet $A_{-,1}$ in Eq. 6.71

$$A_{-,1} = A_{\max,1,-} \cos\left[\frac{\pi}{\tau_p}x + (1 - 2s)\omega_s t - \varphi_{A,-}\right] \tag{6.74}$$

an electromagnetic torque the average value of which is [see also Eq. (6.38)]

$$M = VA_{\max,1,-}B_{-,1}f_{sk,1} \cos\left(\varphi_A - \varphi_B\right) \tag{6.75}$$

An unbalanced rotor generates both positive and negative sequence components of current sheet, which interact with stator flux density. Besides pulsating components, a constant pure electromagnetic torque is generated which is equal to zero at slip $s = 0.5$. This property of negative sequence MMF is called **Görges phenomenon**.

6.5.3 Single-Phase Operation of a Rotating Field Machine

Single-phase synchronous machine in the power range above 100 MVA is a typical source of electrical energy for railroads where only one contact wire is available. In order to avoid too a low winding factor for the fundamental, usually only 2/3 of stator slots are wound. A 2-pole generator usually rotates at 1000 rpm in order to produce voltages at a frequency of 16 2/3 Hz. The rotor is cylindrical with a very strong damping cage, which protects field winding from the 100 % unbalance of the stator MMF.

Single-phase stator winding fed by sinusoidal current creates pulsating current sheet $A_s\ (x,\ t)$

$$A_s(x,t) = \sum_n^{\infty} A_{\max,n}\left[\cos\left(n\frac{\pi}{\tau_p}x - \omega t - \varphi_A\right) + \cos\left(n\frac{\pi}{\tau_p}x + \omega t - \varphi_A\right)\right] \tag{6.76}$$

whereas the field coil rotating at synchronous speed generates air gap flux density which is defined in Eq. 6.50 (Table 6.9).

Table 6.9 Pure electromagnetic torque components created by negative sequence component of stator current sheet in a single-phase synchronous machine at steady state

Spatial order n	Current sheet		Air gap flux density		Torque
	k_A	$\omega_{k,n}/\omega_{1,1}$	k_B^a	$\omega_{k,n}/\omega_{1,1}$	$M/(V \cdot A_{max,n} \cdot B_{max,n})$
1	1	1	1	1	$\cos(2p\Omega t + \varphi_{A,1} - \varphi_{B,1})$
3	1	N.A.	3	1	N.A.
5	1	−1/5	5	1	$\cos(4p\Omega t + \varphi_{A,5} - \varphi_{B,5})$
7	1	1/7	7	1	$\cos(8p\Omega t + \varphi_{A,7} - \varphi_{B,7})$
9	1	N.A.	9	1	N.A.
11	1	−1/11	11	1	$\cos(10p\Omega t + \varphi_{A,11} - \varphi_{B,11})$
13	1	1/13	13	1	$\cos(14p\Omega t + \varphi_{A,13} - \varphi_{B,13})$
15	1	N.A.	15	1	N.A.
17	1	−1/17	17	1	$\cos(16p\Omega t + \varphi_{A,17} - \varphi_{B,17})$
19	1	1/19	19	1	$\cos(20p\Omega t + \varphi_{A,19} - \varphi_{B,19})$

[a]Equivalent order of the time harmonic, which would result in the same angular velocity as in case of rotor rotating at synchronous speed

In addition to the torques created by positive sequence component of stator current sheet and rotor flux density (Table 6.1), pulsating torques are generated in a single-phase synchronous machine as a result of interaction between negative sequence component of stator current sheet and the rotor flux density.

The fundamental negative sequence component of stator current sheet creates with the fundamental component of the rotor flux density an electromagnetic torque, which pulsates at twice the mains frequency. Since the machine is single-phase-fed, its power has a pulsating component besides constant term $U \cdot I \cdot \cos\varphi$.

The single-phase synchronous generator for railroads is unique among all electric machine types because the amplitude of its pulsating torque is larger or equal to the constant, useful torque which it generates. For this reason, the housing of a single-phase synchronous generator is fixed with a series of springs to its fundamental. In addition, the rotor damping cage must be dimensioned for a 100 % unbalanced load.

Single-phase operation of an induction machine at steady state is characterized by stator current sheet

$$A_s = \sum_n^{\infty} A_{max,n}\left[\cos\left(n\frac{\pi}{\tau_p}x - k_s\omega_s t - \varphi_A\right) + \cos\left(n\frac{\pi}{\tau_p}x + k_s\omega_s t - \varphi_A\right)\right] \quad (6.77)$$

which induces voltages after Eq. 2.266 in symmetrical rotor windings rotating at mechanical angular speed Ω. Denoting by k_s the order of stator current time harmonic, the frequency of induced voltage can be expressed by means of Eq. 2.667 as

$$\omega_{i,-} = \omega[k_s - n(1-s)] \tag{6.78}$$

for voltage components introduced by positive sequence of stator current sheet, and

$$\omega_{i,+} = \omega[k_s + n(1-s)] \tag{6.79}$$

for voltage components induced by negative sequence of stator current sheet. The positive sequence of stator current sheet rotates in the same direction as the rotor. The fundamental time and spatial harmonic of the positive sequence component of current sheet, $n = k_s = 1$, induces in rotor conductors voltages with frequency $\omega_{i,-} = s \cdot \omega$, whereas the fundamental time and spatial harmonic of the negative sequence component of current sheet induces in rotor conductors voltages with frequency $\omega_{i,+} = (2 - s) \cdot \omega$.

Induced voltages in a squirrel cage rotor generate currents, which are the source of rotor component of air gap flux density B_r. Considering Eqs. 6.68, one can define for positive rotor mechanical angular speed Ω the positive sequence of the rotor air gap flux density as

$$B^+_{r,+} = B_{max} \cos\left\{n\frac{\pi}{\tau_p}x - [n(1-s)(1-k_r) + k_r k_s]\omega t\right\} \tag{6.80}$$

with k_r standing for the order of rotor current time harmonic. Analogously, the negative sequence of the rotor air gap flux density equals

$$B^+_{r,-} = B_{max} \cos\left\{n\frac{\pi}{\tau_p}x - [n(1-s)(1+k_r) - k_r k_s]\omega t\right\} \tag{6.81}$$

The positive sequence component of the rotor air gap flux density for negative rotor mechanical angular speed Ω is equal to

Table 6.10 Trigonometric functions multiplying terms $V\,A_{max}\,B_{max}$ in the expression for pure electromagnetic torque components in a single-phase induction machine rotating in positive direction at steady state

	n	$B^+_{r,+}$	$B^+_{r,-}$
$A_{s,+}$	1	$\cos(\varphi_{A,1} - \varphi_{B,1})$	$\cos(\varphi_{A,1} - \varphi_{B,1} + 2s\omega t)$
	3	$\cos(\varphi_{A,3} - \varphi_{B,3})$	$\cos[\varphi_{A,3} - \varphi_{B,3} - 2(2-3s)\omega t]$
	5	$\cos(\varphi_{A,5} - \varphi_{B,5})$	$\cos[\varphi_{A,5} - \varphi_{B,5} - 2(4-5s)\omega t]$
	7	$\cos(\varphi_{A,7} - \varphi_{B,7})$	$\cos[\varphi_{A,7} - \varphi_{B,7} - 2(6-7s)\omega t]$
	9	$\cos(\varphi_{A,9} - \varphi_{B,9})$	$\cos[\varphi_{A,9} - \varphi_{B,9} - 2(8-9s)\omega t]$
	11	$\cos(\varphi_{A,11} - \varphi_{B,11})$	$\cos[\varphi_{A,11} - \varphi_{B,11} - 2(10-11s)\omega t]$
	13	$\cos(\varphi_{A,13} - \varphi_{B,13})$	$\cos[\varphi_{A,13} - \varphi_{B,13} - 2(12-13s)\omega t]$
	15	$\cos(\varphi_{A,15} - \varphi_{B,15})$	$\cos[\varphi_{A,15} - \varphi_{B,15} - 2(14-15s)\omega t]$
	17	$\cos(\varphi_{A,17} - \varphi_{B,17})$	$\cos[\varphi_{A,17} - \varphi_{B,17} - 2(16-17s)\omega t]$
	19	$\cos(\varphi_{A,19} - \varphi_{B,19})$	$\cos[\varphi_{A,19} - \varphi_{B,19} - 2(18-19s)\omega t]$

(continued)

Table 6.10 (continued)

	n	$B_{r,+}^{+}$	$B_{r,-}^{+}$
	1	$\cos(\varphi_{A,1} - \varphi_{B,1} + 2\omega t)$	$\cos[\varphi_{A,1} - \varphi_{B,1} + 2(s-1)\omega t]$
	3	$\cos(\varphi_{A,3} - \varphi_{B,3} + 2\omega t)$	$\cos[\varphi_{A,3} - \varphi_{B,3} + 6(s-1)\omega t]$
	5	$\cos(\varphi_{A,5} - \varphi_{B,5} + 2\omega t)$	$\cos[\varphi_{A,5} - \varphi_{B,5} + 10(s-1)\omega t]$
	7	$\cos(\varphi_{A,7} - \varphi_{B,7} + 2\omega t)$	$\cos[\varphi_{A,7} - \varphi_{B,7} + 14(s-1)\omega t]$
$A_{s,-}$	9	$\cos(\varphi_{A,9} - \varphi_{B,9} + 2\omega t)$	$\cos[\varphi_{A,9} - \varphi_{B,9} + 18(s-1)\omega t]$
	11	$\cos(\varphi_{A,11} - \varphi_{B,11} + 2\omega t)$	$\cos[\varphi_{A,11} - \varphi_{B,11} + 22(s-1)\omega t]$
	13	$\cos(\varphi_{A,13} - \varphi_{B,13} + 2\omega t)$	$\cos[\varphi_{A,13} - \varphi_{B,13} + 26(s-1)\omega t]$
	15	$\cos(\varphi_{A,15} - \varphi_{B,15} + 2\omega t)$	$\cos[\varphi_{A,15} - \varphi_{B,15} + 30(s-1)\omega t]$
	17	$\cos(\varphi_{A,17} - \varphi_{B,17} + 2\omega t)$	$\cos[\varphi_{A,17} - \varphi_{B,17} + 34(s-1)\omega t]$
	19	$\cos(\varphi_{A,19} - \varphi_{B,19} + 2\omega t)$	$\cos[\varphi_{A,19} - \varphi_{B,19} + 38(s-1)\omega t]$

$$B_{r,+}^{-} = B_{\max} \cos\left\{ n\frac{\pi}{\tau_p}x + [n(1-s)(1-k_r) - k_r k_s]\omega t \right\} \tag{6.82}$$

and the negative sequence

$$B_{r,-}^{-} = B_{\max} \cos\left\{ n\frac{\pi}{\tau_p}x + [n(1-s)(1+k_r) - k_r k_s]\omega t \right\} \tag{6.83}$$

Table 6.11 Trigonometric functions multiplying terms $V\,A_{\max}\,B_{\max}$ in the expression for pure electromagnetic torque components in a single-phase induction machine rotating in negative direction at steady state

	n	$B_{r,-}^{-}$	$B_{r,+}^{-}$
	1	$\cos(\varphi_{A,1} - \varphi_{B,1})$	$\cos(\varphi_{A,1} - \varphi_{B,1} + 2\,s\omega t)$
	3	$\cos(\varphi_{A,3} - \varphi_{B,3})$	$\cos[\varphi_{A,3} - \varphi_{B,3} - 2(2-3s)\omega t]$
	5	$\cos(\varphi_{A,5} - \varphi_{B,5})$	$\cos[\varphi_{A,5} - \varphi_{B,5} - 2(4-5s)\omega t]$
	7	$\cos(\varphi_{A,7} - \varphi_{B,7})$	$\cos[\varphi_{A,7} - \varphi_{B,7} - 2(6-7s)\omega t]$
$A_{s,-}$	9	$\cos(\varphi_{A,9} - \varphi_{B,9})$	$\cos[\varphi_{A,9} - \varphi_{B,9} - 2(8-9s)\omega t]$
	11	$\cos(\varphi_{A,11} - \varphi_{B,11})$	$\cos[\varphi_{A,11} - \varphi_{B,11} - 2(10-11s)\omega t]$
	13	$\cos(\varphi_{A,13} - \varphi_{B,13})$	$\cos[\varphi_{A,13} - \varphi_{B,13} - 2(12-13s)\omega t]$
	15	$\cos(\varphi_{A,15} - \varphi_{B,15})$	$\cos[\varphi_{A,15} - \varphi_{B,15} - 2(14-15s)\omega t]$
	17	$\cos(\varphi_{A,17} - \varphi_{B,17})$	$\cos[\varphi_{A,17} - \varphi_{B,17} - 2(16-17s)\omega t]$
	19	$\cos(\varphi_{A,19} - \varphi_{B,19})$	$\cos[\varphi_{A,19} - \varphi_{B,19} - 2(18-19s)\omega t]$

(continued)

Table 6.11 (continued)

	n	$B^{-}_{r,-}$	$B^{-}_{r,+}$
$A_{s,+}$	1	$\cos(\varphi_{A,1} - \varphi_{B,1} + 2\omega t)$	$\cos[\varphi_{A,1} - \varphi_{B,1} + 2(s - 1)\omega t]$
	3	$\cos(\varphi_{A,3} - \varphi_{B,3} + 2\omega t)$	$\cos[\varphi_{A,3} - \varphi_{B,3} + 6(s- - 1)\omega t]$
	5	$\cos(\varphi_{A,5} - \varphi_{B,5} + 2\omega t)$	$\cos[\varphi_{A,5} - \varphi_{B,5} + 10(s - 1)\omega t]$
	7	$\cos(\varphi_{A,7} - \varphi_{B,7} + 2\omega t)$	$\cos[\varphi_{A,7} - \varphi_{B,7} + 14(s - 1)\omega t]$
	9	$\cos(\varphi_{A,9} - \varphi_{B,9} + 2\omega t)$	$\cos[\varphi_{A,9} - \varphi_{B,9} + 18(s-1)\omega t]$
	11	$\cos(\varphi_{A,11} - \varphi_{B,11} + 2\omega t)$	$\cos[\varphi_{A,11} - \varphi_{B,11} + 22(s - 1)\omega t]$
	13	$\cos(\varphi_{A,13} - \varphi_{B,13} + 2\omega t)$	$\cos[\varphi_{A,13} - \varphi_{B,13} + 26(s - 1)\omega t]$
	15	$\cos(\varphi_{A,15} - \varphi_{B,15} + 2\omega t)$	$\cos[\varphi_{A,15} - \varphi_{B,15} + 30(s - 1)\omega t]$
	17	$\cos(\varphi_{A,17} - \varphi_{B,17} + 2\omega t)$	$\cos[\varphi_{A,17} - \varphi_{B,17} + 34(s - 1)\omega t]$
	19	$\cos(\varphi_{A,19} - \varphi_{B,19} + 2\omega t)$	$\cos[\varphi_{A,19} - \varphi_{B,19} + 38(s - 1)\omega t]$

Denoting by $A_{s,+}$ the positive and by $A_{s,-}$ the negative sequence components of stator current sheet

$$A_{s,+} = \sum_{n}^{\infty} A_{\max,n} \cos\left(n\frac{\pi}{\tau_p}x - k_s\omega_s t - \varphi_A\right) \tag{6.84}$$

$$A_{s,-} = \sum_{n}^{\infty} A_{\max,n} \cos\left(n\frac{\pi}{\tau_p}x + k_s\omega_s t - \varphi_A\right) \tag{6.85}$$

and considering only fundamental time components of stator and rotor currents, $k_s = k_r = 1$, one can express torque components generated in a single-phase-fed squirrel cage induction machine at positive rotor mechanical speed in the manner shown in Table 6.10. Only positive sequence components of stator current sheet and rotor air gap flux density can create constant torque components at any speed; otherwise, the pure electromagnetic torque is pulsating. Since the negative sequence component of current sheet rotates relative to the positive sequence component of rotor air gap flux density with twice the synchronous speed, the frequency of pulsations of pure electromagnetic torque between the two is always equal to the twice angular frequency of the stator current.

When the rotor is rotating in negative direction, the positive and negative sequence components of stator current sheet and rotor air gap flux density swap their roles resulting in identical structure of torque components as for positive speed of rotation (see Table 6.11).

6.6 Spectral Components of Torque in a Machine with Uneven Air Gap: Slotting, Salient Poles, and Rotor Eccentricity

When air gap width is not constant, the nth harmonic of air gap flux density contains a component created by the current sheet harmonic of the same order, as well as components which result from interaction of slot harmonics and current sheet harmonics of orders different from n. As shown in Eq. 6.37, a current sheet harmonic in a machine with constant air gap width cannot create a pure electromagnetic torque with an air gap flux harmonic of the same order which it created. Therefore, in a machine with constant air gap width, two independent sources—one for current sheet and the other for flux density—are needed in order for electromagnetic torque to be created.

In a machine with N/p slots per pole pair, the nth harmonic of the positive sequence of current sheet wave and the jth harmonic of the positive sequence of air gap flux density wave generate electromagnetic torque due to slotting (reluctance torque) after Eq. 6.53

$$M_{\mathrm{rel}} = pRl_{\mathrm{ax}}f_{\mathrm{sk},n}A_nB_j\int_0^{2\tau_{\mathrm{p}}}\sin\left[n\frac{\pi}{\tau_{\mathrm{p}}}(x+x_{\mathrm{c}})-\omega t\right]\cos\left[j\frac{\pi}{\tau_{\mathrm{p}}}(x+x_{\mathrm{c}})-\omega t\right]\cos\frac{N}{p}\frac{\pi}{\tau_{\mathrm{p}}}x\mathrm{d}x \tag{6.86}$$

where

$$x_{\mathrm{c}} = R(\Omega t+\gamma) \tag{6.87}$$

Here x_c denotes the position of the zero crossing point of current sheet relative to the centerline of the 1. slot, Ω the rotor mechanical speed at which the two harmonics modulated by slotting synchronize with each other and γ the initial rotor angle.

The definite integral in Eq. 6.86 is equal to

$$\begin{aligned} I = {} & p^2\tau_{\mathrm{p}}\frac{(n-j)\cos[(n-j)p(\Omega t-\gamma)]}{2\pi\left[N^2-(j-n)^2p^2\right]}+p^2\tau_{\mathrm{p}}\frac{(n+j)\cos[(j+n)p(\Omega t-\gamma)+2\omega t]}{2\pi\left[N^2-(j+n)^2p^2\right]}+ \\ & +p\tau_{\mathrm{p}}\frac{\cos\{(n+j)p\gamma-2\pi(N/p+n)+[2\omega-(n+j)p\Omega]t\}}{4\pi[N+(j+n)p]}+ \\ & +p\tau_{\mathrm{p}}\frac{\cos\{(n+j)p\gamma-2\pi(N/p-n)+[2\omega-(n+j)p\Omega]t\}}{4\pi[N-(j+n)p]}+ \\ & +p\tau_{\mathrm{p}}\frac{\cos[(n-j)p(\Omega t-\gamma)]}{4\pi[N+(j-n)p]}+p\tau_{\mathrm{p}}\frac{\cos[(n-j)p(\Omega t-\gamma)]}{4\pi[N-(j-n)p]} \end{aligned} \tag{6.88}$$

and its average value is different from zero only for certain combinations of N, j, n, p, and Ω . This mathematical result can be as well physically interpreted: A 2p-pole machine with N slots generates reluctance torque only at certain mechanical angular speeds Ω as a result of interaction of a given jth harmonic of air gap flux density and nth harmonic of air gap current sheet.

At a mechanical angular speed

$$\Omega = 2\frac{\omega}{N} \tag{6.89}$$

the jth harmonic of current sheet and nth harmonic of air gap flux density create a reluctance torque after Eq. 6.86 if and only if the condition is fulfilled

$$j + n = \frac{N}{p} \tag{6.90}$$

The amount of reluctance torque created at mechanical angular speed Ω is

$$M_{\mathrm{rel}} = \frac{1}{2} V \sum_{j=1,3,5,\ldots}^{\frac{N}{p}-1} A_{\max,j} B_{\max,\frac{N}{p}-j} \sin(N\delta) \tag{6.91}$$

which means that all harmonics of current sheet and air gap flux density the orders of which are below N/p contribute to the reluctance torque at mechanical angular speed of Ω.

In a double-slotted air gap of an induction machine besides stator current sheet and rotor air gap flux density, rotor current sheet and stator air gap flux density create reluctance torque(s). Since the frequency of rotor currents varies proportional to the rotor slip, one can express the rotor mechanical angular speed Ω' at which the reluctance torque due to stator slotting is generated as

$$\Omega' = 2\frac{s\omega}{N'} \tag{6.92}$$

with s denoting the corresponding slip and N' the number of stator slots. In the worst case, the two mechanical speeds in Eqs. 6.89 and 6.92 are equal, which occur at slip

$$s = \frac{N'}{N} \tag{6.93}$$

In order to avoid that slip s in Eq. 6.93 falls into normal operating mode ($s < 1$), the number of stator slots N' should be larger than the number of rotor slots N.

In special case $N/p = 2$ (salient pole synchronous machine), the torque is generated at synchronous speed ($\Omega = \omega$) and its amplitude is proportional to the product of the amplitudes of fundamental harmonic of current sheet and air gap flux density.

Reluctance torque after Eq. 6.91 is also generated at standstill, $\Omega = 0$, in which case the orders of current sheet and air gap flux density harmonics satisfy equation

$$|j - n| = \frac{N}{p} \tag{6.94}$$

When the coil does not rotate, the nth harmonic of flux density created by the jth harmonic of current sheet and N/pth harmonic of air gap slotting are at standstill. The reluctance torque after Eq. 6.91 can be expressed as

$$M_{\mathrm{rel}} = \frac{1}{2} V \sum_{j=1,3,5,\ldots}^{\infty} A_{\max,j} B_{\max,\frac{N}{p}+j} \sin(N\delta) \tag{6.95}$$

In case of **salient pole machine**, $N/p = 2$, the mechanical angular speed Ω after Eq. 6.89 at which a constant reluctance torque is generated equals ω/p. Following Eq. 6.90, only the fundamental spatial harmonic of MMF ($n = 1$) and the fundamental spatial harmonic of air gap flux density ($j = 1$) can create the reluctance torque.

Eccentric air gap is a source of both odd and even flux density harmonics in a machine with windings creating odd MMF harmonics only. Reluctance torque acting on eccentric rotor can be evaluated by using equations derived in this section and inserting for $N = 1$. In particular, the rotor mechanical speed Ω after Eq. 6.89 at which a constant reluctance torque due to eccentricity is developed equals 2ω, i.e., twice the synchronous speed of a two-pole machine. According to Eq. 6.90, only the fundamental harmonic of MMF ($n = 1$) and constant term of flux density ($j = 0$) can generate this reluctance torque.

Case Study 6.1: A 6 kV, 60 Hz, 440 kW, 10-pole, cos $\varphi = 0.815$, 3-phase squirrel cage induction machine has 90 stator and 70 rotor slots. Air gap diameter is 675 mm, and stator has rectangular slots with a width of 12.5 mm and a height of 57 mm. Stator winding has a pitch of $y = 6$. The number of stator slots per pole and phase is equal to $q = 90/(2 \cdot 5 \cdot 3) = 3$ and the slot angle $\alpha_{\mathrm{el}} = 5 \cdot 2\pi/90 = \pi/9$. The ratio between locked rotor and rated current is 5.6 and between maximum and rated torque 3.2. The air gap flux at the mechanical speed of $\Omega = \omega_s /7$ equals to 60 % of the rated value.

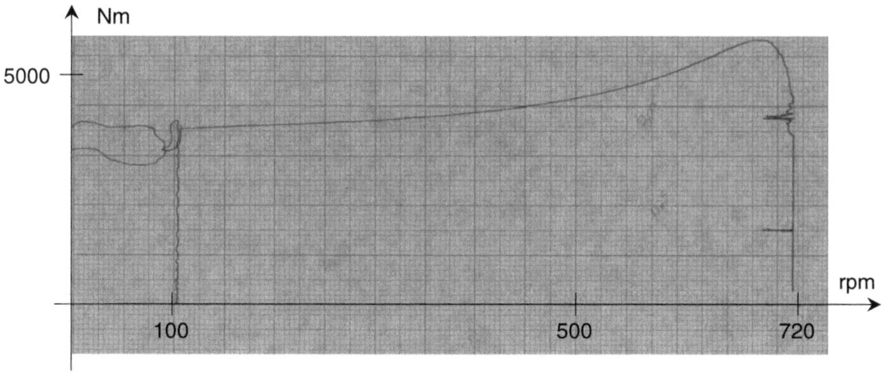

Fig. 6.17 Torque–speed curve of the motor in Case Study 6.1 measured at 50 % of the rated voltage

Table 6.12 Stator winding factors of the machine under study along with relative amplitudes of current sheet and air gap flux density

n	$f_{p,n}$	$f_{z,n}$	$f_{w,n}$	A_n/A_1	B_n/B_1	$A_n\ B_{N/p-n}/A_1\ B_1$ @ Ω
1	0.866	0.96	0.831	1	1	0.794
5	−0.866	0.218	−0.188	0.045	0.12	0.006
7	0.866	−0.177	−0.154	0.026	0.09	0.008
11	−0.866	−0.177	0.154	0.017	0.05	0.01
13	0.866	0.218	0.188	0.017	**0.235**	0.061

Note a substantially increased amplitude of the flux density slot harmonic (n = 13) due to 14 slots per pole pair

Torque–speed curve of the motor recorded during start-up transient is shown in Fig. 6.17. Even unloaded, the motor stalls at a speed of about 103 rpm.

Winding factors along with the amplitudes of current sheet and flux density for the first couple harmonics are evaluated in Table 6.12.

Following Eq. 6.95, the mechanical speed of rotation n_m at which a reluctance torque due to rotor slotting is created can be expressed as

$$n_\mathrm{m} = \frac{30}{\pi}\Omega = \frac{60}{\pi}\frac{\omega}{N} = \frac{60}{\pi}\frac{120\pi}{70} = 102.9\,\mathrm{rpm}$$

The number of rotor slots per pole pair, N/p, is equal to

$$\frac{N}{p} = \frac{70}{5} = 14$$

The ratio of the reluctance torque at 102.9 rpm to rated torque is equal to the sum of all terms in the last column of Table 6.12, i.e.,

$$\frac{M_{\mathrm{rel}}}{M_{\mathrm{rated}}} = 0.879$$

The largest contribution to the reluctance torque comes from the fundamental harmonic of current sheet and the slot harmonic of air gap flux density, the latter being a consequence of the fundamental harmonic of air gap slotting.

Besides reluctance torque, the squirrel cage induction machine in this case study develops pure electromagnetic torques as shown in Table 6.5. The amplitude of the 7. harmonic of stator current sheet equals 2.6 % of the amplitude of the fundamental. At the synchronous speed for the 7. spatial harmonic ω_7

$$\omega_7 = \frac{1}{7}\frac{120\pi}{5} = 10.77\,\mathrm{s}^{-1}$$

and the corresponding rotational speed n_7

$$n_7 = \frac{30}{\pi}\omega_7 = 102.9\,\mathrm{rpm}$$

the machine draws almost the short-circuit current. This helps one estimate the ratio between the maximum motor torque due to 7. harmonic and the rated machine torque as

$$\frac{M_{\mathrm{max},7}}{M_{\mathrm{rated}}} = \frac{M_{\mathrm{max}}}{M_{\mathrm{rated}}}\frac{A_{\mathrm{S.C.},7}}{A_{\mathrm{rated}}}\frac{B_{\mathrm{S.C.},7}}{B_{\mathrm{rated}}} = 3.2 \cdot \frac{0.026 \cdot 5.6}{1}\frac{0.09 \cdot 5.6 \cdot 0.6}{1} = 0.141$$

i.e., 14.1 % of the rated torque. Due to a higher flux level, the maximum generator torque is considerably larger than the maximum motor torque. This makes the total negative torque at and in the vicinity of $n_{\mathrm{m}} = 102.9$ rpm larger than $0.879 + 0.141 = 1.02$ of the rated torque—no wonder that the motor cannot accelerate over the saddle at n_{m}, as recorded in Fig. 6.17.

6.7 Side Effects of Accumulated Magnetic Energy: Radial Air Gap Force, Forces on Conductors in Slots and on Slot Wedges

Both components of electromagnetic torque calculated in the previous sections—the pure electromagnetic and the reluctance torque—add to each other and act in the shaft direction, in the following equation:

$$\vec{M} = \vec{r} \times \vec{F} \tag{6.96}$$

Electromagnetic torque, as a sum of all products of r and F on the air gap circumference, is therefore a global quantity. For this reason, there exists no possibility to access from outside a particular component of torque created by a single pole in a 2p-pole machine.

Besides electromagnetic torque, which is the useful quantity generated in an electric machine, the accumulated magnetic energy creates forces which do not contribute to the electromechanical energy conversion and which are typically a source of unwanted effects, such as vibrations and electromagnetic noise. A typical side effect of air gap flux density is the attractive force between stator and rotor cylindrical surfaces, as in Eq. 6.21. Since the attractive force F_r is proportional to the square of the amplitude of air gap flux density, one can write for the radial air gap force created by the positive sequence component of air gap flux density

$$F_{r,+}(x,z,t) \sim \left\{ \sum_{n=1,3,5,\ldots}^{\infty} \sum_{k=1,2,\ldots}^{\infty} I_k \cos\left[n\frac{\pi}{\tau_p}\left(x - \beta_{sk} R \frac{z}{l_{ax}}\right) - k\omega t \right] \right\}^2 \tag{6.97}$$

As opposed to the electromagnetic torque, radial air gap force is a local quantity, the effects of which act on the particular machine section independent of the amount of radial air gap force at some other place in the machine. Total electromagnetic torque acting on the shaft is equal to the algebraic sum (integral) of torque components over the whole air gap circumference. On the other hand, a vector sum of radial forces along the whole air gap circumference might be equal to zero, which does not mean that there is no locally acting radial force.

Considering only the fundamental time harmonic component of current and the fundamental and the slot harmonic of air gap flux density, one can rewrite Eq. 6.97 as

$$\begin{aligned} F_{r,+}(x,z,t) \sim \Bigg\{ & B_1 \cos\left[\frac{\pi}{\tau_p}\left(x - \beta_{sk} R \frac{z}{l_{ax}}\right) - \omega t \right] \\ & + B_{\frac{N}{p}-1} \cos\left[\left(\frac{N}{p} - 1\right) \frac{\pi}{\tau_p}\left(x - \beta_{sk} R \frac{z}{l_{ax}}\right) - \omega t \right] \Bigg\}^2 \end{aligned} \tag{6.98}$$

or

$$\begin{aligned} F_{r,+}(x,z,t) \sim {} & B_1^2 \cos^2\left[\frac{\pi}{\tau_p}\left(x - \beta_{sk} R \frac{z}{l_{ax}}\right) - \omega t \right] + \\ & + 2 B_1 B_{\frac{N}{p}-1} \cos\left[\frac{\pi}{\tau_p}\left(x - \beta_{sk} R \frac{z}{l_{ax}}\right) - \omega t \right] \cos\left[\left(\frac{N}{p} - 1\right) \frac{\pi}{\tau_p}\left(x - \beta_{sk} R \frac{z}{l_{ax}}\right) - \omega t \right] \end{aligned} \tag{6.99}$$

for

$$B_{\frac{N}{p}-1} \ll B_1 \tag{6.100}$$

Radial force after Eq. 6.99 has a constant term proportional to

$$F_{\mathrm{r,const}} \sim \frac{1}{2}B_1^2 + B_1 B_{\frac{N}{p}-1} \cos\left[\left(\frac{N}{p} - 2\right)\frac{\pi}{\tau_{\mathrm{p}}}\left(x - \beta_{\mathrm{sk}} R \frac{z}{l_{\mathrm{ax}}}\right)\right] \tag{6.101}$$

along with a two-dimensional traveling wave components

$$\begin{aligned} F_{\mathrm{r,wave}} \sim \frac{1}{2}B_1^2 \cos\left[2\frac{\pi}{\tau_{\mathrm{p}}}\left(x - \beta_{\mathrm{sk}} R \frac{z}{l_{\mathrm{ax}}}\right) - 2\omega t\right] \\ + B_1 B_{\frac{N}{p}-1} \cos\left[\frac{N}{p}\frac{\pi}{\tau_{\mathrm{p}}}\left(x - \beta_{\mathrm{sk}} R \frac{z}{l_{\mathrm{ax}}}\right) - 2\omega t\right] \end{aligned} \tag{6.102}$$

the velocities of which in circumferential direction are

$$v_{x,1} = \frac{\omega \tau_{\mathrm{p}}}{\pi}; \quad v_{x,\frac{N}{p}} = 2\frac{p}{N}\frac{\omega \tau_{\mathrm{p}}}{\pi} \tag{6.103}$$

and in axial direction

$$v_{z,1} = \frac{\tau_{\mathrm{p}} \omega l_{\mathrm{ax}}}{\beta_{\mathrm{sk}} R \pi}; \quad v_{z,\frac{N}{p}} = 2\frac{p}{N}\frac{\omega \tau_{\mathrm{p}} l_{\mathrm{ax}}}{\pi \beta_{\mathrm{sk}} R} \tag{6.104}$$

as long as $\beta_{\mathrm{sk}} \neq 0$. The effects of skewing, represented here by the skewing angle β_{sk}, are illustrated in Fig. 6.18, in which radial force waves due to slotting in a 2-pole machine with 24 slots are shown.

If the slots are not skewed, as in Fig. 6.18a, the same amount of force acts in axial direction for a given circumferential coordinate along the whole active part length.

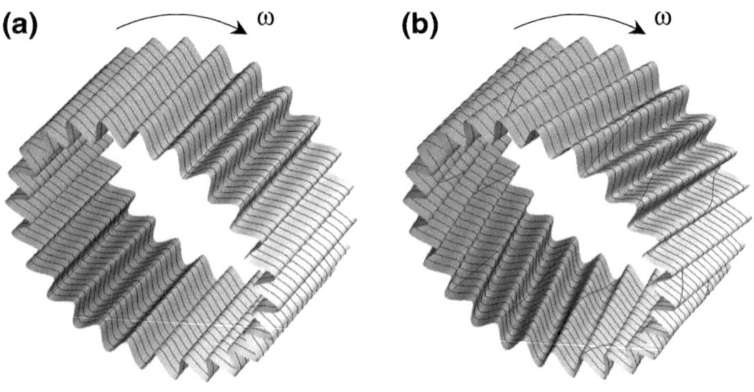

Fig. 6.18 Radial air gap force waves due to slotting in a 2-pole machine with 24 slots: **a** unskewed and **b** skewed for one slot

If the slots are skewed for one slot, as in Fig. 6.18b, the amount of force varies axially along the active part resulting in zero total force for any circumferential coordinate. The effect of slot skewing on radial force is similar to the effect of skewing gears—the operation is smoother and quieter than in a machine with unskewed slots.

Radial air gap forces are always positive, i.e., they tend to decrease the air gap width. Considering mechanically stiff inner structure, radial forces excite mechanically the outer structure of the machine which, as a result, begins to oscillate at various spatial modi [3], as in Fig. 6.19. At each spatial mode, a frequency spectrum of vibrations is generated, which act on the machine fundaments and create electromagnetic noise.

Effects of air gap radial forces can be very strong in machines with thin outer structures, such as high polarity outer rotor permanent magnet machines, characterized by large diameters and extremely thin yokes.

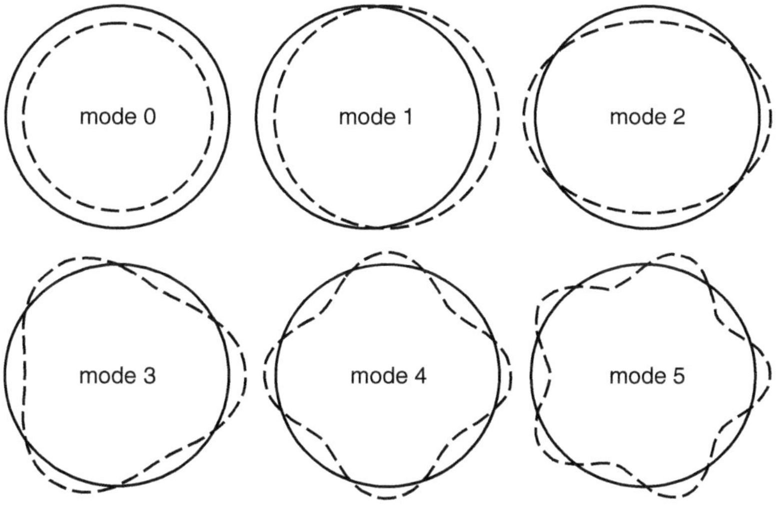

Fig. 6.19 Vibration modi of the outer cylinder of an electric machine produced by air gap radial forces

Case Study 6.2: A 690 V, 2.8 MW, 96-pole, 3-phase outer rotor permanent magnet synchronous generator has 252 slots and a rated speed of 10.2 rpm.

Due to its large outer diameter of 4.8 m and thin yoke (50 mm), the rotor has a structure known as "beer can." Rotor magnets are skewed for 1/8 of the stator slot pitch, following recommendation in [4–6]. During operation at rated speed, tonal sounds at frequencies of 42.8 and 85.6 Hz were measured, the intensity of which exceeded 100 dB (Fig. 6.20). In order to operate in accordance with stringent acoustic emission standards, the machine had to be redesigned and noise level decreased. By applying proper skewing of 1 stator slot pitch, instead of only 1/8, the sound level due to cogging torque was suppressed down to 80 dB, as in Fig. 6.20.

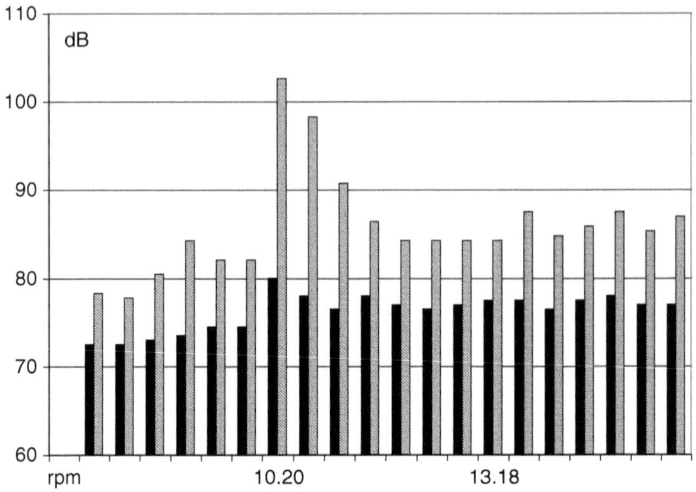

Fig. 6.20 Measured intensity of sound radiated by a 2.8-MW PM wind generator with outer rotor: original magnet skewing of 1/8 of the stator slot pitch after [4–6] (*shaded bars*) and for magnet skewing of 1 stator slot (*black bars*)

6.7.1 Unbalanced Magnetic Pull Caused by Rotor Eccentricity

As shown in Chap. 2, the spectrum of air gap flux density in a machine with eccentric rotor contains besides homopolar flux both even and odd harmonics. Accordingly, the air gap (attractive) force after Eq. 6.97 created only by the fundamental time harmonic of current can be written for a machine with eccentric rotor as

$$F_{\mathrm{r},+}(x,z,t) \sim \left\{ \sum_{n=0,1,2,\ldots}^{\infty} B_n \cos\left[n\frac{\pi}{\tau_\mathrm{p}}\left(x - x_\mathrm{r} - \beta_\mathrm{sk} R \frac{z}{l_\mathrm{ax}} \right) - \omega t \right] \right\}^2 \tag{6.105}$$

Considering the dependence of the amplitudes of flux density on the rotor shift coordinate expressed in Eqs. 2.73–2.74, one can rewrite Eq. 6.104 for a two-pole machine as

$$F_{\mathrm{r},+}(x,z,t) \sim \left\{ B_{0,\max} \cos\frac{\pi}{\tau_\mathrm{p}} x_r + \sum_{n=1,2,\ldots}^{\infty} \left(B_{n,\mathrm{const}} + B_{n,\max} \cos\left(2\frac{\pi}{\tau_\mathrm{p}} x_\mathrm{r} \pm \pi \right) \right) \cos\left[n\frac{\pi}{\tau_\mathrm{p}}\left(x - x_\mathrm{r} - \beta_\mathrm{sk} R \frac{z}{l_\mathrm{ax}} \right) - \omega t \right] \right\}^2 \tag{6.106}$$

Radial force in Eq. 6.106 has a time constant term, the components of which are spatial functions of the rotor shift coordinate x_r, and terms which pulsate with twice the frequency ω, the amplitudes of which are spatial functions of the rotor shift coordinate x_r. The amplitude of resulting radial force in Eq. 6.106 is highest for rotor shift $x_r = 0$, i.e., at the point of minimum air gap width. Since the air gap attractive force is a function of rotor shift x_r, it is often called *unbalanced magnetic pull.*

6.7.2 Radial Forces on Conductors in Slots

Current-carrying conductors in slots of electric machines create leakage flux, the flux lines of which spread tangentially through slots and build radial forces along with current-carrying conductors. Since conductor currents are periodical functions of time, radial forces acting on conductors in slots pulsate with twice the frequency of current. If the conductors are not firmly fixed within slots, the permanently acting pulsating forces can move conductors relative to each other and to the slot wedges and slot insulation, which ultimately causes destruction of insulation. In case of rigid conductors, as in squirrel cage induction machines, the vibrations caused by radial forces on bars can result in their complete destruction.

Consider now a rectangular slot after Fig. 6.21 with two conductors one above another. The lower conductor carries current I_l and the upper I_u. The upper conductor consists of n parallel connected strands, each of which carries current I_u/n.

Magnetic field strength H_j in the jth strand, $j = 1, \ldots, n$ is equal to

$$H_j = \frac{1}{w}\left(I_l + I_u\frac{j-1}{n}\right) \tag{6.107}$$

and the corresponding force on the jth strand:

$$H_j = \frac{\mu_0 l_{ax}}{w}\left[\frac{I_l I_u}{n} + \left(\frac{I_u}{n}\right)^2 (j-1)\right] \tag{6.108}$$

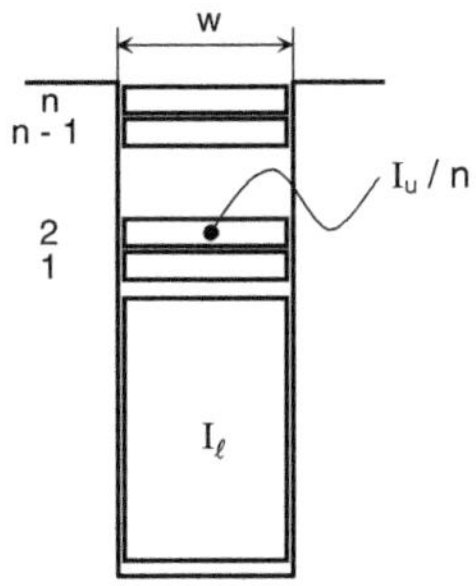

Fig. 6.21 Illustrating radial force on a conductor in a slot

Total force on the upper conductor is equal to the sum of forces acting on its n strands:

$$F_{\mathrm{u}} = \frac{\mu_0 l_{\mathrm{ax}}}{w} I_{\mathrm{u}} \left(I_{\mathrm{l}} + I_{\mathrm{u}} \frac{n-1}{2n} \right) \tag{6.109}$$

If the lower conductor consists of m strands, the total force on it is equal to

$$F_{\mathrm{l}} = \frac{1}{2} \frac{\mu_0 l_{\mathrm{ax}}}{w} I_{\mathrm{l}}^2 \frac{m-1}{m} \tag{6.110}$$

The force on the lower conductor F_ℓ acts always toward slot bottom. The direction of force on the upper conductor F_{u} depends on signs of currents I_{u} and I_ℓ.

In a 3-phase machine, the currents I_{u} and I_ℓ can belong to the same phase (mono slots), or to different phases (mixed slots). The force on the upper conductor in a mono slot acts always in the direction of slot bottom; for the force exerted on the upper conductor in a mixed slot one can write

$$f_{\mathrm{u}}(t) = \frac{\mu_0 l_{\mathrm{ax}}}{w} I^2 \cos \omega t \left[-\cos\left(\omega t - \frac{2\pi}{3} \right) + \frac{1}{2} \cos \omega t \right] \tag{6.111}$$

The extreme values of force after Eq. 6.111 are

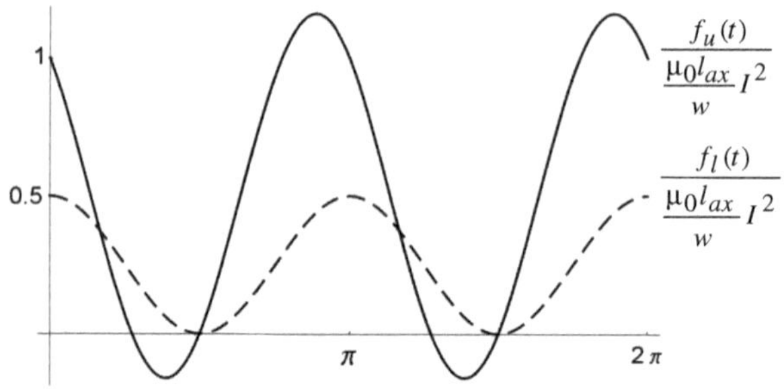

Fig. 6.22 Radial forces on conductors in a mixed slot

$$F_{\max} = \frac{2+\sqrt{7}}{4} \frac{\mu_0 l_{\mathrm{ax}}}{w} I^2; \quad F_{\min} = \frac{2-\sqrt{7}}{4} \frac{\mu_0 l_{\mathrm{ax}}}{w} I^2 \tag{6.112}$$

The force $F_{\min}$ is negative, which means that it acts against the slot wedge, i.e., towards the air gap.

Forces acting on upper and lower conductors in a slot during one period of conductor currents are shown in Fig. 6.22.

6.8 Forces on Conductors in End Winding

Magnetic field in the end winding zone has all three spatial components, created both by the stator and by the rotor windings. Currents in conductors in the end winding zone change their directions as a function of the circumferential coordinate. Total flux density acting on a conductor segment is not only dependent on stator and rotor currents, but also on their mirror images created by the action of stator and rotor iron.

Since the medium in the end winding zone is linear, field distribution can be found by applying Biot–Savart law, as in Fig. 6.23.

The computational procedure starts with evaluation of all 3 spatial components of magnetic field at a given time instant at points within the conductors of end winding as a function of all actual currents in the stator and rotor end winding and of all mirrored currents. In the next step, the three spatial components of forces acting on conductor segments are found (Fig. 6.24).

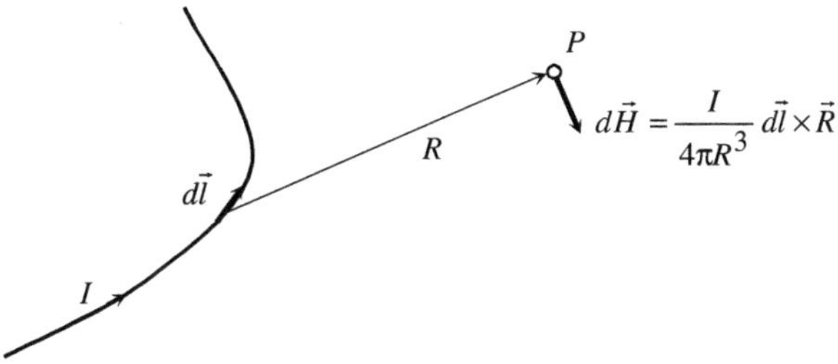

Fig. 6.23 Magnetic field created by current I in segment dl

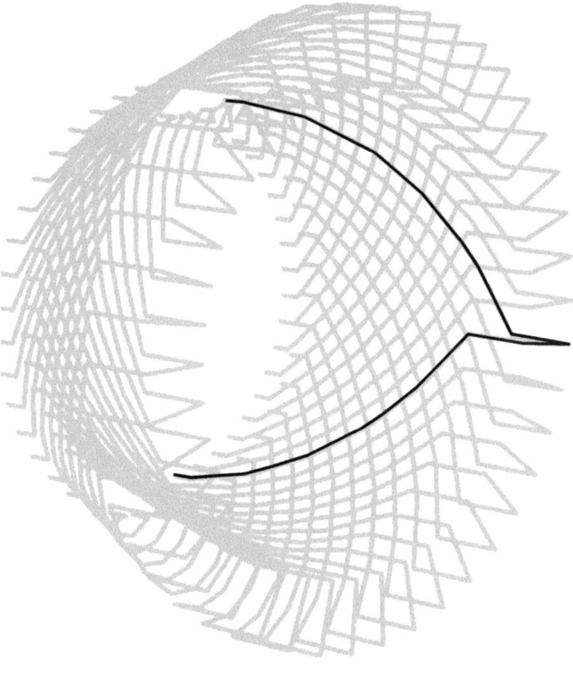

Fig. 6.24 Coils in the stator end winding

6.9 Torque as a Function of Terminal Quantities

One of the most pronounced properties of energy is that it can be stored. Independent of the medium in which it is stored and the form in which it is stored, energy is time invariant. From this point of view, the product $u \cdot i \cdot t$, usually called *electric energy*, cannot be considered energy, but *work* performed by current i and voltage u during time interval t. The dependence of the product $u \cdot i \cdot t$ on time is the reason why this quantity cannot be stored, i.e., left aside for a while and then used as required.

Having this in mind, one can analyze the electromechanical energy conversion in terms of electrical work necessary to change the amount of stored energy in an electric machine, to perform mechanical work, and to cover losses in it [1]

$$\sum_{j=1}^{m} u_j i_j \mathrm{d}t = F\mathrm{d}x + \mathrm{d}W_{\mathrm{mg}} \tag{6.113}$$

Neglecting i^2R losses, one can write

$$\sum_{j=1}^{m} i_j \mathrm{d}\psi_j = F\mathrm{d}x + \mathrm{d}W_{\mathrm{mg}} \tag{6.114}$$

where

$$\mathrm{d}W_{\mathrm{mg}} = \frac{\partial W_{\mathrm{mg}}}{\partial x}\mathrm{d}x + \sum_{j=1}^{m} \frac{\partial W_{\mathrm{mg}}}{\partial \Psi_j}\mathrm{d}\Psi_j \tag{6.115}$$

By substituting Eqs. 6.115 in 6.114, one obtains

$$F = -\frac{\partial W_{\mathrm{mg}}}{\partial x} + \left(\sum_{j=1}^{m} i_j - \sum_{j=1}^{m} \frac{\partial W_{\mathrm{mg}}}{\partial \Psi_j}\right)\frac{\mathrm{d}\Psi_j}{\mathrm{d}x} \tag{6.116}$$

The terms in brackets in Eq. 6.116 multiply quotients of state variables $d\Psi_j/dx$. Since state variables are orthogonal, i.e., independent of each other, one can write

$$\frac{d\Psi_j}{dx} = 0; \quad j = 1, 2, \ldots, m \tag{6.117}$$

Now, one can express the electromagnetic force F in Eq. 6.116 as

$$F = -\frac{\partial W_{mg}}{\partial x} \tag{6.118}$$

Magnetic energy W_{mg} is accumulated in air gap (W_δ), iron core (W_{Fe}), and leakage paths (W_σ) of an electric machine

$$W_{mg} = W_\delta + W_{Fe} + W_\sigma \tag{6.119}$$

When evaluating the electromagnetic force after Eq. 6.118, the stored magnetic energy has to be partially differentiated w.r.t. the circumferential coordinate x, which means that all other state variables (fluxes Ψ_j) have to be kept constant. The magnetic energy W_{Fe} stored in iron is dependent only on the flux level in iron. As long as the flux in iron is kept constant, the accumulated energy W_{Fe} remains unchanged, no matter how the circumferential coordinate (rotor shift) x changes. The same is valid for the accumulated energy in leakage paths

$$\frac{\partial W_{Fe}}{\partial x} = 0; \quad \frac{\partial W_\sigma}{\partial x} = 0 \tag{6.120}$$

Accordingly, one can state that *electromagnetic force is equal to the partial derivative of accumulated magnetic energy in air gap w.r.t. rotor circumferential coordinate x*

$$F = -\frac{\partial W_\delta}{\partial x} \tag{6.121}$$

Similarly, one can write for the electromagnetic torque

$$M = -\frac{\partial W_\delta}{\partial \gamma} \tag{6.122}$$

With γ denoting rotor angle, $\gamma = x_r/R$, with R being the air gap diameter and x_r the rotor shift. Substituting for accumulated magnetic energy in the air gap

$$W_\delta = \int_V H_\delta \mathrm{d}B_\delta = \frac{\mu_0 l_{ax}}{\delta} \int_0^{2R\pi} \Theta_\delta^2 \mathrm{d}x \tag{6.123}$$

where Θ_δ denotes the resulting MMF drop across the air gap, and one can write for the torque

$$M = -R\frac{\mathrm{d}}{\mathrm{d}x_r} W_\delta = -R\frac{\mu_0 l_{ax}}{\delta}\frac{\mathrm{d}}{\mathrm{d}x_r} \int_0^{2R\pi} \Theta_\delta^2 \mathrm{d}x \tag{6.124}$$

Case Study 6.3: A 138 MVA, 10.75 kV, 12.8 kA, 2-pole, 1000 rpm, cos $\varphi = 0.8$, single-phase cylindrical rotor synchronous generator has 142 stator and 52 rotor slots. Single-layer stator winding with 2 parallel circuits and one turn per coil is placed in 48 slots per pole. Rated field current equals to 1330 A. Out of 52 rotor slots, 40 are wound, each of which carrying 11 conductors. Winding distributions are shown in Fig. 6.25. Rotor diameter equals 1550 mm and length 5625 mm.

Generator air gap width is 50 mm. Stator and rotor MMF distributions are presented in Fig. 6.26. Corresponding stator pitch factor is 0.798 and rotor pitch factor of 0.774. Generator no load curve is shown in Fig. 6.27. At rated point, the MMF drop across air gap equals to 38 % of the total winding MMF.

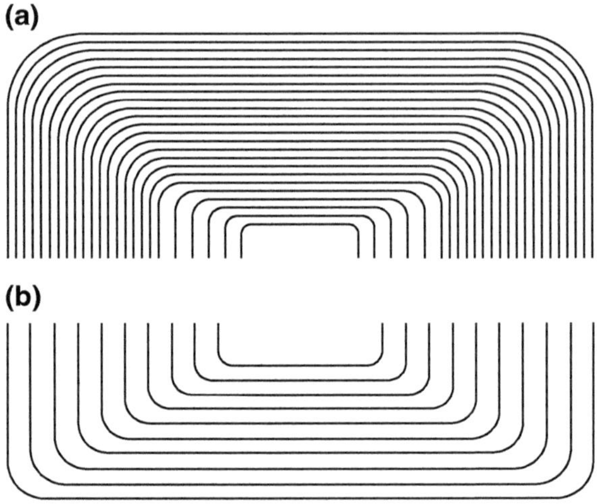

Fig. 6.25 Stator (**a**) and rotor (**b**) winding distribution of the generator in Case Study 6.3

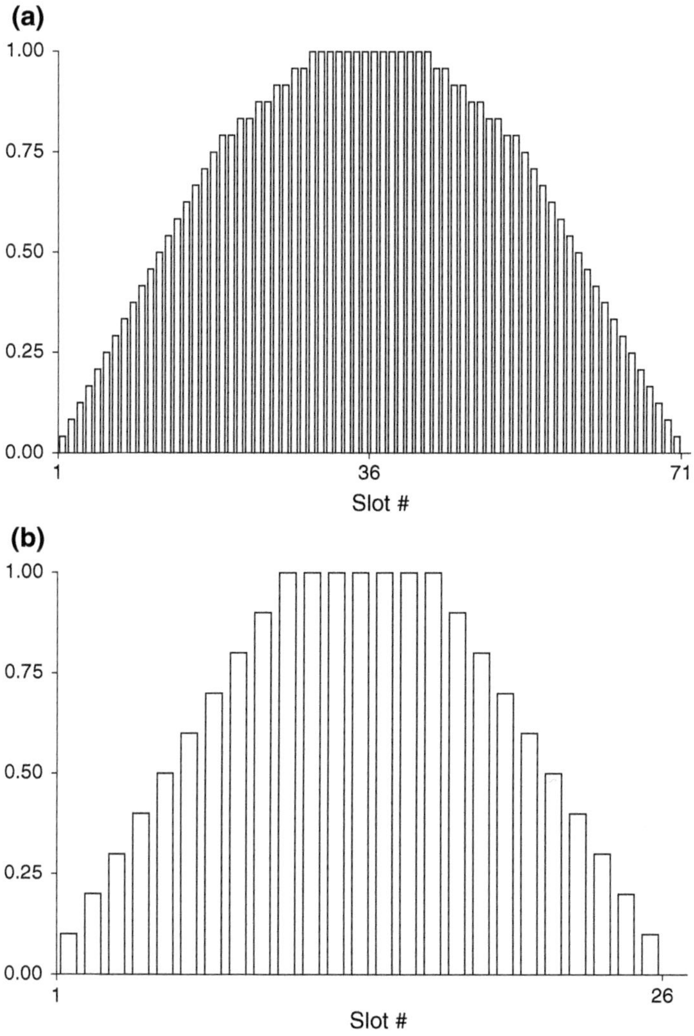

Fig. 6.26 Stator (**a**) and rotor (**b**) MMF distributions in p.u. created by the windings in Fig. 6.25

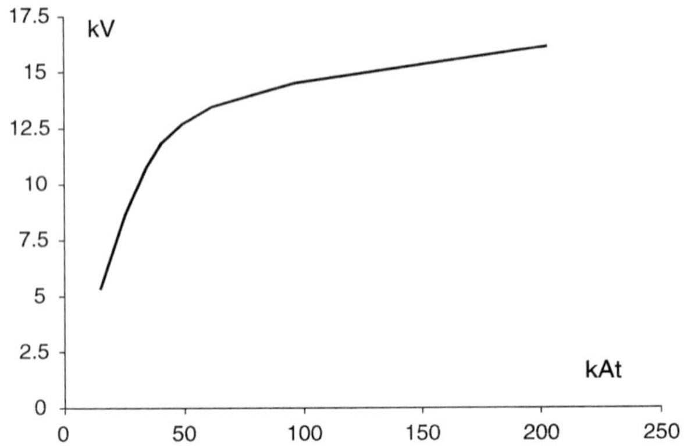

Fig. 6.27 No load curve

Torque created by the fundamental stator MMF wave

$$\Theta_{s,1} = \Theta_{max,s,1} \cos\frac{\pi}{\tau_p} x \sin(\omega t - \varphi)$$

with φ denoting the phase shift between stator voltage and current, and by the fundamental rotor MMF wave

$$\Theta_{r,1} = \Theta_{max,r,1} \cos\frac{\pi}{\tau_p}(x - x_r)$$

with x_r standing for the rotor shift

$$x_r = R\omega_r t$$

and ω_r for the rotor angular speed, can be expressed by using Eq. 6.124 as

$$M = -R\frac{\mu_0 l_{ax}}{\delta}\frac{d}{dx_r}\int_0^{2R\pi}\left[\Theta_{max,s,1}\cos\frac{\pi}{\tau_p}x\sin(\omega t - \varphi) + \Theta_{max,r,1}\cos\frac{\pi}{\tau_p}(x - x_r)\right]^2 dx$$

At the rotor speed of $\omega_r = \omega/p$, the torque can be written as

$$M = \frac{1}{2}Rp\pi\frac{\mu_0 l_{ax}}{\delta}\Theta_{max,s,1}\Theta_{max,r,1}[\cos\varphi - \cos(\varphi - 2\omega t)]$$

and, finally,

$$M = p\frac{I_{max,s,1}w_s f_{w,s} I_{max,r,1} w_r f_{w,r}}{R_{mg,gap,pole}}[\cos\varphi - \cos(\varphi - 2\omega t)]$$

where (Eq. 4.28)

$$R_{mg,gap,pole} = \frac{\pi^2}{16p}\frac{\delta}{\mu_0 l_{ax}\tau_p}$$

After inserting machine dimensions, one obtains

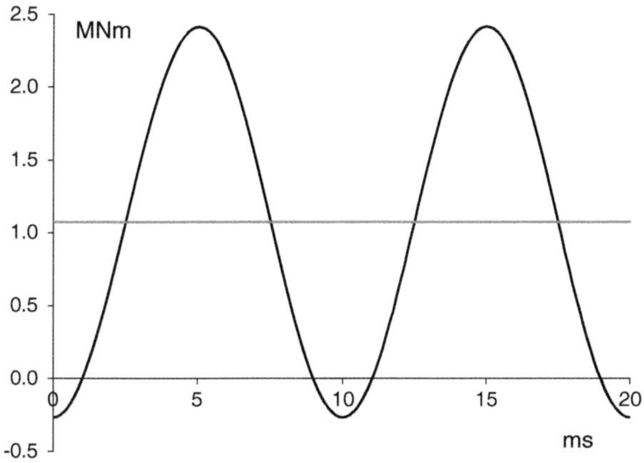

Fig. 6.28 Torque at the rated point of the single-phase synchronous generator 138 MVA, 10.75 kV, 12.8 kA, 2-pole, 1000 rpm, cos φ = 0.8

$$R_{mg,gap,pole} = \frac{\pi^2}{16}\frac{0.05}{\mu_0 \cdot 5.625 \cdot 1.55\pi/2} = 1792H^{-1}$$

and

$$M = \frac{\left(0.35 \cdot 12800 \cdot \sqrt{2}/2\right) \cdot 24 \cdot 0.798 \cdot 0.35 \cdot 1330 \cdot 110 \cdot 0.774}{1792}[0.8 - \cos(\varphi - 2\omega t)]$$

or

$$M = 1.34 \cdot [0.8 - \cos(\varphi - 2\omega t)]\ [\mathrm{MNm}]$$

At rated point, the torque oscillates with twice the mains frequency around the average value of 1.34 · 0.8 = 1.072 MNm. The amplitude of torque oscillations equals to 1.34 MNm, Fig. 6.28.

6.10 A Method for Direct Measurement of Electromagnetic Torque in Large Synchronous Machines

As opposed to the mechanical power, the shaft torque is not a direct function of the rotor speed. Based upon the fundamental equation for electromagnetic force $F = \mathrm{B} \cdot \ell \cdot I$, one concludes that the torque is determined only by the flux level in a machine and currents in its windings. Following this conclusion, one can state that the widely spread method for torque measurement in large electric machines, the so-called back-to-back test [7], becomes superfluous for synchronous and wound rotor induction machines. The back-to-back test requires two identical machines connected electrically and mechanically in such a manner that the total power consumption is used to cover the losses in the system, although each machine operates at its rated point. This is possible because one machine is run as a motor and the other as a generator, as in Fig. 6.29a.

Very often, especially when building a prototype of a large machine, e.g., a PM generator, only one unit is manufactured. Nevertheless, the torque–angle curve of such machine can be recorded by means of the connection in Fig. 6.29b.

The rotor of the tested machine is firmly hold by means of a lever on the other end of which a strain gauge-type force sensor is mounted. Relative position between the (stator) rotating field and (rotor) flux density (load angle) is set by a proper combination of (stator) DC. This way, the stator air gap flux density distribution is fixed in space. The torque is dependent on the amount of all three (stator) currents, which determine both the amount and position of air gap MMF and the amount of the field current i_F. The machine is force-cooled in order to keep the windings temperature at a given level.

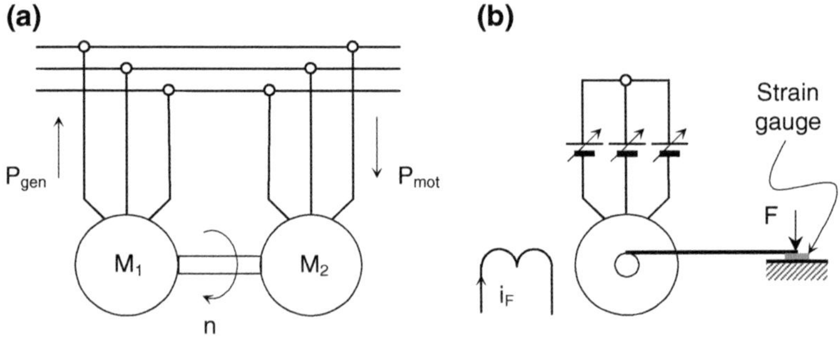

Fig. 6.29 Back-to-back connection of two identical electric machines M_1 and M_2 (**a**) and the torque measurement at standstill of a single machine (**b**)

References

1. Ostovic V (1989) Dynamics of saturated electric machines. Springer, New York
2. Ostovic V (1994) Computer-Aided Analysis of Electric Machines: a Mathematica Approach. Prentice-Hall, London
3. Ostovic V, Boman G (1995) Radial air gap force as a source of audible noise in a sinusoidally fed induction machine. In: Industry applications conference. Conference record of the thirtieth IAS annual meeting, IAS, vol 95, no 1, pp 591–598
4. Zhu ZQ, Howe D (2000) Influence of Design parameters on cogging torque in permanent magnet machines. IEEE Trans Energy Convers 15:407–412
5. Jahns TM, Soong WL (1996) Pulstaing torque minimization techniques for permanent magnet A.C. motor drives—a review. IEEE Trans Power Electron 43:321–330
6. De La Ree J, Boules N (1989) torque production in permanent magnet synchronous motors. IEEE Trans Ind Appl 25:101–112
7. IEC 60034-2: 1996

Chapter 7
Thermal Design of Rotating Field Electric Machines

Contents

Electromechanical energy conversion in an electric machine is accompanied by losses, which increase its temperature. In order to secure smooth operation of a machine, the temperature increase in it has to be kept under certain level, i.e., the losses have to be taken out from the machine. Therefore, cooling of a machine is an inevitable component of its design procedure. Not only the winding and lamination temperatures must stay under allowed level but also inadequate heating at the rated point means poor machine utilization and, ultimately, a too expensive machine.

Besides its reactances, heating is the dominating factor which limits the rated power of an electric machine. Substantial increase of rated power is possible only by applying direct conductor cooling either with gas or fluids (oil and water). An improvement of the machine's torque to volume ratio is possible only with a more efficient machine cooling.

V. Ostović, *The Art and Science of Rotating Field Machines Design*,
DOI 10.1007/978-3-319-39081-9_7

Rotating electric machines are cooled predominantly by the ambient air, primarily because of its availability and simplicity of air-based cooling systems. Besides air, hydrogen, water, and oil are used a coolants in electric machines.

7.1 Types of Cooling

Heat can be taken out from an electric machine in numerous ways, depending on machine size, choice of cooling medium, level of losses, price of the cooling system, etc. As discussed in Introduction, the larger the machine, the more complex its cooling system. The most common types of cooling are

Natural cooling: The machine has no fan; the heat is taken from it by means of free air flow and radiation from its housing. This type of cooling is common for small and micro-machines.

Self-cooling: The machine has its own fan(s). The amount of cooling gas is dependent on the rotor speed. It is suitable for constant speed machines up to the highest power ratings.

External cooling: The machine is cooled by separate fan(s), or by means of cooling medium other than gas. It is used for machines operating at variable speed, or high torque density machines.

Small-size machines are usually cooled on their outer surface, whereas medium- and large-size machines utilize air/gas flowing through its active part.

Open cooling circuit is used in small- and medium-size machines. In an open cooling system the ambient air directly cools the machine heat exchange surfaces. The airborne noise caused by large amounts of cooling air blowing through the machine along with difficulties related to filtering and demoisturizing of the permanently fresh air is the main reason why the open cooling circuit is not used in large machines.

Closed cooling system requires heat exchanger in which the warm gas from the machine transfers its heat to another cooling medium, typically water. The machine is totally enclosed; air or hydrogen are used as primary coolers. Cooling gas takes the heat out from conductors either **directly** (hollow conductors) or **indirectly** by blowing over the conductor insulated outer surfaces.

Closed cooling systems are sometimes combined with **liquid cooling** (oil or water). The coolant usually flows through hollow conductors, or cooling pipes built into the lamination.

Air-cooled machines are characterized by carefully designed cooling circuit in which the air pressure creates optimally distributed air velocities. The availability of cooling air on heat exchange surfaces determines thermal performance of the machine and its overall characteristics. Cooling air paths within the machine are connected in series and in parallel, and the fan is placed either on the inlet (pressure) or outlet (suction) portion of the cooling system. Air pressure in a machine with

suction cooling is lower than the ambient air pressure. In a **pressure**-based cooling, the cooling air is pre-heated by losses created by the fan.

Hydrogen-cooled machines are totally enclosed in order to prevent hydrogen leakage and mixture with air, which could create highly explosive oxyhydrogen gas. Hydrogen as cooling medium, especially at couple of ata overpressure, has a significantly higher heat transfer coefficient and a better thermal conductivity than air.

Cooling gas is set in motion by fan(s). In **axial fans**, the gas inlet and outlet are axial, and the pressure increase is relatively low. In **radial fans**, the gas inlet is axial, whereas the outlet is radial. Radial fans, especially in combination with a **diffusor,** can create a couple of ata pressure difference, necessary for hydrogen cooling.

7.2 Rated Torque and Rated Power

In the previous chapter the dependence of electromagnetic torque on machine geometry, material properties, and current/fluxes was derived. It was shown that the electromagnetic torque has components which are either dependent on stator to rotor angle, or which are constant. The electromagnetic torque was derived from accumulated magnetic energy, which is only one of various energy forms flowing between machine electrical and mechanical terminals.

Another important energy form present in electric machines is losses, which are converted directly into heat. Although losses play a negligible role when determining machine rated torque, they are crucial when defining the machine rated power. In conventional electric machines, the energy of losses is significantly smaller than the energy converted from mechanical to electrical form and vice versa. Nevertheless, the rated power of a machine is determined only by dissipated losses, particularly by that portion of total losses which remain in the machine and increase its temperature. According to international standards, the maximum temperature ϑ_{max} of machine's winding may not exceed the amount determined by its class of insulation. Given temperature of environment ϑ_0, one has the window with the width $\vartheta_{max} - \vartheta_0$ to fill with temperature increase $\Delta\vartheta$ after Eq. 1.20.

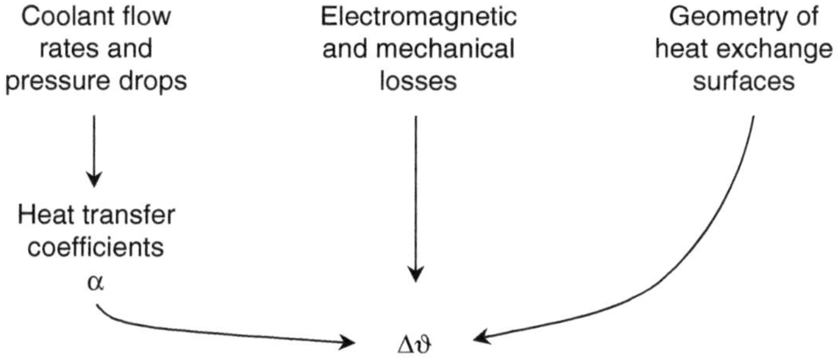

Fig. 7.1 Schematical representation of temperature increase computational procedure

The amount of temperature increase $\Delta\vartheta$ is defined by electrical, fluid dynamical, and geometrical quantities, as visualized in Fig. 7.1. The crucial quantities thereby are the heat transfer coefficients α at machine heat exchange surfaces, as a measure of cooling efficiency: The higher the value of α, the larger the amount of heat that can be taken from a heat exchange surface. Heat transfer coefficient α is a function of the type of coolant (air, hydrogen, water, oil, etc.) and coolant velocity [1].

7.3 Hydraulic Resistances and Fan Curves

Hydraulic properties of cooling ducts and channels in an electric machine can be dealt in terms of sources and resistances and solved by applying rules known from electric circuits in the same manner as in the case of magnetic or thermal networks. The result of such solution procedure is the distribution of coolant velocities in the machine, which are necessary when calculating the heat transfer coefficients.

Hydraulic resistance R_h [kg/m^{4}s] of a portion of machine with length ℓ [m], cross-sectional area S [m^2] and hydraulic diameter d_h [m] is defined by means of pressure drop Δp [Pa] across it and volumetric flow rate Q [m^3/s] through it as

$$R_\mathrm{h} = \frac{\Delta p}{Q} \tag{7.1}$$

where

$$\Delta p = \lambda \frac{l}{d_\mathrm{h}} \frac{Q^2}{S^2} \frac{\rho}{2} + \sum_i \zeta_i \frac{Q^2}{S^2} \frac{\rho}{2} \tag{7.2}$$

with λ denoting the dimensionless *friction factor*, ζ_i the dimensionless *pressure loss coefficient*, and ρ the coolant density [kg/m^3]. Note that the coolant velocity v [m/s] can be expressed as

$$v = \frac{Q}{S} \tag{7.3}$$

The friction factor λ is a function of the roughness of the heat exchange surface and can be expressed by means of transcendental equation [2] as

$$\frac{1}{\sqrt{\lambda}} = -2\ln\left(\frac{2.51}{\mathrm{Re}\sqrt{\lambda}} + \frac{k}{3.72 d_\mathrm{h}}\right) \tag{7.4}$$

with Re standing for the Reynolds number and k for the coefficient of roughness [m]. Since the values of k are far below those typical for hydraulic diameters [3], one may use instead of Eq. 7.4 the definition [4] for λ:

$$\lambda = \frac{0.316}{\sqrt[4]{\text{Re}}} \tag{7.5}$$

The pressure loss coefficient ζ_i of the ith component of hydraulic resistance reflects an increase of pressure drop due to geometry changes or coolant redirection within the resistance of the length ℓ. The values for λ and ζ_i for a particular geometry can be found in [2–7]. Here friction factors and pressure loss coefficients will be discussed, which are characteristic for electric machines geometry, in particular for constructions typical for medium- and large-size electric machines, where cooling plays a more important role than in case of small machines.

7.3.1 Friction Factor for Coolant Expanding in Axial Direction Through Air Gap

Consider coolant inlet into air gap of a rotating electric machine characterized by axial component of velocity only. After entering the air gap, the tangential component of coolant velocity increases, following distribution along radial direction sketched in Fig. 7.2.

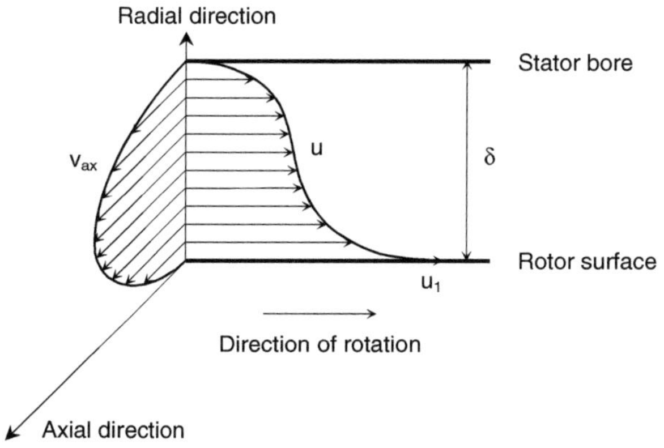

Fig. 7.2 Distribution of tangential and axial velocities in the air gap of a rotating electric machine [7]. Tangential component of coolant velocity u varies from 0 on the stator bore to u_1 on the rotor surface

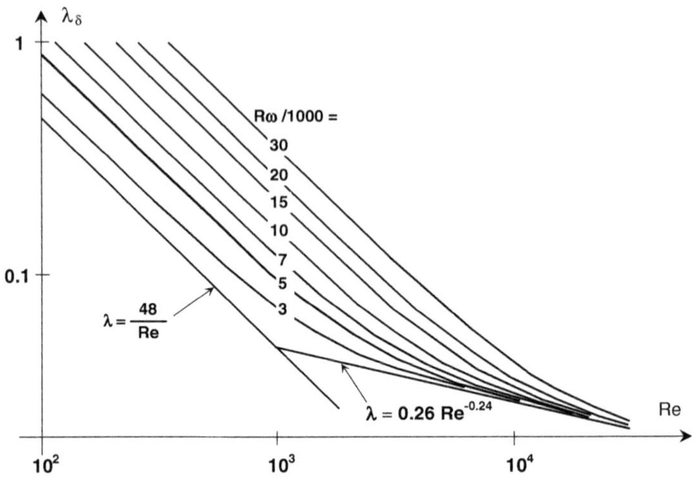

Fig. 7.3 Friction factor for air gap of a rotating electric machine [7]

Denoting by Re the Reynolds number for axial fluid flow defined as

$$\mathrm{Re} = \frac{v_{ax}\delta}{\nu} \tag{7.6}$$

where v_{ax} stands for mean velocity of axial flow, δ for air gap width, and ν for kinematic viscosity of the coolant, as well as by $R\omega$ the Reynolds number for rotating fluid flow, which is defined as

$$R\omega = \frac{u_1\delta}{\nu} \tag{7.7}$$

where u_1 stands for peripheral velocity on the rotor surface, one can write for the friction factor λ_δ of air gap as

$$\lambda_\delta = 0.26\,\mathrm{Re}^{-0.24}\left[1 + \left(\frac{7}{8}\cdot\frac{R\omega}{2\mathrm{Re}}\right)^2\right]^{0.38} \tag{7.8}$$

The dependence of friction factor λ_δ on Reynolds number Re in axial direction and for rotating Reynolds number $R\omega$ as a parameter is shown in Fig. 7.3.

7.3.2 *Pressure Loss Coefficients for Radial Cooling Ducts*

In order to increase the heat exchange surface, lamination in active part of medium- and large-size electric machines is divided in axial direction into subpackages, creating radial cooling ducts that way. Typical width (axially) of a lamination subpackage is about 50 mm and that of a cooling duct is about 10 mm.

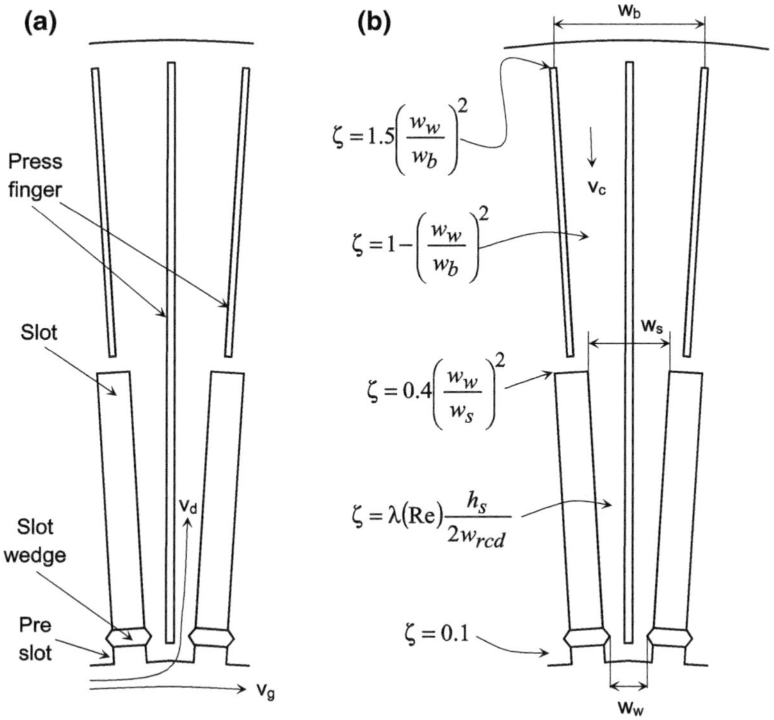

Fig. 7.4 Radial cooling duct with coolant flowing in centrifugal (**a**) and centripetal (**b**) direction [5, 8]. Slot height is denoted by h_s and axial width of radial cooling channel by w_{rcd}

Coolant can flow through a radial cooling duct either in centrifugal, or in centripetal direction (Fig. 7.4), each of which is characterized by its own values of friction factor and pressure loss coefficient [5, 8].

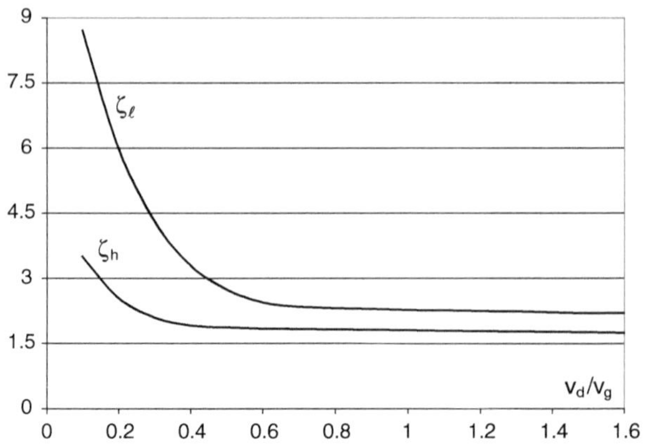

Fig. 7.5 Pressure loss coefficients for centrifugal direction of coolant flow in Fig. 7.4a and for various pre-slot heights [8]

Centrifugal flow (Fig. 7.4a): The pressure loss coefficient is a function of pre-slot height. For the pre-slot height h_{ps} between 5 and 26 mm, the measurements carried out in [5, 8] have shown linear dependence

$$\zeta = \frac{\zeta_l - \zeta_h}{21}\left(26 - h_{ps}\right) + \zeta_h \tag{7.9}$$

with ζ_ℓ denoting the pressure loss coefficient for pre-slot height below 5 mm and ζ_h the pressure loss coefficient for pre-slot height above 26 mm and h_{ps} in mm. The dependencies of ζ_ℓ and ζ_h on the ratio between the duct and gap coolant velocities v_d and v_g are shown in Fig. 7.5.

Centripetal flow (Fig. 7.4b): The pressure loss coefficients at various segments of radial cooling duct in tooth and yoke are given in Fig. 7.4 [5, 8].

7.3.3 Pressure Loss Coefficients for End Winding with Form-wound Coils

Stator end winding acts as a sieve for coolant. As shown in [9], the pressure loss coefficient in zones 1 and 3 of the end winding in Fig. 7.6 can be expressed as

$$\zeta = 0.8\frac{1 - \frac{\Delta}{\tau}}{\left(\frac{\Delta}{\tau}\right)^2} \tag{7.10}$$

In the zone 2 in Fig. 7.6, the loss coefficient can be interpolated as

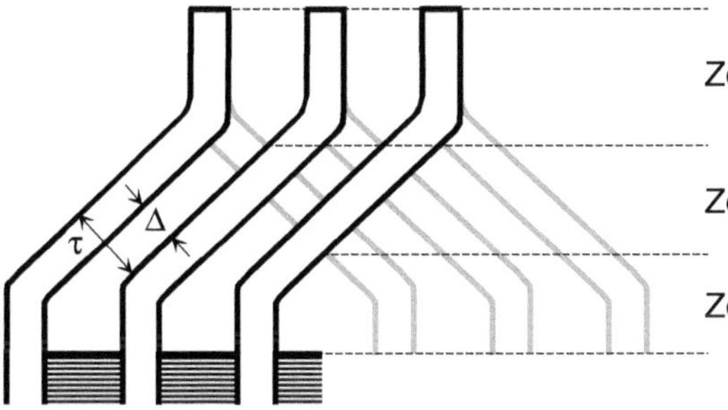

Fig. 7.6 End winding structure from the point of view of coolant distribution

$$\zeta = 275\left(\frac{\Delta}{\tau}\right)^{8.5} \tag{7.11}$$

for $0 \leq \Delta/\tau \leq 0.3$, and

$$\zeta = 175e^{-\frac{\Delta}{0.12\tau}} \tag{7.12}$$

for $\Delta/\tau \geq 0.3$.

7.3.4 Fan Curve

Large majority of electric machines are cooled by their own fans; most of them with one and some with two. A rotating fan creates a pressure increase Δp at a given volumetric flow rate Q and acts as a source in the machine's hydraulic network.

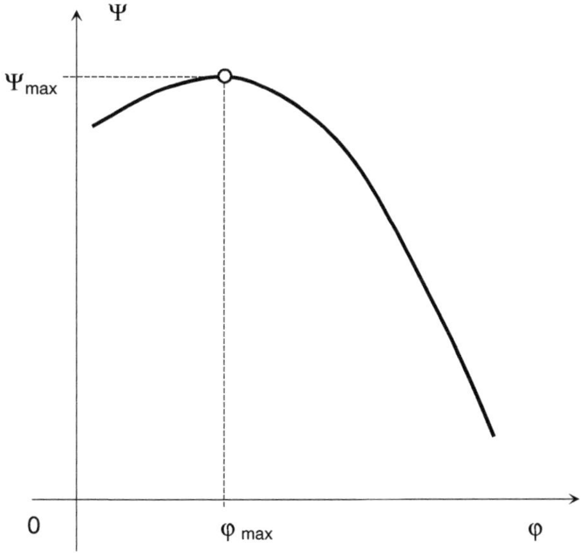

Fig. 7.7 Dimensionless fan curve. The interval of φ below φ_{max} is avoided due to danger of stall and pumping effect [13, 14]

A fan on the machine shaft takes gas axially and blows it out either axially or radially. Instead of the fan curve Δp (Q), a dimensionless dependence of the coefficient of pressure Ψ on the flow coefficient φ is sometimes used, where

$$\Psi = \frac{729.5}{d^2 n^2} \frac{\Delta p}{\rho} \tag{7.13}$$

and

$$\varphi = \frac{24.3}{d^3 n} Q \tag{7.14}$$

with n [rpm] denoting the speed of rotation and d [m] the fan diameter. Typical fan curve is represented in Fig. 7.7.

Besides fan(s), rotating components of an electric machine generate pressure difference (inherent pressure), which also acts as a source in the machine's hydraulic network. The inherent pressure is created by a rotating component as long as the difference between the inlet and outlet speed to the component is different from zero.

7.4 Coolant Distribution in Electric Machines, Pressure, and Volumetric Flow Rate in Elements of Its Hydraulic Network

In the previous sections the nonlinear character of friction factor and pressure loss coefficient of hydraulic resistances in electric machines was discussed. Both λ and ζ are dependent on Reynolds number Re, which on the other hand is a function of the volumetric flow rate Q. Therefore, one can rewrite Eq. 7.2 as

$$\Delta p = f(Q) \tag{7.15}$$

Since the pressure drop Δp is analogous to electrical voltage, and the volumetric flow rate Q to electrical current, one can interpret Eq. 7.15 as an analogue to a nonlinear electrical resistance [10]. As indicated in Table 1.4, one can formulate Kirchhoff's nodal rule as

$$\sum Q = 0 \tag{7.16}$$

and Kirchhoff's mesh rule for a hydraulic network as

$$\sum \Delta p = 0 \tag{7.17}$$

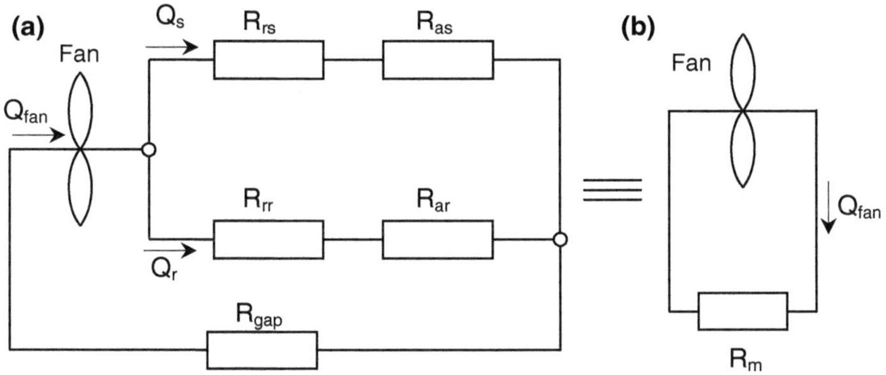

Fig. 7.8 Simplified hydraulic network of an electric machine including fan along with hydraulic resistance of stator radial cooling channels R_{rs}, rotor radial cooling channels R_{rr}, stator yoke back R_{as}, rotor yoke back R_{ar}, and air gap section R_{gap} (**a**), and its equivalent with only one hydraulic resistance R_m (**b**)

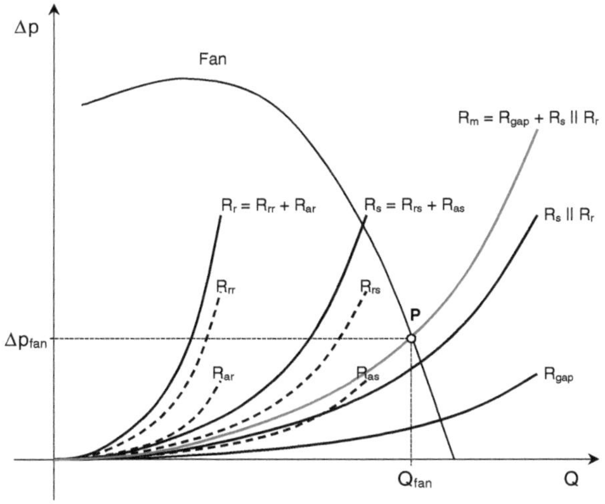

Fig. 7.9 Graphical solution of the simplified hydraulic network of an electric machine in Fig. 7.8. The operating point P of the fan has coordinates (Q_{fan}, Δp_{fan})

Cooling gas in the majority of electric machines is set in motion by only one fan, which extremely simplifies the solution procedure of their hydraulic circuits. Machine hydraulic resistances, as defined in Sect 7.2, are connected in series or in parallel to each other. The hydraulic resistance network can be reduced to a single resulting hydraulic resistance of the machine R_m connected to the fan, as illustrated in Fig. 7.8, in which a simple hydraulic network of an electric machine is converted into a single element connected in parallel/series with the fan.

Graphical solution of the hydraulic network in Fig. 7.8 is represented in Fig. 7.9. One recognizes in this figure the structure of the machine equivalent hydraulic resistance R_m and steps how it is created from hydraulic resistances of the segments of the machine.

Case Study 7.1: The two-pole air-cooled generator as described in [11] has the following data: S = 35.4 MVA, U = 10.5 kV, 50 Hz, rated field current 905 A, operating pressure 1.012 bar, gas inlet temperature 40 °C, 3000 rpm, fan outer diameter 990 mm, fan inner diameter 750 mm, fan stagger angle 29°, rotor diameter 810 mm, active part length 2100 mm, air gap width 37.5 mm, stator yoke outer diameter 1880 mm, 66 stator slots, stator slot height 229 mm, stator slot width 19.4 mm, rotor conductor gas cross-sectional area 105.7 mm^2, 94 field winding turns per rotor pole, shaft diameter below the end bell 400 mm, stator end winding maximum diameter 1600 mm, 49 radial cooling channels, width of a cooling channel 10 mm, and fan curve as shown in Fig. 7.10.

The machine hydraulic network is shown in Fig. 7.11. The elements of hydraulic network in Fig. 7.11 represent the portions of the machine as follows:

- T1–T3: space between stator yoke and housing;
- RCC1–RCC3: radial cooling channels;
- SEW: stator end winding zone;
- PP: channel between press plate and stator lamination;
- RAX: space between fan and air gap;
- REW: rotor end winding;
- GAP1–GAP3: air gap segments in axial direction; and
- RC: rotor hollow conductors.

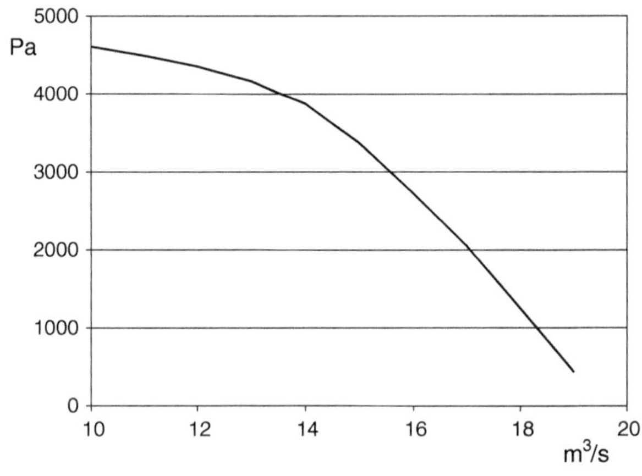

Fig. 7.10 Fan curve

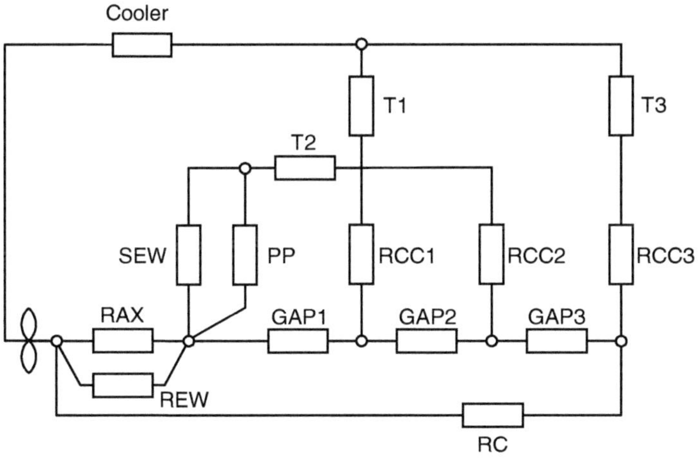

Fig. 7.11 Machine hydraulic network

The solution of the machine hydraulic network [15]shown in Fig. 7.11 is a set of pairs of volumetric flow rates Q and pressure drops Δp for all resistances. Knowing the cross-sectional areas of all elements, one can find corresponding gas velocities and heat transfer coefficients α [1], Table 7.1.

Table 7.1 Volumetric flow rates Q [m^3/s], pressure drops Δp [Pa], coolant velocities v [m/s] and heat transfer coefficients α [W/m^2 K] for hydraulic resistances in Fig. 7.11

Element	Q [m^3/s]	Δp [Pa]	v [m/s]	α [W/m^2 K]
SEW	2.84	27.04	5.66	30.2
PP	0.91	27.76	7.09	35.9
RCC1	4.05	1165.39	38.97	135.8
RCC2	1.81	986.07	34.87	124.5
RCC3	3.06	933.39	33.62	121
GAP1	7.65	311.03	61.03	192.7
GAP2	3.60	179.57	28.74	107.1
GAP3	1.79	52.68	14.29	62.1
REW	1.08	3573.19	30.33	111.7
RC	1.26	2279.88	31.81	115.9

Heat transfer coefficients α [W/m^2 K] for air are evaluated as

$$\alpha = 7.8 \cdot v^{0.78} \tag{7.18}$$

with v in m/s [1].

7.5 Finite Difference Solution of Temperature Distribution in Electric Machines—Thermal Node Potential Equations

Temperature distribution at steady state in a space with heat sources is described by Poisson's equation [12].

$$\lambda^x \frac{\partial^2 \vartheta}{\partial x^2} + \lambda^y \frac{\partial^2 \vartheta}{\partial y^2} + \lambda^z \frac{\partial^2 \vartheta}{\partial z^2} + q(x, y, z) = 0 \tag{7.19}$$

with λ^x, λ^y, and λ^z denoting thermal conductivities in x-, y-, and z-direction, respectively, and q the heat density in W/m^3. If Eq. (7.19) is discretized in the manner shown in Sect. 3.1, i.e., by substituting for

$$\lambda^x \frac{\partial^2 \vartheta}{\partial x^2} = 2\frac{\lambda_{10}}{d_1(d_1 + d_3)}(\vartheta_1 - \vartheta_0) + 2\frac{\lambda_{30}}{d_3(d_1 + d_3)}(\vartheta_3 - \vartheta_0) \tag{7.20}$$

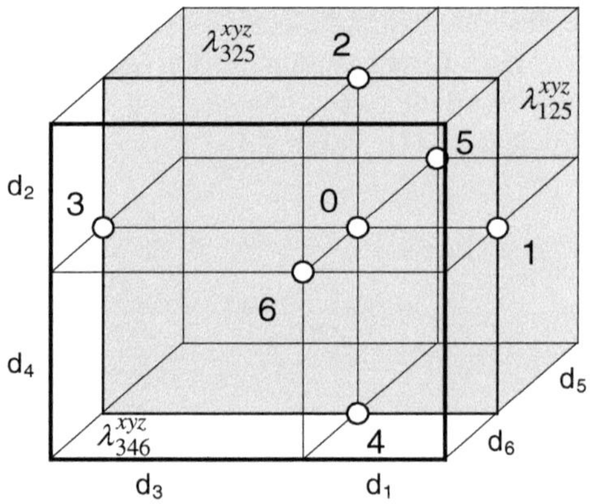

Fig. 7.12 Rectangular grid on which the Poisson's equation is discretized

$$\lambda^y \frac{\partial^2 \vartheta}{\partial y^2} = 2\frac{\lambda_{20}}{d_2(d_2 + d_4)}(\vartheta_2 - \vartheta_0) + 2\frac{\lambda_{40}}{d_4(d_2 + d_4)}(\vartheta_4 - \vartheta_0) \tag{7.21}$$

$$\lambda^z \frac{\partial^2 \vartheta}{\partial z^2} = 2\frac{\lambda_{50}}{d_5(d_5 + d_6)}(\vartheta_5 - \vartheta_0) + 2\frac{\lambda_{60}}{d_6(d_5 + d_6)}(\vartheta_6 - \vartheta_0) \tag{7.22}$$

where

$$\lambda_{10} = \frac{\lambda^x_{125} + \lambda^x_{126} + \lambda^x_{145} + \lambda^x_{146}}{4} \tag{7.23}$$

$$\lambda_{20} = \frac{\lambda^y_{125} + \lambda^y_{126} + \lambda^y_{325} + \lambda^y_{326}}{4} \tag{7.24}$$

$$\lambda_{30} = \frac{\lambda^x_{325} + \lambda^x_{326} + \lambda^x_{345} + \lambda^x_{346}}{4} \tag{7.25}$$

$$\lambda_{40} = \frac{\lambda^y_{145} + \lambda^y_{146} + \lambda^y_{345} + \lambda^y_{346}}{4} \tag{7.26}$$

$$\lambda_{50} = \frac{\lambda^z_{125} + \lambda^z_{145} + \lambda^z_{325} + \lambda^z_{345}}{4} \tag{7.27}$$

$$\lambda_{60} = \frac{\lambda^z_{126} + \lambda^z_{146} + \lambda^z_{326} + \lambda^z_{346}}{4} \tag{7.28}$$

one can write for the rectangular three-dimensional grid shown in Fig. 7.12

$$\begin{aligned}
&(\vartheta_0 - \vartheta_1)\left(\frac{1}{R^x_{125}} + \frac{1}{R^x_{145}} + \frac{1}{R^x_{146}} + \frac{1}{R^x_{126}}\right) + (\vartheta_0 - \vartheta_2)\left(\frac{1}{R^y_{125}} + \frac{1}{R^y_{126}} + \frac{1}{R^y_{326}} + \frac{1}{R^y_{325}}\right) + \\
&(\vartheta_0 - \vartheta_3)\left(\frac{1}{R^x_{325}} + \frac{1}{R^x_{345}} + \frac{1}{R^x_{346}} + \frac{1}{R^x_{326}}\right) + (\vartheta_0 - \vartheta_4)\left(\frac{1}{R^y_{145}} + \frac{1}{R^y_{146}} + \frac{1}{R^y_{346}} + \frac{1}{R^y_{345}}\right) + \\
&(\vartheta_0 - \vartheta_5)\left(\frac{1}{R^z_{125}} + \frac{1}{R^z_{145}} + \frac{1}{R^z_{345}} + \frac{1}{R^z_{325}}\right) + (\vartheta_0 - \vartheta_6)\left(\frac{1}{R^z_{126}} + \frac{1}{R^z_{146}} + \frac{1}{R^z_{346}} + \frac{1}{R^z_{326}}\right) = \\
&= q_{125}d_1d_2d_5 + q_{126}d_1d_2d_6 + q_{325}d_3d_2d_5 + q_{326}d_3d_2d_6 + \\
&+ q_{345}d_3d_4d_5 + q_{346}d_3d_4d_6 + q_{145}d_1d_4d_5 + q_{146}d_1d_4d_6
\end{aligned} \tag{7.29}$$

where

$$R^{xyz}_{ijk} = \frac{1}{\lambda^{xyz}_{ijk}} \cdot \frac{d_i}{d_jd_k} \tag{7.30}$$

Introducing substitutions

$$G_1 = \frac{1}{R^x_{125}} + \frac{1}{R^x_{145}} + \frac{1}{R^x_{146}} + \frac{1}{R^x_{126}} \tag{7.31}$$

$$G_2 = \frac{1}{R^y_{125}} + \frac{1}{R^y_{126}} + \frac{1}{R^y_{326}} + \frac{1}{R^y_{325}} \tag{7.32}$$

$$G_3 = \frac{1}{R^x_{325}} + \frac{1}{R^x_{345}} + \frac{1}{R^x_{346}} + \frac{1}{R^x_{326}} \tag{7.33}$$

$$G_4 = \frac{1}{R^y_{145}} + \frac{1}{R^y_{146}} + \frac{1}{R^y_{346}} + \frac{1}{R^y_{345}} \tag{7.34}$$

$$G_5 = \frac{1}{R^z_{125}} + \frac{1}{R^z_{145}} + \frac{1}{R^z_{345}} + \frac{1}{R^z_{325}} \tag{7.35}$$

$$G_6 = \frac{1}{R^z_{126}} + \frac{1}{R^z_{146}} + \frac{1}{R^z_{346}} + \frac{1}{R^z_{326}} \tag{7.36}$$

along with

$$\begin{aligned} P_0 = {} & q_{125}d_1d_2d_5 + q_{126}d_1d_2d_6 + q_{325}d_3d_2d_5 + q_{326}d_3d_2d_6 + \\ & + q_{345}d_3d_4d_5 + q_{346}d_3d_4d_6 + q_{145}d_1d_4d_5 + q_{146}d_1d_4d_6 \end{aligned} \tag{7.37}$$

one can rewrite the discretized Poisson's Eq. 7.29 as

$$\vartheta_0 \sum_{i=1}^{6} G_i - \sum_{i=1}^{6} \vartheta_i G_i = P_0 \tag{7.38}$$

which is nothing but a regular node potential equation for the node 0! In other words, the discretized Poisson's equation for heat flow and temperature distribution (7.19) can be interpreted in terms of electric circuits as a node potential equation, where electrical potential is analogous to the temperature, and electric current to the heat. In Eq. 7.29 both conduction and convection resistances, as introduced in Table 1.3, can appear. Corresponding boundary conditions have the meaning of fixed temperatures of particular nodes.

Case Study 7.2: The PM generator introduced in the Case Study 1.1 has 288 stator slots with dimensions 23 × 116 mm, mechanical air gap width of 6 mm, and stator yoke height of 46 mm. The stator is indirectly cooled with air blown through 12 radial cooling channels, each of which is 8 mm wide, from air gap toward the back iron.

Machine geometry is discretized, and Eq. 7.38 is applied to each node. The position of nodes in the machine's 3D thermal grid in (r, φ) and (r, z) planes is shown in Fig. 7.13.

Each radial cooling channel is blown with 0.5 m^3 air per second. The resulting temperature increases at selected points of the grid are shown in Fig. 7.14.

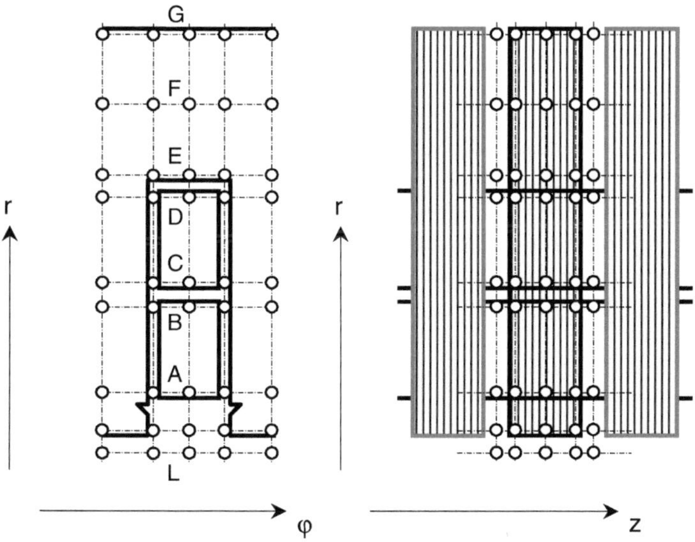

Fig. 7.13 Rectangular grid in (r, φ) and (r, z) planes with points A, G, and L at which temperature increases are shown in Fig. 7.14

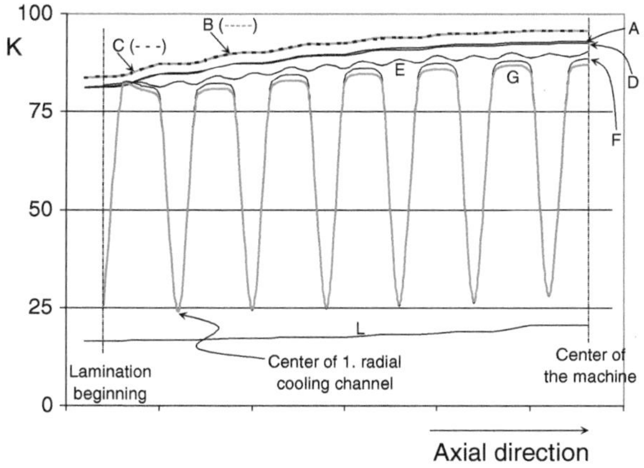

Fig. 7.14 Temperature increase [K] distribution at points A- G, and L in Fig. 7.13 along the machine length, starting at the beginning of the first lamination subpackage and ending at the axial center of the machine

7.6 Thermal Networks of Electric Machines and Methods for Their Solution

In the previous chapter the Poisson's equation was used to derive the node potential equations of the general thermal network, in which the temperatures are analogous to electric potentials, heat to electric currents, and thermal resistances to electrical resistances. It was emphasized that both types of thermal resistance introduced in Table 1.3, i.e., the resistance due to convection and conduction, can appear in the thermal network.

The resistance due to convection is related to the heat transfer from a solid body to fluid or gaseous medium. The value of convection resistance is determined by the heat transfer coefficient α, which is a measure of the amount of heat that warms up the fluid or gaseous medium.

Warming-up of the cooling medium can be taken into account by means of Eq. 1.20. Introducing the inverse of the heat storage capacity, the *coolant heat resistance* R_{ch}, [K/W], defined as

$$R_{\mathrm{ch}} = \frac{1}{c \cdot \rho \cdot \dot{V}} \tag{7.39}$$

one can write for the temperature drop $\Delta\vartheta$ across an element representing cooling medium as:

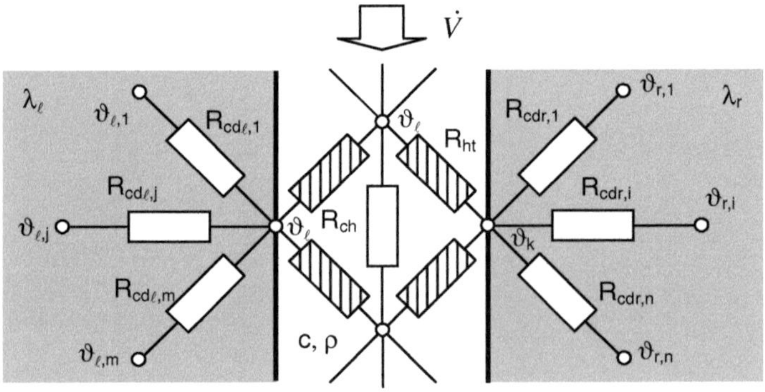

Fig. 7.15 Illustrating the definition of heat transfer resistance R_{ht}

$$\Delta\vartheta = P \cdot R_{\mathrm{ch}} \tag{7.40}$$

with P denoting the amount of losses which increase the cooling medium temperature, such as friction.

Alternatively to the convection resistance R_{cv} introduced in Table 1.3, the temperature drop across a heat exchange surface and cooling medium can be evaluated by means of the thermal conductivity of the heat exchange surface and heat storage capacity of the cooling medium [12]. This approach is illustrated in Fig. 7.15 by the example of channel with cooling medium taking heat from solid surfaces with thermal conductivities λ_ℓ and λ_r.

Temperature difference $\Delta\vartheta$ across the heat transfer resistance R_{ht} in Fig. 7.15 is defined as

$$\Delta\vartheta_{l-k} = \frac{1}{2 \cdot c \cdot \rho \cdot \dot{V}} \left(\sum_{i=1}^{n} \frac{\vartheta_{r,i} - \vartheta_k}{R_{\mathrm{cdr},i}} + \sum_{j=1}^{m} \frac{\vartheta_{l,j} - \vartheta_l}{R_{\mathrm{cdl},j}} \right) \tag{7.41}$$

The value of heat transfer resistance R_{ht} is dependent on temperatures of adjacent nodes. Therefore, R_{ht} is nonlinear and accordingly denoted shaded in Fig. 7.15.

Case Study 7.3: The tooth wound PM surface-mounted generator introduced in the Case Study 2.2 has 72 stator slots with dimensions 70 × 150 mm in which two layer coils are placed, each of which has 8 × 26 conductors. Stator active part is indirectly cooled with water flowing through pipes placed on the stator back iron, as in Fig. 7.16.

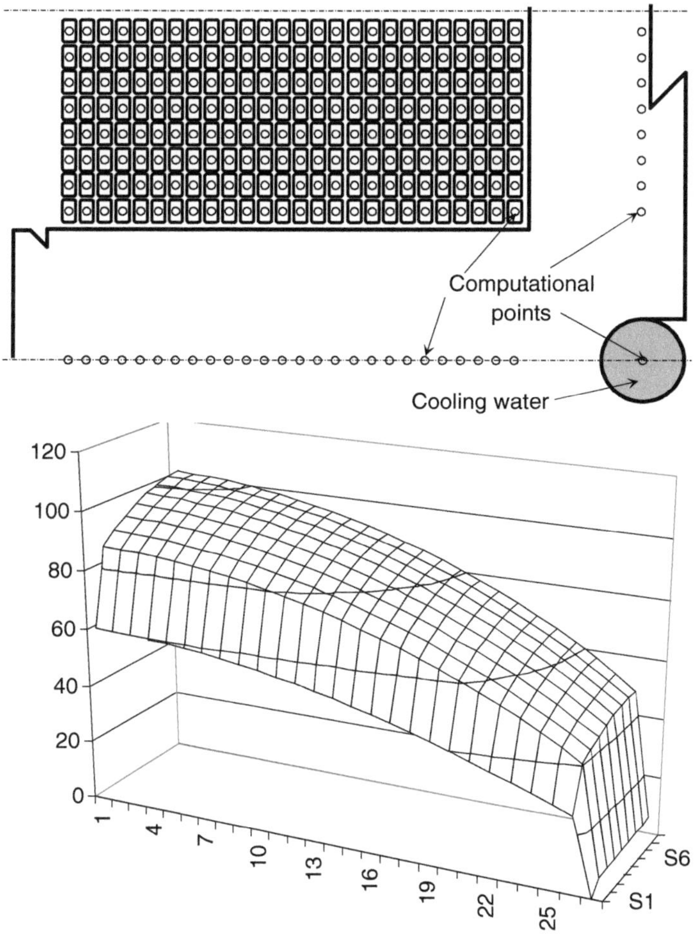

Fig. 7.16 Tooth wound machine geometry (*above*) and temperature rise distribution (*below*)

Temperature rise distribution shown in Fig. 7.16 was evaluated for the stator current density of 1.95 A/mm^2 by using the thermal resistances described in this section. A node of the thermal network is placed in every conductor in Fig. 7.16, as well as along the tooth centerline and in yoke segment. The nodes are connected with each other by means of thermal resistances and fed by generated losses. I^2R losses in each element are updated in each computational step by considering the thermal coefficient of specific resistance.

As one can see in Fig. 7.16, the highest temperature rise takes place in the outermost conductor close to the air gap, which is farthest from the tube with cooling water. One should keep in mind that the analyzed generator rotates at a very slow speed of 24 rpm, at which there exists practically no cooling by means of blowing air. On the other hand, stator iron losses are extremely low due to a low frequency of $24 \cdot 35/60 = 14$ Hz. The overwhelming majority of losses are dissipated in stator winding, and the applied cooling concept is obviously inefficient, because the heat creates too high temperature gradients on its way from coils to water pipes.

Case Study 7.4: A 62.9 MVA, 24-pole hydro generator with 3400 mm stator bore, 120 stator slots, 142 × 35 mm stator slot dimensions, 1860 mm length has directly water cooled stator winding and directly water cooled stator yoke by means of 80 cooling tubes with inner diameter 20 mm.

Circular holes for cooling tubes require special handling in a cylindrical or rectangular grid. Conduction resistance of iron lamination around a circular hole is usually evaluated as a sum of n resistances of rectangular elements with variable widths, as in Fig. 7.17.

Stator current density at rated operating point is 6.2 A/mm^2, average iron core loss density in stator teeth equals to 31.2 kW/m^3, and 28 kW/m^3 in stator yoke. Cooling water velocity is 0.76 m/s, and inlet temperature is 35 °C. Average air temperature in the air gap equals to 45 °C.

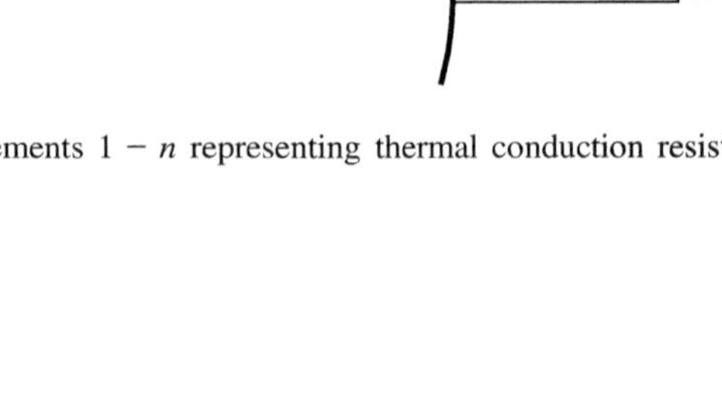

Fig. 7.17 Refined elements $1 - n$ representing thermal conduction resistance around a circular cooling hole

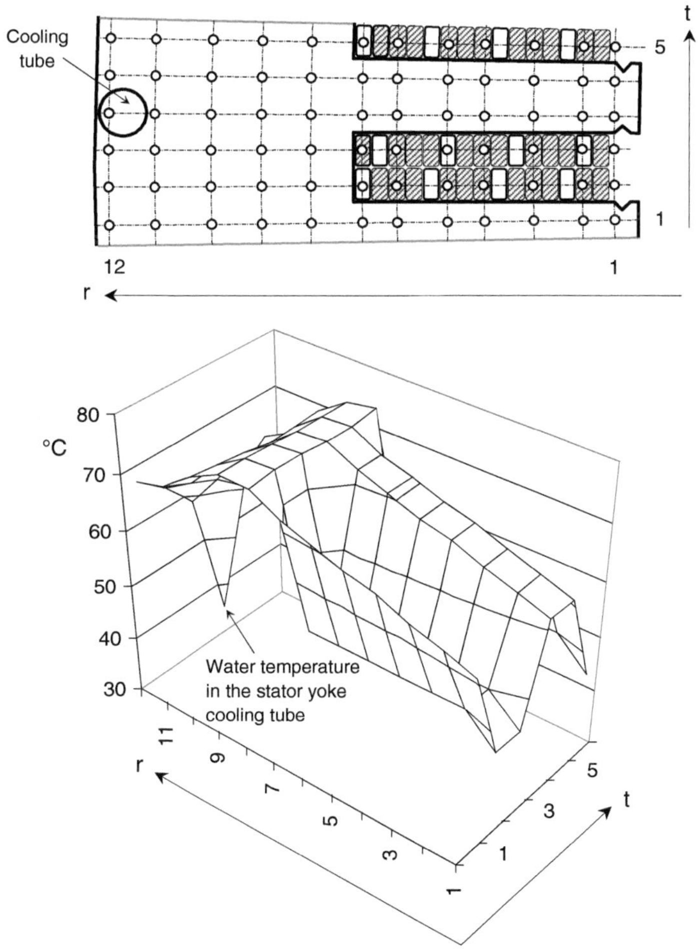

Fig. 7.18 Two dimensional cylindrical grid (*r* radial coordinate; *t* tangential coordinate) in a stator segment of directly water cooled stator of analyzed 62.9 MVA hydro generator (*above*) and temperature distribution over grid points for given current and loss densities (*below*)

As one can see in Fig. 7.18, owing to efficient water cooling of stator conductors and yoke, temperatures at rated point remain far below critical values.

Case Study 7.5: Stator bars of a 688 MVA, two-pole turbogenerator are directly cooled by water flowing at 1.5 m/s through 6 hollow conductors imbedded in 54 strands within each bar, as in Fig. 7.19. Hole dimensions of a hollow conductor are 10.8 × 2 mm, and strand dimensions are 13 × 1.8 mm. Cooling water inlet temperature is 45 °C, and rated current density equals to 6.8 A/mm^2. Total I^2R losses in a bar at rated point amount to 36.4 kW.

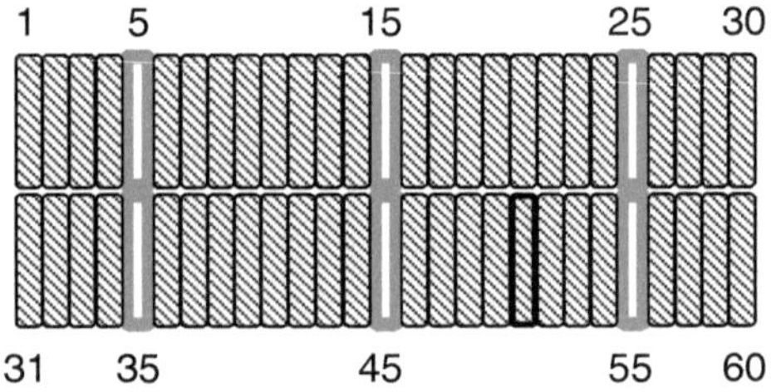

Fig. 7.19 "Green" bar (no bar insulation yet) with a total of 60 strands. Six strands are hollow and carry cooling water. Strand numbers are referred to abscissa in the following 2 figures

Temperature increase distribution at rated point with properly functioning cooling is shown in Fig. 7.20a for strands 1–30 and in Fig. 7.20b for strands 31–60. The lowest temperature increase points at a given axial coordinate z are allocated in hollow conductors and the highest in the strands in the middle between two hollow conductors, at the end of conductors on cooling water outlet (here about 24 K).

If one hollow conductor is blocked, the temperatures of strands around it increase due to worsened cooling. Assume that the hollow conductor denoted by "45" in Fig. 7.19 is blocked at the beginning of the bar, as in Fig. 7.21. Besides immense temperature increase of about 60 % in the blocked hollow conductor (from about 23 K in properly functioning operation to about 37 K in case of blocked conductor), strands in its vicinity are also exposed to higher temperatures. Considering that at the given machine size a bar is typically about 10 m long, an uneven temperature distribution within it could result in relative motion between the strands and possible insulation damage. Therefore, if a hollow conductor is blocked, the machine power must be reduced. Ultimately, the machine must be taken of the mains and the bar must be repaired.

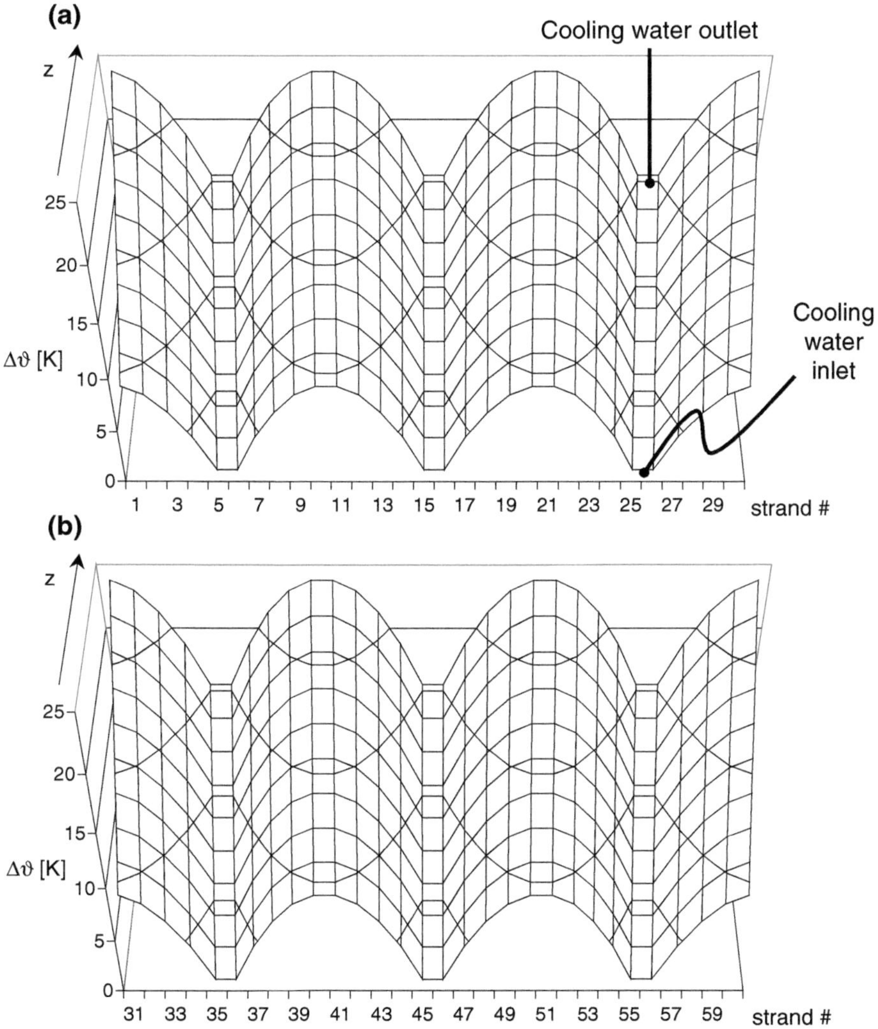

Fig. 7.20 Temperature increase distribution in strands 1–30 (**a**) and 31–60 (**b**) at properly functioning cooling

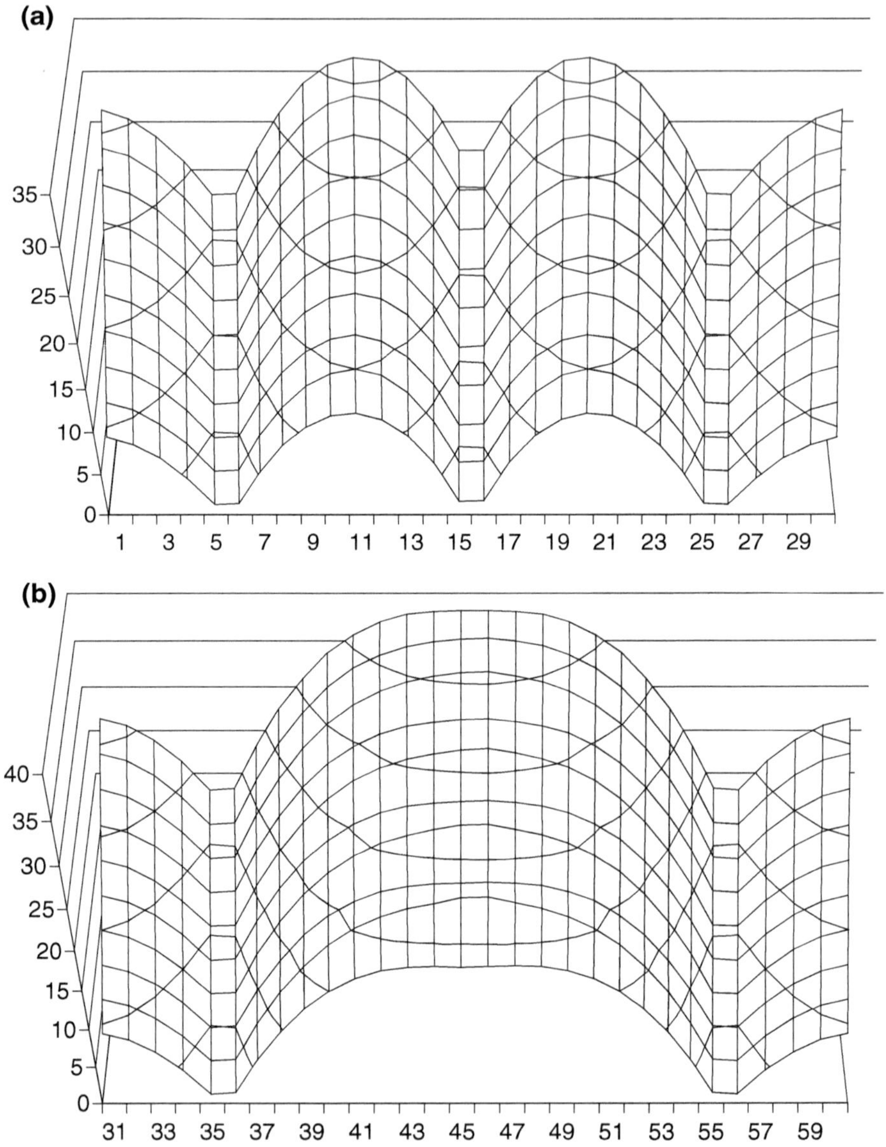

Fig. 7.21 Temperature increase distribution in strands 1–30 (**a**) and 31–60 (**b**) at improperly functioning cooling

Case Study 7.6: Rotor conductors of a 125 MVA, two-pole turbogenerator are directly cooled with air, Fig. 7.22. Cooling air inlet temperature is 40 °C, and rated current density equals to 6.2 A/mm^2. Total rotor I^2R losses at rated field current of 1526 A amount to 210.5 kW. Cooling air inlet is on both driving end (DE) and non-driving end (NDE), and outlet in the middle of the machine (axially). Total rotor conductor length is 3938 mm.

Fig. 7.22 Field winding conductors in a rotor slot. Conductor numbers are referred to abscissa in the following 2 figures

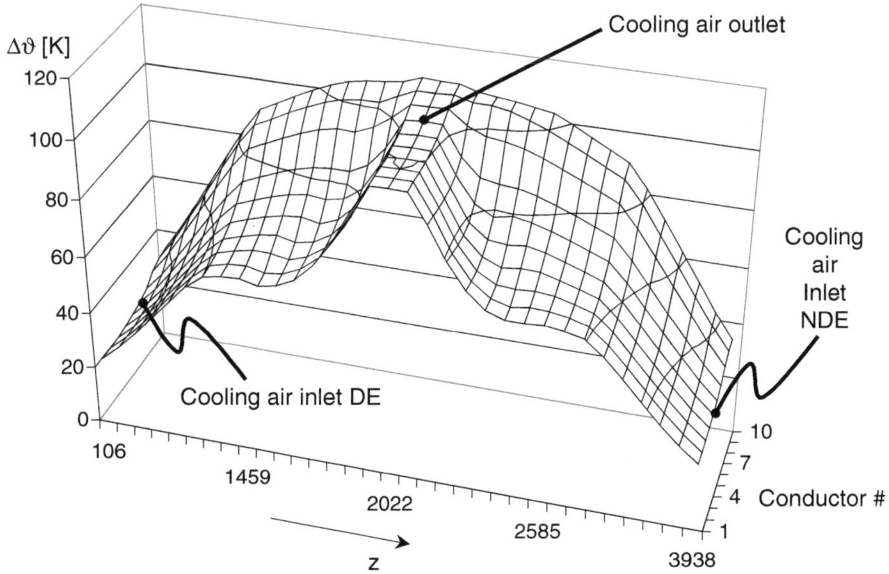

Fig. 7.23 Temperature increase distribution in rotor conductors at rated point and properly functioning cooling

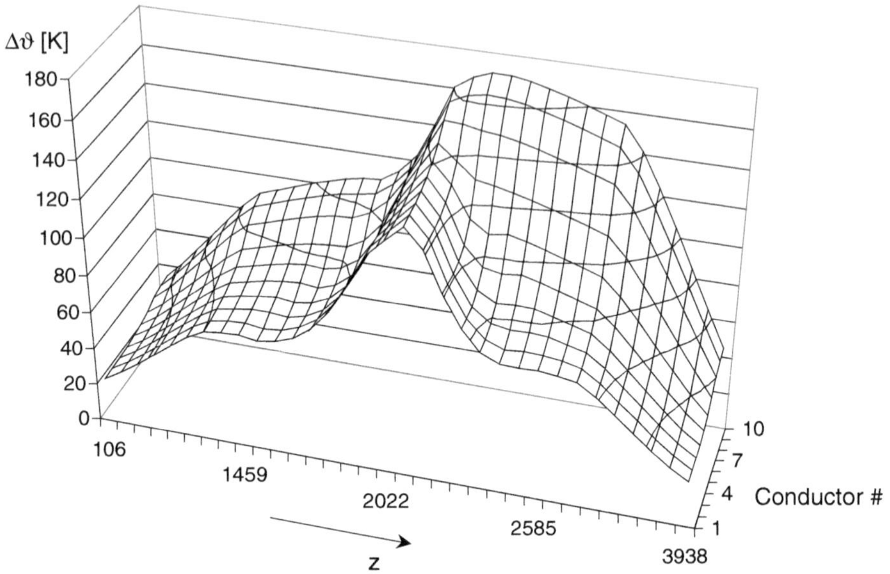

Fig. 7.24 Temperature increase distribution in rotor conductors at rated point and blocked cooling

Temperature increase distribution at properly functioning cooling and at rated field current is shown in Fig. 7.23. If 60 % of the lowermost conductors 8, 9, and 10 on machine's NDE, and 40 % of the conductor 7 are blocked, the temperature increase distribution changes to the one shown in Fig. 7.24. The maximum temperature increase is far above the limit given by the class of insulation, and the machine cannot operate safely at the given load any more.

7.7 Transient Heating of a Hollow Conductor

Highly utilized electric machines are characterized by double-digit current density values of A/mm^2, which create I^2R losses that can be taken out of conductor only by directly cooling them with gas, oil, or water. The combination of parameters determining thermal time constant of a directly cooled conductor is such as to speed up thermal transients. Therefore, not only stationary parameters of the cooling system discussed previously in this chapter are decisive for thermal design of directly cooled conductors, but also their heat storage capability.

Consider a segment of a directly cooled conductor of the length dx in which volume losses $q = P/V = \Gamma^2\ \rho$ [W/m^3] are dissipated as shown in Fig. 7.25.

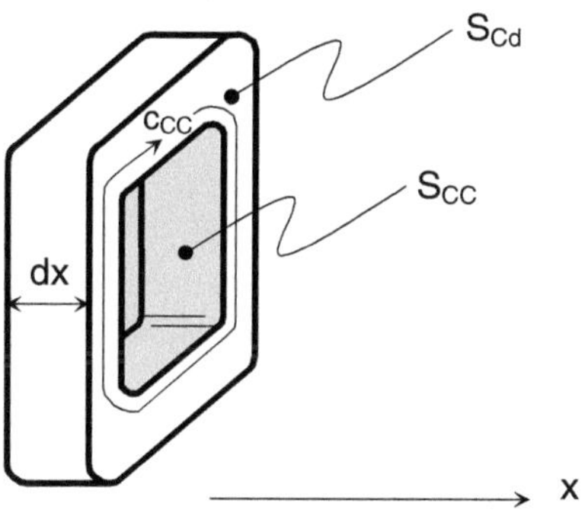

Fig. 7.25 A segment of a directly cooled conductor: S_{cd} conductor cross-sectional area; S_{cc} cooling channel cross-sectional area; c_{cc} cooling channel circumference

Denoting by c_{Cd} the specific heat of the conductor material [Ws/kg K], by c_C the specific heat of the coolant [Ws/kg K], by ρ_{Cd} the mass density of the conductor [kg/m^3], and by ρ_C the mass density of the coolant [kg/m^3], one can write for the differential amount of losses dW_L generated in the conductor element with length dx:

$$dW_L = q\,S_{Cd}\,dx\,dt \tag{7.42}$$

A portion $dW_{S,Cd}$ of generated losses remains accumulated in the conductor, thus increasing its temperature ϑ_{Cd}:

$$dW_{S,Cd} = c_{Cd}\,S_{Cd}\,\rho_{Cd}\frac{\partial\vartheta_{Cd}}{\partial t}dx\,dt \tag{7.43}$$

and the rest dW_α is transferred to the cooling medium

$$dW_\alpha = (\vartheta_{Cd} - \vartheta_C)\alpha_C\,c_{CC}dx\,dt \tag{7.44}$$

with α_C denoting the heat transfer coefficient of the coolant. Based upon the energy balance equation

$$dW_L = dW_{S,Cd} + dW_\alpha \tag{7.45}$$

one can combine Eqs. 7.42–7.44 into a single partial differential equation

$$c_{Cd}\,S_{Cd}\,\rho_{Cd}\frac{\partial\vartheta_{Cd}}{\partial t} + (\vartheta_{Cd} - \vartheta_C)\alpha_C\,c_{CC} = q\,S_{Cd} \tag{7.46}$$

At steady state, $\partial\vartheta_{Cd}/\partial t = 0$ and

$$\left(\vartheta_{Cd,t\to\infty} - \vartheta_{C,t\to\infty}\right) = \frac{q\,S_{Cd}}{\alpha_C\,c_{CC}} = \Delta\vartheta_{CC} \tag{7.47}$$

with $\Delta\vartheta_{CC}$ denoting the *temperature drop across the heat transfer surface at steady state*.

The differential of energy $dW_{S,C}$ accumulated by the coolant is equal to

$$dW_{S,C} = c_C\, S_C\, \rho_C \frac{\partial \vartheta_C}{\partial t} dx\, dt \tag{7.48}$$

and the difference between delivered and dissipated energy differentials in the coolant

$$dW_{diff} = c_C\, \rho_C\, S_{cc}\, v\, \frac{\partial \vartheta_C}{\partial x} dx\, dt \tag{7.49}$$

with v denoting the coolant velocity [m/s].

The energy balance partial differential equation for the coolant is now

$$c_C\, \rho_C\, S_{cc}\, v\, \frac{\partial \vartheta_C}{\partial x} + c_C\, \rho_C\, S_{cc} \frac{\partial \vartheta_C}{\partial t} - (\vartheta_{Cd} - \vartheta_C)\alpha_C\, c_{CC} = 0 \tag{7.50}$$

The solution of the system of partial differential Eqs. 7.46 and 7.50 are the temperature increase distributions for the conductor, $\vartheta_{Cd}(t, x)$, and for the coolant, $\vartheta_C\,(t, x)$. The corresponding initial condition can be expressed as

$$\vartheta_{Cd}(0, x) = 0 \tag{7.51}$$

and boundary condition as

$$\vartheta_C(t, 0) = 0 \tag{7.52}$$

Substituting in Eq. 1.50 for the conductor current $I = 0$, one obtains the *differential thermal time constant of the conductor* $\tau_{th,Cd}$, which is defined as

$$\tau_{th,Cd} = \frac{c_{Cd}\, S_{Cd}\, \rho_{Cd}}{\alpha_C\, c_{CC}} \tag{7.53}$$

Analogously one can define the *differential thermal time constant of the coolant* $\tau_{th,C}$ as

$$\tau_{th,C} = \frac{c_C\, S_C\, \rho_C}{\alpha_C\, c_{CC}} \tag{7.54}$$

Division of Eq. 7.47 by $q\ S_{Cd}$ and Eq. 7.50 by $\alpha_C c_{CC}$ results in the system of partial differential equations for a directly cooled conductor

$$\frac{\tau_{th,Cd}}{\Delta\vartheta_{CC}} \frac{\partial \vartheta_{Cd}}{\partial t} + \frac{\vartheta_{Cd} - \vartheta_C}{\Delta\vartheta_{CC}} = 1 \tag{7.55}$$

$$v\frac{\partial \vartheta_C}{\partial x} + \frac{\partial \vartheta_C}{\partial t} - \frac{\vartheta_{Cd} - \vartheta_C}{\tau_{th,C}} = 0 \tag{7.56}$$

Solution of the system of partial differential Eqs. 7.55–7.56 can be expressed by means of modified Bessel function I_0 as:

$$\vartheta_C(x,t) = \int_0^{\frac{x}{\tau_{th,C}}} \int_0^{\frac{t}{\tau_{th,Cd}}} e^{-\eta} e^{-\xi} I_0\left(\sqrt{4\eta\xi}\right) d\eta \, d\xi \tag{7.57}$$

$$\vartheta_{Cd}(x,t) = \vartheta_C(x,t) + e^{-x} \int_0^{\frac{t}{\tau_{th,Cd}}} e^{-\eta} I_0\left(\sqrt{4\eta x}\right) d\eta \tag{7.58}$$

Solution of the system 7.55–7.56 is shown qualitatively in Fig. 7.26.

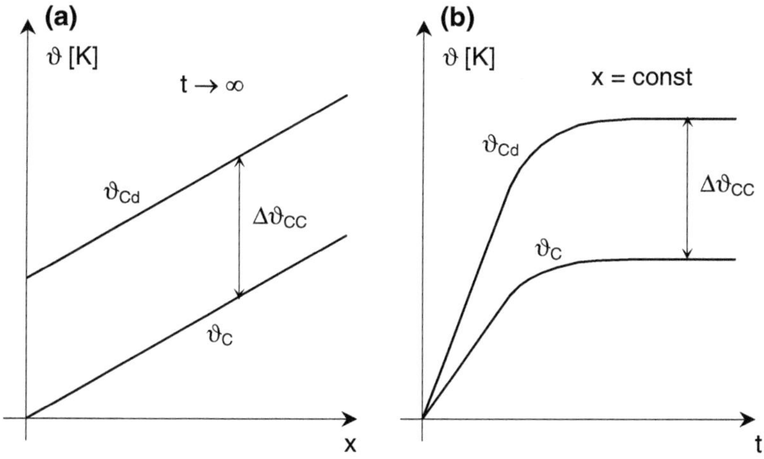

Fig. 7.26 Qualitative representation of the solution of system 7.55–7.56 in spatial (**a**) and time (**b**) domain

Case Study 7.5: Stator bars of a 688 MVA, two-pole turbogenerator are built of 60 strands and 6 water-cooled hollow conductors. Cross-sectional area of copper strands is 1386 mm^2 and that of a stainless steel hollow conductors is 178 mm^2. Stator current density of 6.8 A/mm^2 at rated point creates 19.5 kW I^2R losses in each bar. Bar length is 9.48 m, and hole circumference of a hollow conductor equals to 25.6 mm. Cooling water has a velocity of 1.5 m/s, at which the heat transfer coefficient reaches the value of 8961 W/m^2 K.

The coefficients of partial differential Eqs. 7.55 and 7.56 are as follows:

$$\Delta\vartheta_{CC} = \frac{q\, S_{Cd}}{\alpha_C\, c_{CC}} = \frac{\dfrac{19500\,\mathrm{W}}{9.48\,\mathrm{m}\cdot 0.001386\,\mathrm{m}^2}\,\dfrac{0.001386\,\mathrm{m}^2}{6}}{8961\,\dfrac{\mathrm{W}}{\mathrm{m}^2\mathrm{K}}\cdot 0.0256\,\mathrm{m}} = 1.5\,\mathrm{K}$$

$$\tau_{th,Cd} = \frac{c_{Cd}\, S_{Cd}\, \rho_{Cd}}{\alpha_C\, c_{CC}} = \frac{384.4\,\dfrac{\mathrm{Ws}}{\mathrm{kg\,K}}\,\dfrac{0.001386\,\mathrm{m}^2}{6}\,8960\,\dfrac{\mathrm{kg}}{\mathrm{m}^3}}{8961\,\dfrac{\mathrm{W}}{\mathrm{m}^2\mathrm{K}}\cdot 0.0256\,\mathrm{m}} = 3.47\,\mathrm{s}$$

$$\tau_{th,C} = \frac{c_C\, S_C\, \rho_C}{\alpha_C\, c_{CC}} = \frac{4187\,\dfrac{\mathrm{Ws}}{\mathrm{kg\,K}}\,0.0000216\,\mathrm{m}^2\,1000\,\dfrac{\mathrm{kg}}{\mathrm{m}^3}}{8961\,\dfrac{\mathrm{W}}{\mathrm{m}^2\mathrm{K}}\cdot 0.0256\,\mathrm{m}} = 0.342\,\mathrm{s}$$

At time instant $t = 0$, the generator is loaded with rated current. Both bar and water temperatures increase to their steady-state values within 300–400 s. Such short thermal transients are a consequence of extremely large heat transfer coefficient. Steady-state temperature increase in a conductor at the end of the bar equals 29.5 K and water outlet temperature 28 K.

Graphical representation of solutions is given in Figs. 7.27, 7.28, 7.29, and 7.30.

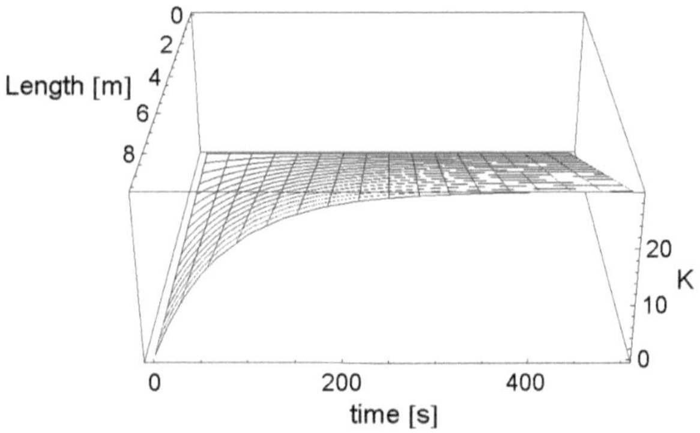

Fig. 7.27 Temperature increase within a conductor

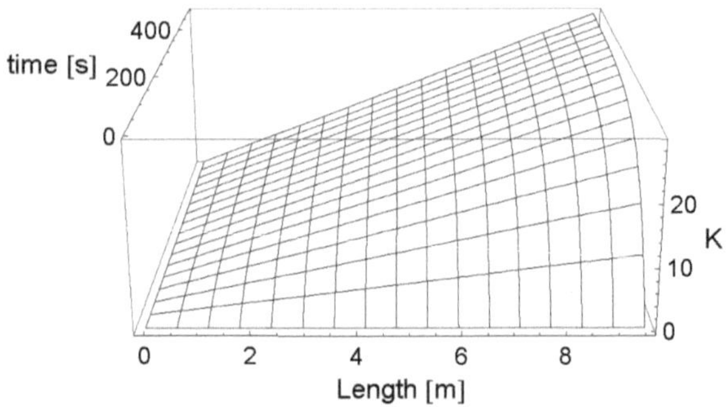

Fig. 7.28 Temperature increase within a conductor (same as in Fig. 7.27, but with rotated abscissa and ordinate)

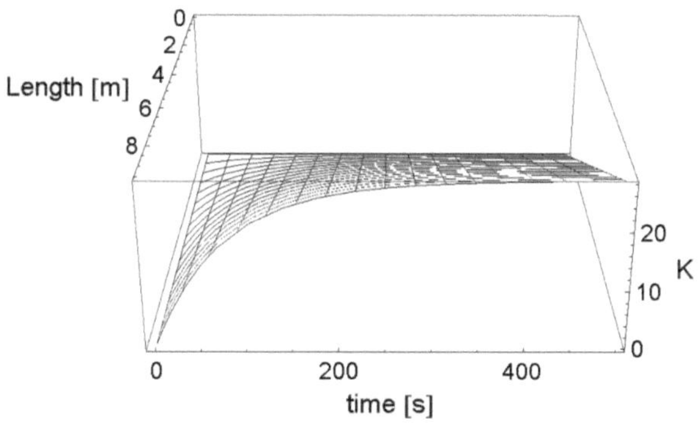

Fig. 7.29 Temperature increase of cooling water in a hollow conductor

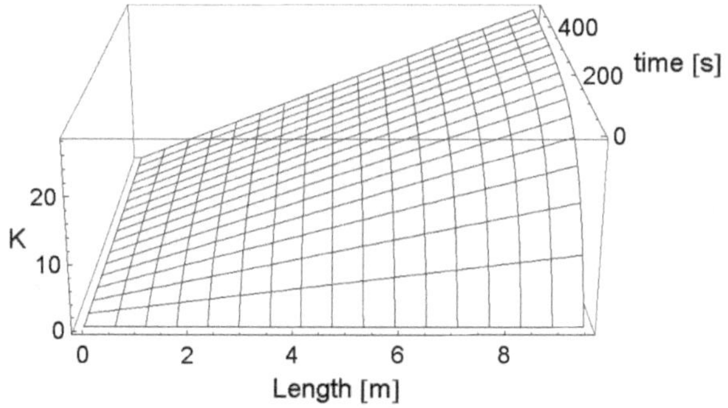

Fig. 7.30 Temperature increase of cooling water in a hollow conductor (same as in Fig. 7.29, but with rotated abscissa and ordinate)

Case Study 7.6: Rotor conductors of a 688 MVA, two-pole turbogenerator are directly cooled with hydrogen at 5 ata. Cross-sectional area of a conductor is 405 mm^2 and that of a cooling hole is 245 mm^2. Rotor current density of 14.7 A/mm^2 at rated point creates loss density of 5.4 MW/m^3 in each bar. The rotor is cooled with two fans, one on each side. Therefore, the rotor conductor length is 5.7/2 = 2.85 m, 5.7 m being the rotor active length. The conductor hole circumference equals to 95 mm. Hydrogen in hollow conductors has a velocity of 59 m/s, at which the heat transfer coefficient equals to 1188 W/m^2 K.

The coefficients of partial differential Eqs. 7.55 and 7.56 are as follows:

$$\Delta\vartheta_{CC} = \frac{q\,S_{Cd}}{\alpha_C\,c_{CC}} = \frac{5400000\,\frac{\mathrm{W}}{\mathrm{m^3}}\,0.000405\,\mathrm{m}^2}{1188\,\frac{\mathrm{W}}{\mathrm{m^2K}}\cdot 0.095\,\mathrm{m}} = 19.4\,\mathrm{K}$$

$$\tau_{th,Cd} = \frac{c_{Cd}\,S_{Cd}\,\rho_{Cd}}{\alpha_C\,c_{CC}} = \frac{384.4\,\frac{\mathrm{Ws}}{\mathrm{kg\,K}}\,0.000405\,\mathrm{m}^2 8960\,\frac{\mathrm{kg}}{\mathrm{m^3}}}{1188\,\frac{\mathrm{W}}{\mathrm{m^2K}}\cdot 0.095\,\mathrm{m}} = 12.36\,\mathrm{s}$$

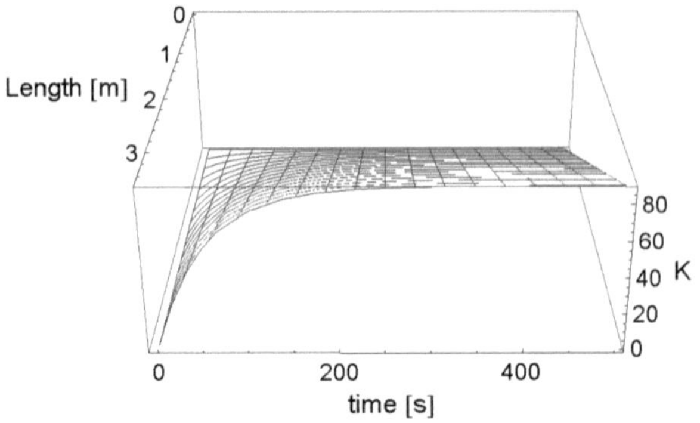

Fig. 7.31 Temperature increase within a conductor

$$\tau_{\mathrm{th,C}} = \frac{c_{\mathrm{C}}\, S_{\mathrm{C}}\, \rho_{\mathrm{C}}}{\alpha_{\mathrm{C}}\, c_{\mathrm{CC}}} = \frac{14400\,\frac{\mathrm{Ws}}{\mathrm{kg\,K}}\,0.000245\,\mathrm{m}^2 \cdot 6 \cdot 0.0899\,\frac{\mathrm{kg}}{\mathrm{m}^3}}{1188\,\frac{\mathrm{W}}{\mathrm{m}^2\mathrm{K}} \cdot 0.095\,\mathrm{m}} = 0.0169\,\mathrm{s}$$

Graphical representation of solutions is given in Figs. 7.31, 7.32, 7.33, and 7.34.

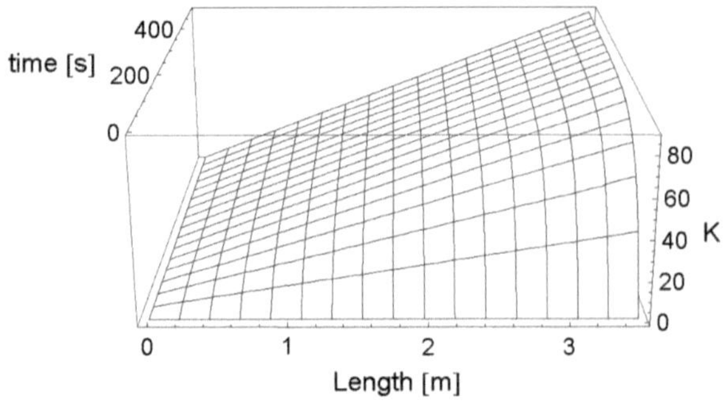

Fig. 7.32 Temperature increase within a conductor (same as in Fig. 7.31, but with rotated abscissa and ordinate)

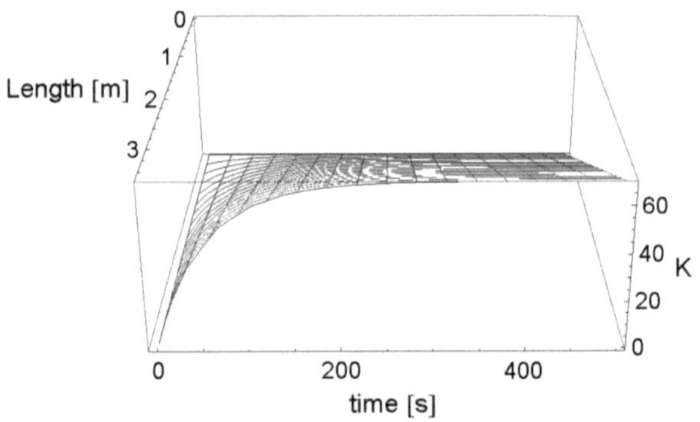

Fig. 7.33 Temperature increase of hydrogen in a hollow conductor

In the middle of the rotor, the steady-state temperature increase in a conductor equals 91 K and hydrogen temperature increase equals 72 K.

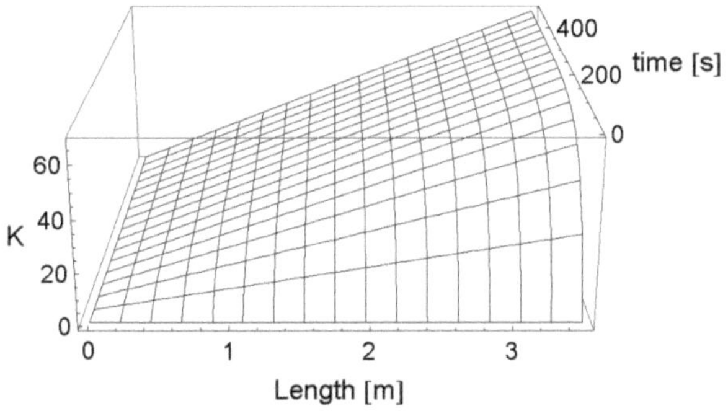

Fig. 7.34 Temperature increase of hydrogen in a hollow conductor (same as in Fig. 7.33, but with rotated abscissa and ordinate)

References

1. Wiedemann E, Kellenberger W (1967) Konstruktion elektrischer Maschinen. Springer, Berlin
2. Colebrook CF, White CM (1937/38) The reduction of carrying capacity of pipes with age. J Inst Civ Eng 7(1):99–118
3. Richter H (1958) Rohrhydraulik. Springer, Berlin
4. Blasius H (1931) Das Aehnlichkeitsgesetz bei Reibungsvorgaengen in Fluessigkeiten. VDI-Forschungsheft (31)
5. Idelchik IE (1960) Spravochnik po gidravlicheskim soprotivleniyam- Koeffitsienty mestnykhsoprotivlenii i soprotivleniya treniya. Gosudarstvennoe Energeticheskoe Izdatel'stvo, Moskva- Leningrad
6. Woods G, Osborne WC, Turner CG (1952) Practical guide to fan engineering. Benham & Co., Ltd., Colchester
7. Yamada Y (1962) Resistance of a flow through an annulus with an inner rotating cylinder. Bulletin of JSME 5(18):302–310
8. Muench B, Zardin H (1964) Rechnerische und experimentelle Behandlung von Ventilationsproblemen an elektrischen Maschinen. Report 115, 20. Congress CIGRE, Paris
9. Wieghardt K (1953) Ueber den Stroemungswiderstand von Sieben. Z Angew Math Mech 33 (8/9):312–314
10. Pohle R (1963) Moeglichkeiten zur Bestimmung der Mengenstrom und Druckverteilung in stark verzweigten Luftleitungssystemen. Heiz.- Lueft.- Haustechnik (14/10):355–360
11. Haase H, Lagardièr H, Suter J (1986) Luftgekuehlte Turbogeneratoren der 200-MVA-Klasse. Brown Boveri Technik (3/73):133–138
12. Gotter G (1954) Erwaermung und Kuehlung elektrischer Maschinen. Springer, Berlin
13. Eckert B, Schnell E (1961) Axial-und Radialkompressoren. Springer, Berlin
14. Eck B (1972) Ventilatoren. Springer, Berlin
15. Burden RL, Faires JD, Reynolds AC (1981) Numerical analysis. PWS Publishers, Boston

Chapter 8
General Principles of AC Machine Design

Contents

Key aspects of electromagnetic design of induction and synchronous machines are discussed starting from equivalent circuit and/or voltage diagram. It is shown how the design criteria, such as short-circuit current, torque overload capability, e.t.c. influence the selection of parameters of an induction machine. The role of synchronous reactance and short-circuit ratio is discussed, along with the influence of air gap geometry of a synchronous machine to its transient parameters.

8.1 Introduction

To design an electric machine means to find an economically optimal solution which satisfies customer specifications and relevant standards, and which is manufacturable with given production facilities and materials. Since in the process of design several independent criteria have to be fulfilled, many of them being opposite to each other, the designer's challenge is to consider all of them in a proper weight and implement simplest and most efficient solutions. Obviously, there exists no best machine which would satisfy given specification, because machine design procedure is not only the number crunching procedure per se, but also a creative activity with attributes of applied arts and industrial design. The designer is free to select the optimal solution, which he reaches by using models and tools in which he has to have absolute confidence. This is especially important in design from scratch, when the limits of existing machine models are shifted and a question is often posed, how reliable is the machine design model. Even if the machine manufacturer has developed his own software for the purpose of machine design, it is hard to predict how accurate it would be when applied on a virgin soil. Very often,

V. Ostović, *The Art and Science of Rotating Field Machines Design*,
DOI 10.1007/978-3-319-39081-9_8

however, the machine manufacturer does not posses such an extensive computational tool. In such cases, a commercial finite element software package turns out to be too a large risk factor, because its vendor would not allow designer to access the source code in order to estimate the software performance in critical cases. Even worse, vendors tend to bind machine designers offering them to design the machine by themselves, however, disclaiming any responsibility for results of their design. The situation becomes precarious when a three-dimensional problem has to be solved, such as eddy current losses in solid components in the end winding zone or in stator wedges on the outer diameter of lamination package. Here methods and procedures for generating a functioning mesh and obtaining reliable results still have an enormous optimization potential.

National and international standards specify, among others, rated and test voltages, current and torque overloads, protection types, classes of insulation, wire shapes, shaft heights, power factor and efficiency intervals, etc. In case of large machines, e.g., turbo- and hydrogenerators, the specifications from standards are usually extended by conditions requested by the customer.

Since every machine manufacturer has his own design philosophy based on specific experience and production facilities, in this chapter only the common principles of machine design will be discussed. As a result, no recipes, but general concepts of sizing AC machines from scratch are developed in the following two sections [8, 9, 10].

8.2 Sizing Equations of an Induction Machine

In order do design an induction machine for a given rated voltage U_r and rated frequency f_r, the following data have to be specified:

- Rated mechanical power P_r (continuous), or its equivalent for intermittent operation;
- Synchronous speed n_s;
- Rotor type (single/ double cage, slip rings) along with the ratio between the starting and the rated torque M_s/M_r;
- Ratio between the starting and the rated current I_s/I_r;
- Torque overload capability M_{max}/M_r;
- Type of construction and mounting arrangements (horizontal or vertical shaft, bearing type, flange, feet, etc.);
- Degree of protection provided by enclosure;
- Cooling method

Furthermore, very often certain machine parameters have to be reached, such as rated and starting torque, starting (short circuit) current, rated slip, efficiency and power factor at partial load, etc.

The design of an induction machine for the fundamental spatial harmonic is based on its per-phase equivalent circuit, Fig. 8.1. Real and imaginary components of impedance in Fig. 8.1 are

$$\mathrm{Re}\{\hat{Z}\} = R_1 + \frac{R_{\mathrm{Fe}} X_{\mathrm{main}}^2 \left(R_2'^2 + sR_2' R_{\mathrm{Fe}} + s^2 X_{2\sigma}'^2 \right)}{2sR_2' R_{\mathrm{Fe}} X_{\mathrm{main}}^2 + R_2'^2 \left(R_{\mathrm{Fe}}^2 + X_{\mathrm{main}}^2 \right) + s^2 \left[X_{\mathrm{main}}^2 X_{2\sigma}'^2 + R_{Fe}^2 \left(X_{\mathrm{main}} + X_{2\sigma}' \right)^2 \right]} \tag{8.1}$$

$$\mathrm{Im}\{\hat{Z}\} = X_{1\sigma} + \frac{R_{\mathrm{Fe}} X_{\mathrm{main}}^2 \left[R_2'^2 + sX_{2\sigma}' \left(X_{\mathrm{main}} + X_{2\sigma}' \right) \right]}{2sR_2' R_{\mathrm{Fe}} X_{\mathrm{main}}^2 + R_2'^2 \left(R_{\mathrm{Fe}}^2 + X_{\mathrm{main}}^2 \right) + s^2 \left[X_{\mathrm{main}}^2 X_{2\sigma}'^2 + R_{\mathrm{Fe}}^2 \left(X_{\mathrm{main}} + X_{2\sigma}' \right)^2 \right]} \tag{8.2}$$

Usually, **iron core losses** are calculated separately, because of their negligible influence on power distribution in Fig. 8.1. This is equivalent to connecting the equivalent iron core resistance R_{Fe} in parallel with the terminal voltage U_1 instead of in parallel with U_i, as shown in Fig. 8.1. The error introduced that way is low, knowing that the ratio between induced and applied voltage in motor mode equals about 0.95 for machines with a number of pole pairs $p \leq 7$ and $0.98 - 0.004p$ for $p > 7$. Accordingly, by setting $R_{\mathrm{Fe}} \to \infty$ in Eqs. 8.1 and 8.2, one obtains

$$\mathrm{Re}\{\hat{Z}\} = R_1 + \frac{X_{\mathrm{main}}^2 s R_2'}{R_2'^2 + s^2 \left(X_{\mathrm{main}} + X_{2\sigma}' \right)^2} \tag{8.3}$$

and

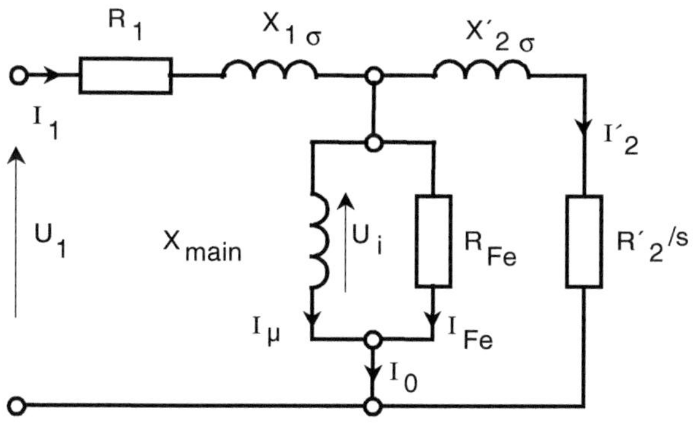

Fig. 8.1 Per-phase equivalent circuit of an induction machine

$$\mathrm{Im}\{\hat{Z}\} = X_{1\sigma} + \frac{X_{\mathrm{main}} \left[R_2'^2 + sX_{2\sigma}' \left(X_{\mathrm{main}} + X_{2\sigma}' \right) \right]}{R_2'^2 + s^2 \left(X_{\mathrm{main}} + X_{2\sigma}' \right)^2} \tag{8.4}$$

Introducing factors σ_1 and σ_2, defined as

$$\sigma_1 = \frac{X_{1\sigma}}{X_{\text{main}}} \approx 0.04 \quad ; \quad \sigma_2 = \frac{X'_{2\sigma}}{X_{\text{main}}} \approx 0.04 \tag{8.5}$$

one can simplify expressions for impedance and write them as

$$R = \text{Re}\{\hat{Z}\} = R_1 + \frac{\frac{R'_2}{s}}{\left(\frac{R'_2}{s \cdot X_{\text{main}}}\right)^2 + (1+\sigma_2)^2} \tag{8.6}$$

$$X = \text{Im}\{\hat{Z}\} = (1+\sigma_1)X_{\text{main}} - \frac{(1+\sigma_2)X_{\text{main}}}{\left(\frac{R'_2}{s \cdot X_{\text{main}}}\right)^2 + (1+\sigma_2)^2} \tag{8.7}$$

At short circuit ($s = 1$) the components of machine impedance are equal to

$$R_{\text{SC}} = R_1 + \frac{R'_2}{\left(\frac{R'_2}{X_{\text{main}}}\right)^2 + (1+\sigma_2)^2} \tag{8.8}$$

$$X_{\text{SC}} = (1+\sigma_1)X_{\text{main}} - \frac{(1+\sigma_2)X_{\text{main}}}{\left(\frac{R'_2}{X_{\text{main}}}\right)^2 + (1+\sigma_2)^2} \tag{8.9}$$

with typical values of

$$\frac{R_1}{X_{\text{SC}}} \approx 0.2; \quad \frac{R'_2}{X_{\text{SC}}} \approx 0.2; \quad \frac{R_{\text{SC}}}{X_{\text{SC}}} \approx 0.4 \tag{8.10}$$

Phase current I_{Ph} can now be expressed by using Eqs. 8.6 and 8.7 as

$$I_{\text{Ph}} = \frac{U_1}{\sqrt{R^2 + X^2}} \tag{8.11}$$

and the short-circuit current I_{SC} as

$$I_{\text{SC}} = \frac{U_1}{\sqrt{R_{\text{SC}}^2 + X_{\text{SC}}^2}} \tag{8.12}$$

or, in p.u.

$$i_{\text{SC}} = \frac{I_{\text{SC}}}{I_{\text{Ph}}} = \frac{U_1}{X_{\text{SC}} I_{\text{rated}}} \frac{1}{\sqrt{1 + \frac{R_{\text{SC}}^2}{X_{\text{SC}}^2}}} = \frac{1}{x_{\text{SC}}} \frac{1}{\sqrt{1 + \frac{R_{\text{SC}}^2}{X_{\text{SC}}^2}}} \tag{8.13}$$

Air gap torque M_δ can be written as

$$M_\delta = \frac{m_1}{\Omega_1} I_1^2 \frac{\frac{R_2'}{s}}{\left(\frac{R_2'}{s \cdot X_{\text{main}}}\right)^2 + (1+\sigma_2)^2} \tag{8.14}$$

or, after substituting for the stator current $I_1 = U_1/Z$:

$$M_\delta = \frac{m_1 U_1^2}{\Omega_1} \frac{1}{2R_1 + \frac{R_2'}{s}(1+\sigma_1)^2 \left[1 + \left(\frac{R_1}{X_{\text{main}} + X_{1\sigma}}\right)^2\right] + \frac{2}{R_2'}\left\{R_1^2(1+\sigma_2)^2 + \left[X_{1\sigma} + X_{2\sigma}'(1+\sigma_1)\right]^2\right\}} \tag{8.15}$$

with Ω_1 denoting mechanical synchronous speed created by stator MMF. Maximum (pullout) torque $M_{\delta,m}$ is equal to

$$M_{\delta,m} = \frac{m_1 U_1^2}{\Omega_1} \frac{1}{R_1 + (1+\sigma_1)^2 \sqrt{R_1^2(1+\sigma_2)^2 + \left[X_{1\sigma} + X_{2\sigma}'(1+\sigma_1)\right]^2} \sqrt{1 + \left(\frac{R_1}{X_{\text{main}} + X_{1\sigma}}\right)^2}} \tag{8.16}$$

and it is created at pullout slip s_m equal to

$$s_m = \frac{R_2'}{\sqrt{R_1^2(1+\sigma_2)^2 + \left[X_{1\sigma} + X_{2\sigma}'(1+\sigma_1)\right]^2}} (1+\sigma_1) \sqrt{1 + \left(\frac{R_1}{X_{\text{main}} + X_{1\sigma}}\right)^2} \tag{8.17}$$

After simplifying, one can write

$$M_{\delta,m} = \frac{m_1 U_1^2}{2\Omega_1 X_{\text{SC}}} \frac{1}{\frac{R_1}{X_{\text{SC}}} + \sqrt{1 + \left(\frac{R_1}{X_{\text{SC}}}\right)^2}} \tag{8.18}$$

as well as

$$s_m = \frac{R_2'}{X_{\text{SC}}} \frac{1}{(1+\sigma_2)^2 \sqrt{1 + \left(\frac{R_1}{X_{\text{SC}}}\right)^2}} \tag{8.19}$$

Now one can express the p.u. pullout torque m_m as

$$m_m = \frac{M_{\delta,m}}{M_{\text{rated}}} = \frac{1}{2x_{SC}} \frac{1 - s_{\text{rated}}}{\eta_{\text{rated}} \cos \varphi_{\text{rated}}} \frac{1}{\frac{R_1}{X_{SC}} + \sqrt{1 + \left(\frac{R_1}{X_{SC}}\right)^2}} \frac{1}{1 + \frac{P_{f,w}}{P_{\text{rated}}}} \tag{8.20}$$

as well as the p.u. starting torque m_{SC}

$$m_{SC} = \frac{M_{SC}}{M_{\text{rated}}} = c_s \left(\frac{I_{SC}}{I_{\text{rated}}}\right)^2 s_{\text{rated}} k \frac{1 + \left(\frac{1}{s_{\text{rated}}} \frac{R_2'}{X_{2\sigma}'}\right)^2}{1 + \left(k \frac{R_2'}{X_{2\sigma}'}\right)^2} \frac{1}{1 + \frac{P_{f,w}}{P_{\text{rated}}} m_{SC}} \tag{8.21}$$

where the ratio $P_{f,w}/P_{\text{rated}}$ varies between 0.75 and 1.5 % for $p = 1$, 0.45 and 0.9 % for $p = 2$, and 0.3 and 0.6 % for $p = 3$ or 4, all for the rated power P_{rated} between 0.5 and 50 kW. Factor c_s in Eq. 8.21 reflects various levels of saturation at short circuit and at rated point, whereas k stands for increase of rotor resistance due to skin effect.

The p.u. magnetizing current i_μ can be expressed as

$$i_\mu = \frac{I_\mu}{I_{\text{rated}}} = \frac{U_i}{X_{\text{main}}} \frac{1}{I_{\text{rated}}} \tag{8.22}$$

with

$$U_i = \omega B_1 l_{\text{ax}} \tau_p w_{1,\text{Ph}} f_{w1} \frac{\sqrt{2}}{\pi} \tag{8.23}$$

$$I_{\text{rated}} = \frac{p\tau_p}{m_1 w_{1,\text{Ph}} f_{w1}} \frac{A_1}{\sqrt{2}} \tag{8.24}$$

and

$$X_{\text{main}} = \omega \frac{m_1}{\pi} \frac{\mu_0 D l_{\text{ax}}}{\delta \cdot k_\delta p^2} w_{1,\text{Ph}}^2 f_{w1}^2 \tag{8.25}$$

and k_δ standing for increase of air gap width due to slotting and saturation in iron. Substituting Eqs. 8.23–8.25 in Eq. 8.22, one obtains

$$i_\mu = \frac{\pi}{\mu_0} \frac{B_1}{A_1} \frac{\delta \cdot k_\delta}{\tau_p} \tag{8.26}$$

Keeping machine dimensions constant, the p.u. magnetizing current changes proportionally to the ratio B_1/A_1. Keeping electromagnetic loading and level of saturation constant, the p.u. magnetizing current changes proportionally to the ratio of δ/τ_p.

One of the basic parameters characterizing an AC machine is the ratio λ between its active length ℓ and pole pitch τ_p

$$\lambda = \frac{l}{\tau_p} \tag{8.27}$$

When starting new design from scratch [9, 10], the ratio λ helps one define machine active volume based upon the apparent power S_i crossing air gap

$$S_i = m_1 \cdot U_i \cdot I_1 = \frac{U_i}{U_1} S_{\text{rated}} \tag{8.28}$$

and synchronous speed n_s

$$n_s = \frac{60 f_1}{p} \tag{8.29}$$

Substituting for the induced voltage U_i in Eq. 8.2

$$U_i = 2\pi \cdot f_1 w_{1,\text{Ph}} f_{w1} \frac{2}{\pi\sqrt{2}} l \tau_p B_1 = 2\sqrt{2} f_1 w_{1,\text{Ph}} f_{w1} l \tau_p B_1 \tag{8.30}$$

as well as for the phase current I_1

$$I_1 = A_1 \frac{p \tau_p}{\sqrt{2} m_1 w_1 f_{w1}} \tag{8.31}$$

one obtains for the apparent power crossing air gap

$$S_i = 2 f_1 p \tau_p^2 l A_1 B_1 = \frac{U_i}{U_1} S_{\text{rated}} \tag{8.32}$$

and the pole pitch τ_p

$$\tau_p = \sqrt[3]{\frac{S_{\text{rated}}}{p\lambda}} \sqrt{\frac{\frac{U_i}{U_1}}{2 f A_1 B_1}} = c_\tau \sqrt[3]{\frac{S_{\text{rated}}}{p\lambda}} \tag{8.33}$$

In the previous considerations, the minimum (integer) number of slots and the insulation thickness were not taken into account. With these details, the pole pitch τ_p can be expressed as

$$\tau_p = a + c_\tau \cdot \sqrt[3]{\frac{S_{\text{rated}}}{p\lambda}} \tag{8.34}$$

with coefficients a and c_τ being usually in the range given in Table 8.1 [1, 2].

Table 8.1 Orientational values of coefficients a and c_τ for induction machines

	Machine type	a [mm]	c_τ [mm/(VA)$^{1/3}$]
Moderately utilized	Squirrel cage, 2–12 poles, 0.1–100 kW	35	6.5
	Slip rings, 2–12 poles, 1–250 kW	35	6.5
Highly utilized	Squirrel cage, 2-pole, low voltage	60	5.2
	Slip rings, 2-pole, low voltage	50	6.0
	Induction machine, 2–12 poles, <500 V	20	5.2
	Induction machine, high voltage	30	5.3

Usual values of **slot pitch** τ_s lie between 15 and 50 mm. The number of **slots per pole and phase** $q = N/(2pm)$ should not be larger than 5, because of no significant improvement of MMF spatial distribution and not smaller than 3, because of too strong MMF harmonics.

The **apparent rotational thrust** σ_A is a quantity with the same dimension as torque density, however, with a slightly different physical meaning. Whereas the torque density has a meaning of torque per volume, the apparent rotational thrust is equal to apparent force per area. The apparent rotational thrust should not be confused with pressure, which has the same dimension [N/m^2], but a different meaning: The pressure is equal to the *normal* component of force acting on a given surface, whereas the apparent rotational thrust is equal to the *tangential* component of force acting on a surface

$$\sigma_A = \frac{F_\mathrm{A}}{A} = \frac{M_\mathrm{A}}{2R^2\pi l_\mathrm{ax}} \tag{8.35}$$

where apparent force F_A and apparent torque M_A correspond to apparent power S

$$M_A = \frac{S_\mathrm{rated}}{\Omega_1}\frac{U_i}{U_1} = \frac{pS_\mathrm{rated}}{\pi^2 D^2 l_\mathrm{ax} f_1}\frac{U_i}{U_1} = \frac{D^2\pi}{4} l_\mathrm{ax} A_1 B_1 \tag{8.36}$$

By applying Eq. 8.36, one can express the product $A_1 \cdot B_1$ as

$$A_1 B_1 = 2\frac{pS_\mathrm{rated}}{\pi^2 D^2 l_\mathrm{ax} f_1}\frac{U_i}{U_1} \tag{8.37}$$

and the apparent rotational thrust σ_A

$$\sigma_\mathrm{A} = \frac{1}{2} A_1 \cdot B_1 \tag{8.38}$$

The p.u. short-circuit reactance x_SC, defined in Eq. 8.13 as

$$x_\mathrm{sc} = X_\mathrm{SC}\frac{I_\mathrm{rated}}{U_1} \tag{8.39}$$

determines both the p.u short-circuit current i_{SC} (Eq. 8.13) and the p.u. pullout torque m_m, Eq. 8.20. Whereas the short-circuit current must not be larger than a given value, the pullout torque must not be smaller than a given value. Accordingly, one can define the limits for the p.u. short-circuit reactance as

$$x_{SC,min} = \frac{1}{i_{SC,max}} \frac{1}{\sqrt{1 + \frac{R_{SC}^2}{X_{SC}^2}}} \tag{8.40}$$

For the ratio $R_{SC}/X_{SC} \approx 0.4$ (Eq. 8.10) one obtains

$$x_{SC,min} \approx \frac{0.93}{i_{SC,max}} \tag{8.41}$$

Analogously, for $x_{SC,max}$ one can write by using Eq. 8.20

$$x_{SC,max} = \frac{1}{m_{m,min}} \frac{1}{2} \frac{1 - s_{rated}}{\eta_{rated} \cos \varphi_{rated}} \frac{1}{\frac{R_1}{X_{SC}} + \sqrt{1 + \left(\frac{R_1}{X_{SC}}\right)^2}} \frac{1}{1 + \frac{M_{f,w}}{M_m}} \tag{8.42}$$

A reasonably well-designed induction machine is characterized by

$$x_{SC,min} \leq x_{SC} \leq x_{SC,max} \tag{8.43}$$

If the condition 8.43 cannot be fulfilled, a slip ring machine should be used instead of a squirrel cage one.

The short-circuit reactance X_{SC} can be expressed by means of the stator and rotor leakage reactances as

$$X_{SC} \approx X_{1\sigma} + \frac{X'_{2\sigma}}{1 + \sigma_2} \tag{8.44}$$

where (Eq. 4.111)

$$X_{1\sigma} = \omega_1 m_1 \mu_0 l_{ax} \frac{w_{1,Ph}^2}{N_1} \lambda_{1,res} \tag{8.45}$$

and

$$X'_{2\sigma} = \omega_1 m_2 \mu_0 l_{ax} \frac{w_{2,Ph}^2}{N_2} \lambda_{2,res} \cdot \left[\frac{m_1}{m_2} \left(\frac{w_{1,Ph} f_{w,1}}{w_{2,Ph} f_{w,2}} \right)^2 \right] \tag{8.46}$$

After some mathematical manipulations, one can write for X_{SC}

$$X_{sc} = \omega_1 m_1 \mu_0 l_{ax} \frac{w_{1,Ph}^2}{N_1} \lambda_{SC} \sim \frac{w_{1,Ph}^2}{N_1} \lambda_{SC} \tag{8.47}$$

where the value of λ_{SC} can be reconstructed from expressions 8.45–8.47.
On the other hand, based upon Eq. 8.39

$$x_{sc} = X_{SC} \frac{I_{rated}}{U_1} = X_{SC} \frac{I_{rated}}{U_{i,1} \frac{U_1}{U_{i,1}}} \tag{8.48}$$

and substituting for

$$U_{i,1} = \omega_1 w_{1,Ph} f_{w,1} \frac{\Phi_1}{\sqrt{2}} \tag{8.49}$$

$$\Phi_1 = \frac{2}{\pi} B_1 l_{ax} \tau_p \tag{8.50}$$

$$I_{rated} = A_1 \frac{p \tau_p}{m_1 w_{1,Ph} f_{w,1} \sqrt{2}} \tag{8.51}$$

one can write

$$x_{sc} = \frac{\pi p \mu_0}{2 f_{w,1}^2} \frac{U_{i,1}}{U_1} \frac{\lambda_{SC}}{N_1} \frac{A_1}{B_1} \sim \frac{A_1}{B_1} \tag{8.52}$$

By combining Eqs. 8.41, 8.42, and 8.52, one can write

$$\left(\frac{B_1}{A_1}\right)_{max} \sim \frac{1}{x_{SC,min}} \sim i_{SC,max} \tag{8.53}$$

and

$$\left(\frac{B_1}{A_1}\right)_{min} \sim \frac{1}{x_{SC,max}} \sim m_{m,min} \tag{8.54}$$

Equations 8.38, 8.53, and 8.54 define the product $P_{AB} = A_1 \cdot B_1$ as well as the quotient $Q_{AB} = B_1/A_1$ as functions of machine rated data. The quantities P_{AB} and Q_{AB} are basic parameters in the machine design procedure.

Number of turns per phase $w_{1,\mathrm{Ph}}$ is calculated after determining the amount of air gap flux density B_1

$$B_1 = \sqrt{P_{AB} Q_{AB}} = \sqrt{(A_1 \cdot B_1)\left(\frac{B_1}{A_1}\right)} \tag{8.55}$$

and for previously selected values of D and ℓ_{ax}. The main flux is then evaluated as

$$\Phi_1 = \frac{2}{\pi} l_{\mathrm{ax}} \tau_p B_1 \tag{8.56}$$

and the induced voltage per turn u_1

$$u_1 = \frac{\omega_1}{\sqrt{2}} f_{w,1} \Phi_1 \tag{8.57}$$

The number of turns per phase is selected in such a manner that the induced voltage U_{i} corresponds to the phase voltage U_1 in the manner discussed at the beginning of this section. Knowing the number of turns per phase $w_{1,\mathrm{Ph}}$, one can determine the number of conductors per slot z_s as

$$z_s = 2 \frac{m_1 a_1}{N_1} w_{1,\mathrm{Ph}} \tag{8.58}$$

where a_1 denotes the number of parallel branches. The number of conductors per slot is an integer value, which in the case of double-layer winding can only be even.

The bigger the machine, the smaller the number of conductors per slot. Whereas in large machines only one conductor per layer (bar) is necessary in order to obtain the proper value of induced voltage, the coils of small machines have typically a double-digit number of turns. Consequently, the requested value of air gap flux density is much easier to reach accurately in small than in large machines. If the discrepancy between induced and applied voltage is too large, one can try to correct the quotient $\lambda_{\mathrm{SC}}/N_1$ in Eq. 8.48, or to decrease the rotational thrust. The latter measure increases the machine size; therefore, at this point is a good idea to rethink the machine's main dimensions

Air gap width δ of conventional induction machines varies typically between 0.2 and 2 mm, as a function of rated speed and active length. Air gap width should be minimal in order to minimize the magnetizing current after Eq. 8.26. Typical ratios δ/τ_{p} vary between 0.0025 and 0.004 for $p = 1$, 0.006 and 0.007 for $p = 6$, and 0.012 and 0.013 for $p = 30$, the first number standing for short, and the second for long machines.

Air gap flux density represented with amplitude of its fundamental component B_1 typically remains below 1 T, in order to avoid too high iron losses in teeth and too strong air gap radial forces. On the other hand, the values of B_1 below 0.65 T are seldom and characteristic for small, low-utilized machines. If the air gap flux density turns out to be too high, the same countermeasures have to be undertaken as in the case of too a large discrepancy between induced and applied voltage.

Winding resistance is evaluated by using standard equation

$$R_w = \rho(\vartheta)\frac{l_w w_{\mathrm{Ph}}}{S_C} f_{\sim} \tag{8.59}$$

with $\rho(\vartheta)$ denoting specific electric resistance at temperature ϑ, S_C the cross-sectional area of a single conductor, and $f_{\sim}$ the factor of resistance increase due to skin effect. Equation 8.59 is valid both for stator phase winding and for rotor slip ring phase winding.

Squirrel cage resistance is evaluated by means of I^2R losses in it

$$P_{\mathrm{loss,cage}} = N_2 R_B I_B^2 + 2R_R I_R^2 \tag{8.60}$$

with R_B and R_R denoting the bar and ring resistance, respectively

$$R_B = \rho_B(\vartheta)\frac{l_B}{S_B} f_{\sim,B} \tag{8.61}$$

$$R_R = \rho_R(\vartheta)\frac{D_R \pi}{S_R} f_{\sim,R} \tag{8.62}$$

and D_R the ring average diameter. Substituting for the ratio between ring and bar current [3]

$$\frac{I_R}{I_B} = \frac{1}{\sin\frac{p\pi}{N_2}} \tag{8.63}$$

one can rewrite Eq. 8.60 as

$$P_{\mathrm{loss,cage}} = N_2 R_B I_B^2 + \frac{2R_R I_B^2}{\sin^2\frac{p\pi}{N_2}} = N_2 R_2 I_B^2 \tag{8.64}$$

with R_2 standing for equivalent bar resistance

$$R_2 = \rho_B(\vartheta)\frac{l_B}{S_B} f_{\sim,B}\left[1 + \frac{\rho_R(\vartheta)}{\rho_B(\vartheta)} \cdot \frac{S_B}{S_R} \cdot \frac{D_R \pi}{l_B N_2} \cdot \frac{1}{2\sin^2\frac{p\pi}{N_2}} \cdot \frac{f_{\sim,R}}{f_{\sim,B}}\right] \tag{8.65}$$

Introducing ratio r defined as

$$r = \frac{\rho_R(\vartheta)}{\rho_B(\vartheta)} \cdot \frac{S_B}{S_R} \cdot \frac{D_R \pi}{l_B N_2} \cdot \frac{1}{2\sin^2\frac{p\pi}{N_2}} \tag{8.66}$$

one can express the equivalent bar resistance R_2 in Eq. 8.66 as

$$R_2 = \rho_B(\vartheta)\frac{l_B}{S_B}f_{\sim,B}\left(1 + r\frac{f_{\sim,R}}{f_{\sim,B}}\right) \tag{8.67}$$

Denoting by k the ratio between equivalent rotor bar resistance at short circuit and at rated point (Eq. 8.21)

$$k = \frac{R_{B,\text{SC}}}{R_{B,\text{rated}}} = \frac{f_{\sim,B,\text{SC}}}{f_{\sim,B,\text{rated}}} \cdot \frac{1 + r\frac{f_{\sim,R,\text{SC}}}{f_{\sim,B,\text{SC}}}}{1 + r\frac{f_{\sim,R,\text{rated}}}{f_{\sim,B,\text{rated}}}} \tag{8.68}$$

one can express the p.u. starting torque (Eq. 8.21) as

$$m_{\text{SC}} = c \cdot s_{\text{rated}}\left(\frac{I_{\text{SC}}}{I_{\text{rated}}}\right)^2 \cdot k \tag{8.69}$$

Obviously, there exists a minimum value of the coefficient k for which the machine still develops a given p.u. starting torque. If this is not the case, a slip ring rotor should be used instead.

Assuming that losses are taken from the rotor only on its lateral area, one can define the loss density as a ratio P_{loss}/S

$$\frac{P_{\text{loss}}}{S} = \frac{N_2 R_2 I_B^2}{D\pi l_{\text{ax}}} \tag{8.70}$$

with a dimension of kW/m^2. Loss density in the rotor of air-cooled squirrel cage machines can reach the value of ~ 10 kW/m^2, which is about 50 % above the loss density on the stator side. The reason for higher allowed loss density in a squirrel cage rotor is the absence of conductor and slot insulation. Usually, the loss density in bars is higher than in rings, in order to avoid mechanical damage of rings: Whereas thermal dilatation of bars in axial direction has almost no mechanical consequences, the expansion of rings can be a reason for their breakage.

Stator conductor cross-sectional area is selected in such a manner as to fulfill the cooling specifications. Slot width w_s is approximately equal to half of the slot pitch τ_s; however, this ratio varies from one machine to another. Slot height h_s is selected in such a manner as to provide enough space for obtaining the allowed current density Γ in conductors

$$h_s = \frac{A_1}{\Gamma}\frac{\tau_S}{w_s}\frac{1}{\sqrt{2}f_{w,1}f_S} \tag{8.71}$$

with f_S denoting the slot fill factor, i.e., the ratio between conductor and slot area. The slot fill factor for windings with round wires can exceed 40 %, and for formed coils it goes up to 70 %. The highest slot fill factor is achieved in squirrel cage machines, where it exceeds 90 %.

Since smaller machines dissipate less losses per heat exchange surface than larger, their current density may go up to 10 A/mm^2. Large air-cooled machines have typical current densities between 3 and 4 A/mm^2. Highest current densities are found in machines with directly (water, oil) cooled conductors, where they reach values of typically 15 A/mm^2. The amplitude of current sheet in smallest machines is about 30–50 kA/m, in medium size machines about 70–90 kA/m, and in largest machines ~350 kA/m. Accordingly, the ratio A_1/Γ in Eq. 8.60 varies between ~3 mm for the smallest, over ~20 mm for medium size up to ~23 mm for the largest machines.

Magnetic circuit at no load can be solved by using the magnetic equivalent circuit method (MEC), as shown in [4]. In the approach described in this reference, nonlinear magnetic circuits are solved in a straightforward manner, without iterations. This way the computational time is significantly reduced, as compared to similar methods, such as finite element analysis, and at the same time, the computational accuracy is increased. As opposed to the finite element method, which tolerates large local errors, the field distribution evaluated with the MEC method is calculated with the same level of accuracy in all elements.

8.3 Sizing Equations of a Synchronous Machine

The largest electric machines ever built are synchronous machines, in particular high-speed turbogenerators. The overwhelming field of applications of a synchronous machine is the generation of active and reactive power, the latter being controlled by the field current.

Whereas conventional synchronous machines are built with a single- or three-phase armature winding on the stator side and field winding on the rotor side, exciters for large synchronous machines are built inside-out: Their armature winding rotates, along with rectifier diodes, and the field winding is placed on the stator side.

The source of air gap flux inducing no-load voltage can be either the field winding, or permanent magnets, Fig. 8.2.

Denoting by $\alpha_1 = 2/\pi$ the ratio between the average and maximum value of sine function, and by $\beta = B_{max}/B_1$ the ratio between the maximum of the air gap flux density, which determines the level of saturation and the amplitude of its fundamental harmonic, one can express the air gap flux Φ_1 created by the fundamental component of flux density as

$$\Phi_1 = \frac{\alpha_1}{\beta} \tau_p l_{ax} B_{max} \tag{8.72}$$

Current sheet A_s created by z_s conductors per slot of the field winding, each carrying the current I, can be expressed as

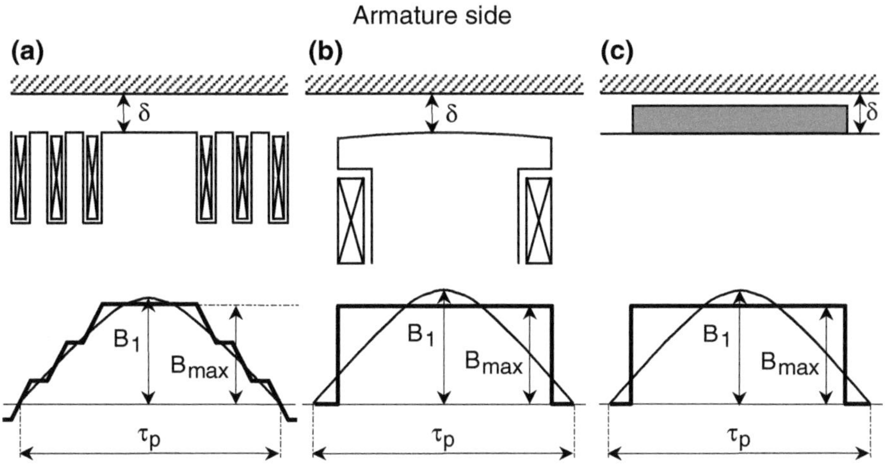

Fig. 8.2 Modi of field generation in cylindrical wound rotor (**a**), salient pole wound rotor (**b**), and surface mounted permanent magnet synchronous machine (**c**)

$$A = \frac{z_s I}{\frac{D\pi}{N}} = \frac{m \cdot w_{\mathrm{Ph}} I}{p\tau_p} \tag{8.73}$$

The amplitude of the fundamental component of current sheet is then

$$A_1 = \sqrt{2}A \cdot f_{w,1} \tag{8.74}$$

Synchronous machines are predominantly used as generators of both active and reactive power. In order to supply reactive power into the power system, a synchronous machine has to be overexcited. Typically, the field MMF at rated operating point is 1.5—3 times the armature MMF, which makes the rotor of a synchronous machine its bottleneck. One should keep in mind that a large field MMF results in increased i^2R losses, which have to be taken out of the rotor.

The **no-load curve** of a synchronous machine is usually evaluated by applying the MEC method [4]. Rotor yoke is split into a network of elements, the reluctances of which determine required MMF for a given flux.

The **short-circuit ratio** k_C is defined as a quotient between the field current required to induce the rated voltage at no load and the field current which generates rated armature current at short circuit. The reciprocal of k_C is equal to the p.u. value of unsaturated synchronous reactance in the d-axis x_d

$$k_C = \frac{\frac{U_{\text{rated}}}{I_{\text{rated}}}}{X_d} = \frac{1}{x_d} \tag{8.75}$$

with X_d being defined as

$$X_d = X_{\text{ad}} + X_{a\sigma} \tag{8.76}$$

where X_{ad} stands for reactance due to armature reaction in the d-axis, as defined in Eq. 8.25, and $X_{a\sigma}$ is the stator leakage reactance

$$X_{a\sigma} = \omega m \mu_0 l_{\text{ax}} \frac{w_{\text{Ph}}^2}{N} \lambda \tag{8.77}$$

and

$$x_{a\sigma} = \frac{X_{a\sigma}}{Z_{\text{rated}}} = \frac{X_{a\sigma}}{\frac{U_{\text{rated}}}{I_{\text{rated}}}} \tag{8.78}$$

By simplifying, the reactance X_d can be written as

$$X_d = X_{\text{ad}}(1 + \sigma_1) \tag{8.79}$$

and the rated voltage U_{rated} and rated current I_{rated} as

$$U_{\text{rated}} = \frac{U_{\text{rated}}}{U_i} \omega w_{\text{Ph},1} f_{w,1} \frac{\sqrt{2}}{\pi} B_1 l_{\text{ax}} \tau_p \tag{8.80}$$

$$I_{\text{rated}} = A_1 \frac{p \tau_p}{m w_{\text{Ph},1} f_{w,1} \sqrt{2}} \tag{8.81}$$

After inserting for U_{rated}, I_{rated} and X_d, one can express the p.u. synchronous reactance x_d as

$$x_d = \frac{\mu_0}{\pi} (1 + \sigma_1) \frac{U_i}{U_{\text{rated}}} \frac{\tau_p}{k_\delta \delta} \frac{A_1}{B_1} \tag{8.82}$$

Since

$$\frac{U_i}{U_{\text{rated}}} \approx 1 + \sigma_1 \tag{8.83}$$

one can express the short-circuit ratio as

$$k_C = \frac{1}{(1+\sigma_1)^2} \frac{\pi}{\mu_0} \frac{k_\delta \delta}{\tau_p} \frac{B_1}{A_1} = \frac{1}{(1+\sigma_1)} \frac{U_{\text{rated}}}{U_i} \frac{\pi}{\mu_0} \frac{k_\delta \delta}{\tau_p} \frac{B_1}{A_1} \tag{8.84}$$

The **armature reactance MMF** $\Theta_{a,\text{rated}}$ per pole pair at rated current is equal to (see Fig. 8.3)

$$\Theta_{a,\text{rated}} = \frac{m_1}{2} \frac{w_{\text{Ph},1} f_{w,1}}{p \cdot f_f} \sqrt{2} I_{\text{rated}} \tag{8.85}$$

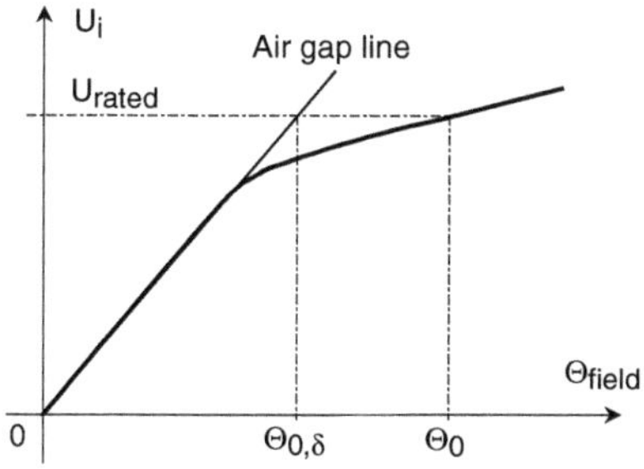

Fig. 8.3 MMF components at armature open circuit

Denoting by $\Theta_{0,\delta}$ the air gap MMF drop across the air gap created by field current at no load, one can define the unsaturated air gap reactance (p.u.) in d-axis as

$$x_{adi} = \frac{\Theta_{a,\text{rated}}}{\Theta_{0,\delta}} \tag{8.86}$$

The p.u. unsaturated air gap reactance in q-axis is

$$x_{aq} = x_{adi} \frac{k_{CRd}}{k_{CRq}} \tag{8.87}$$

with $k_{\text{CR}d}$ and $k_{\text{CR}q}$ denoting the Carter factor in the d- and q-axis, respectively. Saturated air gap reactance in p.u. is defined as

$$x_{adr} = \frac{1}{k_C} - x_{a\sigma}(i_a) \tag{8.88}$$

and unsaturated synchronous reactance in d- and q-axis, respectively, as

$$x_d = x_{adi} + x_{a\sigma}(i_a) \tag{8.89}$$

$$x_q = x_{aq} + x_{a\sigma}(i_a) \tag{8.90}$$

Rotor slots are characterized by specific permeances [3], defined as

$$\sigma_f = p\frac{\pi k_\delta \delta}{2D_{\text{ave}}}\frac{\lambda_f}{N_f f_f^2} \tag{8.91}$$

for the field winding, with D_{ave} denoting the diameter of the center of air gap, N_{f} the number of rotor slots with field winding, and f_{f} the field winding factor;

$$\sigma_{fD} = p\frac{\pi k_\delta \delta}{2D_{\text{ave}}}\frac{\lambda_{fD}}{N_D f_f f_D} \tag{8.92}$$

for the common effects of damper and field winding, where N_{D} denotes the number of rotor slots with damper winding and f_{D} the damper winding factor;

$$\sigma_d = p\frac{\pi k_\delta \delta}{2D_{\text{ave}}}\frac{\lambda_D}{N_D f_d} \tag{8.93}$$

for the damper winding in the d-axis, and

$$\sigma_q = p\frac{\pi k_\delta \delta}{2D_{\text{ave}}}\frac{\lambda_Q}{N_{\text{D}} f_q} \tag{8.94}$$

for the damper winding in the q-axis. The specific permeances of rotor windings help define the p.u. values of characteristic reactances [3] as

$$x_{cD} = \sigma_{fD} x_{adi} \tag{8.95}$$

for the rotor coupling reactance along the d-axis;

$$x_{c\sigma} = \left(\sigma_f - \sigma_{fD}\right) x_{adi} \tag{8.96}$$

for the common leakage reactance of the field and damper winding in the d-axis;

$$x_{cd} = \left(\sigma_d - \sigma_{fD}\right) x_{adi} \tag{8.97}$$

for the ideal leakage reactance of the damper winding in the d-axis, and

$$x_{cq} = \sigma_q x_{aq} \tag{8.98}$$

for the ideal leakage reactance of the damper winding in the q-axis.

More about computation of transient and subtransient machine parameters one can find in [5–7].

References

1. Sirotic Z, Krajzl V (1968) Electromagnetic design of synchronous machines. University of Zagreb, Zagreb
2. Stier P (1960) Entwurf und Berechnung von Drehstrom-Induktionsmaschinen. Technische Hochschule Fridericiana, Karlsruhe
3. Ostovic V (1994) Computer-Aided Analysis of Electric Machines: a Mathematica Approach. Prentice-Hall, London
4. Ostovic V (1989) Dynamics of saturated electric machines. Springer, New York
5. Canay M (1969) Ersatzschemata der Synchronmaschine zur berechnung von Polradgroessen bei nichtstaionaeren Vorgaengen sowie asynchronem Anlauf, Teil 1: Turbogeneratoren. Brown Boveri Mitteilungen 56(2):60–71
6. Canay M (1970) Ersatzschemata der Synchronmaschine zur berechnung von Polradgroessen bei nichtstaionaeren Vorgaengen sowie asynchronem Anlauf, Teil 2: Schenkelpolmaschinen. Brown Boveri Mitteilungen 57(3):135–145
7. Klamt J (1962) Berechnung und Bemessung elektrischer Maschinen. Springer, Berlin
8. Jurkovic B (1970) Electromagnetic design of squirrel cage induction machines. University of Zagreb, Zagreb
9. Jurkovic B (1970) Electromagnetic design of wound rotor induction machines. University of Zagreb, Zagreb
10. Stier P (1959) Entwurf und Berechnung von Drehstrom-Synchronmaschinen. Technische Hochschule Fridericiana, Karlsruhe

Appendix

A.1 Orthogonal Functions

The properties of the most pronounced parameters of an electric machine, such as torque, electric power, and induced voltage, are defined by time and/or spatial orthogonality of physical variables which determine their primordial quantities: mechanical, magnetic, and electric energy.

Mathematically speaking, two functions $f(x)$ and $g(x)$ are orthogonal over the interval $a \leq x \leq b$ if and only if

$$\int_a^b f(x) \cdot g(x)\mathrm{d}x = 0 \tag{A.1.1}$$

Physical quantities, which determine properties of an electric machine, are *periodical* functions of time, space, or both of them. Basic means to handle periodical functions is the Fourier analysis. A periodic function can have:

- a constant term;
- the fundamental harmonic; and
- higher harmonics.

Two periodic functions are *orthogonal* to each other if they have different lengths of periods. If two periodic functions have the same length of period, they are *non-orthogonal*. The fundamental physical meaning of orthogonal functions is that they cannot interact with each other in a sense of producing energy or concatenating fluxes; non-orthogonal functions, on the opposite, are those which create electric power, mechanical torque, mutual inductance, etc.

Three periodic functions $f(x)$, $g(x)$, and $h(x)$, defined as

$$f(x) = F\cos(\alpha \cdot x) \tag{A.1.2}$$

$$g(x) = G\cos(\beta \cdot x) \tag{A.1.3}$$

V. Ostović, *The Art and Science of Rotating Field Machines Design*,
DOI 10.1007/978-3-319-39081-9

$$h(x) = H\cos(\gamma \cdot x) \tag{A.1.4}$$

are non-orthogonal to each other over the interval $a \leq x \leq b$ if and only if

$$\int_a^b \cos(\alpha \cdot x) \cdot \cos(\beta \cdot x) \cdot \cos(\gamma \cdot x) \mathrm{d}x \neq 0 \tag{A.1.5}$$

which is fulfilled for

$$\alpha \pm \beta \pm \gamma = 0 \tag{A.1.6}$$

By using addition theorems, the integrand in Eq. A.1.5 can be further written as

$$\begin{aligned} &\frac{1}{2}\cos(\alpha \cdot x)[\cos(\beta + \gamma)x + \cos(\beta - \gamma)x] \\ &= \frac{1}{4}[\cos(\alpha + \beta + \gamma)x + \cos(\alpha - \beta - \gamma)x \\ &\quad + \cos(\alpha + \beta - \gamma)x + \cos(\alpha - \beta + \gamma)x] \end{aligned} \tag{A.1.7}$$

If the interval length $[a, b]$ is an integer multiple of $(\alpha \pm \beta \pm \gamma)x$, the integral in Eq. A.1.5 will be different from zero if and only if the condition A.1.6 is fulfilled, i.e., if the particular combination of coefficients α, β, and γ yields 0 and one of the four summands in Eq. A.1.7 becomes a constant equal to 1.

Substituting for

$$\alpha = \frac{2}{D}p_F; \quad \beta = \frac{2}{D}p_G; \quad \gamma = \frac{2}{D}p_H \tag{A.1.8}$$

with p_i denoting the number of pole pairs, and $i = F, G, H$, one can rewrite the condition A.1.6 as

$$p_F \pm p_G \pm p_H = 0 \tag{A.1.9}$$

Three air gap distributions are non-orthogonal, i.e., capable of creating torque, if their numbers of pole pairs satisfy condition A.1.9.

Orthogonality of time functions: The length of period T, i.e., time interval after which for a function of time applies $f(t) = f(t + T)$, is a function of angular frequency $\omega = 2\pi f$:

$$T = \frac{2\pi}{\omega} = \frac{1}{f} \tag{A.1.10}$$

with f denoting the frequency [Hz].

Voltage $u(t)$ is defined as

$$u(t) = U_{\max} \sin(n\omega t) \tag{A.1.11}$$

and current $i(t)$ is defined as

$$i(t) = I_{\max} \sin(m\omega t - \varphi_0) \tag{A.1.12}$$

where n and m are positive integers, are orthogonal over interval $-\pi \le \omega t \le \pi$ if the electrical energy W_{el} in the given interval is equal to zero, i.e., if

$$W_{el} = \int_{-\frac{\pi}{\omega}}^{\frac{\pi}{\omega}} U_{\max} \sin(n\omega t) \cdot I_{\max} \sin(m\omega t - \varphi_0) \mathrm{d}t = 0 \tag{A.1.13}$$

which is fulfilled for

$$n \neq m \tag{A.1.14}$$

If both voltage and current have the same frequency, i.e., if

$$n = m \tag{A.1.15}$$

the energy in the given interval is equal to

$$W_{el} = U_{\max} \cdot I_{\max} \cdot \frac{\pi}{\omega} \cdot \cos \varphi_0 \tag{A.1.16}$$

This energy corresponds to an average (active!) power P_{el} of

$$P_{el} = \frac{W_{el}}{T} = \frac{\omega}{2\pi} \cdot U_{\max} \cdot I_{\max} \cdot \frac{\pi}{\omega} \cdot \cos \varphi_0 = U \cdot I \cdot \cos \varphi_0 \tag{A.1.17}$$

with U and I denoting the rms values of voltage and current, respectively.

If an alternating voltage has a different frequency than an alternating current, i.e., if they are orthogonal over a given interval, the energy created by the two is equal to zero. A current must have the same frequency as a voltage in order to create with it electric energy different from zero; the amount of created energy is proportional to the cosine of the phase angle between current and voltage.

If the periodical voltage and current functions contain higher harmonics terms, i.e., if they can be represented as

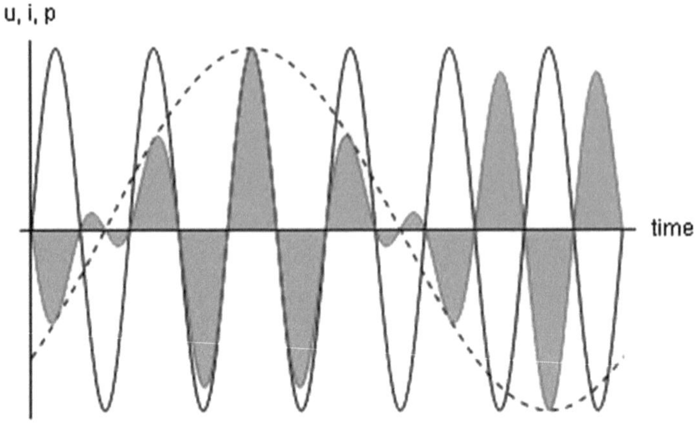

Fig. A.1 1 p.u. voltage $u(t) = \sin(\omega t - \pi/4)$ (*dashed black curve*), 1 p.u. current $i(t) = \sin(6\omega t)$ (*solid black curve*), and their product—instantaneous power $p(t) = u(t) \cdot i(t)$ (*gray curve*). The area between the product of voltage and current represents electrical energy, the average of which in a given interval is equal to zero. In this case, the voltage and the current are orthogonal because of their different frequencies

$$u(t) = \sum_{k=1}^{\infty} U_{\max,k} \sin(k\omega t); \quad i(t) = \sum_{l=1}^{\infty} I_{\max,l} \sin(l\omega t - \varphi_l) \tag{A.1.18}$$

the active electric power is equal to

$$P_{\mathrm{el}} = \sum_{j=1}^{\infty} U_j \cdot I_j \cdot \cos\varphi_j \tag{A.1.19}$$

Only voltage and current harmonics of the same order can generate active electric power.

Graphical interpretation of two orthogonal time functions is given in Fig. A.1, in which a 1 p.u. sinusoidal voltage $v(t) = \sin(\omega t - \pi/4)$ and a 1 p.u. sinusoidal current $i(t) = \sin(6\omega t)$ are shown. The instantaneous power $p(t)$, equal to the product of instantaneous voltage and instantaneous current, oscillates around zero in this case. The integral of power, having a meaning of energy in a given interval, is equal to zero. Graphically, the total area between the product of voltage and current (instantaneous power) and the time axis in the given interval is equal to zero.

Non-orthogonal harmonic functions of time have the same frequency, as shown in Fig. A.2. Here, a p.u. sinusoidal voltage $u(t) = \sin(\omega t - \pi/4)$ and p.u. sinusoidal current $i(t) = \sin(\omega t)$ are shown, along with their product—the instantaneous power $p(t) = u(t) \cdot i(t)$.

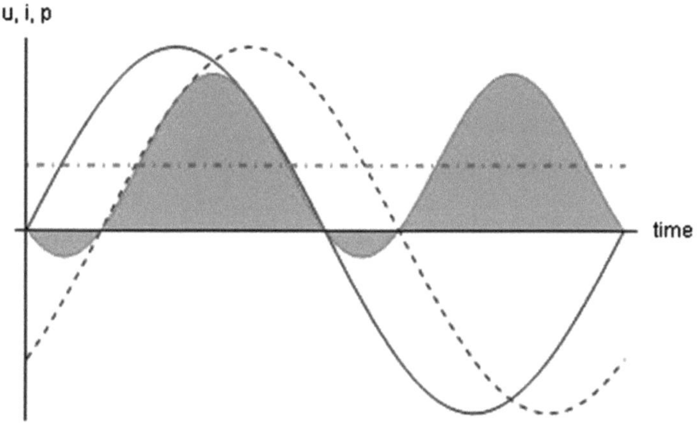

Fig. A.2 1 p.u. voltage $u(t) = \sin(\omega t - \pi/4)$ (*dashed black curve*), 1 p.u. current $i(t) = \sin(\omega t)$ (*solid black curve*), their product—instantaneous power $p(t) = u(t) \cdot i(t)$ (*gray curve*), and the average of the instantaneous power (*gray dash-dotted curve*). The area between the product of voltage and current represents electrical energy, the average of which in a given interval—the active power—monotonically increases, since the voltage and the current with the same frequency are *not* orthogonal ant *not* 90° out of phase

The instantaneous power oscillates around the amount proportional to the cosine of the angle between the voltage and current. The area below the product $p(t) = u(t) \cdot i(t)$ is *not* equal to zero, since the functions are *not* orthogonal.

Orthogonality of spatial functions: The length of period of the fundamental harmonic of a spatial function is twice the machine pole pitch τ_p, defined as

$$\tau_p = \frac{D\pi}{2p} \quad \text{(A.1.20)}$$

with D denoting the air gap diameter and p the number of machine pole pairs. The machine pole pitch τ_p, measured along the circumferential coordinate x in the center of the air gap, equals to the length of interval belonging to one pole. It is important to note that a spatial higher harmonic of the order m has the same meaning as m times more pole pairs, i.e., $m \cdot p$ instead of p.

The orthogonality of spatial functions will be illustrated on the example of torque created by a stator and rotor harmonic of air gap flux density. The mth harmonic of air gap flux density created by a stator winding with p_S pole pairs can be expressed as

$$B_{m,s}(x) = B_{\max,m,s} \sin\left(m\frac{\pi}{\tau_p}x - \alpha_s\right) = B_{\max,m,s} \sin\left(2m \cdot p_s\frac{x}{D} - \alpha_s\right) \quad \text{(A.1.21)}$$

The nth harmonic of current sheet created by a rotor winding with p_r pole pairs can be expressed as

$$A_{n,\mathrm{r}}(x) = A_{\max,n,\mathrm{r}} \sin\left(n\frac{\pi}{\tau_\mathrm{p}}x - \alpha_\mathrm{r}\right) = A_{\max,n,\mathrm{r}} \sin\left(2n \cdot p_\mathrm{r}\frac{x}{D} - \alpha_\mathrm{r}\right) \tag{A.1.22}$$

and the corresponding flux density as

$$B_{n,\mathrm{r}}(x) = \mu_0 \frac{\Theta_{n,\mathrm{r}}(x)}{\delta(x)} \tag{A.1.23}$$

Considering constant air gap width $\delta(x) = \delta$, one can further write

$$B_{n,\mathrm{r}}(x) = \frac{\mu_0}{\delta}\int A_{n,\mathrm{r}}(x)\mathrm{d}x = B_{\max,n,\mathrm{r}} \sin\left(2n \cdot p_\mathrm{r}\frac{x}{D} - \alpha_\mathrm{r} - \frac{\pi}{2}\right) \tag{A.1.24}$$

where the amplitude of the mth harmonic of air gap flux density created by the rotor winding with p_r pole pairs, $B_{\max,n,\mathrm{r}}$, is equal to

$$B_{\max,n,\mathrm{r}} = \frac{\mu_0}{\delta}A_{\max,n,\mathrm{r}}\frac{D}{2n \cdot p_\mathrm{r}} \tag{A.1.25}$$

The accumulated magnetic energy in the air gap is equal to

$$W_\mathrm{mg} = \frac{l\delta}{2\mu_0}\int_0^{D\pi} B^2(x)\mathrm{d}x = \frac{l\delta}{2\mu_0}\int_0^{D\pi} \left[B_{m,\mathrm{s}}(x) + B_{n,\mathrm{r}}(x)\right]^2\mathrm{d}x \tag{A.1.26}$$

or

$$W_\mathrm{mg} = \frac{l\delta}{2\mu_0}\int_0^{D\pi} \left[B_{m,\mathrm{s}}^2(x) + 2B_{m,\mathrm{s}}(x)B_{n,\mathrm{r}}(x) + B_{n,\mathrm{r}}^2(x)\right]\mathrm{d}x \tag{A.1.27}$$

which results in

$$\begin{aligned} W_\mathrm{mg} &= \frac{l\delta}{2\mu_0}\left[\frac{D\pi}{2}B_{\max,m,\mathrm{s}}^2 + \frac{D\pi}{2}B_{\max,n,\mathrm{r}}^2 + 2\int_0^{D\pi} B_{m,\mathrm{s}}(x)B_{n,\mathrm{r}}(x)\mathrm{d}x\right] = \\ &= \frac{D\pi l\delta}{4\mu_0}\left(B_{\max,m,\mathrm{s}}^2 + B_{\max,n,\mathrm{r}}^2\right) + \\ &\quad + \frac{Dl\delta}{4\mu_0}B_{\max,m,\mathrm{s}}B_{\max,n,\mathrm{r}} \cdot \frac{\cos(\alpha_\mathrm{r} - \alpha_\mathrm{s}) - \cos[\alpha_\mathrm{r} - \alpha_\mathrm{s} - 2\pi \cdot (n \cdot p_\mathrm{r} - m \cdot p_\mathrm{s})]}{n \cdot p_\mathrm{r} - m \cdot p_\mathrm{s}} \end{aligned} \tag{A.1.28}$$

Since n, m, p_r, and p_s are all integers, the expression for magnetic energy in case

$$m \cdot p_s \neq n \cdot p_r \tag{A.1.29}$$

i.e., when created by rotor and stator spatial harmonics of different orders, can be written as

$$W_{mg} = \frac{D\pi l\delta}{4\mu_0}\left(B^2_{max,m,s} + B^2_{max,n,r}\right) \tag{A.1.30}$$

Electromagnetic torque is equal to the derivative of the accumulated magnetic energy with respect to angle $\alpha_r - \alpha_s$ between current sheet and flux density

$$M = -\frac{\partial W_{mg}}{\partial(\alpha_r - \alpha_s)} \tag{A.1.31}$$

which in case of spatial harmonics with different orders obviously gives zero as result:

$$M = -\frac{\partial}{\partial(\alpha_r - \alpha_s)}\left[\frac{D\pi l\delta}{4\mu_0}\left(B^2_{max,m,s} + B^2_{max,n,r}\right)\right] = 0 \tag{A.1.32}$$

or *orthogonal spatial harmonics create no torque.*

If the two spatial harmonics are of the same order, i.e.,

$$m \cdot p_s = n \cdot p_r \tag{A.1.33}$$

the expression for magnetic energy created by them becomes

$$W_{mg} = \frac{D\pi l\delta}{4\mu_0}\left(B^2_{max,m,s} + B^2_{max,n,r}\right) - \frac{Dl\pi\delta}{2\mu_0}B_{max,m,s}B_{max,n,r} \cdot \sin(\alpha_r - \alpha_s) \tag{A.1.34}$$

because

$$\begin{aligned} &\lim_{m \to \frac{n \cdot p_r}{p_s}} \int_0^{D\pi} \sin\left(2m \cdot p_s \frac{x}{D} - \alpha_s\right) \sin\left(2n \cdot p_r \frac{x}{D} - \alpha_r - \frac{\pi}{2}\right) dx = \\ &= -\frac{D\pi}{2}\sin(\alpha_r - \alpha_s) \end{aligned} \tag{A.1.35}$$

The derivative of magnetic energy accumulated in the air gap with respect to the angle $\alpha_r - \alpha_s$ is

$$\frac{\partial W_{mg}}{\partial(\alpha_r - \alpha_s)} = -M = \frac{Dl\pi\delta}{2\mu_0}B_{max,m,s}B_{max,n,r} \cdot \cos(\alpha_r - \alpha_s) \tag{A.1.36}$$

After replacing the rotor flux density with current sheet, one becomes the expression for torque created by two harmonics of the same order as:

$$M = V \cdot B_{\max,m,\mathrm{s}} A_{\max,n,\mathrm{r}} \cdot \cos(\alpha_\mathrm{r} - \alpha_\mathrm{s}) \tag{A.1.37}$$

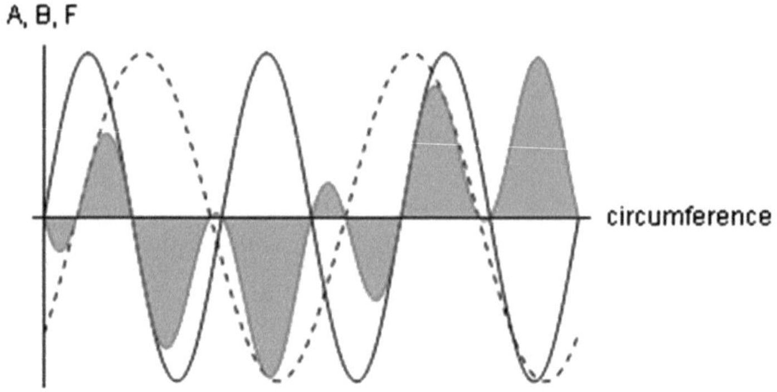

Fig. A.3 1 p.u. current sheet $A(x) = \sin(2\pi x/\tau_\mathrm{p} - \pi/4)$ (*dashed black curve*), 1 p.u. flux density $B(x) = \sin(3\pi x/\tau_\mathrm{p})$ (*solid black curve*), and their product, proportional to the force distribution along the circumference (*gray curve*). The total force created by the two harmonics, equal to the area below the gray curve, is equal to zero, because these harmonics are orthogonal to each another

The stator harmonic of the order $m \cdot p_\mathrm{s}$ is not orthogonal to the rotor harmonic of the order $n \cdot p_\mathrm{r}$ when $m \cdot p_\mathrm{s} = n \cdot p_\mathrm{r}$, and therefore, the two can create an electromagnetic torque. For example, the third harmonic of flux density in a ten-pole stator can create a torque different from zero with the fifth harmonic of flux density in a six-pole rotor. The number of stator pole pairs in electric machines for conventional applications is without exception equal to the number of rotor pole pairs, $p_\mathrm{s} = p_\mathrm{r}$. In that case, only the stator and rotor spatial harmonics of the same order can create a torque, i.e., $m = n$, and all higher spatial harmonics are torqueless.

Orthogonality of spatial functions is illustrated on an example of a 1 p.u. 4-pole current sheet $A(x) = \sin(2\pi x/\tau_\mathrm{p} - \pi/4)$ and a 1 p.u. 6-pole flux density distribution $B(x) = \sin(3\pi x/\tau_\mathrm{p})$, Fig. A.3. The product of the two for a given value of circumferential coordinate is proportional to the force created by them at that particular point. The infinite sum of all products—their integral—is equal to the total force created by the two harmonics. The integral (area below the product of the two functions) along the complete circumference in Fig. A.3 is equal to zero; consequently, any two harmonics of different spatial order cannot create a torque.

If the two harmonics of flux density and current sheet are non-orthogonal, their interaction can result in a force/torque if the spatial shift between them is different from 90°. This is illustrated in Fig. A.4, in which a 1 p.u. 4-pole current sheet

$A(x) = \sin(2\pi x/\tau_p - \pi/4)$ and a 1 p.u. 4-pole flux density distribution $B(x) = \sin(2\pi x/\tau_p)$ are shown, along with their product. The integral of the product of two, equal to the force created by them, is different from zero.

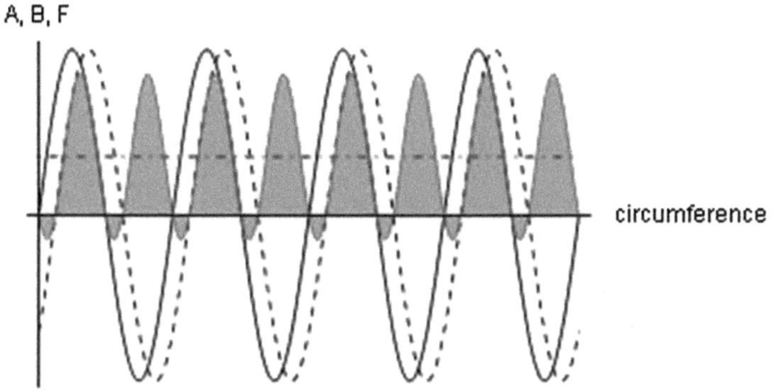

Fig. A.4 1 p.u. current sheet $A(x) = \sin(2\pi x/\tau_p - \pi/4)$ (*dashed black curve*), 1 p.u. flux density $B(x) = \sin(2\pi x/\tau_p)$ (*solid black curve*), and their product, proportional to the force distribution along the circumference (*gray curve*). The total force created by the two harmonics, equal to the area below the gray curve and represented by a dash-dot horizontal line, is different from zero, because these harmonics are non-orthogonal and the spatial angle between them is different from 90°

A.2 Periodic Functions of Time

The complex number $\hat{A} = A \cdot e^{j(\omega t + \varphi_a)}$ with modulus A and time-dependent argument $\omega t + \varphi_a$ is according to Euler's formula

$$A \cdot e^{j(\omega t + \varphi_a)} = A \cdot \cos(\omega t + \varphi_a) + jA \cdot \sin(\omega t + \varphi_a) \tag{A.2.1}$$

equal to the sum of a real

$$\mathrm{Re}\left\{A \cdot e^{j(\omega t + \varphi_a)}\right\} = A \cdot \cos(\omega t + \varphi_a) \tag{A.2.2}$$

and an imaginary component

$$\mathrm{Im}\left\{A \cdot e^{j(\omega t + \varphi_a)}\right\} = A \cdot \sin(\omega t + \varphi_a) \tag{A.2.3}$$

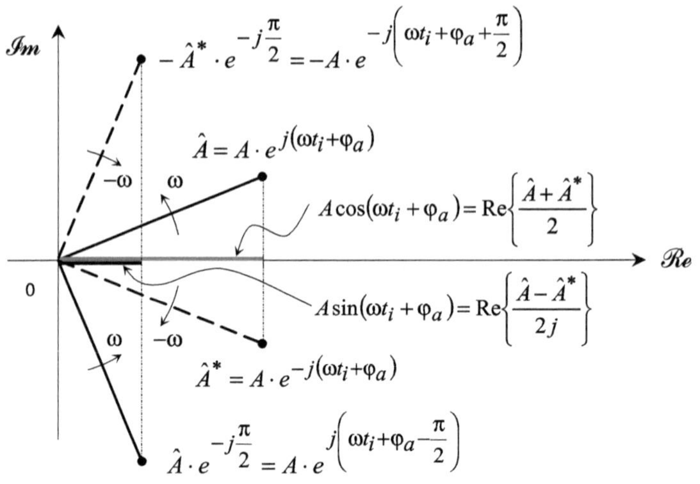

Fig. A.5 Representation of real numbers $A\ \cos(\omega t_i + \varphi_a)$ and $A\ \sin(\omega t_i + \varphi_a)$ at a time instant $t = t_i$ by means of complex conjugates $\hat{A} = A \cdot e^{j(\omega t_i + \varphi_a)}$ and $\hat{A}^* = A \cdot e^{-j(\omega t_i + \varphi_a)}$

Here e denotes the base of natural logarithms, ω is the angular frequency equal to $2\pi/T$, T is the period length in s, and j is the imaginary unit, $j = \sqrt{(-1)}$.

The *complex conjugate* $\hat{A}^*$ of the number $\hat{A} = A \cdot e^{j(\omega t + \varphi_a)}$ is equal to

$$\hat{A}^* = A \cdot e^{-j(\omega t + \varphi_a)} = A \cdot \cos(\omega t + \varphi_a) - jA \cdot \sin(\omega t + \varphi_a) \tag{A.2.4}$$

By introducing a complex number $\hat{Z}$, such that $\mathrm{Im}\{\hat{Z}\} = 0$:

$$\hat{Z} = \frac{\hat{A} + \hat{A}^*}{2} = A\frac{e^{j(\omega t + \varphi_a)} + e^{-j(\omega t + \varphi_a)}}{2} \tag{A.2.5}$$

one can express the real component $a(t) = A\ \cos\ (\omega t + \varphi_a)$ of the complex number $\hat{A} = A \cdot e^{j(\omega t + \varphi_a)}$ as

$$a(t) = A\ \cos(\omega t + \varphi_a) = \mathrm{Re}\{\hat{Z}\} = \hat{Z} \tag{A.2.6}$$

which means that a *time* function $a(t) = A\ \cos\ (\omega t + \varphi_a)$ is identical to a sum of two complex conjugates, the arguments of which are functions of time.

Introduce now a complex number $\hat{D}$, such that $\mathrm{Im}\{\hat{D}\} = 0$:

$$\hat{D} = \frac{\hat{A} - \hat{A}^*}{2j} = -jA\frac{e^{j(\omega t + \varphi_a)} - e^{-j(\omega t + \varphi_a)}}{2} \tag{A.2.7}$$

The imaginary component of the complex number $\hat{A} = A \cdot e^{j(\omega t + \varphi_a)}$ is equal to the real component of $\hat{D}$, because

$$A\,\sin(\omega t+\varphi_{\mathrm{a}})=\frac{\hat{A}-\hat{A}^{*}}{2j}=-jA\frac{e^{j(\omega t+\varphi_{\mathrm{a}})}-e^{-j(\omega t+\varphi_{\mathrm{a}})}}{2}=\mathrm{Re}\{\hat{D}\}=\hat{D}\quad\text{(A.2.8)}$$

Since

$$\frac{1}{j}=-j=e^{-j\frac{\pi}{2}}\quad\text{(A.2.9)}$$

i.e., division through j is identical to the rotation for $-\pi/2$, one can further write

$$A\,\sin(\omega t+\varphi_{\mathrm{a}})=A\cdot e^{-j\frac{\pi}{2}}\frac{e^{j(\omega t+\varphi_{\mathrm{a}})}-e^{-j(\omega t+\varphi_{\mathrm{a}})}}{2}=A\cdot\frac{e^{j\left(\omega t+\varphi_{\mathrm{a}}-\frac{\pi}{2}\right)}-e^{-j\left(\omega t+\varphi_{\mathrm{a}}+\frac{\pi}{2}\right)}}{2}\quad\text{(A.2.10)}$$

Graphical interpretation of previous results in the *complex plane* ($\mathcal{Re}$, $\mathcal{Im}$) at a time instant $t = t_{\mathrm{i}}$ is shown in Fig. A.5. As opposed to vectors, which are characterized by their magnitude and *direction in space*, complex numbers have no direction in the complex plane. Therefore, the usage of arrow for complex numbers is inappropriate. Instead, the bold dot symbol is employed in order to indicate the position of a complex number in the complex plane.

One should note that the argument of the complex number $\hat{A}=A\cdot e^{j(\omega t_{\mathrm{i}}+\varphi_{\mathrm{a}})}$ is proportional to time. This property can be interpreted as *rotation* in the complex plane at an angular frequency ω in mathematically positive direction, as shown in Fig. A.5. The same is valid for its conjugate $\hat{A}^{*}=A\cdot e^{-j(\omega t_{\mathrm{i}}+\varphi_{\mathrm{a}})}$, which rotates at the same angular frequency ω, but in mathematically negative direction.

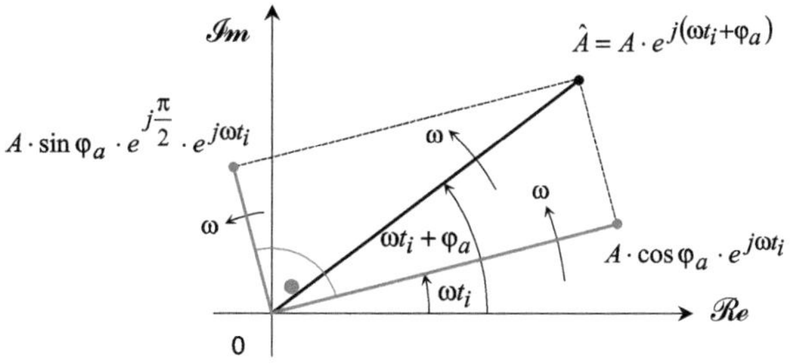

Fig. A.6 Complex number $\hat{A}=A\cdot e^{j(\omega t_{\mathrm{i}}+\varphi_{\mathrm{a}})}$ at a time instant $t = t_{\mathrm{i}}$ represented as a sum of its components $A\cos\varphi_{\mathrm{a}}$ and $A\sin\varphi_{\mathrm{a}}$ along two perpendicular axes

The sum of a complex number and its conjugate, rotating in opposite directions, is a real number which pulsates at an angular frequency ω.

This very interesting property of mathematical objects has a well-known analogy in physics, stating that a pulsating field is equal to the sum of two fields with same amplitudes, rotating in opposite directions.

For the purpose of simplicity it is usual in electrical engineering to represent a time function $A\cos(\omega t + \varphi_a)$ in the complex plane with a single complex number $A \cdot e^{j(\omega t_i + \varphi_a)}$, i.e., without its complex conjugate. A consistent representation, however, requires a complex number *and* its conjugate for a single time function.

The complex number $\hat{A} = A \cdot e^{j(\omega t_i + \varphi_a)}$ can further be expressed as

$$\hat{A} = A \cdot e^{j(\omega t_i + \varphi_a)} = A \cdot e^{j\omega t} \cdot e^{j\varphi_a} = A \cdot \cos\varphi_a \cdot e^{j\omega t} + A \cdot \sin\varphi_a \cdot e^{j\frac{\pi}{2}} \cdot e^{j\omega t} \tag{A.2.11}$$

the graphical interpretation of which at a time instant $t = t_i$ is shown in Fig. A.6.

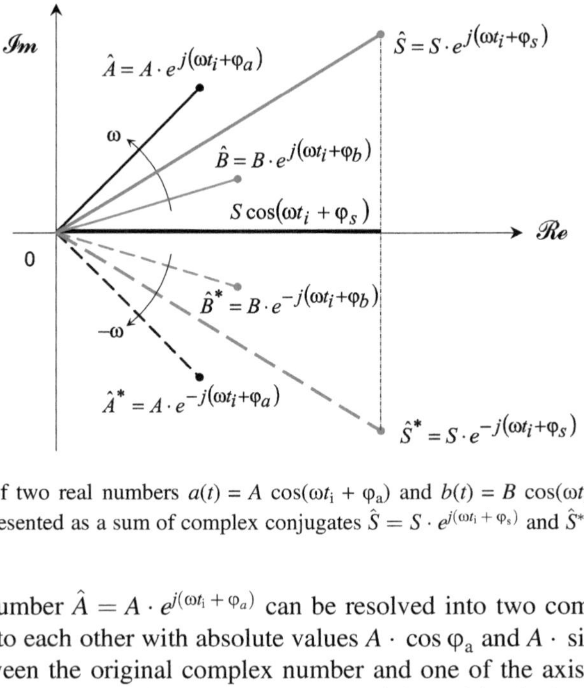

Fig. A.7 Sum of two real numbers $a(t) = A\cos(\omega t_i + \varphi_a)$ and $b(t) = B\cos(\omega t_i + \varphi_b)$ at a time instant $t = t_i$ represented as a sum of complex conjugates $\hat{S} = S \cdot e^{j(\omega t_i + \varphi_s)}$ and $\hat{S}^* = S \cdot e^{-j(\omega t_i + \varphi_s)}$

Complex number $\hat{A} = A \cdot e^{j(\omega t_i + \varphi_a)}$ can be resolved into two complex numbers perpendicular to each other with absolute values $A \cdot \cos\varphi_a$ and $A \cdot \sin\varphi_a$, φ_a being the angle between the original complex number and one of the axis of resolution.

Addition and subtraction in the complex plane follows identical rules as addition and subtraction of vectors in space. This can be illustrated by means of another time-dependent quantity $b(t)$ defined as

$$b(t) = B\cos(\omega t + \varphi_b) \tag{A.2.12}$$

or by means of complex conjugates $\hat{B}$ and $\hat{B}^*$

$$b(t) = \frac{\hat{B} + \hat{B}^*}{2} = B\frac{e^{j(\omega t + \varphi_b)} + e^{-j(\omega t + \varphi_b)}}{2} \tag{A.2.13}$$

The sum $s(t)$ of $a(t)$ and $b(t)$ is equal to

$$s(t) = S\cos(\omega t + \varphi_s) = a(t) + b(t) = e^{j\omega t}\frac{Ae^{j\varphi_a} + Be^{j\varphi_b}}{2} + e^{-j\omega t}\frac{Ae^{-j\varphi_a} + Be^{-j\varphi_b}}{2} \tag{A.2.14}$$

The first term of the sum $s(t)$ in Eq. A.2.14 rotates with an angular frequency ω in positive direction. The second term, its complex conjugate, rotates with the same angular frequency ω in negative direction. The complex sum

$$Ae^{j\varphi_a} + Be^{j\varphi_b} \tag{A.2.15}$$

is evaluated by adding real and imaginary components separately from each other, following rules for vector addition along orthogonal axis $\mathcal{Re}$ and $\mathcal{Im}$. The same is valid for the sum of complex conjugates

$$Ae^{-j\varphi_a} + Be^{-j\varphi_b} \tag{A.2.16}$$

as shown in Fig. A.7.

The analogy between operations on complex numbers and spatial vectors is limited to addition and subtraction only. In case of multiplication and division, completely different rules are valid for spatial vectors than for complex numbers. The result of multiplication of two complex numbers is a complex number again, which can be represented in the complex plane. The result of multiplication of two spatial vectors is either a scalar, or a vector perpendicular to the plane defined by the two multiplicands. In the latter case the product of two spatial vectors is not defined in the plane created by the two vectors. Furthermore, the division of two spatial vectors is not defined, whereas the quotient of two complex numbers is again a complex number.

Consider now the product $p(t)$ of two time-dependent real quantities $u(t)$ and $i(t)$ defined as

$$\begin{aligned} p(t) &= u(t)\cdot i(t) = U\sqrt{2}\cos(\omega t)\cdot I\sqrt{2}\cos(\omega t - \varphi) = \\ &= U\cdot I\cdot[\cos\varphi + \cos(2\omega t - \varphi)] = U\cdot I\cdot\cos\varphi\cdot(1 + \cos 2\omega t) + U\cdot I\cdot\sin\varphi\cdot\sin 2\omega t \end{aligned} \tag{A.2.17}$$

and graphically represented in Fig. A.8. The product $p(t)$ can be interpreted as instantaneous power in an AC circuit with voltage $u(t)$ and current $i(t)$. As known, the instantaneous power pulsates with an amplitude of $S = U \cdot I$ (apparent power) around the average value $P = U \cdot I \cdot \cos\varphi = S \cdot \cos\varphi$ (active power). By further decomposition of the product $u(t) \cdot i(t)$, one obtains that the instantaneous power is equal to a sum of the component $U \cdot I \cdot \cos\varphi$ multiplied by $(1 + \cos 2\omega t)$ and $Q = U \cdot I \cdot \sin\varphi$ (reactive power) multiplied by $\sin 2\omega t$.

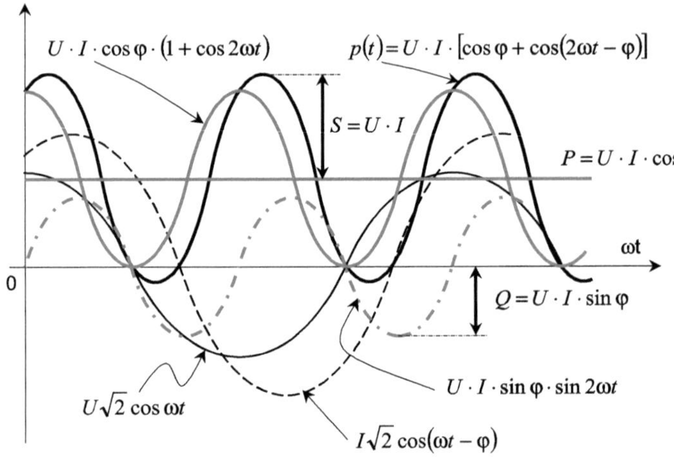

Fig. A.8 Product of two trigonometric functions $u(t) = U\sqrt{2}\cos\omega t$ and $i(t) = I\sqrt{2}\cos(\omega t - \varphi)$ and its components in time domain

One should note in Fig. A.8 that the term $V \cdot I \cdot \cos\varphi \cos 2\omega t$ leads the term $V \cdot I \cdot \sin\varphi \cdot \sin 2\omega t$ for an angle of $\pi/2$.

The result of multiplication of $u(t)$ and $i(t)$ can be as well represented in the complex plane, in which $p(t)$ can be written in terms of complex numbers $\widehat{U} = U\sqrt{2} \cdot e^{j\omega t}$and $\widehat{U}^{*} = V\sqrt{2} \cdot e^{-j\omega t}$, as well as $\hat{I} = I\sqrt{2} \cdot e^{j(\omega t - \varphi)}$ and $\hat{I}^{*} = I\sqrt{2} \cdot e^{-j(\omega t - \varphi)}$ standing for $u(t)$ and $i(t)$, respectively, as

$$\begin{aligned} p(t) &= U\sqrt{2}\frac{e^{j\omega t} + e^{-j\omega t}}{2} I\sqrt{2}\frac{e^{j(\omega t-\varphi)} + e^{-j(\omega t-\varphi)}}{2} = \\ &= \frac{S}{2}\left[e^{j(2\omega t-\varphi)} + e^{-j(2\omega t-\varphi)} + e^{j\varphi} + e^{-j\varphi}\right] \end{aligned} \tag{A.2.18}$$

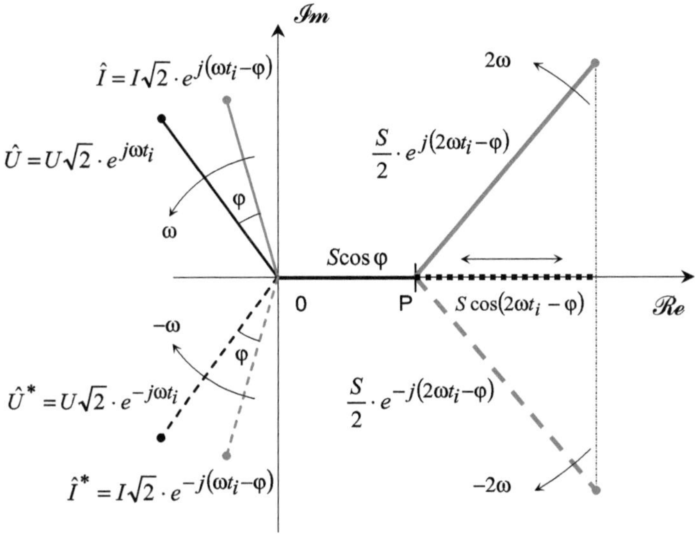

Fig. A.9 Product of two trigonometric functions $u(t) = U\sqrt{2}\cos\omega t$ and $i(t) = I\sqrt{2}\cos(\omega t - \varphi)$ and its components in the complex plane at time instant $t = t_i$

Complex numbers $\hat{U} = U\sqrt{2} \cdot e^{j\omega t}$ and $\hat{I} = I\sqrt{2} \cdot e^{j(\omega t - \varphi)}$, along with their conjugates $\hat{U}^* = U\sqrt{2} \cdot e^{-j\omega t}$ and $\hat{I}^* = I\sqrt{2} \cdot e^{-j(\omega t - \varphi)}$, rotate in the complex plane with angular frequencies ω and −ω, respectively, with the center of rotation at the origin of the complex plane, as shown in Fig. A.9.

The product $p(t)$ has a *constant real termP*:

$$P = S\frac{e^{j\varphi} + e^{-j\varphi}}{2} = S\cos\varphi \tag{A.2.19}$$

and two complex conjugates which rotate at twice the angular frequency 2ω in opposite directions

$$S\frac{e^{j(2\omega t - \varphi)} + e^{-j(2\omega t - \varphi)}}{2} = S\cos(2\omega t - \varphi) \tag{A.2.20}$$

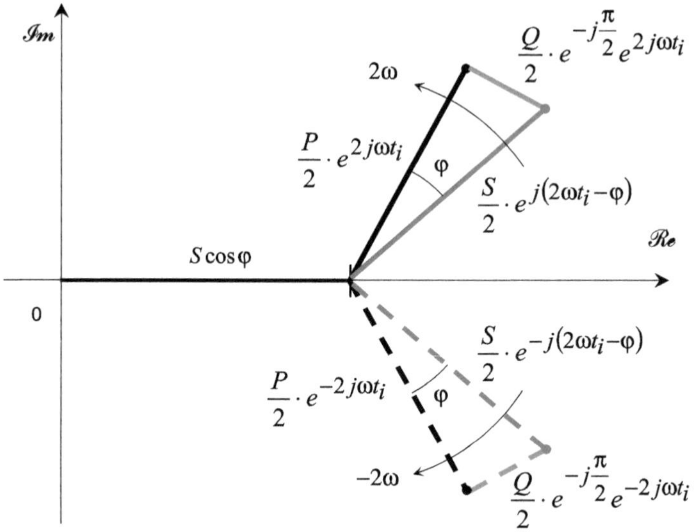

Fig. A.10 Constant and pulsating components of power in the complex plane at $t = t_i$

The product $p(t)$ can be represented in the complex plane in Fig. A.9 as a sum of *steady* (DC) component $S\cos\varphi$ on the real axis and two complex conjugates, $\hat{S} = S/2 \cdot e^{j(2\omega t-\varphi)}$ and $\hat{S}^* = S/2 \cdot e^{-j(2\omega t-\varphi)}$, which rotate around the point P on the real axis with coordinates (S cos φ; 0) at angular frequencies 2ω and −2ω, respectively. Each of the complex conjugates carries 50 % of the pulsating component of power $U \cdot I \cdot \cos(2\omega t - \varphi)$ and can be resolved into two components perpendicular to each other.

The two pulsating components of power can be represented in terms of the power triangle as shown in Fig. A.10.

The projection of $\hat{S} = S/2 \cdot e^{j(2\omega t_i-\varphi)}$ to an axis rotating at an angular frequency 2ω and leading for an angle φ is equal to $P/2 \cdot e^{j2\omega t_i}$; the projection of $\hat{S} = S/2 \cdot e^{j(2\omega t_i-\varphi)}$ to an axis lagging for an angle π/2 – φ is equal to $Q/2 \cdot e^{j\left(2\omega t_i+\frac{\pi}{2}\right)}$, since $P = S\cos\varphi$ and $Q = S\sin\varphi$.

Similarly, the projection of $\hat{S}^* = S/2 \cdot e^{-j(2\omega t_i-\varphi)}$ to an axis rotating at an angular frequency of −2ω and leading for an angle φ is equal to $P/2 \cdot e^{-j2\omega t_i}$, and to an axis lagging for an angle π/2 − φ to $Q/2 \cdot e^{-j\left(2\omega t_i+\frac{\pi}{2}\right)}$, since $Q = S\sin\varphi$. The phase shift between the complex numbers $P/2 \cdot e^{-j2\omega t_i}$ and $Q/2 \cdot e^{-j\left(2\omega t_i+\frac{\pi}{2}\right)}$ is π/2, as already illustrated in the time domain, see Fig. A.8. One should note that *lagging* position between two complex numbers rotating with positive angular frequency means *leading* position for negative angular frequency and vice versa.

In an absolutely symmetrical m-phase system with voltages

$$u_j(t) = U\sqrt{2}\cos\left[\omega t - (j-1)\frac{2\pi}{m}\right]; \quad j = 1, 2, 3, \ldots, m \tag{A.2.21}$$

and currents

$$i_j(t) = I\sqrt{2}\cos\left[\omega t - (j-1)\frac{2\pi}{m} - \varphi\right]; \quad j = 1, 2, 3, \ldots, m \tag{A.2.22}$$

the instantaneous power is equal to

$$p(t) = \sum_{j=1}^{m} u_j(t) \cdot i_j(t) = \sum_{j=1}^{m} U \cdot I \cdot \left\{\cos\varphi + \cos\left[2\omega t - 2(j-1)\frac{2\pi}{m}\right]\right\} \tag{A.2.23}$$

or

$$p(t) = P = m \cdot U \cdot I \cdot \cos\varphi \tag{A.2.24}$$

Instantaneous power in a symmetrical m-phase system is time invariant, as shown in Fig. A.11.

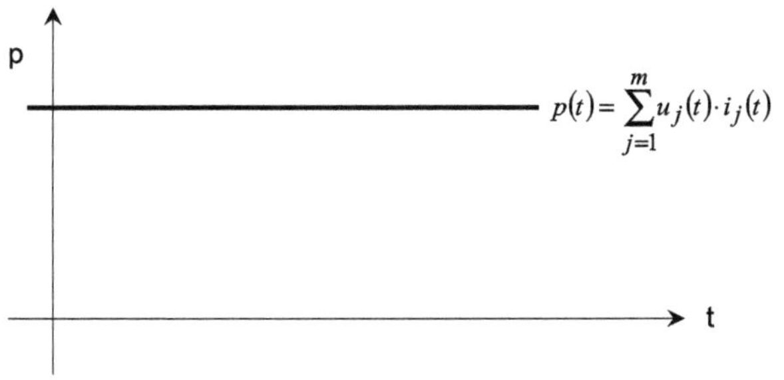

Fig. A.11 Instantaneous power in a symmetrical m-phase system

A symmetrical m-phase system acts in terms of power consumption identically as a resistor fed from a DC source, as shown in Fig. A.12. The conclusion that the total instantaneous power in a symmetrical m-phase system is constant is not contradictory to the fact that the power pulsates in each phase with twice the supply frequency. One should keep in mind that the total instantaneous current in a symmetrical m-phase system is also constant (and equal to zero), independent of how large are the phase currents.

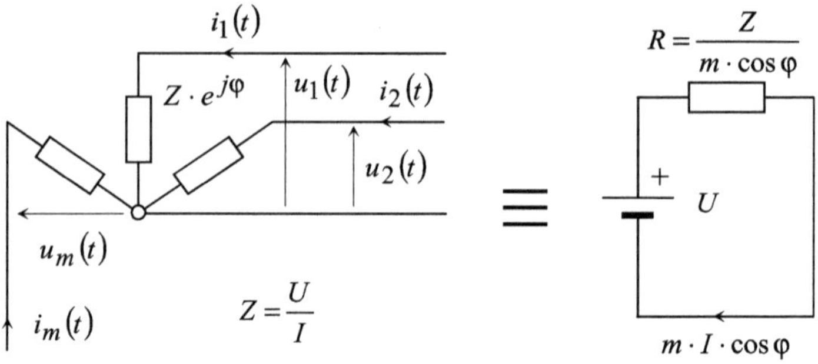

Fig. A.12 Comparison between instantaneous power in a symmetrical m-phase system and in a DC circuit

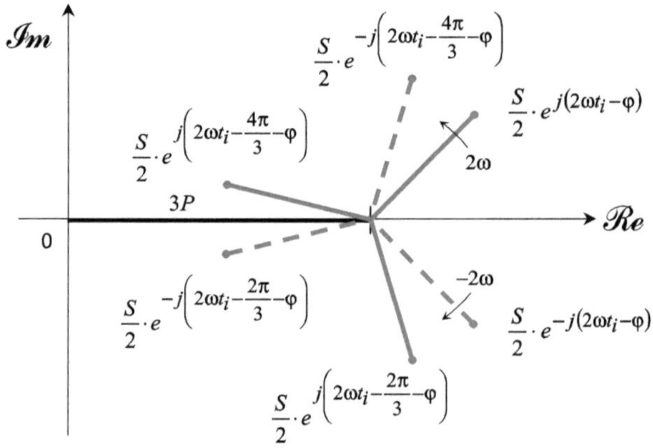

Fig. A.13 Instantaneous power in a symmetrical three-phase system and its components in the complex plane

Graphical interpretation of power in a symmetrical three-phase system is shown in Fig. A.13, in which power components in all three phases are shown along with the resulting power $3P$, with P denoting the active power in one phase. The sum of power components in separate phases for both positive (solid) direction and negative (dashed) direction of rotation is equal to zero at each time instant.

Very often, sinusoidal quantities are represented only with complex numbers rotating at positive angular frequencies and named *phasors*, as shown in Fig. A.14a. The description *phasor* is redundant, because it is nothing but a complex number. As such, it does not carry any information more than a complex number, and therefore, will not be used.

Sinusoidal voltage and current in Fig. A.14a are real numbers, which, accordingly, can be represented only on real axis in the complex plane. There exists no correct physical interpretation for the voltage and current components on imaginary axis in Fig. A.14a.

The only consequent way to represent a real current or voltage in the complex plane is to express them as sums of two complex conjugates, in which case their imaginary components are equal to zero, as shown in Fig. A.14b. If the real current or voltage is a periodical function of time, the two complex conjugates representing them rotate in the complex plane in opposite directions at the same angular frequency.

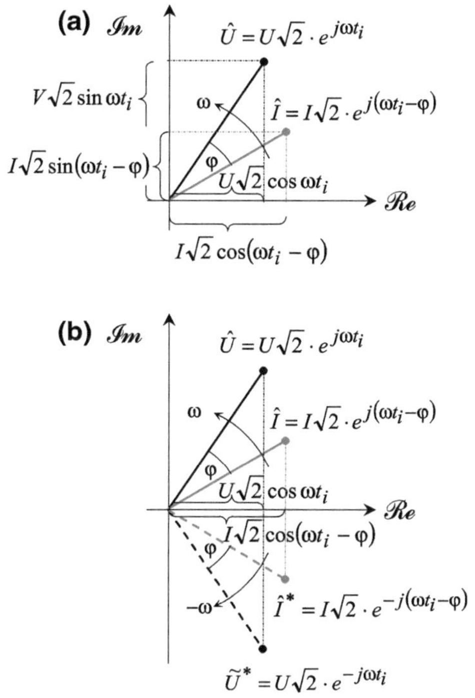

Fig. A.14 Time-dependent periodical functions in the complex plane: incomplete (**a**) and comprehensive (**b**) presentation

Correct representation of periodical quantities in the complex plane becomes extremely important in applications related to rotating field electric machines, where the physical rotation of windings and fields is often confused with fictitious rotation of a complex number in the complex plane. By focusing exclusively on rotation in the complex plane in positive direction, a false impression is obtained that the positive direction of rotation of complex numbers representing machine currents and voltages coincides with physical direction of rotation of the rotor. Neither do the complex numbers representing AC voltages and currents rotate in the machine (r, φ) plane, nor does the rotor along with its windings, air gap MMF, and flux density distributions rotate in the complex plane.

A.3 Power Factor

Power factor of a rotating field machine is an important parameter which not only stands for the level of magnetic energy absorbed or generated by the machine, but also has a strong influence on its I^2R losses. A machine with poor power factor requires an additional component of current in order to build the magnetic field, which increases the stator copper losses.

The only rotating field machine that can generate magnetic energy is the wound rotor synchronous machine. All other machine types, including a permanent magnet machine, require magnetic energy from an external source in order to operate properly at rated point. Consequently, the only machine that can have rated power factor equal to 1 and, therefore, minimum stator copper losses, is the wound rotor synchronous machine.

A.3.1 Field MMF Necessary to Operate Wound Rotor Synchronous Machine at a Given Power Factor

The bottleneck of a wound rotor synchronous machine is its field winding, which on the one hand has to provide the MMF necessary to operate the machine at a given power factor cos φ lagging and, on the other hand, has to be properly cooled in order to keep its temperature within the limit for a given insulation class. The field winding ampere-turns have to override the stator MMF due to armature reaction, so that the rated field current $i_{f,r}$ is typically up to three times larger than the no-load field current i_0, resulting in up to nine times larger rotor i^2R losses at rated point than at no load.

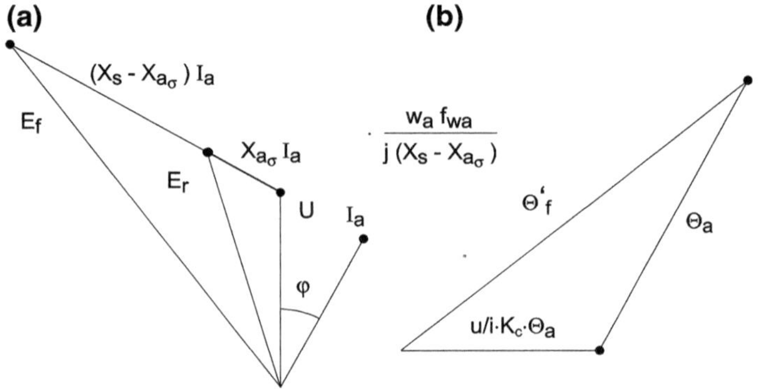

Fig. A.15 Voltage (**a**) and MMF (**b**) diagram of a cylindrical rotor synchronous machine

The ratio between the field and armature MMF at rated point of a wound rotor synchronous machine can be determined on the basis of its voltage and MMF diagram, as shown in Fig. A.15.

E_{f} in Fig. A.15 denotes the excitation voltage, E_{r} the induced voltage, $X_{\mathrm{a}\sigma}$ the stator leakage reactance, and X_{s} the synchronous reactance. The factor K_{c} stands for the no-load short-circuit ratio S.C.R., u for p.u. rated voltage, i for p.u. rated current, $x_{\mathrm{a}\sigma}$ for p.u. stator leakage reactance, Θ_{a} for the armature MMF, and Θ'_f for the field MMF referred to one stator phase. Based upon the two diagrams in Fig. A.15, one can further write

$$U \cdot \frac{w_{\mathrm{a}} \cdot f_{\mathrm{wa}}}{X_{\mathrm{s}} - X_{\mathrm{a}\sigma}} = \frac{\frac{U}{I_{\mathrm{a}}}}{X_{\mathrm{s}}} \cdot w_{\mathrm{a}} \cdot f_{\mathrm{wa}} \cdot I_{\mathrm{a}} \cdot \frac{X_{\mathrm{s}}}{X_{\mathrm{s}} - X_{\mathrm{a}\sigma}} = \frac{u}{i} \cdot K_{\mathrm{c}} \cdot (1 + x_{\mathrm{a}\sigma}) \tag{A.3.1}$$

as well as

$$E_{\mathrm{f}} \cdot \frac{w_{\mathrm{a}} \cdot f_{\mathrm{wa}}}{X_{\mathrm{s}} - X_{\mathrm{a}\sigma}} = \Theta'_{\mathrm{f}} \tag{A.3.2}$$

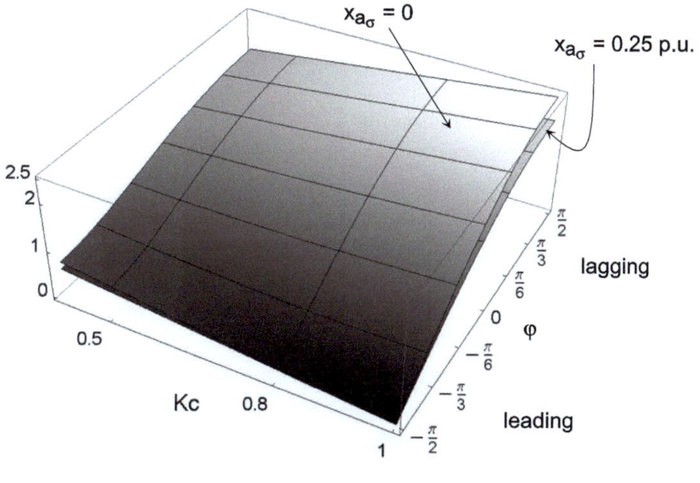

Fig. A.16 Ratio between field and armature MMF necessary to operate a synchronous machine with a given short-circuit ratio K_{c} at rated current and voltage and phase shift φ between them

Using the MMF diagram in Fig. A.15b, one can express the ratio between the (referred) field and armature MMF as

$$\frac{\Theta_{\mathrm{f}}'}{\Theta_{\mathrm{a}}} = (1 + x_{\mathrm{a}\sigma})\sqrt{1 + \left(\frac{u}{i}K_{\mathrm{c}}\right)^2 + 2\frac{u}{i}K_{\mathrm{c}}\sin\varphi} \tag{A.3.3}$$

whereas at the rated point, $i = 1$, $u = 1$, one can write

$$\frac{\Theta_{\mathrm{f}}'}{\Theta_{\mathrm{a}}} = (1 + x_{\mathrm{a}\sigma})\sqrt{1 + K_{\mathrm{c}}^2 + 2K_{\mathrm{c}}\sin\varphi} \tag{A.3.4}$$

The ratio between field and armature MMF as a function of the short-circuit ratio K_{c} and phase shift φ for various values of $x_{\mathrm{a}\sigma}$ and at $u = 1$ p.u., $i = 1$ p.u. is shown in Fig. A.16. As expected, one needs a field MMF stronger than the armature MMF in order to operate a synchronous machine at a given lagging power factor.

A.3.2 *Power Factor and Magnetic Energy Demand of a Permanent Magnet Synchronous Machine*

Permanent magnets are widely used in synchronous machine because they provide air gap flux without dissipating field $I^2 R$ losses. At first sight, this might look as an advantageous property as compared to a wound rotor machine. However, after taking a closer look to physical relationships in a permanent magnet machine, one comes to a quite opposite conclusion.

Usually, the magnet dimensions and magnetic properties are such that the induced voltage at rated speed is equal to the machine rated voltage, in which case the power factor at rated operating point must be leading and, obviously, less than one. This can be easily justified by means of the machine operating chart, as shown in Fig. A.17.

In Fig. A.17, the area within which a regular wound rotor machine can operate is denoted gray. The dashed circle with center at point $(-1/x_{\mathrm{s\ p.u.}}; 0)$ and a radius of $1/x_{\mathrm{s\ p.u.}}$ denotes 1 p.u. excitation (i.e., the no-load voltage) and represents a permanent magnet machine. Operating point of a permanent magnet machine slides along the circle of constant no-load voltage denoted by PM. As shown in Fig. A.17, the maximum power of a permanent magnet excited machine, $P_{\mathrm{max,PM}}$, is proportional to the reciprocal of the p.u. synchronous reactance $x_{\mathrm{s\ p.u.}}$.

Rated *power factor* $\cos\varphi_{\mathrm{r,PM}}$ of a permanent magnet synchronous machine depends on machine topology. It can be as low as 0.7 leading, as is the case in tooth-wound machines, where the high synchronous reactance is a consequence of an excessive air gap leakage flux.

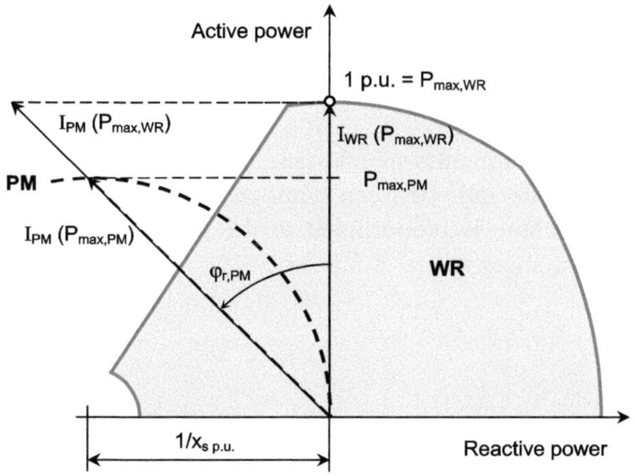

Fig. A.17 Simplified operating chart of a wound rotor synchronous generator (*gray area*) and permanent magnet generator (*dashed circle*). $I_{WR}(P_{max,WR})$ denotes the wound rotor machine armature current at the point of maximum power of wound rotor machine and $I_{PM}(P_{max,PM})$ (shown dashed in figure) the permanent magnet machine armature current at the point of maximum power of permanent magnet machine. $I_{PM}(P_{max,WR})$ is the permanent magnet machine armature current necessary to produce the same maximum power as the wound rotor machine. $P_{max,PM}$ is the maximum power of the PM machine in p.u., and $\varphi_{r,PM}$ denotes its rated power factor

Rated power factor $\cos\varphi_r$ determines the amount of stator copper losses $P_{Cu,S,r}$ as

$$P_{Cu,S,r} = 3I_r^2 R_S = \frac{R_S}{\Omega_r^2 U_r^2} \cdot \frac{M_r^2}{\cos^2 \varphi_r} \tag{A.3.5}$$

with M_r denoting the rated torque at the rated mechanical angular speed Ω_r.

In order to find the ratio between stator copper losses of a PM and wound rotor synchronous machine for the same rated torque and active volume, one should first determine the ratio between the fundamental components of permanent magnet and wound rotor current sheet, A_{PM} and A_{WR}, respectively.

$$\frac{A_{PM}}{A_{WR}} = \frac{B_{WR} \cos \Psi_{WR}}{B_{PM} \cos \Psi_{PM}} \tag{A.3.6}$$

with B_{WR} standing for the wound rotor and B_{PM} for permanent magnet machine air gap flux density. Ψ_{PM} is the spatial angle between fundamental components of stator current sheet and rotor flux density in a permanent magnet machine and Ψ_{WR} ditto for a wound rotor machine. Whereas Ψ_{WR} can be varied by changing the amount of field current, Ψ_{PM} at rated point is constant and can be determined by using the MMF diagram in Fig. A.15b extended for the fundamental of armature current sheet A_a, as shown in Fig. A.19.

Based upon the MMF and current sheet diagram in Fig. A.18, one can express the cosine of angle Ψ_{PM} as

$$\cos\Psi_{PM} = \frac{\frac{u}{i}K_c\cos\varphi_{PM}}{\sqrt{1+\left(\frac{u}{i}K_c\right)^2+2\frac{u}{i}K_c\sin\varphi_{PM}}} \tag{A.3.7}$$

Keeping the machine dimensions constant, current sheet is proportional only to the current. Therefore, the ratio between stator current losses in a permanent magnet and wound rotor machine is proportional to the square of the ratio between the amplitudes of current sheet in Eq. A.3.6, or

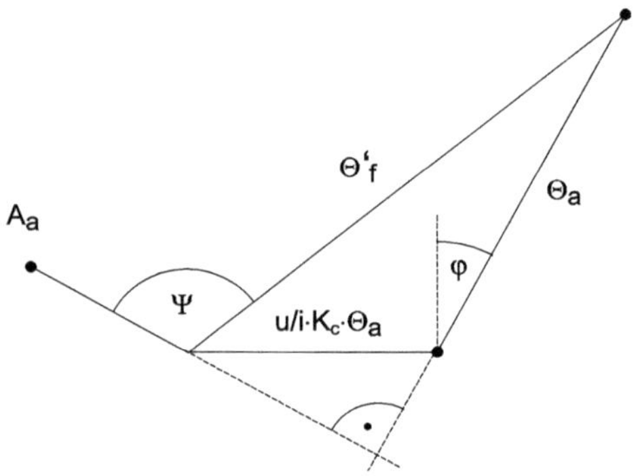

Fig. A.18 Relationship between fundamental components of MMF and current sheet in a synchronous machine

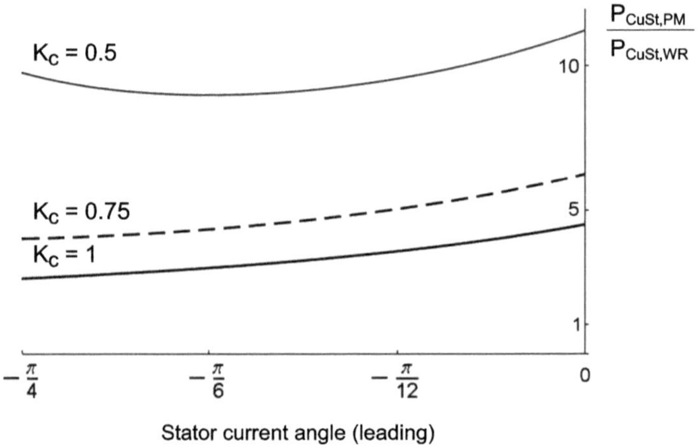

Fig. A.19 Ratio between stator copper losses in a permanent magnet $P_{\mathrm{CuSt,PM}}$ and wound rotor $P_{\mathrm{CuSt,WR}}$ machine with equal rated torques and active volumes as a function of the short-circuit ratio K_c and stator current phase shift at rotor temperature of 20 °C. Typical stator current angle of a permanent magnet machine is 30° leading. The air gap flux of permanent magnet machine is only about 2/3 of the air gap flux of the wound rotor machine, as discussed in Chap. 1

$$\frac{P_{\mathrm{CuSt,PM}}}{P_{\mathrm{CuSt,WR}}}=\frac{1+\left(\frac{u}{i}K_{\mathrm{c}}\right)^{2}+2\frac{u}{i}K_{\mathrm{c}}\sin\varphi_{\mathrm{PM}}}{\left(\frac{u}{i}K_{\mathrm{c}}\cos\varphi_{\mathrm{PM}}\right)^{2}}\left(\frac{B_{\mathrm{WR}}\cos\Psi_{\mathrm{WR}}}{B_{\mathrm{PM}}}\right)^{2} \tag{A.3.8}$$

In Fig. A.19 the ratio between stator copper losses in a permanent magnet and wound rotor machine with equal rated torques and active volumes as a function of the short-circuit ratio K_{c} and stator current phase shift at 20° C rotor temperature is presented. Already at this low rotor temperature, a permanent magnet machine can dissipate an order of magnitude of higher stator copper losses than a wound rotor synchronous machine, depending on the combination of machine parameters K_{c} and $\cos\varphi_{\mathrm{rated}}$.

The most rotors of synchronous machines are built with insulation materials in the class F, but operated in the class B (120° absolute temperature) in order to extend its service life. Considering temperature coefficient of −0.12 %/K for remanent flux density and intrinsic coercive force, one obtains the ratio between stator copper losses in a permanent magnet and wound rotor machine with equal rated torques and active volumes at 120° rotor temperature as shown in Fig. A.20.

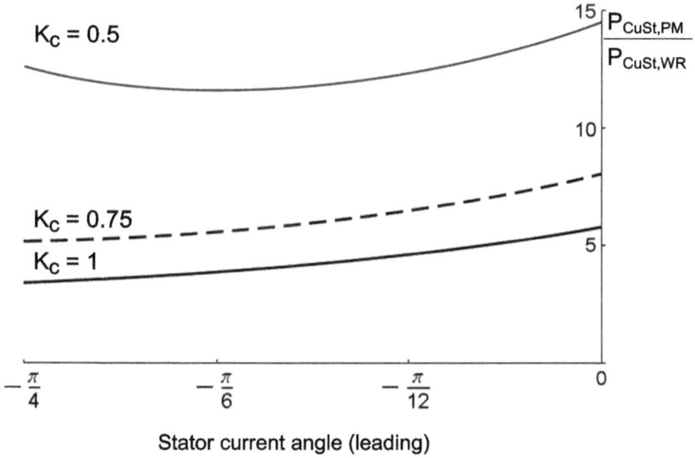

Fig. A.20 Same as in Fig. A.19, however, at rotor temperature of 120 °C. The loss of magnetization of permanent magnets at field winding rated temperature results in a significant increase of stator copper losses and makes permanent magnet machines fully non-competitive to a field wound machine

A permanent magnet machine having the same volume as, and operating in the thermal environment of a wound rotor machine dissipates several times higher stator copper losses, due to:

- Incapability of permanent magnets to generate additional ampere-turns needed to compensate armature reaction and

– Loss of magnetization of permanent magnets at normal operating temperatures of contemporary electric machines.

For example, a permanent magnet machine with $K_c = 1$ and rated power factor $\cos\varphi = 0.866$ dissipates 3.87 times more stator copper losses than a wound rotor machine at a field winding temperature of 120 °C. This means that the field winding of a wound rotor machine placed in the same geometry as the permanent magnet rotor may dissipate 3.87 – 1 = 2.87 times the losses of the armature winding if both machines should have the same efficiency and operate at the same temperature.

A.3.3 Power Factor of an Induction Machine—The Influence of the Number of Machine Poles

As shown in Chap. 4, the main inductance of a winding can be expressed as

$$L_{\mathrm{gap,cyl,1}} = \frac{16}{\pi^2} p \frac{\mu_0 l_{\mathrm{ax}} \tau_{\mathrm{p}}}{\delta} w^2 f_{\mathrm{w,1}}^2$$

with w denoting the number of turns per pole

$$w = \frac{w_{\mathrm{Ph}}}{2p} \tag{A.3.9}$$

and w_{Ph} denoting the number of in-series connected turns per phase. A symmetrically fed, symmetrically built m-phase machine has a main inductance per phase due to fundamental harmonic of MMF L_{main}

$$L_{\mathrm{main}} = \frac{m}{\pi} \frac{\mu_0 D l_{\mathrm{ax}}}{\delta p^2} \left(w_{\mathrm{Ph}} f_{\mathrm{w,1}}\right)^2 = \frac{L_{\mathrm{main,1}}}{p^2} \tag{A.3.10}$$

with $L_{\mathrm{main,1}}$ denoting the main inductance due to fundamental harmonic in a 2-pole machine ($p = 1$). Leakage inductance of one phase can be expressed as

$$L_{\sigma} = \mu_0 l_{\mathrm{ax}} w_{\mathrm{Ph}}^2 \cdot \frac{m}{N} \lambda \tag{A.3.11}$$

with m denoting the number of phases, N the number of slots, and λ the specific leakage permeance.

Keeping machine dimensions and winding parameters constant, one can draw the equivalent circuit of an induction machine as a function of the number of pole pairs p in the manner shown in Fig. A.21. Whereas leakage reactances are independent of the number of pole pairs p, the main reactance decreases as p increases because of an increasing air gap reluctance.

Real and imaginary components of impedance of the equivalent circuit in Fig. A.21 can be expressed as

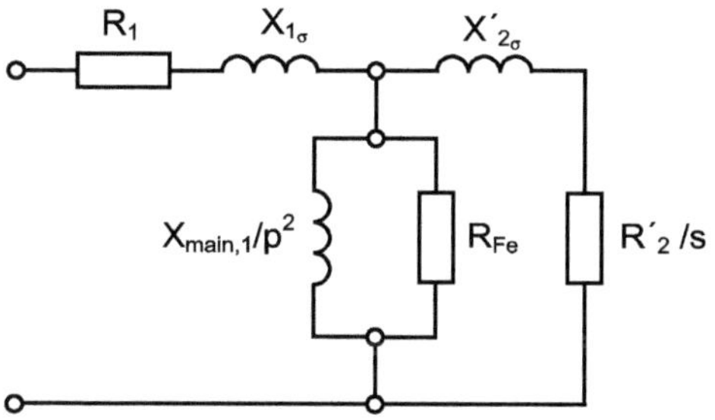

Fig. A.21 Equivalent circuit of an induction machine. Only the main reactance is a function of the number of pole pairs

$$\mathrm{Re}\{\hat{Z}\} = R_1 + \frac{R_2' X_{\mathrm{main},1}^2}{p^4 s\left[\left(\frac{R_2'}{s}\right)^2 + \left(\frac{X_{\mathrm{main},1}}{p^2} + X_{2,\sigma}'\right)^2\right]} \tag{A.3.12}$$

$$\mathrm{Im}\{\hat{Z}\} = X_{1,\sigma} + \frac{R_2'^2 X_{\mathrm{main},1}}{p^2 s^2\left[\left(\frac{R_2'}{s}\right)^2 + \left(\frac{X_{\mathrm{main},1}}{p^2} + X_{2,\sigma}'\right)^2\right]} + \\ + \frac{X_2' X_{\mathrm{main},1}^2}{p^4\left[\left(\frac{R_2'}{s}\right)^2 + \left(\frac{X_{\mathrm{main},1}}{p^2} + X_{2,\sigma}'\right)^2\right]} + \frac{X_2'^2 X_{\mathrm{main},1}}{p^2\left[\left(\frac{R_2'}{s}\right)^2 + \left(\frac{X_{\mathrm{main},1}}{p^2} + X_{2,\sigma}'\right)^2\right]} \tag{A.3.13}$$

Power factor cos φ is now

$$\cos\varphi = \frac{\mathrm{Re}\{\hat{Z}\}}{|\hat{Z}|} \tag{A.3.14}$$

Case Study A.1 A 320 kW, 6 kV, and 2-pole squirrel cage induction machine has the following equivalent circuit parameter: $R_1 = 0.798\ \Omega$, $X_{1,\sigma} = 8.7\ \Omega$, $R'_2 = 0.75\ \Omega$, $X'_{2,\sigma} = 0.75\ \Omega$, $X_{\mathrm{main},1} = 465.5\ \Omega$, and $R_{Fe} = 3658\ \Omega$. The dependence of its power factor on slip is shown in Fig. A.22.

If the machine is wound as a four pole with identical parameters as the original two-pole one, its power factor decreases in the manner shown in Fig. A.22.

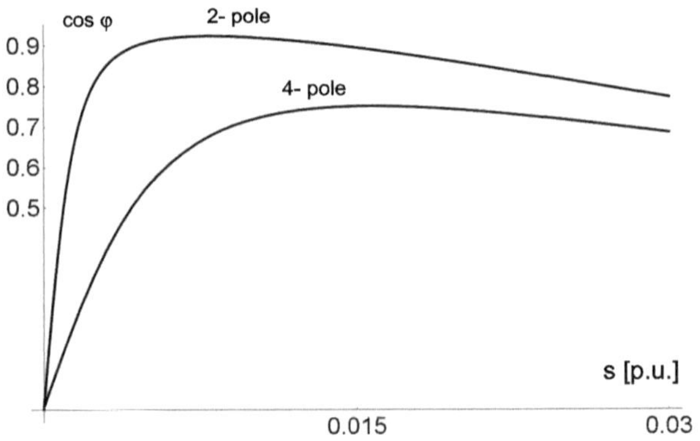

Fig. A.22 Power factor as a function of slip of a squirrel cage induction machine built as a 2-pole (*curve above*) and as a 4-pole (*curve below*)

A.4 Efficiency

Although machine efficiency is a function of numerous parameters, there exists a scaling law for a given cooling art and construction practice, which shows a clear dependence of efficiency on the machine rated power.

Machine electrical losses generated in the active volume V are transferred to the cooling medium on its heat exchange surface S. If the cooling art and insulation class remain unchanged, the heat transfer coefficient remains the same and the losses taken from the heat exchange surface are proportional to its area

$$P_{\text{loss}} \sim S \sim x^2 \tag{A.4.1}$$

with x denoting the machine's linear dimension.

Rated torque and rated power (considering rotational speed constant) are proportional to the machine volume

$$P_{\text{rated}} \sim V \sim x^3 \tag{A.4.2}$$

On the other hand, one can write

$$P_{\text{loss}} = (1 - \eta)P_{\text{rated}} \tag{A.4.3}$$

Substituting expressions A. 4.1–4.2 in A.4.3, one obtains

$$\eta = 1 - c\frac{1}{\sqrt[3]{P_{\text{rated}}}} \tag{A.4.4}$$

with calibrating constant c being evaluated for a given cooling art.

Case Study A.2 The squirrel cage induction machine in Case Study A.1 has a rated efficiency of 91 %.

Taking for P_{rated} in Eq. A.4.4 1 p.u., one can determine the constant c_{IM}

$$c_{\text{IM}} = (1 - \eta)\sqrt[3]{P_{\text{rated}}} = 0.09$$

A low-speed permanent magnet wind generator 3200 kW, 690 V, and $\varphi_{\text{rated}} = -34°$ (leading), 12.75 rpm has a rated efficiency of 93.2 %. The constant c_{PM} in Eq. A.4.4 is

$$c_{\text{PM}} = (1 - \eta)\sqrt[3]{P_{\text{rated}}} = 0.068$$

The efficiency curve as a function of rated power of similarly built squirrel cage and permanent magnet machines is shown in Fig. A.23. In the whole power range from $P_{\text{rated,IM}} = 320$ kW to $P_{\text{rated,PM}} = 3200$ kW, the efficiency of the squirrel cage induction machine is superior to the efficiency of the permanent machine due to drawbacks of permanent magnet excitation.

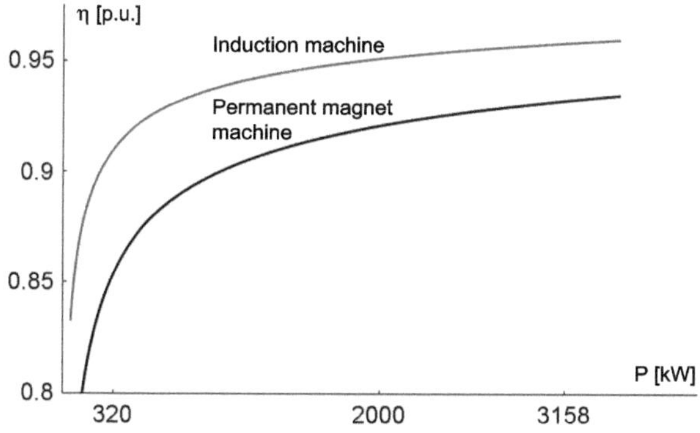

Fig. A.23 Efficiency curve of a squirrel cage induction machine and a permanent magnet machine in a given power range

Index

V. Ostović, *The Art and Science of Rotating Field Machines Design*,
DOI 10.1007/978-3-319-39081-9

MIX
Papier aus verantwortungsvollen Quellen
Paper from responsible sources
FSC® C105338
FSC
www.fsc.org

If you have any concerns about our products,
you can contact us on
ProductSafety@springernature.com

In case Publisher is established outside the EU,
the EU authorized representative is:
Springer Nature Customer Service Center GmbH
Europaplatz 3, 69115 Heidelberg, Germany

Printed by Libri Plureos GmbH
in Hamburg, Germany